QA 278.8 GUP

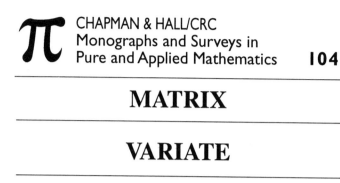

CHAPMAN & HALL/CRC
Monographs and Surveys in
Pure and Applied Mathematics   104

MATRIX

VARIATE

DISTRIBUTIONS

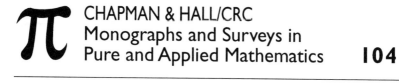

CHAPMAN & HALL/CRC
Monographs and Surveys in
Pure and Applied Mathematics 104

MATRIX

VARIATE

DISTRIBUTIONS

A.K. GUPTA
D.K. NAGAR

CHAPMAN & HALL/CRC

Boca Raton   London   New York   Washington, D.C.

Library of Congress Cataloging-in-Publication Data

Gupta, A. K. (Arjun K.), 1938–
    Matrix variate distributions / A.K. Gupta, D.K. Nagar.
        p. cm. — (Monographs and surveys in pure and applied
mathematics)
    Includes bibliographical references and index.
    ISBN 1-58488-046-5
    1. Distribution (Probability theory) 2. Multivariate analysis.
3. Random matrices. I. Nagar, D. K. II. Title. III. Series:
Chapman & Hall/CRC monographs and surveys in pure and applied
mathematics.
QA273.6.G875 1999
519.2′4—dc21
                                   99-40291
                                   CIP

No claim to original U.S. Government works
International Standard Book Number 1-58488-046-5
Library of Congress Card Number 99-40291
Printed in the United States of America   1 2 3 4 5 6 7 8 9 0
Printed on acid-free paper

To Meera

AKG

To the memory of my mother

DKN

PREFACE

Random matrices play an important role in the study of multivariate statistical methods. They have been found useful in physics, economics, psychology and other fields of scientific investigation. The literature on the subject is widely dispersed throughout statistical journals. A lot of material has accumulated over the years and an analytical review of the material was deemed necessary. At present there is no book available which deals with this subject primarily. This volume presents most of the developments that have taken place in continuous matrix variate distribution theory in a systematic and integrated form. Some new results have also been included. It is hoped that this volume will stimulate further research and help advance the field of multivariate statistical analysis. This book will be especially useful to graduate students, teachers and researchers who are interested in multivariate statistical analysis. The first author presented parts of this book in a one semester course at Bowling Green State University. It can also serve as a source of supplementary reading and reference to many researchers. It is assumed that the reader is familiar with introductory multivariate statistical analysis and matrix algebra.

This work supplements the four volume encyclopaedic work by Johnson and Kotz on the Distributions in Statistics (*Discrete Distributions, Continuous Univariate Distributions 1, Continuous Univariate Distributions 2 and Continuous Multivariate Distributions*) which has been an important contribution to statistical literature. The authors have also benefited by many fine books in multivariate analysis, *e.g.*, Anderson, *An Introduction to Multivariate Statistical Analysis*; Kshirsagar, *Multivariate Analysis*; Srivastava and Khatri, *An Introduction to Multivariate Statistics*; Muirhead, *Aspects of Multivariate Statistical Theory*; and Siotani, Hayakawa and Fujikoshi, *Modern Multivariate Statistical Analysis: A Graduate Course and Handbook*.

After introducing the basic mathematical results from matrix algebra, integration, zonal polynomials and hypergeometric functions in Chapter 1, we study the matrix variate normal and Wishart distributions in Chapters 2 and 3, respectively. We discuss the matrix variate t-distribution in Chapter 4, the matrix variate beta and F-distributions in Chapter 5, and the matrix variate Dirichlet distributions in Chapter 6. The distribution of matrix quadratic forms is given in Chapter 7, and some miscellaneous distributions are presented in Chapter 8. The last chapter, Chapter 9, gives some general families of distributions. Every chapter is followed with a set of problems. Finally, a bibliography, which contains only items cited in the text, has been provided. It is not exhaustive especially as regards to papers on applied topics.

It is a pleasure to express our thanks to all who contributed to the production of this book. Foremost, we would like to record our gratitude to late Professor

C. G. Khatri for many valuable suggestions and for reading parts of the manuscript critically before his untimely death. The authors also benefited greatly from conversations with Professors V. Girko, S. Konishi, S. Kotz, N. Sugiura, and C. G. Troskie. Furthermore, Drs. J. Chen, D. Song, J. Tang and T. Varga helped by reading parts of the manuscript. Thanks are also due to Drs. D. J. de Waal, P. C. N. Groenewald, R. D. Gupta, J. M. Juritz, Ravindra Khattree, N. J. le Roux, D. G. Marx, D. G. Nel, H. M. Nel, J. J. J. Roux, W. Y. Tan and C. A. van der Merwe for providing the preprints and reprints of their work.

The first author would like to acknowledge his special thanks to his wife, Meera, for her continued encouragement and his children Alka, Mita and Nisha, for their support and help in editing the book. However, whatever errors of omission or commission that remain are entirely ours.

The academic environment at the Department of Mathematics and Statistics, Bowling Green State Univeristy and the Department of Mathematics, Universidad de Antioquia, was essential to complete this project. Ms. Mary Lince, CRC Press, has been very cooperative during the publishing phase. Finally, we very much appreciate the help rendered by Ms. Cynthia Patterson in the early stages of the production of this book.

Bowling Green, Ohio   A. K. Gupta
August, 1999.   D. K. Nagar

TABLE OF CONTENTS

CHAPTER 1

PRELIMINARIES

1.1. INTRODUCTION

Multivariate normal distribution plays a pivotal role in the theory of multivariate statistical analysis. Besides mathematical tractability, there are other reasons for this phenomenon. Often the multivariate observations are at least approximately normally distributed. Even when the original data is not multivariate normal, due to the central limit theorem, sampling distributions of certain statistics can be approximated by normal distribution.

The independent multivariate observations are often written in terms of a matrix, which is known as *Sample Observation Matrix* (Roy, 1957). In such a matrix, when sampling from multivariate normal distribution, the columns are distributed independently as multivariate normal with common mean vector and dispersion matrix. The assumption of independence of multivariate observations is not met in multivariate time series, stochastic processes and repeated measurements on multivariate variables. In these cases, the matrix of observations leads to the introduction of the matrix variate normal distribution.

As already stated, the multivariate statistical analysis heavily depends upon multivariate normal distribution. Therefore, the distribution of sample covariance matrix, which has Wishart distribution (Wishart, 1928), plays a central role in almost all multivariate inferential procedures. A distribution closely connected to the Wishart distribution, known as *Matrix Variate Beta* (Khatri, 1959a; Olkin and Rubin, 1964), was introduced by Hsu (1939a) while studying distribution of roots of certain determinantal equation. The matrix variate t-distribution was first obtained by Kshirsagar (1961a), when he proved that the unconditional distribution of the usual estimator of the parameter matrix of regression coefficients has a matrix variate t-distribution.

The subsequent development of the theory of random matrices was brought about by theoretical and practical considerations. Furthermore, multivariate techniques depend upon functions of random matrices such as determinants, traces and characteristic roots. Thus, random matrices are the backbone of multivariate statistical analysis.

Random matrices have found their applications in many fields. Wigner (1967) applied the theory of random matrices to nuclear physics. Treatment of this application and its development are reported by Mehta (1991). Carmeli (1974, 1983), dealing with the statistical theory of energy levels and its relation to random matrices studied the complex Gaussian random matrix and introduced the quaternion random matrix. Girko and Gupta (1996) surveyed distributions of random matrices and their applications in such diverse fields as control theory, stochastic systems, linear stochastic programming, molecular chemistry, experiment planning and ring accelerator. Random matrices have also been used in studies connected with information theory, pattern recognition problems, statistical signal analysis, target detection, identification procedure, and multiple time series. Random matrices are also widely used in experimental studies in various branches such as agriculture, anthropology, biology, cybernetics, economics, education, medicine, and psychology. In these studies the observed random phenomena often can be described by random matrices which include the dependence structure of the relevant random vectors.

Many books on multivariate statistical analysis, *e.g.*, Kshirsagar (1972); Srivastava and Khatri (1979); Muirhead (1982); Anderson (1984); Siotani, Hayakawa and Fujikoshi (1985), give some results on matrix random variables. In particular, all of them cover Wishart distribution. The book by Gupta and Varga (1993) covers most results on matrix variate elliptically contoured distribution. The present volume incorporates most of the results on matrix variates distributions.

1.2.   MATRIX ALGEBRA

This section presents a brief discussion of some of the definitions and theorems from matrix algebra. These results can be found in any book of linear algebra, *e.g.*, Bellman (1970), Graybill (1983), Magnus and Neudecker (1988), or books on multivariate statistical analysis, *e.g.*, Roy (1957), Rao (1973), Muirhead (1982), Anderson (1984), Siotani, Hayakawa and Fujikoshi (1985), and Gupta and Varga (1993).

DEFINITION 1.2.1. *Let $A = (a_{ij})$ be a square matrix of order p. Then, A is called*
  (i) nonsingular if $\det(A) \neq 0$,
  (ii) a diagonal matrix, denoted by $\mathrm{diag}(a_{11}, \ldots, a_{pp})$ if $a_{ij} = 0$, $i \neq j$,
  (iii) an identity matrix, denoted by I_p, if A is diagonal and $a_{ii} = 1$, $i = 1, \ldots, p$,
  (iv) a symmetric matrix if $a_{ij} = a_{ji}$, for $i \neq j$ or equivalently $A = A'$,
  (v) a lower triangular matrix if $a_{ij} = 0$, $i < j$,
  (vi) an upper triangular matrix if $a_{ij} = 0$, $i > j$,
  (vii) an orthogonal matrix if $AA' = A'A = I_p$,
  (viii) an idempotent matrix if $A^2 = A$,
  (ix) a symmetric idempotent matrix if $A = A'$ and $A^2 = A$,
  (x) a positive definite (positive semidefinite) matrix, denoted by $A > 0$ $(A \geq 0)$ if A is symmetric and for every p-dimensional nonzero vector $\boldsymbol{v}$, $\boldsymbol{v}'A\boldsymbol{v} > 0$ $(\boldsymbol{v}'A\boldsymbol{v} \geq 0)$.

THEOREM 1.2.1. *Let A, B be $p \times p$ and C be $q \times p$ matrices. Then, we have the following results.*
  (i) If $A > 0$, then $A^{-1} > 0$.
  (ii) If $A \geq 0$, then $CAC' \geq 0$.

(iii) If $q \leq p$, $A > 0$, and $\operatorname{rank}(C) = q$, then $CAC' > 0$.

(iv) If $A > 0$, $B > 0$ and $A - B > 0$, then $B^{-1} - A^{-1} > 0$.

(v) If $A > 0$, and $B > 0$, then $\det(A + B) \geq \det(A) + \det(B)$.

DEFINITION 1.2.2. *Let A be a $p \times p$ matrix. Then, the roots (with multiplicity) of the equation*

$$\det(A - \lambda I_p) = 0$$

are called characteristic roots or eigenvalues of the matrix A.

THEOREM 1.2.2. *Let $\lambda_1, \ldots, \lambda_p$ be the characteristic roots of $A\,(p \times p)$. Then, the following results hold.*

(i) $\det(A) = \prod_{i=1}^{p} \lambda_i$.

(ii) $\operatorname{tr}(A) = \sum_{i=1}^{p} \lambda_i$.

(iii) $\operatorname{rank}(A) =$ *the number of nonzero characteristic roots.*

(iv) *A is nonsingular if and only if all its characteristic roots are nonzero.*

(v) *Furthermore, if we assume that A is symmetric, then the characteristic roots are real.*

(vi) *A is positive definite (positive semidefinite) if and only if all the characteristic roots of A are positive (non-negative).*

(vii) *A is symmetric idempotent of $\operatorname{rank}(A) = k$ if and only if all the nonzero characteristic roots are unity.*

DEFINITION 1.2.3. *Let $A = (a_{ij})$ be a $p \times q$ matrix. Then, a 2×2 partition of A is defined as*

$$A = \begin{pmatrix} A_{11} & A_{12} \\ A_{21} & A_{22} \end{pmatrix} \begin{matrix} r \\ p - r \end{matrix}$$
$$\begin{matrix} s & q - s \end{matrix}$$

where the submatrices A_{11}, A_{12}, A_{21}, and A_{22} are

$$A_{11} = (a_{ij}),\ i = 1, \ldots, r,\ j = 1, \ldots, s$$
$$A_{12} = (a_{ij}),\ i = 1, \ldots, r,\ j = s + 1, \ldots, q$$
$$A_{21} = (a_{ij}),\ i = r + 1, \ldots, p,\ j = 1 \ldots, s$$

and

$$A_{22} = (a_{ij}),\ i = r + 1, \ldots, p,\ j = s + 1 \ldots, q.$$

Similarly, an $m \times n$ partition of A is defined as

$$A = \begin{pmatrix} A_{11} & A_{12} & \cdots & A_{1n} \\ A_{21} & A_{22} & \cdots & A_{2n} \\ \vdots & & & \\ A_{m1} & A_{m2} & \cdots & A_{mn} \end{pmatrix} \begin{matrix} p_1 \\ p_2 \\ \vdots \\ p_m \end{matrix}$$
$$\begin{matrix} q_1 & q_2 & \cdots & q_n \end{matrix}$$

where $p_1 + \cdots + p_m = p$ and $q_1 + \cdots + q_n = q$. Thus, for $m = 1$, $p_1 = p$, $n = q$, $q_1 = \cdots = q_n = 1$, one can write A as

$$A = (\boldsymbol{a}_1, \ldots, \boldsymbol{a}_q),$$

where $\boldsymbol{a}_1, \ldots, \boldsymbol{a}_q$ are the p-dimensional column vectors of A. Also, when $n = 1$, $q_1 = q$, $m = p$, $p_1 = \cdots = p_m = 1$, we have

$$A = \begin{pmatrix} \boldsymbol{a}_1^{*\prime} \\ \vdots \\ \boldsymbol{a}_p^{*\prime} \end{pmatrix},$$

where $\boldsymbol{a}_1^{*\prime}, \ldots, \boldsymbol{a}_p^{*\prime}$ are the q-dimensional row vectors of A.

THEOREM 1.2.3. *Let A be a nonsingular square matrix of order p. Then,*
  (i) $(kA)^{-1} = k^{-1}A^{-1}$, $k \neq 0$ is a scalar,
  (ii) $(AB)^{-1} = B^{-1}A^{-1}$, B $(p \times p)$ is nonsingular,
  (iii) $A^{-1} = \operatorname{diag}(a_{11}^{-1}, \ldots, a_{pp}^{-1})$ if $A = \operatorname{diag}(a_{11}, \ldots, a_{pp})$, $a_{ii} \neq 0$, $i = 1, \ldots, p$.
  (iv) For B $(p \times q)$, D $(q \times p)$ and nonsingular C $(q \times q)$,

$$(A + BCD)^{-1} = A^{-1} - A^{-1}B(C^{-1} + DA^{-1}B)^{-1}DA^{-1}.$$

  (v) For symmetric A $(p \times p)$ partitioned as $A = \begin{pmatrix} A_{11} & A_{12} \\ A_{21} & A_{22} \end{pmatrix}$,

$$A^{-1} = \begin{pmatrix} A_{11\cdot2}^{-1} & -A_{11\cdot2}^{-1}A_{12}A_{22}^{-1} \\ -A_{22}^{-1}A_{21}A_{11\cdot2}^{-1} & A_{22\cdot1}^{-1} \end{pmatrix}$$

$$= \begin{pmatrix} A_{11\cdot2}^{-1} & -A_{11}^{-1}A_{12}A_{22\cdot1}^{-1} \\ -A_{22\cdot1}^{-1}A_{21}A_{11}^{-1} & A_{22\cdot1}^{-1} \end{pmatrix},$$

where $A_{11\cdot2} = A_{11} - A_{12}A_{22}^{-1}A_{21}$, $A_{22\cdot1} = A_{22} - A_{21}A_{11}^{-1}A_{12}$, assuming A_{11}^{-1}, A_{22}^{-1}, $A_{11\cdot2}^{-1}$ and $A_{22\cdot1}^{-1}$ exist.

THEOREM 1.2.4. *Let A be a $p \times q$ matrix. Then, we have the following results.*
  (i) $0 \leq \operatorname{rank}(A) \leq \min(p, q)$.
  (ii) If $A = 0$, then $\operatorname{rank}(A) = 0$.
  (iii) $\operatorname{rank}(A) = \operatorname{rank}(A') = \operatorname{rank}(AA') = \operatorname{rank}(A'A)$.
  (iv) $\operatorname{rank}(A + B) \leq \operatorname{rank}(A) + \operatorname{rank}(B)$, for A and B of order $p \times q$.
  (v) $\operatorname{rank}(AB) \leq \min(\operatorname{rank}(A), \operatorname{rank}(B))$, for A $(p \times q)$ and B $(q \times r)$.
  (vi) If B $(p \times p)$ and C $(q \times q)$ are nonsingular, then $\operatorname{rank}(BAC) = \operatorname{rank}(A)$.
  (vii) For A $(p \times p)$, $\operatorname{rank}(A) = p$ iff A is nonsingular.

THEOREM 1.2.5. *For the trace function, defined as the sum of the diagonal elements of a square matrix, we have*
  (i) $\operatorname{tr}(A) = \operatorname{tr}(A')$, A $(p \times p)$,

(ii) $\operatorname{tr}(kA) = k \operatorname{tr}(A)$, A $(p \times p)$ and $k \neq 0$ is a scalar,

(iii) $\operatorname{tr}(A + B) - \operatorname{tr}(A) + \operatorname{tr}(B)$, A $(p \times p)$, B $(p \times p)$,

(iv) $\operatorname{tr}(AB) = \operatorname{tr}(BA)$, A $(p \times q)$, B $(q \times p)$,

(v) $\operatorname{tr}(ABC) = \operatorname{tr}(ACB)$, A $(p \times p)$, B $(p \times p)$ and C $(p \times p)$ are symmetric,

(vi) $\operatorname{tr}(HAH') = \operatorname{tr}(A)$, H $(p \times p)$ is orthogonal,

(vii) $\operatorname{tr}(A) = \operatorname{rank}(A)$ if A is idempotent, and

(viii) $\operatorname{tr}(A^k) = \sum_{i=1}^{p} \lambda_i^k$, k is a positive integer, $\lambda_1, \ldots, \lambda_p$ are the characteristic roots of A $(p \times p)$.

THEOREM 1.2.6. *Let A be a symmetric nonsingular matrix of order p. Then, for B $(q \times p)$ and C $(p \times q)$,*

(i) $\det(I_p + CB) = \det(I_q + BC)$.

(ii) *For a partition $A = \begin{pmatrix} A_{11} & A_{12} \\ A_{21} & A_{22} \end{pmatrix}$,*

$$\det(A) = \det(A_{11}) \det(A_{22} - A_{21} A_{11}^{-1} A_{12}), \text{ if } A_{11} \text{ is nonsingular}$$
$$= \det(A_{22}) \det(A_{11} - A_{12} A_{22}^{-1} A_{21}), \text{ if } A_{22} \text{ is nonsingular}.$$

THEOREM 1.2.7. *Let $A = (\boldsymbol{a}_1, \ldots, \boldsymbol{a}_p) = (\boldsymbol{a}_1^*, \ldots, \boldsymbol{a}_p^*)'$ be an orthogonal matrix of order p. Then,*

(i) $A^{-1} = A'$,

(ii) *the characteristic roots of A are either $+1$ or -1,*

(iii) $\det(A) = \pm 1$,

(iv) $\boldsymbol{a}_i' \boldsymbol{a}_j = 0$, $i \neq j$, $\boldsymbol{a}_i' \boldsymbol{a}_i = 1$, $i, j = 1, \ldots, p$,

(v) $\boldsymbol{a}_i^{*\prime} \boldsymbol{a}_j^* = 0$, $i \neq j$, $\boldsymbol{a}_i^{*\prime} \boldsymbol{a}_i^* = 1$, $i, j = 1, \ldots, p$.

THEOREM 1.2.8. *Let A $(p \times p)$ be an idempotent matrix. Then,*

(i) $I_p - A$ is idempotent,

(ii) PAP^{-1} is idempotent, P $(p \times p)$ is a nonsingular matrix,

(iii) HAH' is idempotent, H $(p \times p)$ is orthogonal,

(iv) non-zero characteristic roots of A are unity,

(v) $\operatorname{tr}(A) = \operatorname{rank}(A)$,

(vi) A^k is idempotent, k is a positive integer.

(vii) If $\operatorname{rank}(A) = p$, then $A = I_p$.

DEFINITION 1.2.4. *Let $A = (a_{ij})$ be a square matrix of order p. Then, the sub-matrices $A^{[\alpha]}$ and $A_{[\alpha]}$, $1 \leq \alpha \leq p$, of the matrix A, are defined as*

$$A^{[\alpha]} = \begin{pmatrix} a_{11} & \cdots & a_{1\alpha} \\ \vdots & & \\ a_{\alpha 1} & \cdots & a_{\alpha\alpha} \end{pmatrix}$$

$$= A^{(\alpha)}$$

and

$$A_{[\alpha]} = \begin{pmatrix} a_{p-\alpha+1,p-\alpha+1} & \cdots & a_{p-\alpha+1,p} \\ & \vdots & \\ a_{p,p-\alpha+1} & \cdots & a_{pp} \end{pmatrix}$$

$$= A_{(p-\alpha+1)}$$

respectively.

THEOREM 1.2.9. *Let A and B be two lower (upper) triangular matrices of order p. Then, the following results hold.*
  (i) A^{-1} is lower (upper) triangular.
  (ii) $(AB)^{-1}$ is lower (upper) triangular.
  (iii) $A_{(\alpha)}$ is lower (upper) triangular and $(A_{(\alpha)})^{-1} = (A^{-1})_{(\alpha)}$.
  (iv) $A^{(\alpha)}$ is lower (upper) triangular and $(A^{(\alpha)})^{-1} = (A^{-1})^{(\alpha)}$.
  (v) If A is lower triangular and partitioned as

$$A = \begin{pmatrix} A_{11} & 0 \\ A_{21} & A_{22} \end{pmatrix},$$

where A_{11} and A_{22} are nonsingular, then

$$A^{-1} = \begin{pmatrix} A_{11}^{-1} & 0 \\ -A_{22}^{-1} A_{21} A_{11}^{-1} & A_{22}^{-1} \end{pmatrix}.$$

  (vi) If A is upper triangular and partitioned as

$$A = \begin{pmatrix} A_{11} & A_{12} \\ 0 & A_{22} \end{pmatrix},$$

where A_{11} and A_{22} are nonsingular, then

$$A^{-1} = \begin{pmatrix} A_{11}^{-1} & -A_{11}^{-1} A_{12} A_{22}^{-1} \\ 0 & A_{22}^{-1} \end{pmatrix}.$$

  (vii) $(AB)^{(\alpha)} = A^{(\alpha)} B^{(\alpha)}$.
  (viii) $(AB)_{(\alpha)} = A_{(\alpha)} B_{(\alpha)}$.

THEOREM 1.2.10. (spectral decomposition of a symmetric matrix) *Let $A\,(p \times p)$ be a symmetric matrix. Then, there exists an orthogonal matrix H such that*

$$A = HDH',$$

where D is a diagonal matrix having diagonal elements as the characteristic roots of A.

THEOREM 1.2.11. (square root factorization) *Let* $A\,(p \times p)$ *be a positive definite matrix. Then, there exists a positive definite matrix* $B\,(p \times p)$ *such that* $A = B^2$. *Furthermore, we define square root of* A *as* $A^{\frac{1}{2}} = B$.

THEOREM 1.2.12. (rank factorization) *Let* $A\,(p \times p)$ *be a symmetric matrix of rank* q. *Then, there exists a matrix* $B\,(p \times q)$ *of rank* q *such that* $A = BB'$.

THEOREM 1.2.13. *Let* $A\,(q \times p)$ *be of rank* $q\,(\leq p)$. *Then, there exist an orthogonal matrix* $H\,(p \times p)$ *and a positive definite matrix* $B\,(q \times q)$ *such that*

$$A = B\,(\,I_q \quad 0\,)\,H$$

where 0 *denotes the* $q \times (p - q)$ *null matrix.*

THEOREM 1.2.14. (Cholesky decomposition) *Let* $A\,(p \times p)$ *be a positive definite matrix. Then there exists a unique lower (upper) triangular matrix* $T\,(p \times p)$ *with positive diagonal elements such that*

$$A = TT'.$$

THEOREM 1.2.15. *Let* $A\,(q \times p)$ *be of rank* $q \leq p$. *Then, there exist a lower (upper) triangular matrix* T *with positive diagonal elements, and a semiorthogonal matrix* H_1, $H_1H_1' = I_p$, *such that*

$$A = TH_1.$$

Moreover this representation is unique.

THEOREM 1.2.16. *Let* $A_1, \ldots, A_m$ *be* $p \times p$ *symmetric matrices. Then necessary and sufficient condition for simultaneous diagonalization of* $A_1, \ldots, A_m$ *by an orthogonal matrix* H *is that* $A_iA_j = A_jA_i$, *for every pair* (i, j), $i \neq j$, $i, j = 1, \ldots, m$.

THEOREM 1.2.17. *Let* $A_1, \ldots, A_m$ *be* $p \times p$ *symmetric idempotent matrices and* $A_iA_j = 0$, $i \neq j$. *Then there exists an orthogonal matrix* $H\,(p \times p)$ *such that*

$$H'A_1H = \begin{pmatrix} \Lambda_{r_1} & 0 \\ 0 & 0 \end{pmatrix}, \ H'A_2H = \begin{pmatrix} 0 & 0 & 0 \\ 0 & \Lambda_{r_2} & 0 \\ 0 & 0 & 0 \end{pmatrix}, \ldots, H'A_mH = \begin{pmatrix} 0 & 0 & 0 \\ 0 & \Lambda_{r_m} & 0 \\ 0 & 0 & 0 \end{pmatrix},$$

where $\Lambda_{r_i} = \mathrm{diag}(\lambda_{i1}, \ldots, \lambda_{ir_i})$, λ_{ij}*'s are the characteristic roots of* A_i, $r_i = \mathrm{rank}(A_i)$, $i = 1, \ldots, m$, *and the null matrices are of appropriate orders.*

From Theorem 1.2.10, we can derive the following well-known representation.

THEOREM 1.2.18. (spectral representation) *Let* $A\,(p \times p)$ *be a symmetric matrix. Then the matrix* A *can be written as*

$$A = \sum_{j=1}^{m} \lambda_j A_j$$

where $A_1, \ldots, A_m$ *are symmetric idempotent matrices,* $\mathrm{rank}(A_i) = f_i$, $A_iA_j = 0$,

$i \neq j$, and $\lambda_1, \ldots, \lambda_m$ are the characteristic roots of A with multiplicities $f_1, \ldots, f_m$ respectively, $\lambda_1 > \cdots > \lambda_m$.

THEOREM 1.2.19. Let $A\,(p \times p)$ be a symmetric matrix written as

$$A = \sum_{j=1}^{m} \alpha_j A_j$$

where $\alpha_1, \ldots, \alpha_m$ are positive real numbers, $A_1, \ldots, A_m$ are symmetric idempotent matrices, $A_i A_j = 0$, $i \neq j$, and $\sum_{j=1}^{m} A_j = I_p$. Then the matrix A is positive definite and $A^{-1} = \left(\sum_{j=1}^{m} \alpha_j A_j \right)^{-1} = \sum_{j=1}^{m} \alpha_j^{-1} A_j$.

THEOREM 1.2.20. Let $A_1, \ldots, A_m$ be symmetric matrices of order p and let $A = \sum_{j=1}^{m} A_j$. Consider the following four conditions:
   (i) $A_i^2 = A_i$
   (ii) $A_i A_j = 0$, $i \neq j$
   (iii) $A^2 = A$
   (iv) $\sum_{i=1}^{k} \text{rank}(A_i) = \text{rank}(A)$.
Then (a) any two of the conditions (i), (ii), and (iii) imply the remaining, and (b) conditions (iii) and (iv) imply (i) and (ii).

The above result was given by Graybill and Marsaglia (1957). For an easy proof of this result, the reader is referred to Loynes (1966) and Searle (1971, pp. 62–63).

The next two theorems state some useful results concerning the Kronecker product, also called the direct product.

DEFINITION 1.2.5. The Kronecker product of two matrices $A\,(m \times n) = (a_{ij})$ and $B\,(p \times q) = (b_{ij})$, denoted by $A \otimes B$, is the $mp \times nq$ matrix defined by

$$A \otimes B = \begin{pmatrix} a_{11}B & a_{12}B & \cdots & a_{1n}B \\ a_{21}B & a_{22}B & \cdots & a_{2n}B \\ \vdots & & & \\ a_{m1}B & a_{m2}B & \cdots & a_{mn}B \end{pmatrix}$$

$$= (a_{ij}B).$$

Using the Definition 1.2.5, the following properties of Kronecker product can be easily proved (see Graham, 1981; Graybill, 1983).

THEOREM 1.2.21. (i) For any nonzero scalars α and β,

$$(\alpha A) \otimes (\beta B) = \alpha \beta (A \otimes B).$$

(ii) For $A\,(m \times n)$, $B\,(m \times n)$ and any C,

$$(A + B) \otimes C = (A \otimes C) + (B \otimes C),$$

$$C \otimes (A + B) = (C \otimes A) + (C \otimes B).$$

(iii) $(A \otimes B) \otimes C = A \otimes (B \otimes C)$.

(iv) $(A \otimes B)' = A' \otimes B'$.

(v) For $A\,(m \times m)$ and $B\,(m \times m)$,

$$\mathrm{tr}(A \otimes B) = (\mathrm{tr}\,A)(\mathrm{tr}\,B).$$

(vi) For $A\,(m \times n)$, $B\,(p \times q)$, $C\,(n \times r)$ and $D\,(q \times s)$,

$$(A \otimes B)(C \otimes D) = (AC) \otimes (BD).$$

(vii) For nonsingular matrices A and B

$$(A \otimes B)^{-1} = A^{-1} \otimes B^{-1}.$$

(viii) If P and Q are orthogonal matrices, then $P \otimes Q$ is an orthogonal matrix.

(ix) If P and Q are positive definite matrices, then $P \otimes Q$ is also a positive definite matrix.

(x) For $A\,(m \times m)$ and $B\,(n \times n)$,

$$\det(A \otimes B) = \det(A)^n \det(B)^m.$$

(xi) If the eigenvalues of $A\,(m \times m)$ are a_i, $i = 1, \ldots, m$, and of $B\,(n \times n)$ are b_j, $j = 1, \ldots, n$ then the eigenvalues of $A \otimes B$ are $a_i b_j$, $i = 1, \ldots, m$, $j = 1, \ldots, n$.

DEFINITION 1.2.6. For a matrix $X\,(m \times n)$, $\mathrm{vec}(X)$ is the $mn \times 1$ vector defined as

$$\mathrm{vec}(X) = \begin{pmatrix} \boldsymbol{x}_1 \\ \vdots \\ \boldsymbol{x}_m \end{pmatrix},$$

where $\boldsymbol{x}_i$, $i = 1, \ldots, n$ is the ith column of X.

THEOREM 1.2.22. For $A\,(p \times m)$, $B\,(n \times q)$, $C\,(q \times m)$, $D\,(q \times n)$, $E\,(m \times m)$, and $X\,(m \times n)$, we have

(i) $\mathrm{vec}(AXB) = (B' \otimes A)\,\mathrm{vec}(X)$,

(ii) $\mathrm{tr}(CXB) = (\mathrm{vec}(C'))'(I_q \otimes X)\,\mathrm{vec}(B)$,

(iii) $\mathrm{tr}(DX'EXB) = (\mathrm{vec}(X))'(D'B' \otimes E)\,\mathrm{vec}(X)$
$\qquad\qquad = (\mathrm{vec}(X))'(BD \otimes E')\,\mathrm{vec}(X)$.

Now, we define the commutation matrix, also known as the permutation matrix, which transforms $\mathrm{vec}(A)$ into $\mathrm{vec}(A')$.

The commutation matrix K_{pq} of order $pq \times pq$ is defined as

$$K_{pq} = \sum_{i=1}^{p} \sum_{j=1}^{q} (H_{ij} \otimes H'_{ij}), \tag{1.2.1}$$

where the matrix H_{ij} $(p \times q)$ has a unit element at the $(i,j)^{\text{th}}$ place and zero elsewhere. Note that

$$H_{ij} = \boldsymbol{a}_i \boldsymbol{b}_j' \tag{1.2.2}$$

where $\boldsymbol{a}_i$ $(p \times 1)$, $\boldsymbol{b}_j$ $(q \times 1)$ have unity as the ith and j^{th} element, respectively, together with the remaining elements as zero. Using this representation, the following properties can easily be proved.

(i) $K_{pq} = K_{qp}' = K_{qp}^{-1}$. $\hspace{3cm}$ (1.2.3)

(ii) $K_{pq} \text{vec}(A) = \text{vec}(A')$, $\hspace{2.5cm}$ (1.2.4)

   where A is an $p \times q$ matrix.

(iii) $K_{pq}(A \otimes B)K_{rs} = B \otimes A$, $\hspace{2cm}$ (1.2.5)

   where A is an $q \times r$ matrix and B is an $p \times s$ matrix.

(iv) $\text{tr}\{K_{pq}(A' \otimes B)\} = \text{tr}(A'B) = (\text{vec}(A))' \text{vec}(B)$, $\hspace{1cm}$ (1.2.6)

   for $A\,(p \times q)$, $B\,(p \times q)$.

(v) $\text{vec}(A \otimes B) = (I_q \otimes K_{pr} \otimes I_r)(\text{vec}(A) \otimes \text{vec}(B))$, $\hspace{0.5cm}$ (1.2.7)

   for $A\,(m \times n)$ and $B\,(r \times s)$.

For proof of these results, one can refer to Magnus and Neudecker (1979), and Neudecker and Wansbeek (1983).

Next we define the vec notation for a symmetric matrix (Brown, 1974).

DEFINITION 1.2.7. *For a symmetric matrix $X\,(p \times p)$, $\text{vecp}(X)$ is a $\frac{1}{2}p(p+1)$-dimensional column vector formed from the elements above and including the diagonal, taken columnwise. In other words if*

$$X = \begin{pmatrix} x_{11} & x_{12} & \cdots & x_{1p} \\ x_{21} & x_{22} & \cdots & x_{2p} \\ \vdots & & & \\ x_{p1} & x_{p2} & \cdots & x_{pp} \end{pmatrix}$$

then

$$\text{vecp}(X) = \begin{pmatrix} x_{11} \\ x_{12} \\ x_{22} \\ \vdots \\ x_{1p} \\ \vdots \\ x_{pp} \end{pmatrix}$$

$$= \text{vecp}(X').$$

DEFINITION 1.2.8. *The matrix* B_p *of order* $p^2 \times \frac{1}{2}p(p+1)$, *with typical element*

$$(B_p)_{ij,gh} = \frac{1}{2}(\delta_{ig}\delta_{jh} + \delta_{ih}\delta_{jg}), \ i \leq p, \ j \leq p, \ g \leq h \leq p, \qquad (1.2.8)$$

where δ_{rs} *is the Kronecker's delta, is called the transition matrix.*

It may be noticed that the rank of B_p is $\frac{1}{2}p(p+1)$. The Moore-Penrose inverse of B_p is

$$B_p^+ = (B_p'B_p)^{-1}B_p' \qquad (1.2.9)$$

which is of order $\frac{1}{2}p(p+1) \times p^2$ with typical element

$$
\begin{aligned}
(B_p^+)_{gh,ij} &= (2 - \delta_{gh})(B_p)_{ij,gh}, \ i \leq p, \ j \leq p, \ g \leq h \leq p \\
&= 1, \ ij = gh \text{ or } ij = hg \\
&= 0, \text{ otherwise.}
\end{aligned}
\qquad (1.2.10)
$$

For symmetric matrix X ($p \times p$), the matrices B_p and B_p^+ can be used to express $\mathrm{vec}(X)$ in terms of $\mathrm{vecp}(X)$ and $\mathrm{vecp}(X)$ in terms of $\mathrm{vec}(X)$, respectively,

$$\mathrm{vecp}(X) = B_p' \, \mathrm{vec}(X) \qquad (1.2.11)$$

$$\mathrm{vec}(X) = (B_p^+)' \, \mathrm{vecp}(X). \qquad (1.2.12)$$

The $p^2 \times p^2$ idempotent matrix

$$M_p = B_p B_p^+$$

has the typical element

$$(M_p)_{ij,gh} = \frac{1}{2}(\delta_{ig}\delta_{jh} + \delta_{ih}\delta_{jg}), \ i \leq p, \ j \leq p, \ g \leq p, \ h \leq p. \qquad (1.2.13)$$

It is interesting to note that

$$\frac{1}{2}(I_{p^2} + K_{pp}) = M_p, \qquad (1.2.14)$$

$$M_p B_p = B_p, \qquad (1.2.15)$$

and

$$B_p^+ M_p = B_p^+. \qquad (1.2.16)$$

Further, if Y is a $p \times p$ matrix, then from (1.2.4) and (1.2.14) we get

$$
\begin{aligned}
M_p \, \mathrm{vec}(Y) &= \frac{1}{2}(I_{p^2} + K_{pp}) \, \mathrm{vec}(Y) \\
&= \frac{1}{2}(\mathrm{vec}(Y) + \mathrm{vec}(Y')) \\
&= \mathrm{vec}(X)
\end{aligned}
\qquad (1.2.17)
$$

where $X = \frac{1}{2}(Y + Y') = X'$. Thus for a symmetric Y $(p \times p)$,

$$M_p \operatorname{vec}(Y) = \operatorname{vec}(Y).$$

For a matrix A $(p \times m)$,
$$M_p(A \otimes A) = (A \otimes A)M_m,$$

and for A $(p \times p)$,

$$\det(B_p'(A \otimes A)B_p) = 2^{-\frac{1}{2}p(p-1)} \det(A)^{p+1}. \qquad (1.2.18)$$

In particular if $A = I_p$ then $\det(B_p'B_p) = 2^{-\frac{1}{2}p(p-1)}$.

1.3.   JACOBIANS OF TRANSFORMATIONS

Let X and Y be two matrices having the same number of independent elements $x_1, \ldots, x_p$ and $y_1, \ldots, y_p$ respectively. Consider the matrix transformation $Y = F(X)$. Then the Jacobian of the transformation from X to Y is defined as

$$J(X \to Y) = \operatorname{mod} \det \begin{pmatrix} \frac{\partial x_1}{\partial y_1} & \cdots & \frac{\partial x_1}{\partial y_p} \\ \vdots & & \\ \frac{\partial x_p}{\partial y_1} & \cdots & \frac{\partial x_p}{\partial y_p} \end{pmatrix}.$$

The following results for Jacobians are well known. Their proofs and details are given in Deemer and Olkin (1951), Olkin (1953), Olkin and Roy (1954), Roy (1957), Olkin and Rubin (1964), Perlman (1977), and Rogers (1980).

(i) $dx_1 \cdots dx_p = J(X \to Y)\, dy_1 \cdots dy_p$

(ii)   If $J(X \to Y) \neq 0$, then
$$J(Y \to X) = \{J(X \to Y)\}^{-1}.$$

(iii)   If $Y = F(Z)$ and $Z = G(X)$, then
$$J(X \to Y) = J(X \to Z)J(Z \to Y).$$

(iv)   If $Y = F(X)$ and $Z = G(W)$, then
$$J(X, W \to Y, Z) = J(X \to Y)J(W \to Z).$$

(v) $J(X \to Y) = J((dX) \to (dY))$
where (dZ) is the matrix of differentials of elements of Z.

(vi)   If $y_i = f_i(x_1, \ldots, x_m, x_{m+1}, \ldots, x_{m+n})$, $i = 1, \ldots, m$, where $x_1, \ldots, x_{m+n}$ are subject to n constraints $f_i(x_1, \ldots, x_{m+n}) = 0$, $i = m+1, \ldots, m+n$, then

$$J(y_1, \ldots, y_m \to x_1, \ldots, x_m) = \frac{J(f_1, \ldots, f_{m+n} \to x_1, \ldots, x_{m+n})}{J(f_{m+1}, \ldots, f_{m+n} \to x_{m+1}, \ldots, x_{m+n})}.$$

The Jacobians of certain transformations which are needed in the subsequent chapters are now given, where for simplicity "mod" has been suppressed from their values.

LINEAR TRANSFORMATIONS

(i) For $y\,(p \times 1)$, $x\,(p \times 1)$ and $A\,(p \times p)$ if $y = Ax$, then

$$J(y \to x) = \det(A). \tag{1.3.1}$$

(ii) For $Y\,(p \times q)$, $X\,(p \times q)$, and $A\,(p \times p)$, if $Y = AX$, then

$$J(Y \to X) = \det(A)^q. \tag{1.3.2}$$

(iii) For $Y\,(p \times q)$, $X\,(p \times q)$, and $B\,(q \times q)$, if $Y = XB$, then

$$J(Y \to X) = \det(B)^p. \tag{1.3.3}$$

(iv) For $Y\,(p \times q)$, $X\,(p \times q)$, $A\,(p \times p)$, and $B\,(q \times q)$, if $Y = AXB$, then

$$J(Y \to X) = \det(A)^q \det(B)^p. \tag{1.3.4}$$

(v) For $Y\,(p \times p)$, $X\,(p \times p)$ symmetric and $A\,(p \times p)$ if $Y = AXA'$, then

$$J(Y \to X) = \det(A)^{p+1}. \tag{1.3.5}$$

(vi) For $Y\,(p \times q)$, $X\,(p \times q)$ and a scalar α if $Y = \alpha X$, then

$$J(Y \to X) = \alpha^{pq} \tag{1.3.6}$$

which follows from (ii) by taking $A = \alpha I_p$.

(vii) For lower triangular matrices $Y\,(p \times p)$, $X\,(p \times p)$, and $A = (a_{ij})$, if $Y = AX$ then

$$J(Y \to X) = \prod_{i=1}^{p} a_{ii}^i. \tag{1.3.7}$$

(viii) For upper triangular matrices $Y\,(p \times p)$, $X\,(p \times p)$, and $A = (a_{ij})$, if $Y = AX$ then

$$J(Y \to X) = \prod_{i=1}^{p} a_{ii}^{p-i+1}. \tag{1.3.8}$$

(ix) For lower triangular matrices $Y\,(p \times p)$, $X\,(p \times p)$, $A\,(p \times p) = (a_{ij})$ and $B\,(p \times p) = (b_{ij})$, if $Y = AXB$, then

$$J(Y \to X) = \prod_{i=1}^{p} (a_{ii}^i\, b_{ii}^{p-i+1}). \tag{1.3.9}$$

(x) For a symmetric matrix $Y\,(p \times p)$ and lower triangular matrices $X\,(p \times p)$ and $A\,(p \times p) = (a_{ij})$, if $Y = AX' + XA'$, then

$$J(Y \to X) = 2^p \prod_{i=1}^{p} a_{ii}^{p-i+1}. \tag{1.3.10}$$

(xi) For a symmetric matrix $Y\,(p \times p)$ and upper triangular matrices $X\,(p \times p)$

and $A\,(p \times p) = (a_{ij})$, if $Y = AX' + XA'$, then

$$J(Y \to X) = 2^p \prod_{i=1}^{p} a_{ii}^i. \tag{1.3.11}$$

INVERSE TRANSFORMATIONS

(xii)   For nonsingular matrices $Y\,(p \times p)$ and $X\,(p \times p)$, if $Y = X^{-1}$, then
$$J(Y \to X) = \det(X)^{-2p}. \tag{1.3.12}$$

(xiii) For nonsingular symmetric matrices $Y\,(p \times p)$ and $X\,(p \times p)$, if $Y = X^{-1}$, then
$$J(Y \to X) = \det(X)^{-(p+1)}. \tag{1.3.13}$$

QUADRATIC TRANSFORMATIONS

(xiv)   For a symmetric positive definite matrix $Y\,(p \times p)$ and a lower triangular matrix $T\,(p \times p) = (t_{ij})$, if $Y = TT'$, then

$$J(Y \to T) = 2^p \prod_{i=1}^{p} t_{ii}^{p-i+1} \tag{1.3.14}$$

and if $Y = TAT'$, then

$$J(Y \to T) = 2^p \prod_{i=1}^{p} \left(t_{ii}^{p-i+1} \det(A^{[i]}) \right), \tag{1.3.15}$$

where $A\,(p \times p) = (a_{jk})$ is nonsingular and $A^{[i]} = (a_{jk})$, $j, k = 1, \ldots, i$.

(xv)   For a symmetric positive definite matrix $Y\,(p \times p)$ and an upper triangular matrix $T\,(p \times p) = (t_{ij})$, if $Y = TT'$, then

$$J(Y \to T) = 2^p \prod_{i=1}^{p} t_{ii}^i \tag{1.3.16}$$

and if $Y = TAT'$, then

$$J(Y \to T) = 2^p \prod_{i=1}^{p} \left(t_{ii}^i \det(A_{[i]}) \right), \tag{1.3.17}$$

where $A\,(p \times p) = (a_{jk})$ is nonsingular and $A_{[i]} = (a_{jk})$, $j, k = p-i+1, \ldots, p$.

(xvi)   For symmetric positive definite matrices $Y\,(p \times p)$, $X\,(p \times p)$, $U\,(p \times p)$ and $V\,(p \times p)$, if $U = X + Y$ and $V = U^{-\frac{1}{2}} Y (U^{-\frac{1}{2}})'$ where $U = U^{\frac{1}{2}} (U^{\frac{1}{2}})'$, then
$$J(X, Y \to U, V) = \det(U)^{\frac{1}{2}(p+1)}. \tag{1.3.18}$$

(xvii) For symmetric positive definite matrices $Y\,(p \times p)$ and $X\,(p \times p)$, if
$$Y = (I_p + X)^{-\frac{1}{2}} X ((I_p + X)^{-\frac{1}{2}})' \text{ where } (I_p + X) = (I_p + X)^{\frac{1}{2}} ((I_p + X)^{\frac{1}{2}})',$$

then

$$J(Y \to X) = \det(I_p + X)^{-(p+1)}. \tag{1.3.19}$$

(xviii) For symmetric positive definite matrices Y $(p \times p)$ and X $(p \times p)$, if $Y = X^2$, then Olkin and Rubin (1964) showed that

$$J(Y \to X) = \prod_{i \leq j}^{p} (\delta_i + \delta_j) = h(\delta_1, \dots, \delta_p) \tag{1.3.20}$$

where $\delta_1 \dots, \delta_p$ are the eigenvalues of matrix X. For $p = 2, 3$ and 4, (1.3.20) is simplified as

$$h(\delta_1, \delta_2) = 2^2 a_2 a_1$$
$$h(\delta_1, \delta_2, \delta_3) = 2^3 a_3 (a_1 a_2 - a_3)$$
$$h(\delta_1, \delta_2, \delta_3, \delta_4) = 2^4 a_4 (a_1 a_2 a_3 - a_3^2 - a_1^2 a_4),$$

where a_k is the kth elementary symmetric function of $\delta_1, \dots, \delta_p$. Further $a_k = \text{tr}_k(X)$ where $\text{tr}_k(X)$ is the sum of all k^{th} order principal minors of X.

(xix) For symmetric positive definite matrices Y $(p \times p)$, X $(p \times p)$ and a symmetric matrix B $(p \times p)$, if $Y = XBX$, then (Olkin and Rubin, 1964)

$$J(Y \to X) = \prod_{i \leq j}^{p} (\lambda_i + \lambda_j) \tag{1.3.21}$$

where $\lambda_1, \dots, \lambda_p$ are eigenvalues of $B^{\frac{1}{2}} X B^{\frac{1}{2}}$.

ORTHOGONAL TRANSFORMATIONS

Let the rank of the matrix X $(p \times n)$ be p $(\leq n)$. Then X has the unique representation $X = TH_1$ where T $(p \times p)$ is a lower triangular matrix with positive diagonal elements and H_1 $(p \times n)$ is a semiorthogonal matrix, i.e., $H_1 H_1' = I_p$. Here the matrix X has np variables, T has $\frac{1}{2}p(p+1)$ variables, and the semiorthogonal matrix H_1 has $np - \frac{1}{2}p(p+1)$ functionally independent variables due to the restriction $H_1 H_1' = I_p$. To obtain the Jacobian $J(X \to T, H_1)$, consider the differential form

$$(dX) = (d(TH_1)) = (dT)H_1 + T(dH_1). \tag{1.3.22}$$

Then $J(X \to T, H_1) = J((dX) \to (dT), (dH_1))$. Let $H = \begin{pmatrix} H_1 \\ H_2 \end{pmatrix} \begin{matrix} p \\ n-p \end{matrix}$ be an orthogonal matrix. Then $H_1 H' = (\, I_p \quad 0\,)$ and

$$(d(H_1 H')) = (\,0 \quad 0\,) = (dH_1)H' + H_1 (\, dH_1' \quad dH_2'\,).$$

Also, let $R_1 = (dH_1)H'$ and $R_2 = (dH_1)H_2'$. Then it is easy to see that R_1 $(p \times p)$ is skew-symmetric. Post-multiplying the differential form (1.3.22) by H', we get

$$(dX)H' = (dT)H_1 H' + T(dH_1)H'$$

$$= (dT)\,(\,I_p \quad 0\,) + T\,(\,R_1 \quad R_2\,)$$
$$= (\,W_1 \quad W_2\,)$$

where $W_1 = (dT) + TR_1 = (w_{ij}^1)$ and $W_2 = TR_2$. Thus

$$J(X \to T, H_1) = J((dX) \to (dT), (dH_1))$$
$$= J((dX) \to W)J(W_1 \to (dT), R_1)J(W_2 \to R_2)J(R_1, R_2 \to (dH_1)).$$

Now
$$J((dX) \to W) = \mathrm{mod}\,\det(H')^p = 1,$$
$$J(W_2 \to R_2) = \det(T)^{n-p},$$

and
$$J(R_1, R_2 \to (dH_1)) = g_{n,p}(H_1) \text{ (say)}.$$

Then

$$J(X \to T, H_1) \quad = \quad J(W_1 \to (dT), R_1)\det(T)^{n-p}g_{n,p}(H_1). \qquad (1.3.23)$$

Further, if

$$(dT) = \begin{pmatrix} dt_{11} & 0 & \cdots & 0 \\ dt_{21} & dt_{22} & \cdots & 0 \\ \vdots & \vdots & & \\ dt_{p1} & dt_{p2} & \cdots & dt_{pp} \end{pmatrix}, \text{ and } R_1 = \begin{pmatrix} 0 & r_{12}^1 & r_{13}^1 & \cdots & r_{1p}^1 \\ -r_{12}^1 & 0 & r_{23}^1 & \cdots & r_{2p}^1 \\ \vdots & & & & \\ -r_{1p}^1 & -r_{2p}^1 & -r_{3p}^1 & \cdots & 0 \end{pmatrix}$$

then

$$w_{ij}^1 = dt_{ij} + (TR_1)_{ij}, \; i \geq j$$
$$= (TR_1)_{ij}, \; i < j$$

and

$$J(W_1 \to (dT), R_1) = t_{11}^{p-1}t_{22}^{p-2}\cdots t_{pp}^{p-p} = \prod_{i=1}^{p} t_{ii}^{p-i}. \qquad (1.3.24)$$

Substituting from (1.3.24) in (1.3.23), we finally get

$$J(X \to T, H_1) = \prod_{i=1}^{p} t_{ii}^{n-i}g_{n,p}(H_1), \qquad (1.3.25)$$

where

$$g_{n,p}(H_1) = J((dH_1)\,(\,H_1' \quad H_2'\,) \to (dH_1)). \qquad (1.3.26)$$

Here $g_{n,p}(H_1)\,dH_1$ defines the invariant measure on the Stiefel manifold $O(p,n)$,

$$O(p,n) = \{H_1\,(p \times n) : H_1H_1' = I_p\} \qquad (1.3.27)$$

and is denoted by $[(dH_1)H_1']$. For $p = n$, the Stiefel manifold reduces to the orthogonal group,

$$O(p, p) = O(p) = \{H\,(p \times p) : HH' = I_p\},$$

and the invariant measure on $O(p)$ is $[(dH)H']$. In the next section it is shown that

$$\int_{O(p,n)} [(dH_1)H_1'] = \frac{2^p \pi^{\frac{1}{2}np}}{\Gamma_p(\frac{1}{2}n)}$$

$$= \text{Vol}(O(p, n)). \qquad (1.3.28)$$

For $p = n$,

$$\text{Vol}(O(p)) = \frac{2^p \pi^{\frac{1}{2}p^2}}{\Gamma_p(\frac{1}{2}p)}.$$

Dividing $[(dH_1)H_1']$ by the volume (or the surface area) of the Stiefel manifold, we obtain the unit invariant measure,

$$[dH_1] = \frac{[(dH_1)H_1']}{\text{Vol}(O(p, n))}. \qquad (1.3.29)$$

Thus, the measure (1.3.29) is the normalized surface area of the $np - \frac{1}{2}p(p+1)$ dimensional surface in the np-space defined by (1.3.27). From here the density of H_1 for $n \geq p$ is obtained as

$$\frac{\Gamma_p(\frac{1}{2}n)}{2^p \pi^{\frac{1}{2}np}} g_{n,p}(H_1), \quad H_1 H_1' = I_p.$$

For $p = n$, we have

$$[dH] = \frac{[(dH)H']}{\text{Vol}(O(p))},$$

which is known as the unit invariant Haar measure on the orthogonal group $O(p)$.

Partition $H_1\,(p \times n)$ as

$$H_1 = \begin{pmatrix} V \\ V_1 \end{pmatrix} \begin{matrix} q \\ p - q \end{matrix}.$$

Now choose $Z\,((p-q) \times (n-q))$ and $G(V)\,((n-q) \times n)$ such that $ZZ' = I_{p-q}$, $\begin{pmatrix} V \\ G(V) \end{pmatrix}$ is orthogonal, $V_1 = ZG(V)$, and the relationship between $Z \in O(p-q, n-q)$ and $V_1 \in O(p-q, n)$ is one-to-one. Then the unit invariant measure $[dH_1]$ can be decomposed as the product

$$[dH_1] = [dV][dZ], \qquad (1.3.30)$$

where $[dV]$ and $[dZ]$ are unit invariant measures on $O(q, n)$ and $O(p-q, n-q)$ respectively.

The decomposition (1.3.30) was derived by Chikuse (1990a). She has also given the sequential decomposition of $[dH_1]$ into the product of several invariant measures. For further details the reader is referred to James (1954), Herz (1955), Farrell (1985), Muirhead (1982), and Chikuse (1990a, 1990b).

1.4.   INTEGRATION

Integrals involving functions of matrix arguments are frequently used in this book. In this section, we study such integrals.

Let $f(X)$ be a scalar function of the matrix X. Then

$$\int_R f(X)\,dX$$

is defined as the iterated integral of $f(X)$ with respect to each element of X separately over a region R in the space defined by the simplex bounding the ranges of the elements of X. Evaluation of these integrals is facilitated by the use of Laplace transform discussed in detail in Herz (1955).

DEFINITION 1.4.1. *Let $f(A)$ be a function of $A\,(p \times p) > 0$ and $Z = X + \iota Y$, $\iota = \sqrt{-1}$, be a $p \times p$ complex symmetric matrix. Then, the Laplace transform $g(Z)$ of $f(A)$ is defined as*

$$g(Z) = \int_{A>0} \mathrm{etr}(-ZA)f(A)\,dA,$$

where the integral is assumed to be absolutely convergent in the right half plane $\mathrm{Re}(Z) = X > X_0 > 0$.

The Laplace transform $g(Z)$ of $f(A)$ defined above is an analytic function of Z in the right half plane $\mathrm{Re}(Z) = X > X_0 > 0$. In addition, if

$$\int_{-\infty<Y=Y'<\infty} |g(X + \iota Y)|\,dY < \infty \qquad (1.4.1)$$

for all $X > X_0 > 0$, and

$$\lim_{X\to 0} \int_{-\infty<Y=Y'<\infty} |g(X + \iota Y)|\,dY = 0 \qquad (1.4.2)$$

then the unique inverse Laplace transform $f(A)$ of $g(Z)$ is

$$f(A) = \frac{2^{\frac{1}{2}p(p-1)}}{(2\pi\iota)^{\frac{1}{2}p(p+1)}} \int_{\mathrm{Re}(Z)=X>X_0>0} \mathrm{etr}(ZA)g(Z)\,dZ. \qquad (1.4.3)$$

An important property of Laplace transform is the convolution result. If g_1 and g_2 are the respective Laplace transforms of f_1 and f_2, then $g_1 g_2$ is the Laplace transform of f_3 where

$$f_3(B) = \int_{0<A<B} f_1(B - A)f_2(A)\,dA. \qquad (1.4.4)$$

Some integrals useful in the matrix variate distribution theory are now given.

DEFINITION 1.4.2. *The multivariate gamma function, denoted by $\Gamma_p(a)$, is defined as*

$$\Gamma_p(a) = \int_{A>0} \mathrm{etr}(-A)\det(A)^{a-\frac{1}{2}(p+1)}\,dA, \qquad (1.4.5)$$

where $\mathrm{Re}(a) > \frac{1}{2}(p-1)$, and the integral is over the space of $p \times p$ symmetric positive definite matrices.

The multivariate gamma function $\Gamma_p(a)$ can be expressed as product of ordinary gamma functions as given in the following theorem.

THEOREM 1.4.1. *For* $\operatorname{Re}(a) > \frac{1}{2}(p-1)$,

$$\Gamma_p(a) = \pi^{\frac{1}{4}p(p-1)} \prod_{i=1}^{p} \Gamma\left[a - \frac{1}{2}(i-1)\right].$$

Proof: By definition

$$\Gamma_p(a) = \int_{A>0} \operatorname{etr}(-A) \det(A)^{a-\frac{1}{2}(p+1)} \, dA.$$

Substitute $A = TT'$, where T is a lower triangular matrix with $t_{ii} > 0$, $i = 1, \dots, p$. Then, $\operatorname{tr}(A) = \operatorname{tr}(TT') = \sum_{j \le i} t_{ij}^2$, $\det(A) = \det(TT') = \det(T)^2 = \prod_{i=1}^{p} t_{ii}^2$, and from (1.3.14)

$$J(A \to T) = 2^p \prod_{i=1}^{p} t_{ii}^{p-i+1}.$$

Hence,

$$\Gamma_p(a) = 2^p \int \cdots \int_{\substack{-\infty < t_{ij} < \infty \\ t_{ii} > 0}} \prod_{i=1}^{p} (t_{ii}^2)^{a-\frac{1}{2}i} \exp\left(-\sum_{j \le i} t_{ij}^2\right) \prod_{j \le i} dt_{ij}$$

$$= \left[\prod_{j<i}^{p} \int_{-\infty < t_{ij} < \infty} \exp(-t_{ij}^2) \, dt_{ij}\right]\left[\prod_{i=1}^{p} 2 \int_{t_{ii}>0} \exp(-t_{ii}^2)(t_{ii}^2)^{a-\frac{1}{2}i} \, dt_{ii}\right]$$

$$= \pi^{\frac{1}{4}p(p-1)} \prod_{i=1}^{p} \Gamma\left[a - \frac{1}{2}(i-1)\right]. \quad \blacksquare$$

A particular Laplace transform which is quite useful is

$$\int_{\Lambda>0} \operatorname{etr}(-\Lambda Z) \det(\Lambda)^{a-\frac{1}{2}(p+1)} \, d\Lambda = \Gamma_p(a) \det(Z)^{-a}. \qquad (1.4.6)$$

Herz (1955) proved that the above integral is absolutely convergent for $\operatorname{Re}(Z) > 0$, and $\operatorname{Re}(a) > \frac{1}{2}(p-1)$. Hence, for $\operatorname{Re}(Z) > 0$, substituting $A = Z^{\frac{1}{2}} \Lambda Z^{\frac{1}{2}}$ with the Jacobian $J(\Lambda \to A) = \det(Z)^{-\frac{1}{2}(p+1)}$, in the above integral we get

$$\int_{\Lambda>0} \operatorname{etr}(-\Lambda Z) \det(\Lambda)^{a-\frac{1}{2}(p+1)} \, d\Lambda = \det(Z)^{-a} \int_{A>0} \operatorname{etr}(-A) \det(A)^{a-\frac{1}{2}(p+1)} \, dA$$

$$= \Gamma_p(a) \det(Z)^{-a}.$$

This proves (1.4.6) for real Z. It follows for complex Z by analytic continuation since $\operatorname{Re}(Z) > 0$, $\det(Z) \ne 0$ and $\det(Z)^a$ are well defined by continuation. Using the inversion formulas (1.4.3), after verifying the conditions (1.4.1) and (1.4.2), Herz (1955) gave the inversion of (1.4.6) as

$$\frac{2^{\frac{1}{2}p(p-1)}}{(2\pi\iota)^{\frac{1}{2}p(p+1)}} \int_{\operatorname{Re}(Z)=X>X_0>0} \operatorname{etr}(Z\Lambda) \det(Z)^{-a-\frac{1}{2}(p+1)} \, dZ = \frac{\det(\Lambda)^a}{\Gamma_p\left[a + \frac{1}{2}(p+1)\right]}, \quad \Lambda > 0.$$

DEFINITION 1.4.3. *The multivariate beta function, denoted by $\beta_p(a, b)$, is defined by*

$$\beta_p(a, b) = \int_{0 < A < I_p} \det(A)^{a - \frac{1}{2}(p+1)} \det(I_p - A)^{b - \frac{1}{2}(p+1)} \, dA, \qquad (1.4.7)$$

where $\mathrm{Re}(a) > \frac{1}{2}(p - 1)$ *and* $\mathrm{Re}(b) > \frac{1}{2}(p - 1)$.

The multivariate beta function $\beta_p(a, b)$ can be expressed in terms of multivariate gamma functions.

THEOREM 1.4.2. *For* $\mathrm{Re}(a) > \frac{1}{2}(p - 1)$ *and* $\mathrm{Re}(b) > \frac{1}{2}(p - 1)$,

$$\beta_p(a, b) = \frac{\Gamma_p(a)\Gamma_p(b)}{\Gamma_p(a + b)}$$

$$= \beta_p(b, a). \qquad (1.4.8)$$

Proof: We have

$$\Gamma_p(a)\Gamma_p(b) = \int_{A>0} \mathrm{etr}(-A)\det(A)^{a - \frac{1}{2}(p+1)} \, dA \int_{B>0} \mathrm{etr}(-B)\det(B)^{b - \frac{1}{2}(p+1)} \, dB$$

$$= \int_{A>0} \int_{B>0} \mathrm{etr}\{-(A + B)\}\det(A)^{a - \frac{1}{2}(p+1)} \det(B)^{b - \frac{1}{2}(p+1)} \, dA \, dB.$$

Now making the transformation $W = A + B$, $Z = (A + B)^{-\frac{1}{2}} A (A + B)^{-\frac{1}{2}}$ (where $(A + B)^{\frac{1}{2}}$ is symmetric square root of $A + B$ with the Jacobian $J(A, B \to Z, W) = \det(W)^{\frac{1}{2}(p+1)}$, we get

$$\Gamma_p(a)\Gamma_p(b) = \int_{W>0} \mathrm{etr}(-W)\det(W)^{a + b - \frac{1}{2}(p+1)} \, dW$$

$$\int_{0 < Z < I_p} \det(Z)^{a - \frac{1}{2}(p+1)} \det(I_p - Z)^{b - \frac{1}{2}(p+1)} \, dZ$$

$$= \Gamma_p(a + b)\beta_p(a, b). \qquad \blacksquare$$

Alternatively, Theorem 1.4.2 can be proved by using the convolution formula (1.4.4). Substituting $A = (I_p + B)^{-1}$ in (1.4.7) with Jacobian

$$J(A \to B) = J((dA) \to (dB))$$

$$= \det(I_p + B)^{-(p+1)},$$

we get an equivalent integral representation for the multivariate beta function as

$$\beta_p(a, b) = \int_{B>0} \det(B)^{b - \frac{1}{2}(p+1)} \det(I_p + B)^{-(a+b)} \, dB. \qquad (1.4.9)$$

The incomplete gamma and beta functions corresponding to (1.4.5) and (1.4.7), which are expressible in terms of hypergeometric functions, are defined in Section 1.6. Now we generalize the multivariate beta function.

DEFINITION 1.4.4. *The multivariate Dirichlet function, denoted by* $\beta_p(a_1, \ldots, a_r; b)$, *is defined by*

$$\beta_p(a_1, \ldots, a_r; b) = \int \cdots \int\limits_{\substack{0 < \sum_{i=1}^r Z_i < I_p \\ Z_i > 0}} \prod_{i=1}^r \det(Z_i)^{a_i - \frac{1}{2}(p+1)} \det\left(I_p - \sum_{i=1}^r Z_i\right)^{b - \frac{1}{2}(p+1)} \prod_{i=1}^r dZ_i$$

$$(1.4.10)$$

where $\text{Re}(a_i) > \frac{1}{2}(p-1)$, $i = 1, \ldots, r$, *and* $\text{Re}(b) > \frac{1}{2}(p-1)$.

The relation between the multivariate Dirichlet function and the multivariate gamma function is given in the following theorem.

THEOREM 1.4.3. *For* $\text{Re}(a_i) > \frac{1}{2}(p-1)$, $i = 1, \ldots, r$, *and* $\text{Re}(b) > \frac{1}{2}(p-1)$,

$$\beta_p(a_1, \ldots, a_r; b) = \frac{\Gamma_p(b) \prod_{i=1}^r \Gamma_p(a_i)}{\Gamma_p(a+b)} \tag{1.4.11}$$

where $a = \sum_{i=1}^r a_i$.

Proof: First consider the integral

$$\phi(Z) = \int \cdots \int\limits_{\substack{\sum_{i=1}^r Z_i = Z \\ Z_i > 0}} \prod_{i=1}^r \det(Z_i)^{a_i - \frac{1}{2}(p+1)} \det\left(I_p - \sum_{i=1}^r Z_i\right)^{b - \frac{1}{2}(p+1)} \prod_{i=1}^r dZ_i. \tag{1.4.12}$$

Substituting $\sum_{i=1}^r Z_i = Z$, $W_i = Z^{-\frac{1}{2}} Z_i Z^{-\frac{1}{2}}$, $i = 1, \ldots, r-1$, where $Z^{\frac{1}{2}} Z^{\frac{1}{2}} = Z$, in (1.4.12) with Jacobian

$$J(Z_1, \ldots, Z_{r-1}, Z_r \to W_1, \ldots, W_{r-1}, Z) = \det(Z)^{\frac{1}{2}(r-1)(p+1)}$$

we get

$$\phi(Z) = \det(Z)^{a - \frac{1}{2}(p+1)} \det(I_p - Z)^{b - \frac{1}{2}(p+1)}$$

$$\int \cdots \int\limits_{\substack{\sum_{i=1}^{r-1} W_i < I_p \\ W_i > 0}} \prod_{i=1}^{r-1} \det(W_i)^{a_i - \frac{1}{2}(p+1)} \det\left(I_p - \sum_{i=1}^{r-1} W_i\right)^{a_r - \frac{1}{2}(p+1)} \prod_{i=1}^{r-1} dW_i$$

$$= \det(Z)^{a - \frac{1}{2}(p+1)} \det(I_p - Z)^{b - \frac{1}{2}(p+1)} \beta_p(a_1, \ldots, a_{r-1}; a_r). \tag{1.4.13}$$

Now from (1.4.13) and (1.4.7) we can write

$$\beta_p(a_1, \ldots, a_r; b) = \int_{0 < Z < I_p} \phi(Z) \, dZ$$

$$= \beta_p(a_1, \ldots, a_{r-1}; a_r) \int_{0 < Z < I_p} \det(Z)^{a - \frac{1}{2}(p+1)} \det(I_p - Z)^{b - \frac{1}{2}(p+1)} \, dZ$$

$$= \beta_p(a_1, \ldots, a_{r-1}; a_r) \beta_p(a, b). \tag{1.4.14}$$

From the recurrence relation (1.4.14) we get

$$\beta_p(a_1,\ldots,a_r;b) = \beta_p\Big(\sum_{i=1}^r a_i,b\Big)\beta_p\Big(\sum_{i=1}^{r-1} a_i,a_r\Big)\cdots\beta_p(a_1,a_2). \qquad (1.4.15)$$

Substituting for the multivariate beta functions on the right hand side from (1.4.15) we get the result. ∎

The following result, Olkin (1959), is the matrix variate analog of Liouville's extension of Dirichlet integral.

THEOREM 1.4.4. *Let $f(V)$ be a continuous scalar function of the symmetric matrix V ($p \times p$). Then for B ($p \times p$) $> A$ ($p \times p$) > 0, $\mathrm{Re}(a_i) > \frac{1}{2}(p-1)$, $i = 1,\ldots,r$, and $\sum_{i=1}^r a_i = a$,*

$$\int\cdots\int_{\substack{A<\sum_{i=1}^r Z_i<B\\Z_i>0}} \prod_{i=1}^r \det(Z_i)^{a_i-\frac{1}{2}(p+1)} f\Big(\sum_{i=1}^r Z_i\Big) \prod_{i=1}^r dZ_i$$

$$= \beta_p(a_1,a_2)\beta_p(a_1+a_2,a_3)\cdots\beta_p\Big(\sum_{i=1}^{r-1} a_i,a_r\Big)\int_{A<Z<B} \det(Z)^{a-\frac{1}{2}(p+1)} f(Z)\,dZ.$$

Proof: Making the same transformation as in Theorem 1.4.3, the above left hand side integral becomes

$$\int\cdots\int_{\substack{\sum_{i=1}^{r-1} W_i<I_p\\W_i>0}} \prod_{i=1}^{r-1} \det(W_i)^{a_i-\frac{1}{2}(p+1)}\Big(I_p-\sum_{i=1}^{r-1} W_i\Big)^{a_r-\frac{1}{2}(p+1)} \prod_{i=1}^{r-1} dW_i$$

$$\int_{A<Z<B} \det(Z)^{a-\frac{1}{2}(p+1)} f(Z)\,dZ$$

$$= \beta_p(a_1,\ldots,a_{r-1};a_r)\int_{A<Z<B} \det(Z)^{a-\frac{1}{2}(p+1)} f(Z)\,dZ$$

which is obtained by using (1.4.10). The desired result now follows from (1.4.15). ∎

The following integral is useful in the theory of correlation matrices in multivariate statistical analysis.

THEOREM 1.4.5. *Let $R = (r_{ij})$ with $r_{ij} = r_{ji}$, $i \neq j$, $i,j = 1,\ldots,p$ and $r_{ii} = 1$. Then, for $\mathrm{Re}(a) > \frac{1}{2}(p-1)$, we have*

$$\int_{0<R<I_p} \det(R)^{a-\frac{1}{2}(p+1)}\,dR = \frac{\Gamma_p(a)}{[\Gamma(a)]^p},$$

where $dR = \prod_{i<j}^p dr_{ij}$.

Proof: We have
$$\Gamma_p(a) = \int_{A>0} \mathrm{etr}(-A)\det(A)^{a-\frac{1}{2}(p+1)}\,dA. \qquad (1.4.16)$$

Making the transformation $a_{ij} = \sqrt{a_{ii}}\sqrt{a_{jj}}\, r_{ij}$, $i \neq j$, $i, j = 1, \ldots, p$ and $a_{ii} = a_{ii}$ with the Jacobian

$$J(a_{11}, \ldots, a_{pp}, a_{12}, \ldots, a_{p-1,p} \to a_{11}, \ldots, a_{pp}, r_{12}, \ldots, r_{p-1,p}) = \prod_{i=1}^{p} a_{ii}^{\frac{1}{2}(p-1)},$$

in (1.4.16), we get

$$\Gamma_p(a) = \int_{0 < R < I_p} \det(R)^{a - \frac{1}{2}(p+1)}\, dR \prod_{i=1}^{p} \int_{a_{ii} > 0} a_{ii}^{a-1} \exp(-a_{ii})\, da_{ii}.$$

The result follows since $\int_{a_{ii}>0} a_{ii}^{a-1} \exp(-a_{ii})\, da_{ii} = \Gamma(a)$. ∎

Bellman (1956) generalized the multivariate gamma function as follows.

DEFINITION 1.4.5. *The generalized multivariate gamma function, denoted by* $\Gamma_p^*(a_1, \ldots, a_p)$, *is defined as*

$$\Gamma_p^*(a_1, \ldots, a_p) = \int_{A>0} \frac{\det(A)^{a_p - \frac{1}{2}(p+1)} \operatorname{etr}(-A)}{\prod_{\alpha=1}^{p-1} \det(A^{[\alpha]})^{m_{\alpha}+1}}\, dA,$$

where $a_j = m_1 + \cdots + m_j$ *and* $\operatorname{Re}(a_j) > \frac{1}{2}(j-1)$, $j = 1, \ldots, p$.

THEOREM 1.4.6. *For* $\operatorname{Re}(a_j) > \frac{1}{2}(j-1)$, $j = 1, \ldots, p$,

$$\Gamma_p^*(a_1, \ldots, a_p) = \pi^{\frac{1}{4}p(p-1)} \prod_{j=1}^{p} \Gamma\left[a_j - \frac{1}{2}(j-1)\right].$$

Proof: By definition

$$\Gamma_p^*(a_1, \ldots, a_p) = \int_{A>0} \frac{\det(A)^{a_p - \frac{1}{2}(p+1)} \operatorname{etr}(-A)}{\prod_{\alpha=1}^{p-1} \det(A^{[\alpha]})^{m_{\alpha}+1}}\, dA, \qquad (1.4.17)$$

where $a_j = m_1 + \cdots + m_j$. Let $A = TT'$ where T is a lower triangular matrix with positive diagonal elements and partition A and T as

$$A = \begin{pmatrix} A_{11} & A_{12} \\ A_{21} & A_{22} \end{pmatrix} \begin{matrix} \alpha \\ p-\alpha \end{matrix}, \quad T = \begin{pmatrix} T_{11} & 0 \\ T_{21} & T_{22} \end{pmatrix} \begin{matrix} \alpha \\ p-\alpha \end{matrix}.$$
$$\begin{matrix} \alpha \quad p-\alpha \end{matrix} \qquad \begin{matrix} \alpha \quad p-\alpha \end{matrix}$$

Now it is easy to see that $A^{[\alpha]} = T_{11}T_{11}'$, $\det(A^{[\alpha]}) = \prod_{i=1}^{\alpha} t_{ii}^2$, $\det(A) = \prod_{i=1}^{p} t_{ii}^2$ and $\operatorname{tr}(A) = \sum_{i \geq j}^{p} t_{ij}^2$. From (1.3.14) we have $J(A \to T) = 2^p \prod_{i=1}^{p} t_{ii}^{p-i+1}$. Hence we can write (1.4.17) as

$$\Gamma_p^*(a_1, \ldots, a_p) = 2^p \int \cdots \int_{\substack{-\infty < t_{ij} < \infty \\ t_{ii} > 0}} \frac{\prod_{i=1}^{p} (t_{ii}^2)^{a_p - \frac{1}{2}i} \exp\left(-\sum_{j \leq i}^{p} t_{ij}^2\right)}{\prod_{\alpha=1}^{p-1} \prod_{i=1}^{\alpha} (t_{ii}^2)^{m_{\alpha}+1}} \prod_{j \leq i}^{p} dt_{ij}$$

$$= \left[\prod_{j<i} \int_{-\infty < t_{ij} < \infty} \exp(-t_{ij}^2)\, dt_{ij}\right] \left[\prod_{i=1}^{p} 2 \int_{t_{ii}>0} (t_{ii}^2)^{a_i - \frac{1}{2}i} \exp(-t_{ii}^2)\, dt_{ii}\right].$$

The desired result now follows since $\int_{-\infty < t_{ij} < \infty} \exp(-t_{ij}^2)\, dt_{ij} = \sqrt{\pi}$ and $2\int_{t_{ii} > 0}(t_{ii}^2)^{a_i - \frac{1}{2}i}$ $\exp(-t_{ii}^2)\, dt_{ii} = \Gamma\left[a_i - \frac{1}{2}(i-1)\right]$. ∎

THEOREM 1.4.7. *For* $\mathrm{Re}(a_j) > \frac{1}{2}(j-1)$, $j = 1, \ldots, p$ *and* $B > 0$,

$$\int_{A>0} \frac{\det(A)^{a_p - \frac{1}{2}(p+1)}\, \mathrm{etr}(-BA)}{\prod_{\alpha=1}^{p-1} \det(A^{[\alpha]})^{m_\alpha+1}}\, dA = \Gamma_p^*(a_1, \ldots, a_p) \prod_{\alpha=1}^{p} \det(B_{(\alpha)})^{-m_\alpha}, \qquad (1.4.18)$$

where $a_j = m_1 + \cdots + m_j$.

Proof: Since $B > 0$, let $B = UU'$ where $U = (u_{ij})$ is an upper triangular matrix. Substitute $\Lambda = U'AU$, then $\mathrm{tr}(BA) = \mathrm{tr}(UU'A) = \mathrm{tr}(U'AU) = \mathrm{tr}(\Lambda)$, and $J(A \to \Lambda) = \det(U)^{-(p+1)} = \det(B)^{-\frac{1}{2}(p+1)}$. Now partition U, A and Λ as

$$U = \begin{pmatrix} U_{11} & U_{12} \\ 0 & U_{22} \end{pmatrix} \begin{matrix} \alpha \\ p-\alpha \end{matrix}, \quad A = \begin{pmatrix} A_{11} & A_{12} \\ A_{21} & A_{22} \end{pmatrix} \begin{matrix} \alpha \\ p-\alpha \end{matrix}, \quad \Lambda = \begin{pmatrix} \Lambda_{11} & \Lambda_{12} \\ \Lambda_{21} & \Lambda_{22} \end{pmatrix} \begin{matrix} \alpha \\ p-\alpha \end{matrix}.$$
$$\;\;\;\;\alpha \;\; p-\alpha \qquad\qquad \alpha \;\; p-\alpha \qquad\qquad \alpha \;\; p-\alpha$$

Then, $\Lambda_{11} = \Lambda^{[\alpha]} = U_{11}' A_{11} U_{11}$, and

$$\det(\Lambda^{[\alpha]}) = \det(U_{11}' A_{11} U_{11}) = \det(A_{11}) \prod_{i=1}^{\alpha} u_{ii}^2 = \det(A^{[\alpha]}) \prod_{i=1}^{\alpha} u_{ii}^2.$$

Thus the left hand side of (1.4.18) reduces to

$$\frac{\det(B)^{-a_p}}{\prod_{\alpha=1}^{p-1} \prod_{i=1}^{\alpha} (u_{ii}^{-2})^{m_\alpha+1}} \int_{\Lambda>0} \frac{\det(\Lambda)^{a_p - \frac{1}{2}(p+1)}\, \mathrm{etr}(-\Lambda)}{\prod_{\alpha=1}^{p-1} \det(\Lambda^{[\alpha]})^{m_\alpha+1}}\, d\Lambda = \frac{\Gamma_p^*(a_1, \ldots, a_p)\, \det(B)^{-a_p}}{\prod_{\alpha=1}^{p-1} \prod_{i=1}^{\alpha} (u_{ii}^{-2})^{m_\alpha+1}}.$$

Now the result follows by noting that $\prod_{i=1}^{\alpha} u_{ii}^2 = \frac{\det(B)}{\det(B_{(\alpha+1)})}$. ∎

The proofs of above theorems are due to Olkin (1959). He has also generalized the multivariate beta integral as given in the next theorem.

DEFINITION 1.4.6. *The generalized multivariate beta function, denoted by* $\beta_p^*(a_1, \ldots, a_p; b_1, \ldots, b_p)$, *is defined by*

$$\beta_p^*(a_1, \ldots, a_p; b_1, \ldots, b_p) = \int_{0<A<I_p} \frac{\det(A)^{a_p - \frac{1}{2}(p+1)} \det(I_p - A)^{b_p - \frac{1}{2}(p+1)}}{\prod_{\alpha=1}^{p-1} \{\det(A^{[\alpha]})^{m_\alpha+1} \det((I_p - A)^{[\alpha]})^{k_\alpha+1}\}}\, dA,$$

where $a_j = \sum_{i=1}^{j} m_i$, $b_j = \sum_{i=1}^{j} k_i$, $\mathrm{Re}(a_j) > \frac{1}{2}(j-1)$, *and* $\mathrm{Re}(b_j) > \frac{1}{2}(j-1)$, $j = 1, \ldots, p$.

THEOREM 1.4.8. *For* $\mathrm{Re}(a_j) > \frac{1}{2}(j-1)$, *and* $\mathrm{Re}(b_j) > \frac{1}{2}(j-1)$, $j = 1, \ldots, p$,

$$\beta_p^*(a_1, \ldots, a_p; b_1, \ldots, b_p) = \frac{\Gamma_p^*(a_1, \ldots, a_p)\Gamma_p^*(b_1, \ldots, b_p)}{\Gamma_p^*(a_1 + b_1, \ldots, a_p + b_p)}.$$

Proof: We have, by Definition 1.4.5,

$$\Gamma_p^*(a_1, \ldots, a_p)\Gamma_p^*(b_1, \ldots, b_p)$$

$$= \int_{A>0} \int_{B>0} \frac{\det(A)^{a_p - \frac{1}{2}(p+1)} \det(B)^{b_p - \frac{1}{2}(p+1)} \operatorname{etr}(-A - B)}{\prod_{\alpha=1}^{p-1} \{\det(A^{[\alpha]})^{m_\alpha+1} \det(B^{[\alpha]})^{k_\alpha+1}\}} \, dA \, dB \quad (1.4.19)$$

where $a_j = \sum_{i=1}^{j} m_i$, $b_j = \sum_{i=1}^{j} k_i$, $\operatorname{Re}(a_j) > \frac{1}{2}(j-1)$, and $\operatorname{Re}(b_j) > \frac{1}{2}(j-1)$, $j = 1, \ldots, p$. Now, let $A + B = TT'$ where T is a lower triangular matrix with positive diagonal elements and $W = T^{-1}A(T')^{-1}$. Then $A = TWT'$, $B = T(I_p - W)T'$, $\det(B^{[\alpha]}) = \det((I_p - W)^{[\alpha]}) \prod_{i=1}^{\alpha} t_{ii}^2$, $\det(A^{[\alpha]}) = \det(W^{[\alpha]}) \prod_{i=1}^{\alpha} t_{ii}^2$, and $\operatorname{tr}(A + B) = \sum_{j \leq i}^{p} t_{ij}^2$. The Jacobian of transformation from (1.3.14) and (1.3.18) is given by $J(A, B \to W, T) = 2^p \prod_{i=1}^{p} t_{ii}^{2p-i+2}$. Hence (1.4.19) can be written as

$$\Gamma_p^*(a_1, \ldots, a_p)\Gamma_p^*(b_1, \ldots, b_p) = \int_{0<W<I_p} \frac{\det(W)^{a_p - \frac{1}{2}(p+1)} \det(I_p - W)^{b_p - \frac{1}{2}(p+1)}}{\prod_{\alpha=1}^{p-1} \{\det(W^{[\alpha]})^{m_\alpha+1} \det((I_p-W)^{[\alpha]})^{k_\alpha+1}\}} \, dW$$

$$2^p \int \cdots \int_{\substack{-\infty < t_{ij} < \infty \\ t_{ii} > 0}} \frac{\prod_{i=1}^{p} (t_{ii}^2)^{a_p+b_p - \frac{1}{2}i} \exp(-\sum_{j \leq i}^{p} t_{ij}^2)}{\prod_{\alpha=1}^{p-1} \prod_{i=1}^{\alpha} (t_{ii}^2)^{m_\alpha+1+k_\alpha+1}} \prod_{j \leq i}^{p} dt_{ij}$$

$$= \beta_p^*(a_1, \ldots, a_p; b_1, \ldots, b_p)\Gamma_p^*(a_1 + b_1, \ldots, a_p + b_p).$$

The last equality follows from Definition 1.4.6 and Theorem 1.4.6. ∎

In matrix variate distribution theory quite often we transform

$$Y = TH_1 \quad (1.4.20)$$

where Y $(p \times n)$ has rank $p \leq n$, T $(p \times p) = (t_{ij})$ is a lower triangular matrix with $t_{ii} > 0$, $i = 1, \ldots, p$, and H_1 $(p \times n)$ is a semiorthogonal matrix, $H_1 H_1' = I_p$. The Jacobian of this transformation, given in (1.3.25), is

$$J(Y \to T, H_1) = g_{n,p}(H_1) \prod_{i=1}^{p} t_{ii}^{n-i}, \quad (1.4.21)$$

where $g_{n,p}(H_1) \, dH_1$ defines the invariant measure on the Stiefel manifold $O(p, n)$. The following integral, which involves $g_{n,p}(H_1)$, is used to derive the distribution of certain transformed matrices.

THEOREM 1.4.9. *For $n \geq p$,*

$$\int_{H_1 H_1' = I_p} g_{n,p}(H_1) \, dH_1 = \frac{2^p \pi^{\frac{1}{2}np}}{\Gamma_p\left(\frac{1}{2}n\right)}.$$

Proof: Let y_{ij}, $i = 1, \ldots, p$, $j = 1, \ldots, n$ be np $(p \leq n)$ independent standard normal variates. The joint density of these variates is

$$(2\pi)^{-\frac{1}{2}np} \exp\left\{-\frac{1}{2} \sum_{i=1}^{p} \sum_{j=1}^{n} y_{ij}^2\right\}, \quad -\infty < y_{ij} < \infty. \quad (1.4.22)$$

Define

$$Y = \begin{pmatrix} y_{11} & \cdots & y_{1n} \\ \vdots & & \\ y_{p1} & \cdots & y_{pn} \end{pmatrix}.$$

Then the density (1.4.22) in matrix notation is

$$(2\pi)^{-\frac{1}{2}np} \operatorname{etr}\left(-\frac{1}{2}YY'\right), \, Y \in \mathbb{R}^{p \times n}.$$

Since (1.4.22) is a density function, we have

$$(2\pi)^{-\frac{1}{2}np} \int_{Y \in \mathbb{R}^{p \times n}} \operatorname{etr}\left(-\frac{1}{2}YY'\right) dY = 1. \qquad (1.4.23)$$

Making the transformation (1.4.20) with Jacobian (1.4.21) in (1.4.23) and noting

$$\operatorname{tr}(YY') = \operatorname{tr}(TT') = \sum_{j \leq i}^{p} t_{ij}^2,$$

we have

$$1 = (2\pi)^{-\frac{1}{2}np} \int_{\substack{-\infty < t_{ij} < \infty \\ t_{ii} > 0}} \cdots \int \int_{H_1 H_1' = I_p} \prod_{i=1}^{p} t_{ii}^{n-i} \exp\left(-\frac{1}{2}\sum_{j \leq i}^{p} t_{ij}^2\right) g_{n,p}(H_1) \, dH_1 \prod_{j \leq i}^{p} dt_{ij}$$

$$= (2\pi)^{-\frac{1}{2}np} \left[\prod_{j < i} \int_{-\infty < t_{ij} < \infty} \exp\left(-\frac{1}{2}t_{ij}^2\right) dt_{ij}\right]\left[\prod_{i=1}^{p} \int_{t_{ii} > 0} t_{ii}^{n-i} \exp\left(-\frac{1}{2}t_{ii}^2\right) dt_{ii}\right]$$

$$\int_{H_1 H_1' = I_p} g_{n,p}(H_1) \, dH_1.$$

Now using

$$\int_{-\infty < t_{ij} < \infty} \exp\left(-\frac{1}{2}t_{ij}^2\right) dt_{ij} = \sqrt{2\pi},$$

and

$$\int_{t_{ii} > 0} t_{ii}^{n-i} \exp\left(-\frac{1}{2}t_{ii}^2\right) dt_{ii} = 2^{\frac{1}{2}(n-i-1)} \Gamma\left[\frac{1}{2}(n-i+1)\right],$$

and Theorem 1.4.1 the result follows. ∎

Since $g_{n,p}(H_1) \, dH_1$ defines the invariant measure on $O(p, n)$, the above theorem gives the surface area or volume of the Stiefel manifold $O(p, n)$. That is

$$\operatorname{Vol}(O(p, n)) = \int_{O(p,n)} [(dH_1)H_1']$$

$$= \int_{H_1 H_1' = I_p} g_{n,p}(H_1) \, dH_1$$

$$= \frac{2^p \pi^{\frac{1}{2}np}}{\Gamma_p\left(\frac{1}{2}n\right)}.$$

The following theorem (Hsu, 1940) is useful in deriving the distribution of quadratic forms.

THEOREM 1.4.10. *Let Y $(p \times n)$ be of rank $p (\le n)$ and $f( )$ be a function of Y which depends on Y through YY' only. Then,*

$$\int_{YY'=A} f(YY')\, dY = \frac{\pi^{\frac{1}{2}np}}{\Gamma_p\left(\frac{1}{2}n\right)} \det(A)^{\frac{1}{2}(n-p-1)} f(A). \tag{1.4.24}$$

Proof: Note that

$$\int_{YY'=A} f(YY')\, dY = f(A) \int_{YY'=A} dY.$$

Now transform $Y = TH_1$ and $A = TT'$, where T is a triangular matrix with positive diagonal elements, and H_1 $(p \times n)$ is a semiorthogonal matrix. The Jacobian of this transformation is

$$J(Y \to A, H_1) = J(Y \to T, H_1) J(T \to A)$$

$$= \prod_{i=1}^{p} t_{ii}^{n-i} g_{n,p}(H_1) \left\{ 2^p \prod_{i=1}^{p} t_{ii}^{p-i+1} \right\}^{-1}$$

$$= 2^{-p} \det(A)^{\frac{1}{2}(n-p-1)} g_{n,p}(H_1).$$

Thus, we have

$$\int_{YY'=A} dY = 2^{-p} \det(A)^{\frac{1}{2}(n-p-1)} \int_{H_1 H_1' = I_p} g_{n,p}(H_1)\, dH_1$$

$$= \frac{\pi^{\frac{1}{2}np}}{\Gamma_p\left(\frac{1}{2}n\right)} \det(A)^{\frac{1}{2}(n-p-1)}. \tag{1.4.25}$$

This last step is derived by using Theorem 1.4.9. ∎

THEOREM 1.4.11. *For Y $(p \times n)$ and $\mathrm{Re}(m) > n + p - 1$,*

$$\int_{Y \in \mathbb{R}^{p \times n}} \det(I_p + YY')^{-\frac{1}{2}m}\, dY = \frac{\pi^{\frac{1}{2}np} \Gamma_p\left[\frac{1}{2}(m-n)\right]}{\Gamma_p\left(\frac{1}{2}m\right)}.$$

Proof: We first prove the theorem when $\mathrm{rank}(Y) = p \le n$. In this case, using Theorem 1.4.10, we get

$$\int_{Y \in \mathbb{R}^{p \times n}} \det(I_p + YY')^{-\frac{1}{2}m}\, dY = \int_{A>0} \int_{YY'=A} \det(I_p + YY')^{-\frac{1}{2}m}\, dY\, dA$$

$$= \frac{\pi^{\frac{1}{2}np}}{\Gamma_p\left(\frac{1}{2}n\right)} \int_{A>0} \det(A)^{\frac{1}{2}(n-p-1)} \det(I_p + A)^{-\frac{1}{2}m}\, dA$$

$$= \frac{\pi^{\frac{1}{2}np}}{\Gamma_p\left(\frac{1}{2}n\right)} \beta_p\left(\frac{1}{2}n, \frac{1}{2}(m-n)\right). \tag{1.4.26}$$

The last step follows from (1.4.9). Further simplification of beta function in (1.4.26), using Theorem 1.4.2, gives the desired result in this case.

For the case rank$(Y) = n \leq p$, writing

$$\det(I_p + YY') = \det(I_n + Y'Y),$$

and applying the above result we get

$$
\begin{aligned}
\int_{Y \in \mathbb{R}^{p \times n}} \det(I_p + YY')^{-\frac{1}{2}m} \, dY &= \int_{Y \in \mathbb{R}^{p \times n}} \det(I_n + Y'Y)^{-\frac{1}{2}m} \, dY \\
&= \frac{\pi^{\frac{1}{2}np} \Gamma_n\left[\frac{1}{2}(m - p)\right]}{\Gamma_n\left(\frac{1}{2}m\right)} \\
&= \frac{\pi^{\frac{1}{2}np} \Gamma_p\left[\frac{1}{2}(m - n)\right]}{\Gamma_p\left(\frac{1}{2}m\right)}. \quad \blacksquare
\end{aligned}
$$

Another useful result on integration is a generalization of Sverdrup's lemma (Sverdrup, 1947) by Kabe (1965) and Khatri (1965).

THEOREM 1.4.12. *Let Y $(p \times n)$ be of rank $p \leq n$, D $(q \times n)$ be of rank $q \leq n$ and C $(n \times n)$ be a symmetric positive definite matrix. Then for $p + q \leq n$,*

$$
\int_{\substack{YC^{-1}Y'=A \\ DY'=B'}} f(YC^{-1}Y', DY') \, dY = \frac{\pi^{\frac{1}{2}(n-q)p}}{\Gamma_p\left[\frac{1}{2}(n-q)\right]} \det(C)^{\frac{1}{2}p} \det(DCD')^{-\frac{1}{2}p}
$$

$$
\det(A - B(DCD')^{-1}B')^{\frac{1}{2}(n-p-q-1)} f(A, B').
$$

Proof: Let $C^{\frac{1}{2}}$ be the unique symmetric positive definite root of C. Since $DC^{\frac{1}{2}}$ is of rank q, we can find a matrix L $((n - q) \times n)$ such that

$$
G = \begin{pmatrix} DC^{\frac{1}{2}} \\ L \end{pmatrix}, \text{ and } GG' = \begin{pmatrix} DCD' & 0 \\ 0 & I_{n-q} \end{pmatrix}.
$$

Now let

$$
Y = X(G^{-1})'C^{\frac{1}{2}} = \begin{pmatrix} X_1 & X_2 \end{pmatrix}(G^{-1})'C^{\frac{1}{2}} \tag{1.4.27}
$$

where X_1 and X_2 are of order $p \times q$ and $p \times (n - q)$ respectively. The Jacobian of transformation is $J(Y \rightarrow X) = \det(DCD')^{-\frac{1}{2}p} \det(C)^{\frac{1}{2}p}$. From (1.4.27) we get

$$
YC^{-1}Y' = X_1(DCD')^{-1}X_1' + X_2X_2' \tag{1.4.28}
$$

and

$$
X_1 = YD' = B. \tag{1.4.29}
$$

Substituting (1.4.29) in (1.4.28) we get

$$
YC^{-1}Y' = B(DCD')^{-1}B' + X_2X_2', \tag{1.4.30}
$$

and

$$\int\limits_{\substack{YC^{-1}Y'=A \\ DY'=B'}} f(YC^{-1}Y', DY')\, dY = \int\limits_{X_2 X_2'=V} f_1(X_2 X_2')\, dX_2 \qquad (1.4.31)$$

where $V = A - B(DD')^{-1}B'$ and

$$f_1(X_2 X_2') = f(X_2 X_2' + B(DCD')^{-1}B', B').$$

Finally, using Theorem 1.4.10 to evaluate (1.4.31) we get

$$\int\limits_{X_2 X_2'=V} f_1(X_2 X_2')\, dX_2 = \frac{\pi^{\frac{1}{2}(n-q)p}}{\Gamma_p\left[\frac{1}{2}(n-q)\right]} \det(V)^{\frac{1}{2}(n-q-p-1)} f_1(V). \qquad (1.4.32)$$

Substituting for V in (1.4.32) gives the desired result. ∎

COROLLARY 1.4.12.1. *For $p = 1$, $Y = \boldsymbol{y}'$, $A = a$, $B' = \boldsymbol{b}$, the above integral becomes*

$$\int\limits_{\substack{\boldsymbol{y}'C^{-1}\boldsymbol{y}=a \\ D\boldsymbol{y}=\boldsymbol{b}}} f(\boldsymbol{y}'C^{-1}\boldsymbol{y}, D\boldsymbol{y})\, d\boldsymbol{y} = \frac{\pi^{\frac{1}{2}(n-q)}}{\Gamma\left[\frac{1}{2}(n-q)\right]} \det(C)^{\frac{1}{2}} \det(DCD')^{-\frac{1}{2}}$$

$$(a - \boldsymbol{b}'(DCD')^{-1}\boldsymbol{b})^{\frac{1}{2}(n-q-2)} f(a, \boldsymbol{b}). \qquad (1.4.33)$$

The result (1.4.33) was derived by Sverdrup (1947).

1.5. ZONAL POLYNOMIALS

In this and subsequent sections we give a brief description of zonal polynomials and hypergeometric functions of matrix arguments developed by Herz (1955), Hua (1959), and James (1960, 1961b, 1964).

Let $S\,(p \times p)$ be a symmetric matrix and V_k be the vector space of homogeneous polynomials $\phi(S)$ of degree k in $\frac{1}{2}p(p+1)$ distinct elements of S. The space V_k can be decomposed into a direct sum of irreducible invariant subspaces V_κ where $\kappa = (k_1, \ldots, k_p)$, $k_1 + \cdots + k_p = k$, $k_1 \geq \cdots \geq k_p \geq 0$. Then the polynomial $(\operatorname{tr} S)^k \in V_k$ has the unique decomposition into polynomials $C_\kappa(S) \in V_\kappa$ as

$$(\operatorname{tr} S)^k = \sum_\kappa C_\kappa(S). \qquad (1.5.1)$$

Thus we have

DEFINITION 1.5.1. *The zonal polynomial $C_\kappa(S)$ is the component of $(\operatorname{tr} S)^k$ in the subspace V_κ.*

The zonal polynomial $C_\kappa(S)$ is defined for all k and p, but for a partition κ of k into more than p parts, it is identically zero. These polynomials are invariant under orthogonal transformation, *i.e.*,

$$C_\kappa(S) = C_\kappa(HSH'), \ H \in O(p). \tag{1.5.2}$$

Hence $C_\kappa(S)$ is a symmetric homogeneous polynomial in the characteristic roots of S. Also if R is a symmetric positive definite matrix, then

$$C_\kappa(RS) = C_\kappa(R^{\frac{1}{2}} S R^{\frac{1}{2}}) \tag{1.5.3}$$

where $R^{\frac{1}{2}}$ is the unique symmetric positive definite square root of R. Khatri (1971), has shown that

$$|C_\kappa(S)| \le C_\kappa(S_0) \tag{1.5.4}$$

where $S_0 = \text{diag}(|s_1|, \dots, |s_p|)$, and s_i, $i = 1, \dots, p$, are the characteristic roots of S. If $S = I_p$, and the partition κ of k has r nonzero parts, then Constantine (1963) and James (1964) have shown that

$$C_\kappa(I_p) = \frac{2^{2k} k! \left(\frac{1}{2}p\right)_\kappa \prod_{i<j}^r (2k_i - 2k_j - i + j)}{\prod_{i=1}^r (2k_i + r - i)!} \tag{1.5.5}$$

where $(\frac{1}{2}p)_\kappa = \prod_{i=1}^r (\frac{1}{2}(p - i + 1))_{k_i}$ with $(a)_k = a(a+1) \cdots (a + k - 1)$, $(a)_0 = 1$. James (1964) has tabulated $C_\kappa(S)$ up to $k = 6$ and Parkhurst and James (1974) have extended these tables up to $k = 12$.

Next we define the *generalized hypergeometric coefficient* which frequently occurs in integrals involving zonal polynomials. Let $\kappa = (k_1, \dots, k_p)$, $k_1 \ge \cdots \ge k_p \ge 0$, $k_1 + \cdots + k_p = k$. Then

$$(a)_\kappa = \prod_{j=1}^p \left(a - \frac{1}{2}(j-1)\right)_{k_j}. \tag{1.5.6}$$

Using the notation

$$\Gamma_p(a, \kappa) = \pi^{\frac{1}{4}p(p-1)} \prod_{j=1}^p \Gamma\left[a + k_j - \frac{1}{2}(j-1)\right], \ \text{Re}(a) \ge \frac{1}{2}(p-1) - k_p, \tag{1.5.7}$$

so that $\Gamma_p(a, 0) = \Gamma_p(a)$, we can write (1.5.6) as

$$(a)_\kappa = \frac{\Gamma_p(a, \kappa)}{\Gamma_p(a)}. \tag{1.5.8}$$

Khatri (1966) introduced the notation

$$\Gamma_p(a, -\kappa) = \pi^{\frac{1}{4}p(p-1)} \prod_{j=1}^p \Gamma\left[a - k_j - \frac{1}{2}(p-j)\right], \ \text{Re}(a) \ge \frac{1}{2}(p-1) + k_1,$$

which is also used here. Alternatively we can write

$$\Gamma_p(a, -\kappa) = \frac{(-1)^k \Gamma_p(a)}{\left(-a + \frac{1}{2}(p+1)\right)_\kappa}. \tag{1.5.9}$$

Having defined the zonal polynomials, we now give certain integrals involving them. We will have several occasions to use these results.

LEMMA 1.5.1.

$$\int_{O(p)} \{\text{tr}(RH)\}^{2k} \, [dH] = \sum_{\kappa} \frac{(\frac{1}{2})_k}{(\frac{1}{2}p)_\kappa} C_\kappa(RR'), \tag{1.5.10}$$

$$\int_{O(p)} C_\kappa(RHSH') \, [dH] = \frac{C_\kappa(R)C_\kappa(S)}{C_\kappa(I_p)}, \tag{1.5.11}$$

where [dH] is the unit invariant Haar measure on the orthogonal group $O(p)$.

Proof: See James (1961b) for (1.5.10) and James (1960) for (1.5.11). ∎

LEMMA 1.5.2. *Let $Z \, (p \times p)$ be a complex symmetric matrix of which real part is positive definite and $T \, (p \times p)$ be a complex symmetric matrix. Then*

$$\int_{S>0} \text{etr}(-ZS) \det(S)^{a-\frac{1}{2}(p+1)} C_\kappa(TS) \, dS$$

$$= \Gamma_p(a, \kappa) \det(Z)^{-a} C_\kappa(TZ^{-1}), \ \text{Re}(a) > \frac{1}{2}(p-1), \tag{1.5.12}$$

and

$$\int_{S>0} \text{etr}(-ZS) \det(S)^{a-\frac{1}{2}(p+1)} C_\kappa(TS^{-1}) \, dS$$

$$= \Gamma_p(a, -\kappa) \det(Z)^{-a} C_\kappa(TZ), \ \text{Re}(a) > \frac{1}{2}(p-1) + k_1. \tag{1.5.13}$$

Proof: See Constantine (1963) for (1.5.12) and Khatri (1966) for (1.5.13). ∎

It may be noted that (1.5.12) is the Laplace transform of $\det(S)^{a-\frac{1}{2}(p+1)}$. Thus, using inverse Laplace transform, we get

$$\frac{2^{\frac{1}{2}p(p-1)}}{(2\pi\iota)^{\frac{1}{2}p(p+1)}} \int_{\text{Re}(Z)>0} \text{etr}(ZS) \det(Z)^{-a} C_\kappa(TZ^{-1}) \, dZ$$

$$= \frac{1}{\Gamma_p(a, \kappa)} \det(S)^{a-\frac{1}{2}(p+1)} C_\kappa(TS), \ \text{Re}(a) > \frac{1}{2}(p-1). \tag{1.5.14}$$

Similarly, from (1.5.13) we get

$$\frac{2^{\frac{1}{2}p(p-1)}}{(2\pi\iota)^{\frac{1}{2}p(p+1)}} \int_{\text{Re}(Z)>0} \text{etr}(ZS) \det(Z)^{-a} C_\kappa(TZ) \, dZ$$

$$= \frac{1}{\Gamma_p(a, -\kappa)} \det(S)^{a-\frac{1}{2}(p+1)} C_\kappa(TS^{-1}), \ \text{Re}(a) > \frac{1}{2}(p-1) + k_1. \tag{1.5.15}$$

LEMMA 1.5.3. *Let $R\,(p \times p)$ be a symmetric matrix, then*

$$\int_{0<S<I_p} \det(S)^{a-\frac{1}{2}(p+1)} \det(I_p - S)^{b-\frac{1}{2}(p+1)} C_\kappa(RS)\,dS$$

$$= \frac{\Gamma_p(a,\kappa)\Gamma_p(b)}{\Gamma_p(a+b,\kappa)} C_\kappa(R),\ \operatorname{Re}(a) > \frac{1}{2}(p-1),\ \operatorname{Re}(b) > \frac{1}{2}(p-1) \quad (1.5.16)$$

and

$$\int_{0<S<I_p} \det(S)^{a-\frac{1}{2}(p+1)} \det(I_p - S)^{b-\frac{1}{2}(p+1)} C_\kappa(RS^{-1})\,dS$$

$$= \frac{\Gamma_p(a,-\kappa)\Gamma_p(b)}{\Gamma_p(a+b,-\kappa)} C_\kappa(R),\ \operatorname{Re}(a) > \frac{1}{2}(p-1) + k_1,\ \operatorname{Re}(b) > \frac{1}{2}(p-1). \quad (1.5.17)$$

Proof: See Constantine (1963) for (1.5.16) and Khatri (1966) for (1.5.17). ∎

LEMMA 1.5.4. *Let $T\,(p \times p)$ be a complex symmetric matrix, then*

$$\int_{S>0} \det(S)^{a-\frac{1}{2}(p+1)} \det(I_p + S)^{-(a+b)} C_\kappa(TS)\,dS$$

$$= \frac{\Gamma_p(a,\kappa)\Gamma_p(b,-\kappa)}{\Gamma_p(a+b)} C_\kappa(T),\ \operatorname{Re}(a) > \frac{1}{2}(p-1),\ \operatorname{Re}(b) > \frac{1}{2}(p-1) + k_1, \quad (1.5.18)$$

and

$$\int_{S>0} \det(S)^{a-\frac{1}{2}(p+1)} \det(I_p + S)^{-(a+b)} C_\kappa(TS^{-1})\,dS$$

$$= \frac{\Gamma_p(a,-\kappa)\Gamma_p(b,\kappa)}{\Gamma_p(a+b)} C_\kappa(T),\ \operatorname{Re}(a) > \frac{1}{2}(p-1) + k_1,\ \operatorname{Re}(b) > \frac{1}{2}(p-1). \quad (1.5.19)$$

Proof: See Khatri (1966). ∎

LEMMA 1.5.5. *Let $T\,(p \times p)$ be a complex symmetric matrix, then*

$$\int_{S>0} \operatorname{etr}(-S) \det(S)^{a-\frac{1}{2}(p+1)} (\operatorname{tr} S)^j C_\kappa(TS)\,dS$$

$$= \frac{\Gamma_p(a,\kappa)\Gamma(pa+j+k)}{\Gamma(pa+k)} C_\kappa(T),\ \operatorname{Re}(a) > \frac{1}{2}(p-1), \quad (1.5.20)$$

and

$$\int_{S>0} \operatorname{etr}(-S) \det(S)^{a-\frac{1}{2}(p+1)} (\operatorname{tr} S)^j C_\kappa(TS^{-1})\,dS$$

$$= \frac{\Gamma_p(a,-\kappa)\Gamma(pa+j-k)}{\Gamma(pa-k)} C_\kappa(T),\ \operatorname{Re}(a) > \frac{1}{2}(p-1) + k_1, \quad (1.5.21)$$

where $j = 0, 1, 2, \ldots, \ldots$.

Proof: See Khatri (1966). ∎

In the following theorem we give a generalization of Lemma 1.5.3.

THEOREM 1.5.1. *Let $R(p \times p)$ be a symmetric matrix. Then,*

$$
\int\limits_{\substack{\sum_{i=1}^{r} Z_i < I_p \\ Z_i > 0}} \prod_{i=1}^{r} \det(Z_i)^{a_i - \frac{1}{2}(p+1)} \det\left(I_p - \sum_{i=1}^{r} Z_i\right)^{b - \frac{1}{2}(p+1)} C_\kappa\left(R \sum_{i=1}^{r} Z_i\right) \prod_{i=1}^{r} dZ_i
$$

$$
= \frac{\Gamma_p(a, \kappa)}{\Gamma_p(a+b, \kappa)} \cdot \frac{\Gamma_p(b) \prod_{i=1}^{r} \Gamma_p(a_i)}{\Gamma_p(a)} C_\kappa(R), \quad \mathrm{Re}(a_j) > \frac{1}{2}(p-1), \ j = 1, \ldots, p,
$$

$$
\mathrm{Re}(b) > \frac{1}{2}(p-1), \tag{1.5.22}
$$

and

$$
\int\limits_{\substack{\sum_{i=1}^{r} Z_i < I_p \\ Z_i > 0}} \prod_{i=1}^{r} \det(Z_i)^{a_i - \frac{1}{2}(p+1)} \det\left(I_p - \sum_{i=1}^{r} Z_i\right)^{b - \frac{1}{2}(p+1)} C_\kappa\left(R\left(\sum_{i=1}^{r} Z_i\right)^{-1}\right) \prod_{i=1}^{r} dZ_i
$$

$$
= \frac{\Gamma_p(a, -\kappa)}{\Gamma_p(a+b, -\kappa)} \cdot \frac{\Gamma_p(b) \prod_{i=1}^{r} \Gamma_p(a_i)}{\Gamma_p(a)} C_\kappa(R), \quad \mathrm{Re}(a_j) > \frac{1}{2}(p-1), \ j = 1, \ldots, p,
$$

$$
\mathrm{Re}(a) > \frac{1}{2}(p-1) + k_1,
$$

$$
\mathrm{Re}(b) > \frac{1}{2}(p-1), \tag{1.5.23}
$$

where $a = \sum_{i=1}^{r} a_i$.

Proof: Here we give proof of (i). The proof of (ii) is similar. The integral on the left hand side of (1.5.22), using (1.4.12) and (1.4.13), can be written as

$$
\int\limits_{\substack{\sum_{i=1}^{r} Z_i < I_p \\ Z_i > 0}} \prod_{i=1}^{r} \det(Z_i)^{a_i - \frac{1}{2}(p+1)} \det\left(I_p - \sum_{i=1}^{r} Z_i\right)^{b - \frac{1}{2}(p+1)} C_\kappa\left(R \sum_{i=1}^{r} Z_i\right) \prod_{i=1}^{r} dZ_i
$$

$$
= \int_{0 < Z < I_p} \phi(Z) C_\kappa(RZ) \, dZ
$$

$$
= \beta_p(a_1, \ldots, a_{r-1}; a_r) \int_{0 < Z < I_p} \det(Z)^{a - \frac{1}{2}(p+1)} \det(I_p - Z)^{b - \frac{1}{2}(p+1)} C_\kappa(RZ) \, dZ
$$

$$
= \beta_p(a_1, \ldots, a_{r-1}; a_r) \frac{\Gamma(a, \kappa) \Gamma_p(b)}{\Gamma(a+b, \kappa)} C_\kappa(R).
$$

The last step is derived using (1.5.16). Now substituting for $\beta_p(a_1, \ldots, a_{r-1}; a_r)$ from (1.4.11) and simplifying gives the desired result. ∎

For many other results on zonal polynomials and integrals involving zonal polynomials, the reader is referred to Subrahmaniam (1976), and Muirhead (1982).

1.6.   HYPERGEOMETRIC FUNCTIONS OF MATRIX ARGUMENT

Distributional results of random matrices are often derived in terms of hypergeometric functions of matrix arguments. Bochner (1952) defined the Bessel function of matrix argument as the inverse Laplace transform of the exponential function. Herz (1955) introduced the hypergeometric function of matrix argument using Laplace and inverse Laplace transforms. Constantine (1963) gave the power series representation in series involving zonal polynomials as given below.

DEFINITION 1.6.1. *The hypergeometric function of matrix argument is defined by*

$$
{}_mF_n(a_1, \ldots, a_m; b_1, \ldots, b_n; S) = \sum_{k=0}^{\infty} \sum_{\kappa} \frac{(a_1)_\kappa \cdots (a_m)_\kappa}{(b_1)_\kappa \cdots (b_n)_\kappa} \frac{C_\kappa(S)}{k!}, \qquad (1.6.1)
$$

where a_i, $i = 1, \ldots, m$; b_j, $j = 1, \ldots, n$ are arbitrary complex numbers, $S\,(p \times p)$ is a complex symmetric matrix and $\sum_\kappa$ denotes summation over all partitions κ.

Conditions for convergence of the series (1.6.1) are:

(i) none of the b_j is zero, an integer or half integer less than or equal to $\frac{1}{2}(p-1)$,

(ii) if a_i is a negative integer, say $-r$, then the function reduces to a finite polynomial of degree pr,

(iii) the series converges for all $S\,(p \times p)$ if $m < n + 1$,

(iv) if $m = n + 1$, the series converges for all $S\,(p \times p)$ such that $\|S\| < 1$ where the norm $\|S\|$ denotes the maximum absolute value of the characteristic roots of S,

(v) unless the series terminates, it diverges for all $S \neq 0$ if $m > n + 1$.

From Definition 1.6.1 it follows that

$$
\begin{aligned}
{}_0F_0(S) &= \sum_{k=0}^{\infty} \sum_{\kappa} \frac{C_\kappa(S)}{k!} \\
&= \sum_{k=0}^{\infty} \frac{(\operatorname{tr} S)^k}{k!} \\
&= \operatorname{etr}(S).
\end{aligned}
$$

DEFINITION 1.6.2. *The hypergeometric function of two symmetric matrices $S\,(p \times p)$ and $T\,(p \times p)$ is defined by*

$$
{}_mF_n^{(p)}(a_1, \ldots, a_m; b_1, \ldots, b_n; S, T) = \sum_{k=0}^{\infty} \sum_{\kappa} \frac{(a_1)_\kappa \cdots (a_m)_\kappa}{(b_1)_\kappa \cdots (b_n)_\kappa} \frac{C_\kappa(S)C_\kappa(T)}{C_\kappa(I_p)\, k!}. \qquad (1.6.2)
$$

Conditions for convergence of (1.6.2) are similar to the conditions for the convergence of (1.6.1) except that for $m = n+1$ the series converges for $\|S\| < 1$ or $\|T\| < 1$. If both $S\,(p \times p)$ and $T\,(p \times p)$ are such that $\|S\| < 1$ and $\|T\| < 1$, then the series will converge more rapidly.

It is clear from the Definition 1.6.2 that the order of S and T is unimportant and if one of the arguments is identity matrix, this function reduces to the hypergeometric function of one matrix argument.

By averaging the hypergeometric function of one matrix argument over the orthogonal group $O(p)$, one can obtain the hypergeometric function of two matrices as follows.

THEOREM 1.6.1. *If $S\,(p \times p)$ is a symmetric positive definite matrix and $T\,(p \times p)$ is a symmetric matrix, then*

$$\int_{O(p)} {}_mF_n(a_1, \ldots, a_m; b_1, \ldots, b_n; SHTH')\,[dH]$$

$$= {}_mF_n^{(p)}(a_1, \ldots, a_m; b_1, \ldots, b_n; S, T). \qquad (1.6.3)$$

Proof: The result follows by expanding the integrand using (1.6.1) and then integrating term by term using (1.5.11). ∎

Some of the results given in Section 1.5 can be extended for hypergeometric functions.

THEOREM 1.6.2. *Let $Z\,(p \times p)$ be a complex symmetric matrix of which real part is positive definite and $T\,(p \times p)$ be a complex symmetric matrix. Then*

$$\int_{S>0} \operatorname{etr}(-ZS) \det(S)^{a-\frac{1}{2}(p+1)} {}_mF_n(a_1, \ldots, a_m; b_1, \ldots, b_n; ST)\,dS$$

$$= \Gamma_p(a) \det(Z)^{-a} {}_{m+1}F_n(a_1, \ldots, a_m, a; b_1, \ldots, b_n; Z^{-1}T), \ \operatorname{Re}(a) > \frac{1}{2}(p-1), \ (1.6.4)$$

and

$$\int_{S>0} \operatorname{etr}(-ZS) \det(S)^{a-\frac{1}{2}(p+1)} {}_mF_n^{(p)}(a_1, \ldots, a_m; b_1, \ldots, b_n; ST, R)\,dS$$

$$= \Gamma_p(a) \det(Z)^{-a} {}_{m+1}F_n^{(p)}(a_1, \ldots, a_m, a; b_1, \ldots, b_n; Z^{-1}T, R), \ \operatorname{Re}(a) > \frac{1}{2}(p-1), \ (1.6.5)$$

where $R\,(p \times p)$ is a symmetric matrix.

Proof: Expanding the hypergeometric function in the integrand and integrating term by term using Lemma 1.5.2 gives the desired result. ∎

COROLLARY 1.6.2.1. *For $\|Z\| < 1$,*

$$_1F_0(a; Z) = \det(I_p - Z)^{-a}.$$

Proof: From (1.6.4), letting $T = I_p$ and replacing Z by Z^{-1}, we get

$$_1F_0(a; Z) = \frac{\det(Z)^{-a}}{\Gamma_p(a)} \int_{S>0} \operatorname{etr}(-Z^{-1}S) \det(S)^{a-\frac{1}{2}(p+1)} {}_0F_0(S)\,dS.$$

Now substituting $Z^{-\frac{1}{2}}SZ^{-\frac{1}{2}} = A$ with Jacobian $J(S \to A) = \det(Z)^{\frac{1}{2}(p+1)}$ and using ${}_0F_0(ZA) = \text{etr}(ZA)$, we get

$$
\begin{aligned}
{}_1F_0(a; Z) &= \frac{1}{\Gamma_p(a)} \int_{A>0} \text{etr}\{-A(I_p - Z)\} \det(A)^{a-\frac{1}{2}(p+1)} \, dA \\
&= \det(I_p - Z)^{-a}.
\end{aligned}
$$

which follows from (1.4.6). ∎

THEOREM 1.6.3. *Let $R(p \times p)$ be a symmetric matrix, then*

$$
\int_{0<S<I_p} \det(S)^{a-\frac{1}{2}(p+1)} \det(I_p - S)^{b-\frac{1}{2}(p+1)} {}_mF_n(a_1, \ldots, a_m; b_1, \ldots, b_n; RS) \, dS
$$

$$
= \frac{\Gamma_p(a)\Gamma_p(b)}{\Gamma_p(a+b)} {}_{m+1}F_{n+1}(a_1, \ldots, a_m, a; b_1, \ldots, b_n, a+b; R). \tag{1.6.6}
$$

Proof: The result follows by expanding the hypergeometric function in the integrand and integrating term by term using Lemma 1.5.3. ∎

COROLLARY 1.6.3.1. *For $\text{Re}(\alpha) > \frac{1}{2}(p-1)$, $\text{Re}(\beta) > \frac{1}{2}(p-1)$ and $\text{Re}(\beta - \alpha) > \frac{1}{2}(p-1)$ and symmetric $R\,(p \times p)$,*

$$
{}_1F_1(\alpha; \beta; R) = \frac{\Gamma_p(\beta)}{\Gamma_p(\alpha)\Gamma_p(\beta - \alpha)} \int_{0<S<I_p} \det(S)^{\alpha-\frac{1}{2}(p+1)} \det(I_p - S)^{\beta-\alpha-\frac{1}{2}(p+1)}
$$

$$
\text{etr}(RS) \, dS. \tag{1.6.7}
$$

Proof: Substituting $m = n = 0$, $a = \alpha$, and $b = \beta - \alpha$ in (1.6.6), we get

$$
{}_1F_1(\alpha; \beta; R) = \frac{\Gamma_p(\beta)}{\Gamma_p(\alpha)\Gamma_p(\beta - \alpha)} \int_{0<S<I_p} \det(S)^{\alpha-\frac{1}{2}(p+1)} \det(I_p - S)^{\beta-\alpha-\frac{1}{2}(p+1)}
$$

$$
{}_0F_0(RS) \, dS, \quad \text{Re}(\alpha) > \frac{1}{2}(p - 1), \quad \text{Re}(\beta - \alpha) > \frac{1}{2}(p - 1).
$$

The result follows by using ${}_0F_0(RS) = \text{etr}(RS)$. ∎

COROLLARY 1.6.3.2. *For $\text{Re}(\gamma) > \frac{1}{2}(p-1)$, $\text{Re}(\gamma - \alpha) > \frac{1}{2}(p-1)$ and symmetric $R\,(p \times p)$ where $\text{Re}(R) < I_p$,*

$$
{}_2F_1(\alpha, \beta; \gamma; R) = \frac{\Gamma_p(\gamma)}{\Gamma_p(\alpha)\Gamma_p(\gamma - \alpha)} \int_{0<S<I_p} \det(S)^{\alpha-\frac{1}{2}(p+1)} \det(I_p - S)^{\gamma-\alpha-\frac{1}{2}(p+1)}
$$

$$
\det(I_p - RS)^{-\beta} \, dS. \tag{1.6.8}
$$

Proof: Substituting $m = 1$, $n = 0$, $a_1 = \beta$, $a = \alpha$, and $b = \gamma - \alpha$ in (1.6.6), we get

$$_2F_1(\alpha, \beta; \gamma; R) = \frac{\Gamma_p(\gamma)}{\Gamma_p(\alpha)\Gamma_p(\gamma - \alpha)} \int_{0 < S < I_p} \det(S)^{\alpha - \frac{1}{2}(p+1)} \det(I_p - S)^{\gamma - \alpha - \frac{1}{2}(p+1)}$$

$$_1F_0(\beta; RS)\, dS, \; \mathrm{Re}(\alpha) > \frac{1}{2}(p - 1), \; \mathrm{Re}(\gamma - \alpha) > \frac{1}{2}(p - 1).$$

Now using Corollary 1.6.2.1 the result follows. ∎

The integral representations (1.6.7) and (1.6.8) are generalizations of the classical confluent hypergeometric functions $_1F_1$ and Gauss hypergeometric function $_2F_1$ respectively, and are due to Herz (1955). He also generalized Kummer's and Euler's relations for classical $_1F_1$ and $_2F_1$ functions to the matrix argument.

The hypergeometric functions $_1F_1$ and $_2F_1$ satisfy the following relations (Herz, 1955).

$$_1F_1(\alpha; \gamma; S) = \mathrm{etr}(S)\, _1F_1(\gamma - \alpha; \gamma; -S) \tag{1.6.9}$$

$$_2F_1(\alpha, \beta; \gamma; S) = \det(I_p - S)^{-\beta}\, _2F_1(\gamma - \alpha, \beta; \gamma; -S(I_p - S)^{-1})$$

$$= \det(I_p - S)^{\gamma - \alpha - \beta}\, _2F_1(\gamma - \alpha, \gamma - \beta; \gamma; S). \tag{1.6.10}$$

Subrahmaniam (1973) proved (1.6.10) using the partial differential equation for $_2F_1$. Using zonal polynomial expansion it is easy to establish the confluence relations

$$\lim_{\alpha \to \infty} {}_1F_1\left(\alpha; \gamma; \frac{1}{\alpha}S\right) = {}_0F_1(\gamma; S) \tag{1.6.11}$$

and

$$\lim_{\alpha \to \infty} {}_2F_1\left(\alpha, \beta; \gamma; \frac{1}{\alpha}S\right) = {}_1F_1(\alpha; \gamma; S). \tag{1.6.12}$$

From Theorem 1.6.2 it is seen that $_{m+1}F_n$ function can be obtained from $_mF_n$ by means of a Laplace transform. Conversely $_mF_n$ function can be obtained from $_{m+1}F_n$ function by using an inverse Laplace transform. There is also an inverse Laplace transform which enables the $_mF_{n+1}$ function to be obtained from $_mF_n$ function (see Herz, 1955, p. 485). It has already been shown that $_0F_0(S) = \mathrm{etr}(S)$ and $_1F_0(a; S) = \det(I_p - S)^{-a}$.

The hypergeometric functions $_0F_1$ and $_1F_1$ have the following integral representations given by Herz (1955) and James (1961a).

THEOREM 1.6.4. *Let X ($p \times n$), $p \le n$ be a real matrix and $H = \begin{pmatrix} H_1 \\ H_2 \end{pmatrix} \in O(n)$ where H_1 is $p \times n$. Then*

$$\int_{O(p,n)} \mathrm{etr}(XH_1')\, [dH_1] = {}_0F_1\left(\frac{1}{2}n; \frac{1}{4}XX'\right)$$

where $[dH_1]$ denotes the unit invariant measure on $O(p,n)$.

THEOREM 1.6.5. *Let $H_1 \in O(p,n)$, i.e., H_1 is $p \times n$ and $H_1H_1' = I_p$. Further let $[dH_1]$ be the normalized invariant measure on $O(p,n)$ so that $\int_{O(p,n)} [dH_1] = 1$. If X ($n \times n$) is positive definite matrix, then*

$$\int_{O(p,n)} \mathrm{etr}(XH_1'H_1)\, [dH_1] = {}_1F_1\left(\frac{1}{2}p; \frac{1}{2}n; X\right).$$

Next, by using Theorem 1.6.4, we derive an integral useful in the study of noncentral density of Wishart matrix, and the theory of quadratic forms.

THEOREM 1.6.6. *For X $(p \times n)$ of rank $p \leq n$ and L $(p \times n)$,*

$$\int_{XX'=A} \operatorname{etr}(LX') \, dX = \frac{\pi^{\frac{1}{2}np}}{\Gamma_p(\frac{1}{2}n)} \det(A)^{\frac{1}{2}(n-p-1)} \, {}_0F_1\left(\frac{1}{2}n; \frac{1}{4}LL'A\right).$$

Proof: Transform $X = TH_1$, where H_1 is $p \times n$, $H_1 H_1' = I_p$, and T $(p \times p)$ is a lower triangular matrix with positive diagonal elements, with Jacobian, from (1.3.25),

$$J(X \to T, H_1) = \prod_{i=1}^{p} t_{ii}^{n-i} g_{n,p}(H_1),$$

where $g_{n,p}(H_1) \, dH_1$ defines the invariant measure on $O(p, n)$. Then

$$
\begin{aligned}
\int_{XX'=A} \operatorname{etr}(LX') \, dX &= \int_{TT'=A} \prod_{i=1}^{p} t_{ii}^{n-i} \int_{H_1 \in O(p,n)} \operatorname{etr}(T'LH_1') g_{n,p}(H_1) \, dH_1 \, dT \\
&= \frac{2^p \pi^{\frac{1}{2}np}}{\Gamma_p(\frac{1}{2}n)} \int_{TT'=A} \prod_{i=1}^{p} t_{ii}^{n-i} \int_{H_1 \in O(p,n)} \operatorname{etr}(T'LH_1') \, [dH_1] \, dT \\
&= \frac{2^p \pi^{\frac{1}{2}np}}{\Gamma_p(\frac{1}{2}n)} \int_{TT'=A} \prod_{i=1}^{p} t_{ii}^{n-i} \, {}_0F_1\left(\frac{1}{2}n; \frac{1}{4}LL'TT'\right) dT. \qquad (1.6.13)
\end{aligned}
$$

The expression (1.6.13) has been obtained by using Theorem 1.6.4. Further transforming $TT' = S$, with the Jacobian $J(T \to S) = 2^{-p} \prod_{i=1}^{p} t_{ii}^{-(p-i+1)}$ we get the final result. ∎

There is yet another type of confluent hypergeometric function, Ψ, of matrix argument defined below (Muirhead, 1970).

DEFINITION 1.6.3. *The confluent hypergeometric function Ψ of symmetric matrix R $(p \times p)$ is defined by*

$$\Psi(a, c; R) = \frac{1}{\Gamma_p(a)} \int_{S>0} \operatorname{etr}(-RS) \det(S)^{a-\frac{1}{2}(p+1)} \det(I_p + S)^{c-a-\frac{1}{2}(p+1)} \, dS, \quad (1.6.14)$$

where $\operatorname{Re}(R) > 0$, and $\operatorname{Re}(a) > \frac{1}{2}(p-1)$.

Using (1.6.8), it can easily be proved that the confluent hypergeometric function Ψ can also be obtained as a limit of Gauss hypergeometric function,

$$\lim_{c \to \infty} {}_2F_1(a, b; c; I_p - cR^{-1}) = \det(R)^b \Psi\left(b, b - a + \frac{1}{2}(p+1); R\right).$$

The Whittaker's function of matrix argument has been studied by Abdi (1968). The Bessel functions of matrix argument are defined as follows.

DEFINITION 1.6.4. *The Bessel function (type one Bessel function of Herz) of matrix argument, denoted by $A_\gamma(S)$ is defined as*

$$A_\gamma(S) = \frac{1}{\Gamma_p[\gamma + \frac{1}{2}(p+1)]} \sum_{k=0}^{\infty} \sum_{\kappa} \frac{C_\kappa(-S)}{(\gamma + \frac{1}{2}(p+1))_\kappa \, k!}$$

$$= \frac{1}{\Gamma_p[\gamma + \frac{1}{2}(p+1)]} \,{}_0F_1\left(\gamma + \frac{1}{2}(p+1); -S\right), \qquad (1.6.15)$$

where $\text{Re}(\gamma) > -1$.

It can be easily shown that the Bessel function defined above has the integral representation

$$A_\gamma(S) = \frac{2^{\frac{1}{2}p(p-1)}}{(2\pi\iota)^{\frac{1}{2}p(p+1)}} \int_{\text{Re}(Z)>0} \text{etr}(Z - SZ^{-1}) \det(Z)^{-\gamma - \frac{1}{2}(p+1)} \, dZ. \qquad (1.6.16)$$

The result (1.6.16) can be proved by expanding $\text{etr}(-SZ^{-1})$ in zonal polynomials and using (1.5.14).

The Laplace transform of $\det(S)^\gamma A_\gamma(S)$ is derived from (1.6.15) as

$$\int_{S>0} \text{etr}(-SZ) \det(S)^\gamma A_\gamma(S) \, dS = \text{etr}(-Z^{-1}) \det(Z)^{-\gamma - \frac{1}{2}(p+1)}. \qquad (1.6.17)$$

For $p = 1$, the relation between $A_\gamma(\cdot)$ and the ordinary Bessel function, $J_\gamma(\cdot)$, (Luke, 1969; p. 212) is given by

$$J_\gamma(t) = A_\gamma\left(\frac{1}{4}t^2\right)\left(\frac{1}{2}t\right)^\gamma.$$

DEFINITION 1.6.5. *The type two Bessel function of Herz of matrix argument, B_δ, is defined as*

$$B_\delta(WZ) = \det(W)^{-\delta} \int_{S>0} \text{etr}(-SW) \, \text{etr}(-S^{-1}Z) \det(S)^{-\delta - \frac{1}{2}(p+1)} \, dS, \qquad (1.6.18)$$

where $\text{Re}(W) > 0$ *and* $\text{Re}(Z) > 0$.

By changing variables from S to S^{-1}, we note that

$$B_{-\delta}(Z) = B_\delta(Z) \det(Z)^\delta$$

and we can write

$$B_\delta(Z) = \int_{S>0} \text{etr}(-SZ) \, \text{etr}(-S^{-1}) \det(S)^{\delta - \frac{1}{2}(p+1)} \, dS. \qquad (1.6.19)$$

For $p = 1$, the relation between $B_\delta(\cdot)$, and the Bessel function of the third kind of imaginary argument $K_\delta(\cdot)$, (Luke, 1969; p. 212) is given by

$$K_\delta(t) = \frac{1}{2} B_\delta\left(\frac{1}{4}t^2\right)\left(\frac{1}{2}t\right)^\delta.$$

Next we define incomplete gamma and beta functions of matrix argument.

DEFINITION 1.6.6. *For* $\mathrm{Re}(a) > \frac{1}{2}(p-1)$, *the incomplete gamma function is defined by*

$$\gamma_p(a, B) = \int_{0 < A < B} \det(A)^{a - \frac{1}{2}(p+1)} \, \mathrm{etr}(-A) \, dA. \tag{1.6.20}$$

THEOREM 1.6.7. *For* $\mathrm{Re}(a) > \frac{1}{2}(p-1)$,

$$\gamma_p(a, B) = \det(B)^a \frac{\Gamma_p(a)\Gamma_p[\frac{1}{2}(p+1)]}{\Gamma_p[a + \frac{1}{2}(p+1)]} \, {}_1F_1\Big(a; a + \frac{1}{2}(p+1); -B\Big). \tag{1.6.21}$$

Proof: Substituting $S = B^{-\frac{1}{2}} A B^{-\frac{1}{2}}$, with Jacobian $J(A \to S) = \det(B)^{\frac{1}{2}(p+1)}$, in (1.6.20) we get

$$\gamma_p(a, B) = \det(B)^a \int_{0 < S < I_p} \det(S)^{a - \frac{1}{2}(p+1)} \, \mathrm{etr}(-BS) \, dS$$

$$= \det(B)^a \frac{\Gamma_p(a)\Gamma_p[\frac{1}{2}(p+1)]}{\Gamma_p[a + \frac{1}{2}(p+1)]} \, {}_1F_1\Big(a; a + \frac{1}{2}(p+1); -B\Big).$$

The last equality is obtained from Corollary 1.6.3.1. ∎

DEFINITION 1.6.7. *For* $\mathrm{Re}(a) > \frac{1}{2}(p-1)$, *and* $\mathrm{Re}(b) > \frac{1}{2}(p-1)$, *the incomplete beta function is defined by*

$$\beta_p(a, b, B) = \int_{0 < A < B} \det(A)^{a - \frac{1}{2}(p+1)} \det(I_p - A)^{b - \frac{1}{2}(p+1)} \, dA \tag{1.6.22}$$

where $0 < B < I_p$.

THEOREM 1.6.8. *For* $\mathrm{Re}(a) > \frac{1}{2}(p-1)$, $\mathrm{Re}(b) > \frac{1}{2}(p-1)$ *and* $0 < B < I_p$,

$$\beta_p(a, b, B) = \frac{\Gamma_p(a)\Gamma_p[\frac{1}{2}(p+1)]}{\Gamma_p[a + \frac{1}{2}(p+1)]} \det(B)^a$$

$$\quad {}_2F_1\Big(a, -b + \frac{1}{2}(p+1); a + \frac{1}{2}(p+1); B\Big). \tag{1.6.23}$$

Proof: Substituting $S = B^{-\frac{1}{2}} A B^{-\frac{1}{2}}$ with Jacobian $J(A \to S) = \det(B)^{\frac{1}{2}(p+1)}$ in (1.6.23) we get

$$\beta_p(a, b, B) = \det(B)^a \int_{0 < S < I_p} \det(S)^{a - \frac{1}{2}(p+1)} \det(I_p - BS)^{b - \frac{1}{2}(p+1)} \, dS$$

$$= \det(B)^a \frac{\Gamma_p(a)\Gamma_p[\frac{1}{2}(p+1)]}{\Gamma_p[a + \frac{1}{2}(p+1)]} \, {}_2F_1\Big(a, -b + \frac{1}{2}(p+1); a + \frac{1}{2}(p+1); -B\Big).$$

The last equality is obtained from Corollary 1.6.3.2. ∎

1.7. LAGUERRE POLYNOMIALS

Laguerre polynomials of matrix argument were introduced by Herz (1955). Constantine (1966) modified his definition and gave the following integral representation.

DEFINITION 1.7.1. *The Laguerre polynomial $L_\kappa^\gamma(S)$ of a symmetric matrix $S\,(p \times p)$ corresponding to the partition κ of k is defined as*

$$L_\kappa^\gamma(S) = \text{etr}(S) \int_{R>0} \text{etr}(-R)\det(R)^\gamma C_\kappa(R) A_\gamma(RS)\, dR, \qquad (1.7.1)$$

where $A_\gamma(R)$ is the Bessel function and $\text{Re}(\gamma) > -1$.

Substituting for $A_\gamma(RS)$ from (1.6.15) in (1.7.1), changing the order of integration and integrating with respect to R we get

$$L_\kappa^\gamma(S) = \frac{2^{\frac{1}{2}p(p-1)}}{(2\pi\iota)^{\frac{1}{2}p(p+1)}} \Gamma_p\left(\gamma + \frac{1}{2}(p+1), \kappa\right)$$

$$\int_{\text{Re}(Z)>0} \text{etr}(Z)\det(Z)^{-\gamma-\frac{1}{2}(p+1)} C_\kappa(I_p - SZ^{-1})\, dZ. \qquad (1.7.2)$$

Further, write

$$\frac{C_\kappa(I_p - SZ^{-1})}{C_\kappa(I_p)} = \sum_{t=0}^{k} \sum_\tau \binom{\kappa}{\tau} \frac{C_\tau(-SZ^{-1})}{C_\tau(I_p)} \qquad (1.7.3)$$

where $\binom{\kappa}{\tau}$ is the generalized binomial coefficient (Constantine, 1966), and τ is a partition of t. Substituting (1.7.3) in (1.7.2) we get

$$L_\kappa^\gamma(S) = \frac{2^{\frac{1}{2}p(p-1)}}{(2\pi\iota)^{\frac{1}{2}p(p+1)}} \Gamma_p\left(\gamma + \frac{1}{2}(p+1), \kappa\right) C_\kappa(I_p)$$

$$\sum_{t=0}^{k} \sum_\tau \binom{\kappa}{\tau} \frac{1}{C_\tau(I_p)} \int_{\text{Re}(Z)>0} \text{etr}(Z)\det(Z)^{-\gamma-\frac{1}{2}(p+1)} C_\kappa(-SZ^{-1})\, dZ.$$

Now using (1.5.14), we obtain the series representation for $L_\kappa^\gamma(S)$ as

$$L_\kappa^\gamma(S) = \left(\gamma + \frac{1}{2}(p+1)\right)_\kappa C_\kappa(I_p) \sum_{t=0}^{k} \sum_\tau \binom{\kappa}{\tau} \frac{C_\tau(-S)}{\left(\gamma + \frac{1}{2}(p+1)\right)_\tau C_\tau(I_p)}. \qquad (1.7.4)$$

Clearly $L_\kappa^\gamma(S)$ is a symmetric polynomial of degree k in the eigenvalues of S and

$$L_\kappa^\gamma(0) = \left(\gamma + \frac{1}{2}(p+1)\right)_\kappa C_\kappa(I_p), \qquad (1.7.5)$$

$$|L_\kappa^\gamma(S)| \le \left(\gamma + \frac{1}{2}(p+1)\right)_\kappa C_\kappa(I_p)\,\text{etr}(S), \quad \gamma > -1.$$

Next we give the Laplace transform of $\det(S)^\gamma L_\kappa^\gamma(S)$, which is useful in the theory of quadratic forms.

THEOREM 1.7.1. *Let* Z $(p \times p)$, *and* T $(p \times p)$ *be complex symmetric matrices,* $\mathrm{Re}(Z) > 0$. *Then*

$$\int_{S>0} \mathrm{etr}(-ZS) \det(S)^\gamma L_\kappa^\gamma(TS)\, dS = \left(\gamma + \frac{1}{2}(p+1)\right)_\kappa \Gamma_p\left[\gamma + \frac{1}{2}(p+1)\right]$$

$$\det(Z)^{-\gamma-\frac{1}{2}(p+1)} C_\kappa(I_p - Z^{-1}T). \quad (1.7.6)$$

Proof: Substituting from (1.7.4) in the left hand side of (1.7.6), and using Lemma 1.5.2, we get

$$\left(\gamma + \frac{1}{2}(p+1)\right)_\kappa C_\kappa(I_p) \sum_{t=0}^{k} \sum_\tau \binom{\kappa}{\tau} \frac{1}{\left(\gamma + \frac{1}{2}(p+1)\right)_\tau C_\tau(I_p)}$$

$$\int_{S>0} \mathrm{etr}(-ZS) \det(S)^\gamma C_\tau(-ST)\, dS$$

$$= \left(\gamma + \frac{1}{2}(p+1)\right)_\kappa \Gamma_p\left[\gamma + \frac{1}{2}(p+1)\right] C_\kappa(I_p) \sum_{t=0}^{k} \sum_\tau \binom{\kappa}{\tau} \frac{C_\tau(-Z^{-1}T)}{C_\tau(I_p)}.$$

Now the result follows from (1.7.3). ∎

The generating function for the Laguerre polynomial $L_\kappa^\gamma(S)$ is

$$\sum_{k=0}^{\infty} \sum_\kappa L_\kappa^\gamma(S) \frac{C_\kappa(Z)}{C_\kappa(I_p)k!} = \det(I_p - Z)^{-\gamma-\frac{1}{2}(p+1)} \int_{O(p)} \mathrm{etr}\{-SHZ(I_p - Z)^{-1}H'\}\,[dH],$$

$$\|Z\| < 1,\ S > 0, \quad (1.7.7)$$

which can be proved by multiplying both sides by $\det(S)^\gamma$ and showing that their Laplace transforms are equal.

1.8.   GENERALIZED HERMITE POLYNOMIALS

In this section we define the generalized Hermite polynomial and its extensions. These functions of matrix arguments play an important role in the study of the distribution of quadratic forms.

Hayakawa (1969) modified the definition given by Herz (1955) and defined the Hermite polynomial of matrix argument as

$$H_\kappa(T) = \pi^{-\frac{1}{2}np} \mathrm{etr}(TT') \int_U \mathrm{etr}(-UU' - 2\iota TU') C_\kappa(-UU')\, dU, \quad (1.8.1)$$

where T $(p \times n)$ is a real matrix, and $C_\kappa(\cdot)$ is a zonal polynomial. In 1972 he extended the above definition by introducing $C_\kappa(-UAU')$ in place of $C_\kappa(UU')$ in (1.8.1) where A $(n \times n)$ is a real symmetric matrix. He denoted these polynomials by $P_\kappa(T, A)$, and studied several of its properties. He also calculated expressions for $P_\kappa(T, A)$ up to

$k = 4$. Crowther (1975) called these polynomials Hayakawa polynomials and further extended them to $P_\kappa(T, A, B)$ where T $(p \times n)$ is a complex matrix, and A $(n \times n)$, B $(p \times p)$ are real symmetric matrices:

$$
\begin{aligned}
P_\kappa(T, A, B) &= \pi^{-\frac{1}{2}np} \operatorname{etr}(TT') \int_U \operatorname{etr}(-UU' - 2\iota TU') C_\kappa(-BUAU') \, dU \\
&= \pi^{-\frac{1}{2}np} \int_U \operatorname{etr}\{-(U + \iota T)(U + \iota T)'\} C_\kappa(-BUAU') \, dU \\
&= E[C_\kappa(-B(V - \iota T)A(V - \iota T)')],
\end{aligned} \tag{1.8.2}
$$

where $C_\kappa(S)$ is a zonal polynomial and expectation is with respect to the p.d.f.

$$
\pi^{-\frac{1}{2}np} \operatorname{etr}(-VV').
$$

From (1.8.2) it is easily seen that $P_\kappa(T, A, B) = P_\kappa(-T, A, B)$. For $B = I_p$, and T real,

$$
P_\kappa(T, A, I_p) = P_\kappa(T, A).
$$

For $T = 0$, by using invariance property and integrating over $O(n)$, we get

$$
\begin{aligned}
P_\kappa(0, A, B) &= E[C_\kappa(-BVAV')] \\
&= \pi^{-\frac{1}{2}np} \int_{\Lambda > 0} \int_{VV' = \Lambda} \operatorname{etr}(-VV') C_\kappa(-BVAV') \, dV \, d\Lambda \\
&= \frac{C_\kappa(A)}{\Gamma_p(\frac{1}{2}n) C_\kappa(I_n)} \int_{\Lambda > 0} \operatorname{etr}(-\Lambda) \det(\Lambda)^{\frac{1}{2}(n-p-1)} C_\kappa(-B\Lambda) \, d\Lambda \\
&= \left(\frac{1}{2}n\right)_\kappa \frac{C_\kappa(A) C_\kappa(-B)}{C_\kappa(I_n)},
\end{aligned} \tag{1.8.3}
$$

and hence

$$
\begin{aligned}
P_\kappa(0, A) &= \left(\frac{1}{2}n\right)_\kappa \frac{C_\kappa(A) C_\kappa(-I_p)}{C_\kappa(I_n)} \\
&= \left(\frac{1}{2}p\right)_\kappa C_\kappa(-A).
\end{aligned}
$$

An upper bound for $|P_\kappa(T, A, B)|$ can be obtained as

$$
\begin{aligned}
|P_\kappa(T, A, B)| &\leq \operatorname{etr}(TT') P_\kappa(0, A, B) \\
&= \left(\frac{1}{2}n\right)_\kappa \operatorname{etr}(TT') \frac{C_\kappa(A) C_\kappa(-B)}{C_\kappa(I_n)}.
\end{aligned}
$$

Crowther (1975) has calculated the polynomials $P_\kappa(T, A, B)$ for $\kappa = (1)$, (2), $(1, 1)$, (3), $(2, 1)$, and $(1, 1, 1)$.

1.9. NOTION OF RANDOM MATRIX

In this section we define basic concepts related to random matrices. The format of this section corresponds to standard treatment of the univariate case and its step by step generalization (le Roux, 1978; Anderson, 1984; Hogg and Craig, 1994).

A matrix random phenomenon is an observable phenomenon which can be represented in a matrix form which under repeated observations yields different outcomes which are not deterministically predictable. Instead the outcomes obey certain conditions of statistical regularity. The set of descriptions of all possible outcomes which may occur on observing a matrix random phenomenon is the sample space $\mathcal{S}$.

A matrix event is a subset of the sample space $\mathcal{S}$. A measure of the degree of certainty with which a given matrix event will occur when observing a matrix random phenomenon can be found by defining a probability function on subsets of the sample space, $\mathcal{S}$, which assigns a probability to every matrix event according to the three postulates of Kolmogorov (Rao, 1973).

DEFINITION 1.9.1. *A matrix X $(p \times n)$ consisting of np elements $x_{11}(\cdot), x_{12}(\cdot), \ldots,$ $x_{pn}(\cdot)$ which are real valued functions defined on the sample space $\mathcal{S}$ is a real random matrix if the range $\mathbb{R}^{p \times n}$ of*

$$\begin{pmatrix} x_{11}(\cdot) & \cdots & x_{1n}(\cdot) \\ \vdots & & \\ x_{p1}(\cdot) & \cdots & x_{pn}(\cdot) \end{pmatrix},$$

consists of Borel sets of np-dimensional real space and if, for each Borel set B of real np-tuples, arranged in a matrix,

$$\begin{pmatrix} x_{11} & \cdots & x_{1n} \\ \vdots & & \\ x_{p1} & \cdots & x_{pn} \end{pmatrix},$$

in $\mathbb{R}^{p \times n}$, the set

$$\left\{ s \in \mathcal{S} : \begin{pmatrix} x_{11}(s_{11}) & \cdots & x_{1n}(s_{1n}) \\ \vdots & & \\ x_{p1}(s_{p1}) & \cdots & x_{pn}(s_{pn}) \end{pmatrix} \in B \right\}$$

is an event in $\mathcal{S}$.

Now that we have defined a random matrix, let us define its probability density function. Throughout this book we shall consider only real continuous random matrices. Furthermore, no distinction will be made between a random matrix and its realization.

DEFINITION 1.9.2. *A scalar function $f_X(X)$ such that*
   (i) $f_X(X) \geq 0$

(ii) $\int_X f_X(X) \, dX = 1$

and

(iii) $P(X \in A) = \int_A f_X(X) \, dX$

where A is a subset of the space of realizations of X, defines the probability density function (p.d.f.) of the random matrix X.

DEFINITION 1.9.3. *A scalar function $f_{X,Y}(X,Y)$ such that*

(i) $f_{X,Y}(X,Y) \geq 0$

(ii) $\int_Y \int_X f_{X,Y}(X,Y) \, dX \, dY = 1$

and

(iii) $P((X,Y) \in A) = \int \int_A f_{X,Y}(X,Y) \, dX \, dY$

where A is a subset of the space of realizations of (X,Y), defines the joint (bimatrix variate) p.d.f. of X and Y.

DEFINITION 1.9.4. *Let the random matrices $X\,(p \times n)$ and $Y\,(r \times s)$ have the joint p.d.f. $f_{X,Y}(X,Y)$. Then*

(i) the marginal p.d.f. of X is defined by

$$f_X(X) = \int_Y f_{X,Y}(X,Y) \, dY,$$

and

(ii) the conditional p.d.f. of X given Y is defined by

$$f_{X|Y}(X|Y) = \frac{f_{X,Y}(X,Y)}{f_Y(Y)}, \quad f_Y(Y) > 0$$

where $f_Y(Y)$ is the marginal p.d.f. of Y.

Likewise, one can define the marginal p.d.f. of Y, and the conditional p.d.f. of Y given X.

Two random matrices $X\,(p \times n)$ and $Y\,(r \times s)$ are independently distributed if and only if

$$f_{X,Y}(X,Y) = f_X(X) f_Y(Y)$$

where $f_X(X)$ and $f_Y(Y)$ are the marginal densities of X and Y respectively.

DEFINITION 1.9.5. *The moment generating function (m.g.f.) of the random matrix $X\,(p \times n)$ is defined as*

$$M_X(Z) = \int_X \operatorname{etr}(ZX') f_X(X) \, dX$$

where $Z\,(p \times n)$ is a real arbitrary matrix.

A function $M_X(X)$ is a m.g.f. if and only if it is positive and continuous in a neighborhood of $Z = 0$, where $M_X(0) = 1$. In this case, the p.d.f. is determined uniquely by the m.g.f.

The characteristic function (c.f.) of a random matrix X $(p \times n)$ is defined by

$$\phi(Z) = M_X(\iota Z).$$

The m.g.f. of a bimatrix variate distribution is defined by

$$M_{X_1,X_2}(Z_1, Z_2) = E[\exp\{\text{tr}(Z_1 X_1') + \text{tr}(Z_2 X_2')\}]$$

$$= \int_{X_1} \int_{X_2} \exp\{\text{tr}(Z_1 X_1') + \text{tr}(Z_2 X_2')\} f_{X_1,X_2}(X_1, X_2)\, dX_1\, dX_2.$$

The function $M_{X_1,X_2}(Z_1, Z_2)$ is a m.g.f. if and only if it is positive and continuous in the neighborhood of $Z_1 = 0$ and $Z_2 = 0$, where $M_{X_1,X_2}(0,0) = 1$. The m.g.f. of the marginal distributions of X_j, $j = 1, 2$, are given by

$$M_{X_1}(Z_1) = M_{X_1,X_2}(Z_1, 0)$$

and

$$M_{X_2}(Z_2) = M_{X_1,X_2}(0, Z_1)$$

respectively. In this case the joint p.d.f. $f_{X_1,X_2}(X_1, X_2)$ is determined uniquely.

Let X $(p \times n)$ be a random matrix and $h(X) = (h_{ij}(X))$ where $h_{ij} : \mathbb{R}^{p \times n} \to \mathbb{R}$, $i = 1, \ldots, r$, $j = 1, \ldots, s$. Then the expected value of the function $h(X)$ is a $r \times s$ matrix defined by

$$E[h(X)] = (E(h_{ij}(X)))$$

when $E(h_{ij}(X))$ exists.

From above it is an easy consequence that
(i) $E(A) = A$, A constant matrix,
(ii) for A $(p \times r)$ and B $(s \times q)$

$$E[Ah(X)B] = AE[h(X)]B,$$

(iii) for $h_1(X)$ and $h_2(X)$ of the same order

$$E\{h_1(X) + h_2(X)\} = E\{h_1(X)\} + E\{h_2(X)\}.$$

Thus for the random matrix X $(p \times n)$, the mean matrix is given by

$$E(X) = (E(X_{ij})).$$

The $pn \times rs$ covariance matrix of the random matrices X $(p \times n)$ and Y $(r \times s)$ is defined by

$$\begin{aligned}
\text{cov}(X, Y) &= \text{cov}(\text{vec}(X'), \text{vec}(Y')) \\
&= E\{(\text{vec}(X') - E\,\text{vec}(X'))(\text{vec}(Y') - E\,\text{vec}(Y'))'\} \\
&= E\{\text{vec}(X')(\text{vec}(Y'))'\} - E\{\text{vec}(X')\}E\{(\text{vec}(Y'))'\} \\
&= \begin{pmatrix}
\text{cov}(\boldsymbol{x}_1^*, \boldsymbol{y}_1^*) & \text{cov}(\boldsymbol{x}_1^*, \boldsymbol{y}_2^*) & \cdots & \text{cov}(\boldsymbol{x}_1^*, \boldsymbol{y}_r^*) \\
\vdots & & & \\
\text{cov}(\boldsymbol{x}_p^*, \boldsymbol{y}_1^*) & \text{cov}(\boldsymbol{x}_p^*, \boldsymbol{y}_2^*) & \cdots & \text{cov}(\boldsymbol{x}_p^*, \boldsymbol{y}_r^*)
\end{pmatrix}.
\end{aligned}$$

where $\boldsymbol{x}_i^{*\prime}$ and $\boldsymbol{y}_j^{*\prime}$ are the ith and j^{th} rows of the matrices X and Y respectively, $i = 1, \ldots, p$ and $j = 1, \ldots, r$.

As a special case of above we get the covariance matrix of X as

$$
\begin{aligned}
\text{cov}(X) &= \text{cov}(\text{vec}(X')) \\
&= E\{\text{vec}(X')(\text{vec}(X'))'\} - E\{\text{vec}(X')\}E\{(\text{vec}(X'))'\} \\
&= \begin{pmatrix}
\text{cov}(\boldsymbol{x}_1^*) & \text{cov}(\boldsymbol{x}_1^*, \boldsymbol{x}_2^*) & \cdots & \text{cov}(\boldsymbol{x}_1^*, \boldsymbol{x}_p^*) \\
\vdots & & & \\
\text{cov}(\boldsymbol{x}_p^*, \boldsymbol{x}_1^*) & \text{cov}(\boldsymbol{x}_p^*, \boldsymbol{x}_2^*) & \cdots & \text{cov}(\boldsymbol{x}_p^*)
\end{pmatrix}
\end{aligned}
$$

We have given most of the results needed in the book. Several other results, in addition to these, will be given in the text, with relevant references.

PROBLEMS

1.1.   Prove that

$$
\beta_p(a,b) = \pi^{\frac{1}{2}(p-1)} \frac{\Gamma(a)\Gamma[b - \frac{1}{2}(p+1)]}{\Gamma(a+b)} \beta_{p-1}\left(a - \frac{1}{2}, b\right).
$$

1.2.   For $B\,(p \times p) > 0$, prove that

$$
\int_{A>0} \frac{\det(A)^{b_p - \frac{1}{2}(p+1)} \text{etr}(-BA)}{\prod_{\alpha=2}^{p} \det(A_{(\alpha)})^{k_\alpha-1}}\, dA = \Gamma_p^*(b_1, \ldots, b_p) \prod_{\alpha=1}^{p} \det(B^{[\alpha]})^{-k_\alpha},
$$

where $b_j = k_{p-j+1} + \cdots + k_p$, $b_j > \frac{1}{2}(j-1)$, $j = 1, \ldots, p$.

(Olkin, 1959)

1.3.   Show that

(i) $\displaystyle \int_{A>0} \frac{\det(A)^{a_p - \frac{1}{2}(p+1)} \text{etr}(-A)(\text{tr}\,A)^t}{\prod_{\alpha=1}^{p-1} \det(A^{[\alpha]})^{m_\alpha+1}}\, dA = \left(\sum_{j=1}^{p} a_j\right)_t \Gamma_p^*(a_1, \ldots, a_p),$

where $a_j = m_1 + \cdots + m_j$, $a_j > \frac{1}{2}(j-1)$, $j = 1, \ldots, p$, and

(ii) $\displaystyle \int_{A>0} \frac{\det(A)^{b_p - \frac{1}{2}(p+1)} \text{etr}(-A)(\text{tr}\,A)^t}{\prod_{\alpha=2}^{p} \det(A_{(\alpha)})^{k_\alpha-1}}\, dA = \left(\sum_{j=1}^{p} b_j\right)_t \Gamma_p^*(b_1, \ldots, b_p),$

where $b_j = k_{p-j+1} + \cdots + k_p$, $b_j > \frac{1}{2}(j-1)$, $j = 1, \ldots, p$.

1.4.   Let $R = (r_{ij})$ be the matrix of correlations and $dR = \prod_{i<j}^{p} dr_{ij}$. Then, show that

(i) $\displaystyle \int_{0<R<I_p} \frac{\det(R)^{a_p - \frac{1}{2}(p+1)}}{\prod_{\alpha=1}^{p-1} \det(R^{[\alpha]})^{m_\alpha+1}}\, dR = \frac{\Gamma_p^*(a_1, \ldots, a_p)}{\prod_{i=1}^{p} \Gamma(a_i)},$

where $a_j = m_1 + \cdots + m_j$, $a > \frac{1}{2}(j-1)$, $j = 1, \ldots, p$ and

(ii) $\displaystyle\int_{0<R<I_p} \frac{\det(R)^{b_p-\frac{1}{2}(p+1)}}{\prod_{\alpha=2}^p \det(R_{(\alpha)})^{k_\alpha-1}}\, dR = \frac{\Gamma_p^*(b_1, \ldots, b_p)}{\prod_{i=1}^p \Gamma(b_i)},$

where $b_j = k_{p-j+1} + \cdots + k_p$, $b_j > \frac{1}{2}(j-1)$, $j = 1, \ldots, p$.

(Olkin, 1959)

1.5.   Prove Theorem 1.4.2 using triangular decomposition of matrix A in (1.4.7).

1.6.   Show that

$$\int_{0<A<I_p} \frac{\det(A)^{a_p-\frac{1}{2}(p+1)} \det(I_p - A)^{b_p-\frac{1}{2}(p+1)}}{\prod_{\alpha=2}^p \left\{ \det(A_{(\alpha)})^{m_\alpha-1} \det((I_p - A)_{(\alpha)})^{k_\alpha-1} \right\}}\, dA$$

$$= \beta_p^*(a_1, \ldots, a_p; b_1, \ldots, b_p)$$

where $a_j = \sum_{i=p-j+1}^p m_i$, $b_j = \sum_{i=p-j+1}^p k_i$, $\mathrm{Re}(a_j) > \frac{1}{2}(j-1)$, and $\mathrm{Re}(b_j) > \frac{1}{2}(j-1)$, $j = 1, \ldots, p$.

1.7.   Show that

$$\int_{S>0} \det(S)^{\frac{1}{2}(m-p-1)} \left(1 + \frac{1}{n}\, \mathrm{tr}(\Sigma^{-1}S)\right)^{-\frac{1}{2}(n+mp)} dS$$

$$= \frac{n^{\frac{1}{2}mp}\Gamma(\frac{1}{2}n)\Gamma_p(\frac{1}{2}m)}{\Gamma[\frac{1}{2}(n+mp)]}\, \det(\Sigma)^{\frac{1}{2}m}.$$

1.8.   Show that for $m \geq p$, $n_j \geq p$, $j = 1, \ldots, k$ and $n = \sum_{j=1}^k n_j$,

(a) $\displaystyle\int_{U_1>0} \cdots \int_{U_k>0} \frac{\prod_{j=1}^k \det(U_j)^{\frac{1}{2}(n_j-p-1)} \det(I_p + \sum_{j=1}^k U_j)^{-\frac{1}{2}(m+n-p-1)}}{\prod_{\alpha=1}^p \det((I_p + \sum_{j=1}^k U_j)^{[\alpha]})} \prod_{j=1}^k dU_j$

$$= \beta_p\Big(\frac{1}{2}n_1, \ldots, \frac{1}{2}n_k; \frac{1}{2}m\Big)$$

and

(b) $\displaystyle\int_{U_1>0} \cdots \int_{U_k>0} \frac{\prod_{j=1}^k \det(U_j)^{\frac{1}{2}(n_j-p-1)} \det(I_p + \sum_{j=1}^k U_j)^{-\frac{1}{2}(m+n-p-1)}}{\prod_{\alpha=1}^p \det((I_p + \sum_{j=1}^k U_j)_{[\alpha]})} \prod_{j=1}^k dU_j$

$$= \beta_p\Big(\frac{1}{2}n_1, \ldots, \frac{1}{2}n_k; \frac{1}{2}m\Big)$$

(Olkin and Rubin, 1964)

1.9.   Show that

$$\int_{Y\in\mathbb{R}^{p\times n}} \prod_{\alpha=1}^p \det((I_p + YY')_{(\alpha)})^{-m_\alpha} \det(I_p + YY')^{-\frac{1}{2}n}\, dY$$

$$= (2\pi)^{\frac{1}{2}np}\, \frac{\Gamma_p^*(a_1, \ldots, a_p)}{\Gamma_p^*(a_1 + \frac{1}{2}n, \ldots, a_p + \frac{1}{2}n)},$$

where $a_j = m_1 + \ldots + m_j$, and $\mathrm{Re}(a_j) > \frac{1}{2}(j-1)$, $j = 1, \ldots, p$.

1.10. Show that for symmetric $R\,(p \times p)$,

$$
\int_{\substack{0 < \sum_{i=1}^r Z_i < I_p \\ Z_i > 0}} \cdots \int \prod_{i=1}^r \det(Z_i)^{a_i - \frac{1}{2}(p+1)} \det\left(I_p - \sum_{i=1}^r Z_i\right)^{b - \frac{1}{2}(p+1)}
$$

$$
{}_mF_n\Big(\alpha_1, \ldots, \alpha_m; \beta_1, \ldots, \beta_n; R\big(I_p - \sum_{i=1}^r Z_i\big)\Big) \prod_{i=1}^r dZ_i
$$

$$
= \frac{\prod_{i=1}^r \Gamma_p(a_i)\Gamma_p(b)}{\Gamma_p(\sum_{i=1}^r a_i + b)} \;{}_{m+1}F_{n+1}\Big(\alpha_1, \ldots, \alpha_m, b; \beta_1, \ldots, \beta_n, \sum_{i=1}^r a_i + b; R\Big)
$$

where $\mathrm{Re}(b) > \frac{1}{2}(p-1)$, and $\mathrm{Re}(a_i) > \frac{1}{2}(p-1)$, $i = 1, \ldots, r$.

1.11. Let $f(V)$ be a continuous scalar function of the symmetric matrix $V\,(p \times p)$, $a_i > \frac{1}{2}(p-1)$, $i = 1, \ldots, k$ and $b_j > \frac{1}{2}(p-1)$, $j = 1, \ldots, \ell$. Then show that

$$
\int_{\substack{0 < \sum_{i=1}^k V_i + \sum_{j=1}^\ell W_j < B \\ V_i > 0, i=1,\ldots,k \\ W_j > 0, j=1,\ldots,\ell}} \cdots \int \prod_{i=1}^k \det(V_i)^{a_i - \frac{1}{2}(p+1)} \prod_{j=1}^\ell \det(W_j)^{b_j - \frac{1}{2}(p+1)}
$$

$$
f\Big(\sum_{i=1}^k V_i\Big) \prod_{i=1}^k dV_i \prod_{j=1}^\ell dW_j
$$

$$
= \beta_p(a_1, \ldots, a_{k-1}; a_k)\beta_p(b_1, \ldots, b_{\ell-1}; b_\ell)\beta_p\Big(\sum_{j=1}^\ell b_j, \frac{1}{2}(p+1)\Big)
$$

$$
\int_{0 < Z < B} \det(Z)^{\sum_{i=1}^k a_i - \frac{1}{2}(p+1)} \det(B - Z)^{\sum_{j=1}^\ell b_j} f(Z)\,dZ.
$$

$$
\text{(Olkin, 1979)}
$$

1.12. Let f and g be continuous scalar functions of a symmetric matrix $V\,(p \times p)$, $a_i > \frac{1}{2}(p-1)$, $i = 1, \ldots, k$ and $b_j > \frac{1}{2}(p-1)$, $j = 1, \ldots, \ell$. Then show that

$$
\int_{\substack{0 < \sum_{i=1}^k V_i + \sum_{j=1}^\ell W_j < B \\ V_i > 0, i=1,\ldots,k \\ W_j > 0, j=1,\ldots,\ell}} \cdots \int \prod_{i=1}^k \det(V_i)^{a_i - \frac{1}{2}(p+1)} \prod_{j=1}^\ell \det(W_j)^{b_j - \frac{1}{2}(p+1)}
$$

$$
f\Big(\sum_{i=1}^k V_i\Big)g\Big(\sum_{j=1}^\ell W_j\Big) \prod_{i=1}^k dV_i \prod_{j=1}^\ell dW_j
$$

$$
= \beta_p(a_1, \ldots, a_{k-1}; a_k) \int_{\substack{0 < X + \sum_{j=1}^\ell W_j < B \\ X > 0 \\ W_j > 0, j=1,\ldots,\ell}} \cdots \int \det(X)^{\sum_{i=1}^k a_i - \frac{1}{2}(p+1)}
$$

$$\prod_{j=1}^{\ell} \det(W_j)^{b_j - \frac{1}{2}(p+1)} f(X) g\Big(\sum_{j=1}^{\ell} W_j\Big) \, dX \prod_{j=1}^{\ell} dW_j$$

$$= \beta_p(a_1, \ldots, a_{k-1}; a_k)\beta_p(b_1, \ldots, b_{\ell-1}; b_\ell)$$

$$\int \cdots \int_{\substack{0 < X+Y < B \\ X > 0 \\ Y > 0}} \det(X)^{\sum_{i=1}^{k} a_i - \frac{1}{2}(p+1)} \det(Y)^{\sum_{j=1}^{\ell} b_j - \frac{1}{2}(p+1)} f(X)g(Y) \, dX \, dY.$$

(Olkin, 1979)

1.13. For $\mathrm{Re}(t) > \frac{1}{2}(p-1)$, prove that

(i) $\displaystyle\int_{Y>0} \mathrm{etr}(-Y) \det(Y)^{t-1} \prod_{i=1}^{p-1} \det(Y^{[i]})^{-1} C_\kappa(Y) \, dY = \Gamma_p(t, \kappa) C_\kappa(I_p)$

and

(ii) $\displaystyle\int_{Y>0} \mathrm{etr}(-Y) \det(Y)^{t-1} \prod_{i=2}^{p} \det(Y_{(i)})^{-1} C_\kappa(Y) \, dY = \Gamma_p(t, \kappa) C_\kappa(I_p).$

(Gupta and Nagar, 1998)

1.14. For $\mathrm{Re}(\gamma) > \frac{1}{2}(p-1)$ and $\mathrm{Re}(\gamma - \alpha - \beta) > \frac{1}{2}(p-1)$, prove that

$$_2F_1(\alpha, \beta; \gamma; I_p) = \frac{\Gamma_p(\gamma)\Gamma_p(\gamma - \alpha - \beta)}{\Gamma_p(\gamma - \alpha)\Gamma_p(\gamma - \beta)}.$$

1.15. Prove that

$$\int_{A>0} \det(A)^{\gamma - \frac{1}{2}(p+1)} \det(I_p + A)^{-\rho} \, _2F_1(\alpha, \beta; \gamma; -A) \, dA$$

$$= \frac{\Gamma_p(\gamma)\Gamma_p(\alpha + \rho - \gamma)\Gamma_p(\beta + \rho - \gamma)}{\Gamma_p(\rho)\Gamma_p(\alpha + \beta + \rho - \gamma)},$$

where $\mathrm{Re}(\gamma) > \frac{1}{2}(p-1)$ and $\mathrm{Re}(\rho + \alpha - \gamma) > \frac{1}{2}(p-1)$.
[HINT: Use (1.6.10) and transform $U = A(I_p + A)^{-1}$.]

(Subrahmaniam, 1973)

1.16. Prove that

$$\int_{A>0} \det(A)^{\gamma - \frac{1}{2}(p+1)} \det(I_p - A)^{\rho - \frac{1}{2}(p+1)} C_\kappa((I_p - A)B) \, _2F_1(\alpha, \beta; \gamma; A) \, dA$$

$$= \frac{\Gamma_p(\gamma)\Gamma_p(\rho, \kappa)\Gamma_p(\gamma + \rho - \alpha - \beta, \kappa)}{\Gamma_p(\rho + \gamma - \alpha, \kappa)\Gamma_p(\gamma + \rho - \beta, \kappa)}.$$

(Subrahmaniam, 1973; Kabe, 1979)

1.17. For B $(p \times p)$ symmetric positive definite matrix, show that

(i) $\displaystyle\int_{0<A<B} \det(A)^{a-\frac{1}{2}(p+1)} \det(I_p + CA)^{-a-b} \, dA$

$$= \beta_p\left(a, \frac{1}{2}(p+1)\right) \det(B)^a \, {}_2F_1\left(a, a+b; a+\frac{1}{2}(p+1); -BC\right),$$

where $\mathrm{Re}(-BC) < I_p$ and $\mathrm{Re}(a) > \frac{1}{2}(p-1)$, and

(ii) $\displaystyle\int_{A>B} \det(A)^{a-\frac{1}{2}(p+1)} \det(I_p + CA)^{-a-b} \, dA$

$$= \beta_p\left(b, \frac{1}{2}(p+1)\right) \det(C)^{-a-b} \det(B)^{-b}$$

$$\quad {}_2F_1\left(b, a+b; b+\frac{1}{2}(p+1); -(BC)^{-1}\right),$$

where $\mathrm{Re}(-(BC)^{-1}) < I_p$, and $\mathrm{Re}(b) > \frac{1}{2}(p-1)$.

1.18. Prove that

$$\int_{A>0} \det(A)^{a-\frac{1}{2}(p+1)} \det(I_p + A)^{-b} \det(I_p + BA)^{-c} \, dA$$

$$= \beta_p(a, b+c-a) \det(B)^{-c} \, {}_2F_1(b+c-a, c; b+c; I_p - B^{-1}),$$

where $\mathrm{Re}(I_p - B^{-1}) < I_p$, $\mathrm{Re}(b+c-a) > \frac{1}{2}(p-1)$, and $\mathrm{Re}(a) > \frac{1}{2}(p-1)$.

1.19. For C $(p \times p)$ symmetric positive definite matrix, prove that

(i) $\displaystyle\int_{0<A<B} \det(A)^{a-\frac{1}{2}(p+1)} \, \mathrm{etr}(-AC) \, dA = \det(C)^{-a} \gamma_p(a, C^{\frac{1}{2}} BC^{\frac{1}{2}})$

where $\mathrm{Re}(a) > \frac{1}{2}(p-1)$, and

(ii) $\displaystyle\int_{0<A<B} \det(A)^{a-\frac{1}{2}(p+1)} \det(I_p - CA)^{b-\frac{1}{2}(p+1)} \, dA = \det(C)^{-a} \beta_p(a, b, C^{\frac{1}{2}} BC^{\frac{1}{2}})$

where $\mathrm{Re}(a) > \frac{1}{2}(p-1)$, $\mathrm{Re}(b) > \frac{1}{2}(p-1)$, and $C^{\frac{1}{2}} BC^{\frac{1}{2}} < I_p$.

1.20. For $\mathrm{Re}(a) > \frac{1}{2}(p-1)$, $\mathrm{Re}(b) > \frac{1}{2}(p-1)$, and $0 < B < I_p$ prove that

$$\int_{0<A<I_p} \det(A)^{a-\frac{1}{2}(p+1)} \det(I_p - A)^{b-\frac{1}{2}(p+1)} {}_2F_1(\alpha, \beta; \gamma; AB) \, dA$$

$$= \beta_p(a, b) \, {}_3F_2(a, \alpha, \beta; a+b, \gamma; B)$$

(Subrahmaniam, 1973)

1.21. For Y $(p \times n)$ and X $(p \times n)$, $\mathrm{rank}(X) = p \le n$, show that

$$\int_{X \in \mathbb{R}^{p \times n}} \mathrm{etr}(XY' - XX') \det(XX')^{c-\frac{1}{2}(p+1)} \Psi(a, c; XX') \, dX$$

$$= \frac{\pi^{\frac{1}{2}np} \Gamma_p[c+\frac{1}{2}(n-p-1)]}{\Gamma_p(a+\frac{1}{2}n)} {}_1F_1\left(c+\frac{1}{2}(n-p-1); a+\frac{1}{2}n; \frac{1}{4}YY'\right),$$

where $\text{Re}(a) > \frac{1}{2}(p-1)$ and $\text{Re}(c) > -\frac{1}{2}n + m$.

1.22. Prove that

$$\int_{Y>0} \text{etr}(-XY) \det(Y)^{b-\frac{1}{2}(p+1)} {}_2F_1\Big(a, a-c+\frac{1}{2}(p+1); b; -Y\Big)$$
$$= \Gamma_p(b) \det(X)^{b-a} \Psi(a,c;X).$$

1.23. Prove that

$$\int_{Y>0} \det(Y)^{b-\frac{1}{2}(p+1)} \text{etr}(-XY) \Psi(a,c;Y) \, dY$$
$$= \frac{\Gamma_p(b)\Gamma_p[b-c+\frac{1}{2}(p+1)]}{\Gamma_p[a+b-c+\frac{1}{2}(p+1)]}$$
$$\quad {}_2F_1\Big(b, b-c+\frac{1}{2}(p+1); a+b-c+\frac{1}{2}(p+1); I_p-X\Big),$$

where $\text{Re}(X) > 0$, $\text{Re}(b-c) > -1$, and $\text{Re}(a) > \frac{1}{2}(p-1)$.

1.24. Prove that

$$\int_{Y>0} \text{etr}(-XY) \det(Y)^{b-\frac{1}{2}(p+1)} {}_1F_1(a;c;AY) \, dY$$
$$= \Gamma_p(b) \det(X)^{-b} {}_2F_1(a,b;c;AX^{-1}),$$

where $\text{Re}(AX^{-1}) < I_p$.

1.25. Show that for $S \in \mathbb{R}^{m \times n}$, $n \geq m$ and $p < q$,

$$\int_{S \in \mathbb{R}^{m \times n}} \text{etr}(-SX' - XX') {}_pF_q(a_1, \ldots, a_p; b_1, \ldots, b_p; XX') \, dX$$
$$= \pi^{\frac{1}{2}mn} \text{etr}\Big(\frac{1}{4}SS'\Big) \sum_{k=0}^{\infty} \frac{(a_1)_\kappa \cdots (a_p)_\kappa}{(b_1)_\kappa \cdots (b_q)_\kappa} \frac{L_\kappa^\gamma(-\frac{1}{4}SS')}{k!},$$

where $\gamma = \frac{1}{2}(n-m-1)$.

1.26. Let the elements of a matrix A be functions of a random variable x. Let A be symmetric positive definite for all values of x. Then prove that $E(A^{-1}) - \{E(A)\}^{-1}$ is positive semidefinite, provided $E(A^{-1})$ and $E(A)$ exist.

(Groves and Rothenberg, 1969)

1.27. Let $X \, (p \times p) > 0$ be a random matrix with p.d.f. $f_X(X)$. Show that

$$E[\det(X)] = \int_{\det(X)>0} \det(X) f_{\det(X)}(\det(X)) d(\det(X))$$

(le Roux, 1978)

1.28. Let $X \in \mathbb{R}^{p \times n}$ and $Y \in \mathbb{R}^{q \times m}$ be random matrices with joint p.d.f. $f(X, Y)$. Let $g_1(X)$ and $g_2(Y)$ denote the marginal densities and $h_1(X|Y)$ and $h_2(Y|X)$ be the conditional densities. Assume $f(X, Y)$, $g_1(X)$, $g_2(Y)$, $h_1(X|Y)$, and $h_2(Y|X)$ are defined for all $X \in \mathbb{R}^{p \times n}$ and $Y \in \mathbb{R}^{q \times m}$. Suppose there exists $Y_0 \in \mathbb{R}^{q \times m}$ such that $h_2(Y_0|X) \neq 0$ for all $X \in \mathbb{R}^{p \times n}$. Then show that

$$f(X, Y) = k \frac{h_2(Y|X) h_1(X|Y_0)}{h_2(Y_0|X)},$$

where k is a constant.

(Gupta and Varga, 1992)

CHAPTER 2

MATRIX VARIATE NORMAL DISTRIBUTION

2.1. INTRODUCTION

The random variable x, with the p.d.f.

$$(2\pi\sigma^2)^{-\frac{1}{2}} \exp\left\{-\frac{1}{2\sigma^2}(x-\mu)^2\right\}, \ x \in \mathbb{R}, \tag{2.1.1}$$

where $\mu \in \mathbb{R}$, is said to have a normal distribution with mean μ and variance σ^2. The multivariate generalization of (2.1.1) for $\boldsymbol{x} = (x_1, \ldots, x_p)'$ is

$$(2\pi)^{-\frac{1}{2}p} \det(\Sigma)^{-\frac{1}{2}} \operatorname{etr}\left\{-\frac{1}{2}\Sigma^{-1}(\boldsymbol{x}-\boldsymbol{\mu})(\boldsymbol{x}-\boldsymbol{\mu})'\right\}, \ \boldsymbol{x} \in \mathbb{R}^p, \ \boldsymbol{\mu} \in \mathbb{R}^p, \ \Sigma > 0, \tag{2.1.2}$$

and the random vector $\boldsymbol{x}$ is said to have a multivariate normal distribution, denoted by $\boldsymbol{x} \sim N_p(\boldsymbol{\mu}, \Sigma)$, with mean vector $\boldsymbol{\mu}$ and covariance matrix Σ. This distribution has been studied extensively and plays a key role in multivariate statistical analysis. In this chapter, we discuss its matrix variate generalization, *i.e.*, matrix variate normal distribution, which is one of the most important matrix variate distributions.

2.2. DENSITY FUNCTION

DEFINITION 2.2.1. *The random matrix X $(p \times n)$ is said to have a matrix variate normal distribution with mean matrix M $(p \times n)$ and covariance matrix $\Sigma \otimes \Psi$ where Σ $(p \times p) > 0$ and Ψ $(n \times n) > 0$, if $\operatorname{vec}(X') \sim N_{pn}(\operatorname{vec}(M'), \Sigma \otimes \Psi)$.*

We shall use the notation $X \sim N_{p,n}(M, \Sigma \otimes \Psi)$. We now derive the density of the random matrix X.

THEOREM 2.2.1. *If $X \sim N_{p,n}(M, \Sigma \otimes \Psi)$, then the p.d.f. of X is given by*

$$(2\pi)^{-\frac{1}{2}np} \det(\Sigma)^{-\frac{1}{2}n} \det(\Psi)^{-\frac{1}{2}p} \operatorname{etr}\left\{-\frac{1}{2}\Sigma^{-1}(X-M)\Psi^{-1}(X-M)'\right\},$$

$$X \in \mathbb{R}^{p \times n}, \ M \in \mathbb{R}^{p \times n}. \tag{2.2.1}$$

Proof: Let $x = \mathrm{vec}(X')$ and $m = \mathrm{vec}(M')$. Then, according to the Definition 2.2.1, $x \sim N_{pn}(m, \Sigma \otimes \Psi)$, and its p.d.f. is

$$(2\pi)^{-\frac{1}{2}np} \det(\Sigma \otimes \Psi)^{-\frac{1}{2}} \, \mathrm{etr} \left\{ -\frac{1}{2}(\Sigma \otimes \Psi)^{-1}(x - m)(x - m)' \right\}.$$

Using Theorems 1.2.21 and 1.2.22, we get

$$\det(\Sigma \otimes \Psi)^{-\frac{1}{2}} = \det(\Sigma)^{-\frac{1}{2}n} \det(\Psi)^{-\frac{1}{2}p}, \qquad (2.2.2)$$

$$\mathrm{tr}\{(\Sigma \otimes \Psi)^{-1}(x - m)(x - m)'\} = \mathrm{tr}\{(\Sigma^{-1} \otimes \Psi^{-1})(x - m)(x - m)'\}$$
$$= \mathrm{tr}\{\Sigma^{-1}(X - M)\Psi^{-1}(X - M)'\}. \qquad (2.2.3)$$

Now, from (2.2.2) and (2.2.3), the result (2.2.1) is easily established. ∎

This distribution belongs to the class of matrix variate elliptically contoured distributions studied in Chapter 9. In particular for $M = 0$, the distribution belongs to (i) the class of right spherical distributions if $\Psi = I_n$, (ii) the class of left spherical distributions if $\Sigma = I_p$, and (iii) to the class of spherical distributions if $\Psi = I_n$ and $\Sigma = I_p$.

The matrix variate normal distribution arises when sampling from multivariate normal population. Let $x_1, \ldots, x_N$ be a random sample of size N from $N_p(\mu, \Sigma)$. Define the observation random matrix (*e.g.*, see Roy, 1957; Siotani, Hayakawa and Fujikoshi, 1985), as

$$X = \begin{pmatrix} x_{11} & \cdots & x_{1N} \\ x_{21} & \cdots & x_{2N} \\ \vdots & & \vdots \\ x_{p1} & \cdots & x_{pN} \end{pmatrix} = (x_1, \ldots, x_N) = \begin{pmatrix} x_1^{*\prime} \\ \vdots \\ x_p^{*\prime} \end{pmatrix}, \qquad (2.2.4)$$

then $X' \sim N_{N,p}(e\mu', I_N \otimes \Sigma)$, where $e \, (N \times 1) = (1, \ldots, 1)'$.

2.3. PROPERTIES

In this section, we study various properties of matrix variate normal distribution.

THEOREM 2.3.1. *If $X \sim N_{p,n}(M, \Sigma \otimes \Psi)$, then $X' \sim N_{n,p}(M', \Psi \otimes \Sigma)$.*

Proof: It suffices to prove that the exponents occurring in the densities of $\mathrm{vec}(X')$ and $\mathrm{vec}(X)$ are equal. This, however, follows easily from Theorem 1.2.22. ∎

THEOREM 2.3.2. *If $X \sim N_{p,n}(M, \Sigma \otimes \Psi)$, then the characteristic function of X is*

$$\phi_X(Z) = \mathrm{etr} \left(\iota Z'M - \frac{1}{2}Z'\Sigma Z\Psi \right). \qquad (2.3.1)$$

Proof: We have

$$\phi_X(Z) = E\{\mathrm{etr}(\iota X Z')\}, \ \iota = \sqrt{-1}$$
$$= E[\exp\{\iota(\mathrm{vec}(X'))' \, \mathrm{vec}(Z')\}].$$

Now we know that $\mathrm{vec}(X') \sim N_{pn}(\mathrm{vec}(M'), \Sigma \otimes \Psi)$. Hence, from the characteristic function of a multivariate normal distribution, we get

$$\phi_X(Z) = \exp\left\{\iota(\mathrm{vec}(M'))' \, \mathrm{vec}(Z') - \frac{1}{2}(\mathrm{vec}(Z'))'(\Sigma \otimes \Psi) \, \mathrm{vec}(Z')\right\}$$
$$= \mathrm{etr}\left(\iota Z'M - \frac{1}{2}Z'\Sigma Z\Psi\right).$$

The last equality follows from Theorem 1.2.22. ∎

THEOREM 2.3.3. *Let* $X \sim N_{p,n}(M, \Sigma \otimes \Psi)$*, and* $M = (m_{ij})$*,* $\Sigma = (\sigma_{ti})$*,* $\Psi = (\psi_{jk})$*. Then,*

(i) $E(x_{i_1 j_1} x_{i_2 j_2}) = \sigma_{i_1 i_2} \psi_{j_1 j_2} + m_{i_1 j_1} m_{i_2 j_2}$

(ii) $E(x_{i_1 j_1} x_{i_2 j_2} x_{i_3 j_3}) = m_{i_1 j_1} \sigma_{i_2 i_3} \psi_{j_2 j_3} + m_{i_2 j_2} \sigma_{i_1 i_3} \psi_{j_1 j_3} + m_{i_3 j_3} \sigma_{i_1 i_2} \psi_{j_1 j_2}$
$$+ m_{i_1 j_1} m_{i_2 j_2} m_{i_3 j_3}$$

and

(iii) $E(x_{i_1 j_1} x_{i_2 j_2} x_{i_3 j_3} x_{i_4 j_4}) = \sigma_{i_1 i_4} \psi_{j_1 j_4} \sigma_{i_2 i_3} \psi_{j_2 j_3} + \sigma_{i_1 i_2} \psi_{j_1 j_2} \sigma_{i_4 i_3} \psi_{j_4 j_3}$
$$+ \sigma_{i_1 i_3} \psi_{j_1 j_3} \sigma_{i_4 i_2} \psi_{j_4 j_2} + m_{i_1 j_1} m_{i_2 j_2} \sigma_{i_4 i_3} \psi_{j_4 j_3}$$
$$+ m_{i_1 j_1} m_{i_3 j_3} \sigma_{i_4 i_2} \psi_{j_4 j_2} + m_{i_2 j_2} m_{i_3 j_3} \sigma_{i_1 i_4} \psi_{j_1 j_4}$$
$$+ m_{i_1 j_1} m_{i_4 j_4} \sigma_{i_2 i_3} \psi_{j_2 j_3} + m_{i_4 j_4} m_{i_2 j_2} \sigma_{i_1 i_3} \psi_{j_1 j_3}$$
$$+ m_{i_4 j_4} m_{i_3 j_3} \sigma_{i_1 i_2} \psi_{j_1 j_2} + m_{i_1 j_1} m_{i_2 j_2} m_{i_3 j_3} m_{i_4 j_4}.$$

Proof: From Theorem 2.3.2, the characteristic function of X is

$$\phi_X(Z) = \exp\{h(Z)\}, \tag{2.3.2}$$

where

$$h(Z) = \iota \sum_{i=1}^{p} \sum_{j=1}^{n} m_{ij} z_{ij} - \frac{1}{2} \sum_{i=1}^{p} \sum_{t=1}^{p} \sum_{k=1}^{n} \sum_{j=1}^{n} z_{ij} \psi_{jk} z_{tk} \sigma_{ti} \tag{2.3.3}$$

and $Z = (z_{ij})$. Now, from (2.3.2)

$$\frac{\partial}{\partial z_{i_1 j_1}} \phi_X(Z) = \exp\{h(Z)\} \frac{\partial}{\partial z_{i_1 j_1}} h(Z) \tag{2.3.4}$$

and

$$\frac{\partial}{\partial z_{i_1 j_1}} h(Z) = \frac{\partial}{\partial z_{i_1 j_1}}\left[\iota \sum_{i=1}^{p} \sum_{j=1}^{n} m_{ij} z_{ij} - \frac{1}{2}\left\{\sum_{i=1}^{p} \sum_{j=1}^{n} z_{ij}^2 \psi_{jj} \sigma_{ii} + \sum_{i=1}^{p} \sum_{\substack{k=1 \\ \neq j}}^{n} \sum_{j=1}^{n} z_{ij} \psi_{jk} z_{ik} \sigma_{ii}\right.\right.$$
$$\left.\left. + \sum_{i=1}^{p} \sum_{\substack{t=1 \\ \neq i}}^{p} \sum_{j=1}^{n} z_{ij} \psi_{jj} z_{tj} \sigma_{ti} + \sum_{i=1}^{p} \sum_{\substack{t=1 \\ \neq i}}^{p} \sum_{\substack{k=1 \\ \neq j}}^{n} \sum_{j=1}^{n} z_{ij} \psi_{jk} z_{tk} \sigma_{ti}\right\}\right]$$

$$= \iota m_{i_1 j_1} - \left[z_{i_1 j_1} \psi_{j_1 j_1} \sigma_{i_1 i_1} + \sum_{\substack{k=1 \\ \neq j_1}}^{n} \psi_{j_1 k} z_{i_1 k} \sigma_{i_1 i_1} \right.$$

$$\left. + \sum_{\substack{t=1 \\ \neq i_1}}^{p} \psi_{j_1 j_1} z_{t j_1} \sigma_{t i_1} + \sum_{\substack{t=1 \\ \neq i_1}}^{p} \sum_{\substack{k=1 \\ \neq j_1}}^{n} \psi_{j_1 k} z_{t k} \sigma_{t i_1} \right]. \tag{2.3.5}$$

Substituting from (2.3.5) in (2.3.4), we get

$$E(x_{i_1 j_1}) = \frac{1}{\iota} \frac{\partial}{\partial z_{i_1 j_1}} \phi_X(Z) \Big|_{Z=0} = m_{i_1 j_1}.$$

Now differentiating (2.3.4) with respect to $z_{i_2 j_2}$, we get

$$\frac{\partial^2}{\partial z_{i_1 j_1} \partial z_{i_2 j_2}} \phi_X(Z) = \exp\{h(Z)\} \left[\frac{\partial}{\partial z_{i_1 j_1}} h(Z) \cdot \frac{\partial}{\partial z_{i_2 j_2}} h(Z) + \frac{\partial^2}{\partial z_{i_1 j_1} \partial z_{i_2 j_2}} h(Z) \right], \tag{2.3.6}$$

where $\frac{\partial}{\partial z_{i_2 j_2}} h(Z)$ is obtained from (2.3.5) by replacing i_1 by i_2 and j_1 by j_2. Further differentiating (2.3.5) with respect to $z_{i_2 j_2}$,

$$\frac{\partial^2}{\partial z_{i_1 j_1} \partial z_{i_2 j_2}} h(Z) = -\sigma_{i_1 i_2} \psi_{j_1 j_2}. \tag{2.3.7}$$

Hence,

$$E(x_{i_1 j_1} x_{i_2 j_2}) = \frac{1}{\iota^2} \frac{\partial^2}{\partial z_{i_1 j_1} \partial z_{i_2 j_2}} \phi_X(Z) \Big|_{Z=0} = \sigma_{i_1 i_2} \psi_{j_1 j_2} + m_{i_1 j_1} m_{i_2 j_2}.$$

Now, from (2.3.6)

$$\frac{\partial^3}{\partial z_{i_1 j_1} \partial z_{i_2 j_2} \partial z_{i_3 j_3}} \phi_X(Z) = \exp\{h(Z)\} \left[\frac{\partial}{\partial z_{i_1 j_1}} h(Z) \cdot \frac{\partial}{\partial z_{i_2 j_2}} h(Z) \cdot \frac{\partial}{\partial z_{i_3 j_3}} h(Z) \right.$$

$$+ \frac{\partial^2}{\partial z_{i_1 j_1} \partial z_{i_2 j_2}} h(Z) \cdot \frac{\partial}{\partial z_{i_3 j_3}} h(Z)$$

$$+ \frac{\partial^2}{\partial z_{i_1 j_1} \partial z_{i_3 j_3}} h(Z) \cdot \frac{\partial}{\partial z_{i_2 j_2}} h(Z)$$

$$+ \frac{\partial^2}{\partial z_{i_2 j_2} \partial z_{i_3 j_3}} h(Z) \cdot \frac{\partial}{\partial z_{i_1 j_1}} h(Z)$$

$$\left. + \frac{\partial^3}{\partial z_{i_1 j_1} \partial z_{i_2 j_2} \partial z_{i_3 j_3}} h(Z) \right]. \tag{2.3.8}$$

Using (2.3.5), (2.3.17), and

$$\frac{\partial^3}{\partial z_{i_1 j_1} \partial z_{i_2 j_2} \partial z_{i_3 j_3}} h(Z) = 0$$

in (2.3.8), we get

$$E(x_{i_1j_1}x_{i_2j_2}x_{i_3j_3}) = \frac{1}{i^3}\frac{\partial^3}{\partial z_{i_1j_1}\partial z_{i_2j_2}\partial z_{i_3j_3}}\phi_X(Z)\Big|_{Z=0}$$

$$= m_{i_1j_1}\sigma_{i_2i_3}\psi_{j_2j_3} + m_{i_2j_2}\sigma_{i_1i_3}\psi_{j_1j_3}$$

$$+ m_{i_3j_3}\sigma_{i_1i_2}\psi_{j_1j_2} + m_{i_1j_1}m_{i_2j_2}m_{i_3j_3}.$$

Continuing this procedure one can also establish (iii). ∎

COROLLARY 2.3.3.1. *Let* $X \sim N_{p,n}(0, \Sigma \otimes \Psi)$, *then*

(i) $E(x_{i_1j_1}x_{i_2j_2}) = \sigma_{i_1i_2}\psi_{j_1j_2}$

(ii) $E(x_{i_1j_1}x_{i_2j_2}x_{i_3j_3}) = 0$

and

(iii) $E(x_{i_1j_1}x_{i_2j_2}x_{i_3j_3}x_{i_4j_4}) = \sigma_{i_1i_2}\psi_{j_1j_2}\sigma_{i_3i_4}\psi_{j_3j_4} + \sigma_{i_1i_3}\psi_{j_1j_3}\sigma_{i_2i_4}\psi_{j_2j_4}$

$$+ \sigma_{i_1i_4}\psi_{j_1j_4}\sigma_{i_2i_3}\psi_{j_2j_3}.$$

Proof: Substitute $M = 0$ in Theorem 2.3.3. ∎

van der Merwe (1980) has derived expectations of the traces of certain functions of X. Some of these are given in the next theorem.

THEOREM 2.3.4. *Let* $X \sim N_{p,n}(M, \Sigma \otimes \Psi)$ *and* $\Sigma = (\sigma_{ij})$, $\Psi = (\psi_{ij})$. *Then, for any constant matrix* A ($p \times n$) *and* $a = 0, 1, 2, \ldots, \ldots$, *we have*

(i) $E(\text{tr}(XX')) = \text{tr}(\Sigma)\text{tr}(\Psi)$

(ii) $E(\text{tr}(XX'(AA')^a)) = \text{tr}(\Sigma(AA')^a)\text{tr}(\Psi)$

(iii) $E(\text{tr}^2(XA')) = \text{tr}(\Sigma A\Psi A')$

(iv) $E(\text{tr}(XA')^2) = \text{tr}(\Sigma A\Psi A')$

(v) $E(\text{tr}(XA')\text{tr}(XA'(AA')^a)) = \text{tr}(\Sigma A\Psi A'(AA')^a)$.

Proof: Here we give the proof for (i) and (iii); the others can be similarly derived.

(i) $E(\text{tr}(XX')) = E\Big(\sum_{i=1}^{p}\sum_{j=1}^{n}x_{ij}x_{ij}\Big) = \sum_{i=1}^{p}\sum_{j=1}^{n}E(x_{ij}x_{ij})$

$$= \sum_{i=1}^{p}\sum_{j=1}^{n}\sigma_{ii}\psi_{jj} = \text{tr}(\Sigma)\text{tr}(\Psi).$$

(iii) $E(\text{tr}^2(XA')) = E(\text{tr}^2\Psi)$, where $Y = XA' \sim N_{p,p}(0, \Sigma \otimes (A\Psi A'))$

$$= E\Big(\sum_{i=1}^{p}y_{ii}\Big)^2 = E\Big(\sum_{i=1}^{p}\sum_{j=1}^{p}y_{ii}y_{jj}\Big)$$

$$= \sum_{i=1}^{p}\sum_{j=1}^{p}E(y_{ii}y_{jj}) = \sum_{i=1}^{p}\sum_{j=1}^{p}\sigma_{ij}\psi_{ij}^*; \text{ (where } A\Psi A' = (\psi_{ij}^*))$$

$$= \text{tr}(\Sigma A\Psi A'). \blacksquare$$

H. M. Nel (1977) has derived expectations of certain matrix valued functions of X, some of which are given in the next five theorems.

THEOREM 2.3.5. *Let* $X \sim N_{p,n}(M, \Sigma \otimes \Psi)$, *then*
  (i) $E(X'AX) = \text{tr}(\Sigma A')\Psi + M'AM$, $A\,(p \times p)$
  (ii) $E(XAX') = \text{tr}(A'\Psi)\Sigma + MAM'$, $A\,(n \times n)$
  (iii) $E(XAX) = \Sigma A'\Psi + MAM$, $A\,(n \times p)$
  (iv) $E(\text{tr}(AX)X) = \Sigma A'\Psi + \text{tr}(AM)M$, $A\,(n \times p)$
  (v) $E(\text{tr}(AX)X') = \Psi A\Sigma + \text{tr}(AM)M'$, $A\,(n \times p)$
  (vi) $E(\text{tr}(AX')X) = \Sigma A\Psi + \text{tr}(AM')M$, $A\,(p \times n)$
  (vii) $E(\text{tr}(AX')X') = \Psi A'\Sigma + \text{tr}(AM')M'$, $A\,(p \times n)$.

Proof: (i) Let $X = (x_{ij})$ and $A = (a_{ij})$, then the $(i,j)^{\text{th}}$ element of $X'AX$ is $\sum_{k=1}^{p}\sum_{t=1}^{p} x_{ti}a_{tk}x_{kj}$, and from Theorem 2.3.3 we get

$$E(X'AX) = E\Big(\Big(\sum_{k=1}^{p}\sum_{t=1}^{p} x_{ti}a_{tk}x_{kj}\Big)\Big)$$

$$= \Big(\sum_{k=1}^{p}\sum_{t=1}^{p} a_{tk}(\sigma_{tk}\psi_{ij} + m_{ti}m_{kj})\Big)$$

$$= \Big(\psi_{ij}\sum_{k=1}^{p}\sum_{t=1}^{p} a_{tk}\sigma_{tk} + \sum_{k=1}^{p}\sum_{t=1}^{p} a_{tk}m_{ti}m_{kj}\Big)$$

$$= \text{tr}(A'\Sigma)\Psi + M'AM.$$

(ii) From Theorem 2.3.1, $X' \sim N_{n,p}(M', \Psi \otimes \Sigma)$. Therefore, result (ii) follows from result (i).

(iii) The $(i,j)^{\text{th}}$ element of XAX is $\sum_{t=1}^{p}\sum_{k=1}^{n} x_{ik}a_{kt}x_{tj}$ and hence using Theorem 2.3.3, we get

$$E(XAX) = E\Big(\Big(\sum_{t=1}^{p}\sum_{k=1}^{n} x_{ik}a_{kt}x_{tj}\Big)\Big)$$

$$= \Big(\sum_{t=1}^{p}\sum_{k=1}^{n} a_{kt}(\sigma_{it}\psi_{kj} + m_{ik}m_{tj})\Big)$$

$$= \Sigma A'\Psi + MAM.$$

(iv) The $(i,j)^{\text{th}}$ element of $\text{tr}(AX)X$ is $x_{ij}\sum_{k=1}^{n}\sum_{t=1}^{p} a_{kt}x_{tk}$ and hence from Theorem 2.3.3, we get

$$E(\text{tr}(AX)X) = E\Big(\Big(x_{ij}\sum_{k=1}^{n}\sum_{t=1}^{p} a_{kt}x_{tk}\Big)\Big)$$

$$= \Big(\sum_{k=1}^{n}\sum_{t=1}^{p} a_{kt}(\sigma_{it}\psi_{jk} + m_{ij}m_{tk})\Big)$$

$$= \Sigma A'\Psi + \text{tr}(AM)M.$$

(v)–(vii) Using Theorem 2.3.1, the result (iv) and noting that $\text{tr}(AX) = \text{tr}(AX)' = \text{tr}(A'X')$ the results follow. ∎

It may be noted that some of the results given in Theorem 2.3.4 can be derived from Theorem 2.3.5.

THEOREM 2.3.6. Let $X \sim N_{p,n}(M, \Sigma \otimes \Psi)$, then

(i) $E(XAXBX) = MA\Sigma B'\Psi + \Sigma B'M'A'\Psi + \Sigma A'\Psi BM + MAMBM$,
$\qquad A(n \times p), B(n \times p)$,

(ii) $E(X'AXBX) = M'A\Sigma B'\Psi + \text{tr}(\Sigma B'M'A')\Psi + \text{tr}(A\Sigma)\Psi BM$
$\qquad + M'AMBM, A(p \times p), B(n \times p)$,

(iii) $E(X'AX'BX) = \text{tr}(\Sigma B')M'A\Psi + \text{tr}(AM'B\Sigma)\Psi + \Psi A'\Sigma BM$
$\qquad + M'AM'BM, A(p \times n), B(p \times p)$,

(iv) $E(X'AXBX') = \text{tr}(B\Psi)M'A\Sigma + \Psi B'M'A'\Sigma + \text{tr}(A\Sigma)\Psi BM'$
$\qquad + M'AMBM', A(p \times p), B(n \times n)$,

(v) $E(XAX'BX') = MA\Psi B'\Sigma + \text{tr}(AM'B\Psi)\Sigma + \text{tr}(A\Psi)\Sigma BM'$
$\qquad + MAM'BM', A(n \times n), B(p \times p)$,

(vi) $E(X'AX'BX') = M'A\Psi B'\Sigma + \Psi B'MA'\Sigma + \Psi A'\Sigma BM' + M'AM'BM'$,
$\qquad A(p \times n), B(p \times n)$,

(vii) $E(XAX'BX) = \text{tr}(B\Sigma)MA\Psi + \Sigma B'MA'\Psi + \text{tr}(A\Psi)\Sigma BM$
$\qquad + MAM'BM, A(n \times n), B(p \times p)$.

Proof: (i) The $(i,j)^{\text{th}}$ element of $XAXBX$ is $\sum_{g=1}^{p}\sum_{\ell=1}^{n}\sum_{t=1}^{p}\sum_{k=1}^{n} x_{ik}a_{kt}x_{t\ell}b_{\ell g}x_{gj}$. Hence,

$$E(XAXBX) = \left(\sum_{g=1}^{p}\sum_{\ell=1}^{n}\sum_{t=1}^{p}\sum_{k=1}^{n} a_{kt}b_{\ell g}E(x_{ik}x_{t\ell}x_{gj})\right)$$

$$= \left(\sum_{g=1}^{p}\sum_{\ell=1}^{n}\sum_{t=1}^{p}\sum_{k=1}^{n} a_{kt}b_{\ell g}(m_{ik}\sigma_{tg}\psi_{\ell j} + m_{t\ell}\sigma_{ig}\psi_{kj} + m_{gj}\sigma_{it}\psi_{k\ell} + m_{ik}m_{t\ell}m_{gj})\right)$$

$$= MA\Sigma B'\Psi + \Sigma B'M'A'\Psi + \Sigma A'\Psi BM + MAMBM.$$

The proofs of (ii), (iii), and (iv) follow similar steps. For the proof of (v), (vi), and (vii), notice that $X' \sim N_{n,p}(M', \Psi \otimes \Sigma)$ and use the results (ii), (i), and (iv) respectively. ∎

THEOREM 2.3.7. Let $X \sim N_{p,n}(M, \Sigma \otimes \Psi)$, then

(i) $E(\text{tr}(X'AXB)X) = \text{tr}(A'\Sigma)\text{tr}(B'\Psi)M + \Sigma A'MB'\Psi + \Sigma AMB\Psi$,
$\qquad + \text{tr}(M'AMB)M, A(p \times p), B(n \times n)$,

(ii) $E(\text{tr}(AX)XBX) = MB\Sigma A'\Psi + \Sigma A'\Psi BM + \text{tr}(AM)\Sigma B'\Psi$
$\qquad + \text{tr}(AM)MBM, A(n \times p), B(n \times p)$,

(iii) $E(\text{tr}(AX)X'BX) = M'B\Sigma A'\Psi + \Psi A\Sigma BM + \text{tr}(AM)\text{tr}(B\Sigma)\Psi$
$\qquad + \text{tr}(AM)M'BM, A(n \times p), B(p \times p)$.

Proof: (i) The $(r,s)^{\text{th}}$ element of $\text{tr}(X'AXB)X$ is

$$\sum_{i=1}^{n}\sum_{j=1}^{n}\sum_{k=1}^{p}\sum_{t=1}^{p} x_{ti}x_{kj}x_{rs}a_{tk}b_{ji}.$$

Hence,

$$E(\mathrm{tr}(X'AXB)X)$$

$$= \Big(\sum_{i=1}^{n}\sum_{j=1}^{n}\sum_{k=1}^{p}\sum_{t=1}^{p} a_{tk}b_{ji}(m_{rs}\sigma_{tk}\psi_{ij} + m_{ti}\sigma_{rk}\psi_{sj} + m_{kj}\sigma_{rt}\psi_{si} + m_{rs}m_{ti}m_{kj})\Big)$$

$$= \mathrm{tr}(A'\Sigma)\,\mathrm{tr}(B'\Psi)M + \Sigma A'MB'\Psi + \Sigma AMB\Psi + \mathrm{tr}(M'AMB)M.$$

(ii) The $(r,s)^{\text{th}}$ element of $\mathrm{tr}(AX)XBX$ is

$$\sum_{t=1}^{p}\sum_{k=1}^{n}\sum_{i=1}^{n}\sum_{j=1}^{p} a_{ij}b_{kt}x_{rk}x_{ts}x_{ji}.$$

Hence,

$$E(\mathrm{tr}(AX)XBX)$$

$$= \Big(\sum_{t=1}^{p}\sum_{k=1}^{n}\sum_{i=1}^{n}\sum_{j=1}^{p} a_{ij}b_{kt}(m_{rk}\sigma_{tj}\psi_{si} + m_{ts}\sigma_{rj}\psi_{ki} + m_{ji}\sigma_{rs}\psi_{ks} + m_{rk}m_{ts}m_{ji})\Big)$$

$$= MB\Sigma A'\Psi + \Sigma A'\Psi BM + \mathrm{tr}(AM)\Sigma B'\Psi + \mathrm{tr}(AM)MBM.$$

(iii) The derivation is similar to (ii). ■

THEOREM 2.3.8. *Let* $X \sim N_{p,n}(M, \Sigma \otimes \Psi)$, *then*

(i) $E(XAXBXCX) = \Sigma C'\Psi B\Sigma A'\Psi + \Sigma A'\Psi B\Sigma C'\Psi + \mathrm{tr}(A\Sigma C'\Psi)\Sigma B'\Psi$
$\quad + MAMB\Sigma C'\Psi + MA\Sigma C'M'B'\Psi + \Sigma C'M'B'M'A'\Psi + MA\Sigma B'\Psi CM$
$\quad + \Sigma B'M'A'\Psi CM + \Sigma A'\Psi BMCM + MAMBMCM$, $A\,(n \times p)$,
$\quad B\,(n \times p)$, $C\,(n \times p)$,

(ii) $E(X'AXBXCX) = \mathrm{tr}(\Sigma C'\Psi B\Sigma A')\Psi + \mathrm{tr}(A\Sigma)\Psi B\Sigma C'\Psi + \Psi C\Sigma A'\Sigma B'\Psi$
$\quad + M'AMB\Sigma C'\Psi + M'A\Sigma C'M'B'\Psi + \mathrm{tr}(AMBMC\Sigma)\Psi + M'A\Sigma B'\Psi CM$
$\quad + \mathrm{tr}(AMB\Sigma)\Psi CM + \mathrm{tr}(A\Sigma)\Psi BMCM + M'AMBMCM$, $A\,(p \times p)$,
$\quad B\,(n \times p)$, $C\,(n \times p)$,

(iii) $E(XAX'BXCX) = \mathrm{tr}(B\Sigma)\Sigma C'\Psi A'\Psi + \mathrm{tr}(A\Psi)\Sigma B\Sigma C'\Psi + \Sigma B'\Sigma C'\Psi A\Psi$
$\quad + MAM'B\Sigma C'\Psi + \mathrm{tr}(MC\Sigma B)MA\Psi + \Sigma C'M'B'MA'\Psi$
$\quad + \mathrm{tr}(\Sigma B)MA\Psi CM + \Sigma B'MA'\Psi CM + \mathrm{tr}(A\Psi)\Sigma BMCM$
$\quad + MAM'BMCM$, $A\,(n \times n)$, $B\,(p \times p)$, $C\,(n \times p)$,

(iv) $E(X'AX'BXCX) = \mathrm{tr}(\Sigma C'\Psi A')\,\mathrm{tr}(B\Sigma)\Psi + \Psi A'\Sigma B\Sigma C'\Psi + \Psi C\Sigma B\Sigma A\Psi$
$\quad + M'AM'B\Sigma C'\Psi + \mathrm{tr}(MC\Sigma B)M'A\Psi + \mathrm{tr}(AM'BMC\Sigma)\Psi$
$\quad + \mathrm{tr}(B\Sigma)M'A\Psi CM + \mathrm{tr}(AM'B\Sigma)\Psi CM + \Psi A'\Sigma BMCM$
$\quad + M'AM'BMCM$, $A\,(p \times n)$, $B\,(p \times p)$, $C\,(n \times p)$,

(v) $E(X'AXBX'CX) = \mathrm{tr}(\Sigma C'\Sigma A')\,\mathrm{tr}(B\Psi)\Psi + \mathrm{tr}(A\Sigma)\,\mathrm{tr}(C\Sigma)\Psi B\Psi$
$\quad + \mathrm{tr}(A\Sigma C'\Sigma)\Psi B'\Psi + \mathrm{tr}(\Sigma C)M'AMB\Psi + M'A\Sigma C'MB'\Psi$
$\quad + \mathrm{tr}(AMBM'C\Sigma)\Psi + \mathrm{tr}(B\Psi)M'A\Sigma CM + \Psi B'M'A'\Sigma CM$
$\quad + \mathrm{tr}(A\Sigma)\Psi BM'CM + M'AMBM'CM$, $A\,(p \times p)$, $B\,(n \times n)$, $C\,(p \times p)$,

(vi) $E(X'AXBXCX') = \Psi C'\Psi B\Sigma A'\Sigma + \mathrm{tr}(A\Sigma)\,\mathrm{tr}(C\Psi)\Psi B\Sigma + \Psi C\Psi B\Sigma A\Sigma$
$\quad + \mathrm{tr}(\Psi C)M'AMB\Sigma + \mathrm{tr}(BMC\Psi)M'A\Sigma + \Psi C'M'B'M'A'\Sigma$
$\quad + M'A\Sigma B'\Psi CM' + \mathrm{tr}(\Sigma AMB)\Psi CM' + \mathrm{tr}(A\Sigma)\Psi BMCM'$
$\quad + M'AMBMCM'$, $A\,(p \times p)$, $B\,(n \times p)$, $C\,(n \times n)$.

Proof: As in the proof of Theorem 2.3.7, the results (i)–(vi) can be derived by taking the $(i,j)^{\text{th}}$ element of the random matrix, substituting its expected value from Theorem 2.3.3, and converting the resulting expression in matrix form. ∎

THEOREM 2.3.9. Let $X \sim N_{p,n}(M, \Sigma \otimes \Psi)$, then

(i) $E(\text{tr}(XBXCX')XA) = \Sigma^2 B'\Psi C\Psi A + \text{tr}(C\Psi)\Sigma^2 B'\Psi A + \text{tr}(\Sigma)\Sigma B'\Psi C'\Psi A$
   $+ \text{tr}(MB\Sigma)\text{tr}(C\Psi)MA + \text{tr}(MC\Psi B)\text{tr}(\Sigma)MA + \Sigma MBMC\Psi A$
   $+ \text{tr}(MC'\Psi B\Sigma)MA + \Sigma B'M'MC'\Psi A + \Sigma MC'M'B'\Psi A$
   $+ \text{tr}(MBMCM')MA, \ A\,(n \times n), \ B\,(n \times p), \ C\,(n \times n),$

(ii) $E(\text{tr}(BXCX')XAX) = \Sigma B\Sigma A'\Psi C'\Psi + \text{tr}(B\Sigma)\text{tr}(C\Psi)\Sigma A'\Psi$
   $+ \Sigma B'\Sigma A'\Psi C\Psi + \text{tr}(B\Sigma)\text{tr}(C\Psi)MAM + MA\Sigma BMC\Psi + \Sigma BMC\Psi AM$
   $+ MA\Sigma B'MC'\Psi + \Sigma B'MC'\Psi AM + \text{tr}(MC'MB')\Sigma A'\Psi$
   $+ \text{tr}(BMCM')MAM, \ A\,(n \times p), \ B\,(n \times p), \ C\,(n \times n),$

(iii) $E(\text{tr}(BXCX')X'AX) = \text{tr}(\Sigma B\Sigma A')\Psi C'\Psi + \text{tr}(A\Sigma)\text{tr}(B\Sigma)\text{tr}(C\Psi)\Psi$
   $+ \text{tr}(\Sigma B'\Sigma A')\Psi C\Psi + \text{tr}(B\Sigma)\text{tr}(C\Psi)M'AM + M'A\Sigma BMC\Psi$
   $+ \Psi C'M'B'\Sigma AM + M'A\Sigma B'MC'\Psi + \Psi CM'B\Sigma AM$
   $+ \text{tr}(A\Sigma)\text{tr}(MCM'B)\Psi + \text{tr}(BMCM')M'AM, \ A\,(p \times p),$
   $B\,(p \times p), \ C\,(n \times n),$

(iv) $E(\text{tr}(AX)XBXCX) = \Sigma B'\Psi C\Sigma A'\Psi + \Sigma A'\Psi B\Sigma C'\Psi + \Sigma C'\Psi A\Sigma B'\Psi$
   $+ \text{tr}(AM)MB\Sigma C'\Psi + \text{tr}(AM)\Sigma C'M'B'\Psi + MBMC\Sigma A'\Psi$
   $+ \text{tr}(MA)\Sigma B'\Psi CM + MB\Sigma A'\Psi CM + \Sigma A\Psi BM\Psi M$
   $+ \text{tr}(AM)MBM\Psi M, \ A\,(p \times p), \ B\,(n \times p), \ C\,(n \times p),$

(v) $E(\text{tr}(AX)X'BXCX) = \text{tr}(B\Sigma)\Psi C\Sigma A'\Psi + \Psi A\Sigma B\Sigma C'\Psi$
   $+ \text{tr}(\Sigma B'\Sigma C'\Psi A)\Psi + \text{tr}(MA)M'B\Sigma C'\Psi + \text{tr}(MA)\text{tr}(MC\Sigma B)\Psi$
   $+ M'BMC\Psi A'\Psi + \text{tr}(B\Sigma)\text{tr}(MA)\Psi CM + M'B\Sigma A'\Psi CM$
   $+ \Psi A\Sigma BMCM + \text{tr}(MA)M'BMCM, \ A\,(p \times p), \ B\,(p \times p), \ C\,(n \times p),$

(vi) $E(\text{tr}(AX)X'BX'CX) = \Psi B'\Sigma C\Psi A'\Psi + \text{tr}(\Sigma C)\Psi A\Sigma B\Psi$
   $+ \text{tr}(\Sigma C\Sigma B\Psi A)\Psi + \text{tr}(MA)\text{tr}(\Sigma C)M'B\Psi + \text{tr}(MA)\text{tr}(MB'\Sigma C')\Psi$
   $+ M'BM'C\Sigma A'\Psi + \text{tr}(MA)\Psi B'\Sigma CM + M'B\Psi A\Sigma CM$
   $+ \Psi A\Sigma BMCM + \text{tr}(AM)M'BM'CM, \ A\,(p \times p), \ B\,(p \times p), C\,(p \times p),$

(vii) $E(\text{tr}(AX)X'BXCX') = \text{tr}(\Psi C)\Psi A\Sigma B\Sigma + \text{tr}(B\Sigma)\Psi C\Psi A\Sigma$
   $+ \Psi C'\Psi A\Sigma B'\Sigma + M'BMC\Psi A\Sigma + M'B\Sigma A'\Psi CM' + \Psi A\Sigma BMCM'$
   $+ \text{tr}(\Psi C)\text{tr}(AM)M'B\Sigma + \text{tr}(AM)\Psi C'M'B'\Sigma + \text{tr}(AM)\text{tr}(B\Sigma)\Psi CM'$
   $+ \text{tr}(AM)M'BMCM', \ A\,(n \times p), \ B\,(p \times p), \ C\,(n \times n).$

Proof: The results can be derived by using the procedure described above. ∎

It should be noted that Theorems 2.3.5–2.3.9 are sufficiently general to cover many cases.

Neudecker and Wansbeek (1987) have given an alternative method of derivation of (v) of Theorem 2.3.8. They (and von Rosen, 1988b) have also given expectation of $X \otimes X \otimes X \otimes X$, and the $\text{cov}(\text{vec}(XAX'), \text{vec}(XBX'))$ (derived in Chapter 7).

Further let $X \sim N_{p,p}(M, I_p \otimes I_p)$ and define

$$\mu_k = E(AX)^k, \ k = 2, 3, \ldots, \ldots$$

and $B = AA'$, where $A\,(p \times p)$ is a constant matrix. Then, Hudak and Richter (1996), besides many other results, have proved that

$$\mu_{2k-1} = 0,$$

and

$$\mu_{2k} = \text{tr}(\mu_{2k-2})B + (2k - 2)B\mu_{2k-2},\ k = 2, 3, \ldots, \ldots$$

with $\mu_2 = B$. From this recurrence formula it can be deduced that

$$E(X^2) = I_p,$$

and

$$E(X^{2k}) = (p + 2k - 2)(p + 2k - 4)\cdots(p + 2)I_p,\ k = 2, 3, \ldots, \ldots.$$

THEOREM 2.3.10. *If* $X \sim N_{p,n}(M, \Sigma \otimes \Psi)$, $D\,(m \times p)$ *is of rank* $m \leq p$ *and* $C\,(n \times t)$ *is of rank* $t \leq n$, *then* $DXC \sim N_{m,t}(DMC, (D\Sigma D') \otimes (C'\Psi C))$.

Proof: The characteristic function of DXC is

$$\phi_{DXC}(Z) = E[\text{etr}(\iota DXCZ')]$$

$$= E[\text{etr}(\iota XZ_1')],\ Z_1' = CZ'D.$$

Now, from Theorem 2.3.2, we get

$$\phi_{DXC}(Z) = \text{etr}\left(\iota Z_1'M - \frac{1}{2}Z_1'\Sigma Z_1\Psi\right)$$

$$= \text{etr}\left\{\iota Z'(DMC) - \frac{1}{2}Z'(D\Sigma D')Z(C'\Psi C)\right\}. \tag{2.3.9}$$

Since (2.3.9) is the characteristic function of a matrix variate normal distribution with mean DMC and covariance matrix $(D\Sigma D') \otimes (C'\Psi C)$, the proof is complete. ∎

COROLLARY 2.3.10.1. *In the above theorem, let* $m = t = 1$, $D \equiv d'\,(1 \times p)$ *and* $C \equiv c\,(n \times 1)$, *then*

$$d'Xc \sim N(d'Mc, (d'\Sigma d)(c'\Psi c)).$$

Furthermore,

$$\frac{\{d'(X - M)c\}^2}{(d'\Sigma d)(c'\Psi c)} \sim \chi_1^2.$$

COROLLARY 2.3.10.2. *In the above theorem,*
   (i) if $m = p$, *and* $D = \Sigma^{-\frac{1}{2}}$, *then*

$$\Sigma^{-\frac{1}{2}}XC \sim N_{p,t}(\Sigma^{-\frac{1}{2}}MC, I_p \otimes (C'\Psi C)),$$

and
   (ii) if $t = n$, *and* $C = \Psi^{-\frac{1}{2}}$, *then*

$$DX\Psi^{-\frac{1}{2}} \sim N_{m,n}(DM\Psi^{-\frac{1}{2}}, (D\Sigma D') \otimes I_n).$$

THEOREM 2.3.11. *Let $X \sim N_{p,n}(M, \Sigma \otimes \Psi)$, and partition X, M, Σ, and Ψ as*

$$X = \begin{pmatrix} X_{11} & X_{12} \\ X_{21} & X_{22} \end{pmatrix} \begin{matrix} m \\ p-m \end{matrix}, \quad M = \begin{pmatrix} M_{11} & M_{12} \\ M_{21} & M_{22} \end{pmatrix} \begin{matrix} m \\ p-m \end{matrix},$$
$$\begin{matrix} t & n-t \end{matrix} \qquad\qquad \begin{matrix} t & n-t \end{matrix}$$

$$\Sigma = \begin{pmatrix} \Sigma_{11} & \Sigma_{12} \\ \Sigma_{21} & \Sigma_{22} \end{pmatrix} \begin{matrix} m \\ p-m \end{matrix}, \quad \Psi = \begin{pmatrix} \Psi_{11} & \Psi_{12} \\ \Psi_{21} & \Psi_{22} \end{pmatrix} \begin{matrix} t \\ n-t \end{matrix}.$$
$$\begin{matrix} m & p-m \end{matrix} \qquad\qquad \begin{matrix} t & n-t \end{matrix}$$

Then, $X_{11} \sim N_{m,t}(M_{11}, \Sigma_{11} \otimes \Psi_{11})$.

Proof: The result follows by taking

$$D = (\, I_m \quad 0 \,) \text{ and } C' = (\, I_t \quad 0 \,)$$

in Theorem 2.3.10. ∎

THEOREM 2.3.12. *Let $X \sim N_{p,n}(M, \Sigma \otimes \Psi)$, and partition X, M, Σ, and Ψ as*

$$X = \begin{pmatrix} X_{1r} \\ X_{2r} \end{pmatrix} \begin{matrix} m \\ p-m \end{matrix} = (\, X_{1c} \quad X_{2c} \,)$$
$$\begin{matrix} t & n-t \end{matrix}$$

$$M = \begin{pmatrix} M_{1r} \\ M_{2r} \end{pmatrix} \begin{matrix} m \\ p-m \end{matrix} = (\, M_{1c} \quad M_{2c} \,)$$
$$\begin{matrix} t & n-t \end{matrix}$$

$$\Sigma = \begin{pmatrix} \Sigma_{11} & \Sigma_{12} \\ \Sigma_{21} & \Sigma_{22} \end{pmatrix} \begin{matrix} m \\ p-m \end{matrix} \text{ and } \Psi = \begin{pmatrix} \Psi_{11} & \Psi_{12} \\ \Psi_{21} & \Psi_{22} \end{pmatrix} \begin{matrix} t \\ n-t \end{matrix}.$$
$$\begin{matrix} m & p-m \end{matrix} \qquad\qquad \begin{matrix} t & n-t \end{matrix}$$

Then, (i) $X_{1r} \sim N_{m,n}(M_{1r}, \Sigma_{11} \otimes \Psi)$, $X_{1c} \sim N_{p,t}(M_{1c}, \Sigma \otimes \Psi_{11})$,
(ii) $X_{2r}|X_{1r} \sim N_{p-m,n}(M_{2r} + \Sigma_{21}\Sigma_{11}^{-1}(X_{1r} - M_{1r}), \Sigma_{22\cdot1} \otimes \Psi)$, and $X_{2c}|X_{1c} \sim N_{p,n-t}(M_{2c} + (X_{1c} - M_{1c})\Psi_{11}^{-1}\Psi_{12}, \Sigma \otimes \Psi_{22\cdot1})$, where $\Sigma_{22\cdot1} = \Sigma_{22} - \Sigma_{21}\Sigma_{11}^{-1}\Sigma_{12}$ and $\Psi_{22\cdot1} = \Psi_{22} - \Psi_{21}\Psi_{11}^{-1}\Psi_{12}$.

Proof: (i) In Theorem 2.3.11 substitute $t = n$ to get the density of X_{1r}, and $m = p$ to get the density of X_{1c}.

(ii) Let $\Sigma^{-1} = \begin{pmatrix} \Sigma^{11} & \Sigma^{12} \\ \Sigma^{21} & \Sigma^{22} \end{pmatrix}$, then $\Sigma^{11} = \Sigma_{11\cdot2}^{-1}$, $\Sigma^{22} = \Sigma_{22\cdot1}^{-1}$, $\Sigma^{12} = -\Sigma_{11\cdot2}^{-1}\Sigma_{12}\Sigma_{22}^{-1} = -\Sigma_{11}^{-1}\Sigma_{12}\Sigma_{22\cdot1}^{-1} = (\Sigma^{21})'$, and

$$(X - M)'\Sigma^{-1}(X - M)$$

$$= (\, (X_{1r} - M_{1r})' \quad (X_{2r} - M_{2r})' \,) \begin{pmatrix} \Sigma^{11} & \Sigma^{12} \\ \Sigma^{21} & \Sigma^{22} \end{pmatrix} \begin{pmatrix} X_{1r} - M_{1r} \\ X_{2r} - M_{2r} \end{pmatrix}$$

$$= (X_{1r} - M_{1r})'\Sigma^{11}(X_{1r} - M_{1r}) + (X_{1r} - M_{1r})'\Sigma^{12}(X_{2r} - M_{2r})$$

$$\quad + (X_{2r} - M_{2r})'\Sigma^{21}(X_{1r} - M_{1r}) + (X_{2r} - M_{2r})'\Sigma^{22}(X_{2r} - M_{2r})$$

$$= (X_{1r} - M_{1r})'(\Sigma^{11} - \Sigma^{12}(\Sigma^{22})^{-1}\Sigma^{21})(X_{1r} - M_{1r})$$

$$+ (X_{1r} - M_{1r})'\Sigma^{12}(\Sigma^{22})^{-1}\Sigma^{21}(X_{1r} - M_{1r}) + (X_{1r} - M_{1r})'\Sigma^{12}(X_{2r} - M_{2r})$$

$$+ (X_{2r} - M_{2r})'\Sigma^{21}(X_{1r} - M_{1r}) + (X_{2r} - M_{2r})'\Sigma^{22}(X_{2r} - M_{2r})$$

$$= (X_{1r} - M_{1r})'\Sigma_{11}^{-1}(X_{1r} - M_{1r})$$

$$+ (X_{2r} - M_{2r} - \Sigma_{21}\Sigma_{11}^{-1}(X_{1r} - M_{1r}))'\Sigma_{22\cdot1}^{-1}(X_{2r} - M_{2r} - \Sigma_{21}\Sigma_{11}^{-1}(X_{1r} - M_{1r})).$$

Thus, the density of X can be written as

$$f(X) = (2\pi)^{-\frac{1}{2}np} \det(\Sigma)^{-\frac{1}{2}n} \det(\Psi)^{-\frac{1}{2}p} \operatorname{etr}\left[-\frac{1}{2}(X - M)'\Sigma^{-1}(X - M)\Psi^{-1}\right]$$

$$= (2\pi)^{-\frac{1}{2}nm} \det(\Sigma_{11})^{-\frac{1}{2}n} \det(\Psi)^{-\frac{1}{2}m} \operatorname{etr}\left[-\frac{1}{2}(X_{1r} - M_{1r})'\Sigma_{11}^{-1}(X_{1r} - M_{1r})\Psi^{-1}\right]$$

$$\cdot (2\pi)^{-\frac{1}{2}n(p-m)} \det(\Sigma_{22\cdot1})^{-\frac{1}{2}n} \det(\Psi)^{-\frac{1}{2}(p-m)} \operatorname{etr}\left[-\frac{1}{2}(X_{2r} - M_{2r}\right.$$

$$\left. - \Sigma_{21}\Sigma_{11}^{-1}(X_{1r} - M_{1r}))'\Sigma_{22\cdot1}^{-1}(X_{2r} - M_{2r} - \Sigma_{21}\Sigma_{11}^{-1}(X_{1r} - M_{1r}))\Psi^{-1}\right].$$

Hence, $X_{2r}|X_{1r} \sim N_{p-m,n}(M_{2r} + \Sigma_{21}\Sigma_{11}^{-1}(X_{1r} - M_{1r}), \Sigma_{22\cdot1} \otimes \Psi)$. Since from Theorem 2.3.1, $X' = \begin{pmatrix} X'_{1c} \\ X'_{2c} \end{pmatrix} \sim N_{n,p}\left(\begin{pmatrix} M'_{1c} \\ M'_{2c} \end{pmatrix}, \Psi \otimes \Sigma\right)$, the above result gives

$$X'_{2c}|X'_{1c} \sim N_{n-t,p}(M'_{2c} + \Psi_{21}\Psi_{11}^{-1}(X'_{1c} - M'_{1c}), \Psi_{22\cdot1} \otimes \Sigma).$$

Therefore,

$$X_{2c}|X_{1c} \sim N_{p,n-t}(M_{2c} + (X_{1c} - M_{1c})\Psi_{11}^{-1}\Psi_{12}, \Sigma \otimes \Psi_{22\cdot1}). \quad \blacksquare$$

THEOREM 2.3.13. If $X \sim N_{p,n}(M, \Sigma \otimes \Psi)$, then

$$\operatorname{tr}\{\Sigma^{-1}(X - M)\Psi^{-1}(X - M)'\} \sim \chi_{np}^2.$$

Proof: The result follows by noting that

$$\operatorname{tr}\{\Sigma^{-1}(X - M)\Psi^{-1}(X - M)'\} = \operatorname{tr}(YY').$$

where $Y = \Sigma^{-\frac{1}{2}}(X - M)\Psi^{-\frac{1}{2}}$ which, according to Corollary 2.3.10.2, is distributed as $N_{p,n}(0, I_p \otimes I_n)$. $\quad \blacksquare$

THEOREM 2.3.14. Let $X \sim N_{p,n}(M, \Sigma \otimes \Psi)$, and B $(n \times t)$, and D $(n \times s)$ be given matrices. Then XB and XD are independent if and only if $B'\Psi D = 0$.

Proof: Without loss of generality, assume that $M = 0$. The matrix of covariances between XB and XD is given by

$$\operatorname{cov}(XB, XD) = \operatorname{cov}(\operatorname{vec}(XB)', \operatorname{vec}(XD)')$$

$$= E\{\mathrm{vec}(XB)'(\mathrm{vec}(XD)')'\}$$
$$= E\{(I_p \otimes B')\,\mathrm{vec}(X')(\mathrm{vec}(X'))'(I_p \otimes D)\}$$
$$= (I_p \otimes B')E\{\mathrm{vec}(X')(\mathrm{vec}(X'))'\}(I_p \otimes D)$$
$$= (I_p \otimes B')(\Sigma \otimes \Psi)(I_p \otimes D)$$
$$= \Sigma \otimes (B'\Psi D). \tag{2.3.10}$$

It follows that $\mathrm{cov}(XB, XD) = 0$ if and only if $B'\Psi D = 0$. This completes the proof of the theorem. ■

In (2.3.10), by taking $t = 1$, $B = e_i$ ($n \times 1$), $s = 1$, and $D = e_j$ ($n \times 1$), we get

$$\mathrm{cov}(x_i, x_j) = \psi_{ij}\Sigma, \ i, j = 1, \ldots, n,$$

where x_i is the ith column of the matrix X. Further, it can be shown that

$$\mathrm{cov}(x_i^*, x_j^*) = \sigma_{ij}\Psi, \ i, j = 1, \ldots, p,$$

where $x_i^{*'}$ is the ith row of matrix X.

THEOREM 2.3.15. *Let* $X \sim N_{p,n}(M, \Sigma \otimes \Psi)$, $A\,(r \times p)$, *and* $C\,(q \times p)$ *be given matrices. Then* AX *and* CX *are independent if and only if* $A\Sigma C' = 0$.

Proof: The proof is similar to the proof of Theorem 2.3.14. ■

By combining the results of Theorems 2.3.14 and 2.3.15, we get the following.

THEOREM 2.3.16. *Let* $X \sim N_{p,n}(M, \Sigma \otimes \Psi)$, $A\,(r \times p)$, $B\,(n \times t)$, $C\,(q \times p)$, *and* $D\,(n \times s)$ *be given matrices. Then* AXB *and* CXD *are independent if and only if either* $A\Sigma C' = 0$ *or* $B'\Psi D = 0$.

Now, we generalize a result of Basu and Khatri (1969) for the matrix variate normal case.

THEOREM 2.3.17. *Let* $X \sim N_{p,n}(M, \Sigma_1 \otimes \Sigma_2)$, *and* $f_{ij}(X)$ *be a real-valued function of* X, $i = 1, \ldots, r$, $j = 1, \ldots, s$. *If* $F(X) = (f_{ij}(X)) \sim N_{r,s}(\mu, \Psi_1 \otimes \Psi_2)$ *for every* $M \in \mathbb{R}^{p \times n}$, $\Sigma_1 > 0$ *and* $\Sigma_2 > 0$, *then* $F(X) = AXB + C$ *almost everywhere, where* A, B *and* C *do not depend on* M, Σ_1 *and* Σ_2.

Proof: Noting the fact that $\mathrm{vec}(X')$ is multivariate normal, the proof follows from Basu and Khatri (1969). ■

THEOREM 2.3.18. *Let* $X \sim N_{p,n}(0, I_p \otimes I_n)$, *and* $X = TL$, *where* $T\,(p \times p) = (t_{ij})$, $t_{ii} > 0$ *is a lower triangular matrix and* $L\,(p \times n)$ *is a semiorthogonal matrix,* $LL' = I_p$. *Then* T *and* L *are independently distributed. The p.d.f. of* T *is*

$$\left\{2^{\frac{1}{2}(n-2)p}\,\Gamma_p\left(\tfrac{1}{2}n\right)\right\}^{-1} \prod_{i=1}^{p} t_{ii}^{n-i} \, \mathrm{etr}\left(-\tfrac{1}{2}TT'\right), \tag{2.3.11}$$

that is t_{ij}'s are independently distributed, $1 \leq j \leq i \leq p$, $t_{ii}^2 \sim \chi_{n-i+1}^2$, $i = 1, \ldots, p$, and $t_{ij} \sim N(0, 1)$, $1 \leq j < i \leq p$.

Proof: The p.d.f. of X is given by

$$(2\pi)^{-\frac{1}{2}np} \operatorname{etr}\left(-\frac{1}{2}XX'\right).$$

Now using the transformation $X = TL$ with Jacobian, from (1.3.25), $J(X \to T, L) = g_{n,p}(L) \prod_{i=1}^{p} t_{ii}^{n-i}$, where $g_{n,p}(L)$ is a function of L only, we get the joint density of T and L as

$$(2\pi)^{-\frac{1}{2}np} \prod_{i=1}^{p} t_{ii}^{n-i} \operatorname{etr}\left(-\frac{1}{2}TT'\right) g_{n,p}(L). \qquad (2.3.12)$$

From (2.3.12), it follows that T and L are independent and the density of T, using Theorem 1.4.9, is given by

$$\frac{2^{-\frac{1}{2}np+p}}{\Gamma_p(\frac{1}{2}n)} \prod_{i=1}^{p} t_{ii}^{n-i} \operatorname{etr}\left(-\frac{1}{2}TT'\right). \qquad (2.3.13)$$

This completes the proof of the theorem. ∎

A result more general than Theorem 2.3.18 is proven in the following theorem.

THEOREM 2.3.19. Let Y $(p \times n)$ be a random matrix with $\operatorname{rank}(Y) = p \leq n$, and p.d.f. $f(YY')$. If $Y = TL$, where $T = (t_{ij})$, $t_{ii} > 0$ is a lower triangular matrix and L is a semiorthogonal matrix, $LL' = I_p$, then T and L are independently distributed and the p.d.f. of T is

$$\frac{2^p \pi^{\frac{1}{2}np}}{\Gamma_p(\frac{1}{2}n)} \prod_{i=1}^{p} t_{ii}^{n-i} f(TT'). \qquad (2.3.14)$$

Proof: As in the proof of Theorem 2.3.18, the joint density of T and L is now given by

$$f(TT') \prod_{i=1}^{p} t_{ii}^{n-i} g_{n,p}(L). \qquad (2.3.15)$$

From (2.3.15), it follows that T and L are independent. Integrating (2.3.15) with respect to L, by using Theorem 1.4.9, we get (2.3.14). ∎

The random matrix L, in Theorems 2.3.18 and 2.3.19, has uniform distribution over the Stiefel manifold $O(p, n) = \{L : LL' = I_p\}$, which will be studied in Chapter 8.

2.4.   SINGULAR MATRIX VARIATE NORMAL DISTRIBUTION

The density (2.2.1) of X does not exist if $\Sigma \otimes \Psi$ is positive semidefinite. In this case, X is said to have singular normal distribution which we now define.

DEFINITION 2.4.1. *Let X $(p \times n)$ be a random matrix with $E(X) = M$ and $\text{cov}(X) = \Sigma \otimes \Psi$, where Σ $(p \times p)$ and Ψ $(n \times n)$ are positive semidefinite with ranks p_1 $(< p)$ and n_1 $(< n)$ respectively. Then X is said to have singular matrix variate normal distribution if there exist matrices H $(p \times p_1)$ and R $(n_1 \times n)$ of ranks p_1 and n_1 respectively such that $X = HYR + M$ for some random matrix $Y \sim N_{p_1,n_1}(0, P \otimes Q)$, P $(p_1 \times p_1) > 0$ and Q $(n_1 \times n_1) > 0$.*

We will denote this by $X \sim N_{p,n}(M, \Sigma \otimes \Psi | p_1, n_1)$. From Theorem 2.3.10, it follows that $\Sigma = HPH'$ and $\Psi = R'QR$. It may be noted that if either (i) $p_1 = p$, and $n_1 < n$ or (ii) $p_1 < p$, and $n_1 = n$, then also $\Sigma \otimes \Psi$ is positive semidefinite and the random matrix X has a singular matrix variate normal distribution.

THEOREM 2.4.1. *Let $X \sim N_{p,n}(M, \Sigma \otimes \Psi | p_1, n_1)$, then*

$$\phi_X(Z) = \text{etr}\left(\iota Z'M - \frac{1}{2}Z'\Sigma Z\Psi\right).$$

Proof: By definition,

$$
\begin{aligned}
\phi_X(Z) &= E[\text{etr}(\iota X Z')] \\
&= E[\text{etr}\{\iota(HYR + M)Z'\}] \\
&= \text{etr}(\iota M Z')\phi_Y(H'ZR') \\
&= \text{etr}(\iota M Z')\,\text{etr}\left(-\frac{1}{2}PH'ZR'QRZ'H\right) \\
&= \text{etr}\left\{\iota Z'M - \frac{1}{2}Z'(HPH')Z(R'QR)\right\}. \quad \blacksquare
\end{aligned}
$$

THEOREM 2.4.2. *If $X \sim N_{p,n}(M, \Sigma \otimes \Psi | p_1, n_1)$, D $(m \times p)$, and C $(n \times t)$, then $DXC \sim N_{m,t}(DMC, D\Sigma D') \otimes (C'\Psi C) | m_1, t_1)$, where $m_1 = \text{rank}(D\Sigma D')$ and $t_1 = \text{rank}(C'\Psi C)$.*

Proof: The characteristic function of DXC is

$$
\begin{aligned}
\phi_{DXC}(Z) &= E[\text{etr}(\iota DXCZ')] \\
&= \phi_X(D'ZC') \\
&= \text{etr}\left(\iota MCZ'D - \frac{1}{2}CZ'D\Sigma D'ZC'\Psi\right) \\
&= \text{etr}\left\{\iota DMCZ' - \frac{1}{2}(D\Sigma D')Z(C'\Psi C)Z'\right\},
\end{aligned}
$$

from which the result follows. $\blacksquare$

THEOREM 2.4.3. *Let $X \sim N_{p,n}(M, \Sigma \otimes \Psi | p_1, n_1)$, and partition X, M, Σ, and Ψ as*

$$
X = \begin{pmatrix} X_{11} & X_{12} \\ X_{21} & X_{22} \end{pmatrix} \begin{matrix} m \\ p-m \end{matrix}, \quad M = \begin{pmatrix} M_{11} & M_{12} \\ M_{21} & M_{22} \end{pmatrix} \begin{matrix} m \\ p-m \end{matrix},
$$
$$
\begin{matrix} t & n-t \end{matrix} \qquad\qquad\qquad \begin{matrix} t & n-t \end{matrix}
$$

$$\Sigma = \begin{pmatrix} \Sigma_{11} & \Sigma_{12} \\ \Sigma_{21} & \Sigma_{22} \end{pmatrix} \begin{matrix} m \\ p-m \end{matrix}, \quad \Psi = \begin{pmatrix} \Psi_{11} & \Psi_{12} \\ \Psi_{21} & \Psi_{22} \end{pmatrix} \begin{matrix} t \\ n-t \end{matrix}.$$
$$\qquad\quad\; m \quad p-m \qquad\qquad\qquad t \quad n-t$$

Then $X_{11} \sim N_{m,t}(M_{11}, \Sigma_{11} \otimes \Psi_{11}|m_1, t_1)$ where $m_1 = \mathrm{rank}(\Sigma_{11})$ and $t_1 = \mathrm{rank}(\Psi_{11})$.

Proof: In Theorem 2.4.2, let $D = (\, I_m \quad 0\,)$ and $C' = (\, I_t \quad 0\,)$ so that $DXC = X_{11}$, $D\Sigma D' = \Sigma_{11}$ and $C'\Psi C = \Psi_{11}$. ∎

2.5. SYMMETRIC MATRIX VARIATE NORMAL DISTRIBUTION

Let $Y\,(p \times p) \sim N_{p,p}(N, \Sigma \otimes \Psi)$, then $\mathrm{vec}(Y') \sim N_{p^2}(\mathrm{vec}(N'), \Sigma \otimes \Psi)$ and $\mathrm{vec}(Y) \sim N_{p^2}(\mathrm{vec}(N), \Psi \otimes \Sigma)$. Now, using the transformations

$$\mathrm{vec}(X') = M_p \,\mathrm{vec}(Y'), \qquad\qquad\qquad (2.5.1)$$

and

$$\mathrm{vec}(X) = M_p \,\mathrm{vec}(Y), \qquad\qquad\qquad (2.5.2)$$

where the matrix M_p is defined in Section 1.2, we have

$$\mathrm{vec}(X') \sim N_{p^2}(M_p \,\mathrm{vec}(N'), M_p(\Sigma \otimes \Psi)M_p), \qquad\qquad (2.5.3)$$

and

$$\mathrm{vec}(X) \sim N_{p^2}(M_p \,\mathrm{vec}(N), M_p(\Psi \otimes \Sigma)M_p). \qquad\qquad (2.5.4)$$

From (1.2.17), note that $X = X'$. Therefore, from (2.5.3) and (2.5.4),

$$M_p(\Sigma \otimes \Psi)M_p = M_p(\Psi \otimes \Sigma)M_p,$$

which is satisfied if $\Sigma\Psi = \Psi\Sigma$. The characteristic function of $\mathrm{vec}(X')$ is

$$
\begin{aligned}
\phi_{\mathrm{vec}(X')}(\mathrm{vec}(T')) &= E[\exp\{\iota(\mathrm{vec}(T'))'\,\mathrm{vec}(X')\}],\; T\,(p \times p) = T' \\
&= E[\exp\{\iota(\mathrm{vec}(T'))'M_p\,\mathrm{vec}(Y')\}] \\
&= E[\exp\{\iota(M_p\,\mathrm{vec}(T'))'\,\mathrm{vec}(Y')\}] \\
&= \mathrm{etr}\left[\iota(M_p\,\mathrm{vec}(N'))'\,\mathrm{vec}(T') - \frac{1}{2}(M_p\,\mathrm{vec}(T'))'(\Sigma \otimes \Psi)(M_p\,\mathrm{vec}(T'))\right] \\
&= \mathrm{etr}\left[\iota(\mathrm{vec}(M'))'\,\mathrm{vec}(T) - \frac{1}{2}(\mathrm{vec}(T))'(\Sigma \otimes \Psi)(\mathrm{vec}(T))\right] \\
&= \mathrm{etr}\left[\iota TM - \frac{1}{2}T\Sigma T\Psi\right]. \qquad\qquad\qquad (2.5.5)
\end{aligned}
$$

where $\mathrm{vec}(M') = M_p\,\mathrm{vec}(N')$ gives $M = M'$. Since $X\,(p \times p)$ is a symmetric matrix, it contains only $\frac{1}{2}p(p+1)$ distinct elements and therefore the covariance matrix of X

should be a matrix of order $\frac{1}{2}p(p+1) \times \frac{1}{2}p(p+1)$. To obtain this covariance matrix, we derive the characteristic function of $\mathrm{vecp}(X)$,

$$
\begin{aligned}
\phi_{\mathrm{vecp}(X)}(\mathrm{vecp}(T)) &= E[\exp\{\iota(\mathrm{vecp}(T))'\,\mathrm{vecp}(X)\}] \\
&= E[\exp\{\iota(B_p\,\mathrm{vecp}(T))'\,\mathrm{vec}(X)\}] \\
&= \exp\Big\{\iota(\mathrm{vec}(M))'B_p\,\mathrm{vecp}(T) - \frac{1}{2}(B_p\,\mathrm{vecp}(T))'(\Sigma \otimes \Psi)B_p\,\mathrm{vecp}(T)\Big\} \\
&= \exp\Big\{\iota(B_p'\,\mathrm{vec}(M))'\,\mathrm{vecp}(T) - \frac{1}{2}(\mathrm{vecp}(T))'B_p'(\Sigma \otimes \Psi)B_p\,\mathrm{vecp}(T)\Big\} \\
&= \exp\Big\{\iota(\mathrm{vecp}(M))'\,\mathrm{vecp}(T) - \frac{1}{2}(\mathrm{vecp}(T))'B_p'(\Sigma \otimes \Psi)B_p\,\mathrm{vecp}(T)\Big\},
\end{aligned}
$$

where $\mathrm{vecp}(X)$, and B_p have been defined in Section 1.2. Thus, we define the symmetric matrix variate normal distribution as follows.

DEFINITION 2.5.1. *Let $X\ (p \times p)$ be a symmetric random matrix and M, Σ, and Ψ be constant symmetric $p \times p$ matrices such that $\Sigma\Psi = \Psi\Sigma$. If the $\frac{1}{2}p(p+1) \times 1$ vector $\mathrm{vecp}(X)$ formed from X is distributed as $N_{\frac{1}{2}p(p+1)}(\mathrm{vecp}(M), B_p'(\Sigma \otimes \Psi)B_p)$, then X is said to have symmetric matrix variate normal distribution, with mean matrix M and covariance matrix $B_p'(\Sigma \otimes \Psi)B_p$, and is denoted as $X = X' \sim SN_{p,p}(M, B_p'(\Sigma \otimes \Psi)B_p)$.*

From the Definition 2.5.1, the probability density function of X, in terms of $\mathrm{vecp}(X)$, is

$$
(2\pi)^{-\frac{1}{4}p(p+1)} \det(B_p'(\Sigma \otimes \Psi)B_p)^{-\frac{1}{2}} \exp\Big[-\frac{1}{2}(\mathrm{vecp}(X) - \mathrm{vecp}(M))'
$$
$$
\cdot B_p^+(\Sigma \otimes \Psi)^{-1}B_p^{+'}(\mathrm{vecp}(X) - \mathrm{vecp}(M))\Big]. \tag{2.5.6}
$$

Using (1.2.12), (2.5.6) can be written in terms of $\mathrm{vec}(X)$ as

$$
(2\pi)^{-\frac{1}{4}p(p+1)} \det(B_p'(\Sigma \otimes \Psi)B_p)^{-\frac{1}{2}} \exp\Big[-\frac{1}{2}(\mathrm{vec}(X) - \mathrm{vec}(M))'
$$
$$
\cdot (\Sigma \otimes \Psi)^{-1}(\mathrm{vec}(X) - \mathrm{vec}(M))\Big] \tag{2.5.7}
$$

which, applying Theorem 1.2.22, can be rewritten as

$$
(2\pi)^{-\frac{1}{4}p(p+1)} \det(B_p'(\Sigma \otimes \Psi)B_p)^{-\frac{1}{2}} \mathrm{etr}\Big[-\frac{1}{2}\Sigma^{-1}(X - M)\Psi^{-1}(X - M)\Big]. \tag{2.5.8}
$$

We now derive the product moments of the elements of the random matrix X, given by H. M. Nel (1977) and D. G. Nel (1978).

THEOREM 2.5.1. *Let $X = X' \sim SN_{p,p}(M, B_p'(\Sigma \otimes \Psi)B_p)$, then*
(i) $E(x_{ij}) = m_{ij}$
(ii) $\mathrm{cov}(x_{ij}, x_{k\ell}) = \frac{1}{4}(\sigma_{ik}\psi_{j\ell} + \sigma_{jk}\psi_{i\ell} + \sigma_{i\ell}\psi_{jk} + \sigma_{j\ell}\psi_{ik})$

(iii) $E(x_{ij}x_{k\ell}x_{rs}) = m_{ij}\operatorname{cov}(x_{k\ell}, x_{rs}) + m_{k\ell}\operatorname{cov}(x_{ij}, x_{rs}) + m_{rs}\operatorname{cov}(x_{ij}, x_{k\ell})$
$$+ m_{ij}m_{k\ell}m_{rs}$$

and

(iv) $E(x_{ij}x_{k\ell}x_{rs}x_{tq}) = \operatorname{cov}(x_{ij}, x_{tq})\operatorname{cov}(x_{k\ell}, x_{rs}) + \operatorname{cov}(x_{ij}, x_{k\ell})\operatorname{cov}(x_{rs}, x_{tq})$
$$+ \operatorname{cov}(x_{ij}, x_{rs})\operatorname{cov}(x_{k\ell}, x_{tq}) + m_{ij}m_{tq}\operatorname{cov}(x_{k\ell}, x_{rs})$$
$$= m_{ij}m_{k\ell}\operatorname{cov}(x_{rs}, x_{qt}) + m_{ij}m_{rs}\operatorname{cov}(x_{k\ell}, x_{tq})$$
$$+ m_{k\ell}m_{rs}\operatorname{cov}(x_{ij}, x_{tq}) + m_{rs}m_{qt}\operatorname{cov}(x_{ij}, x_{k\ell})$$
$$+ m_{k\ell}m_{tq}\operatorname{cov}(x_{ij}, x_{rs}) + m_{ij}m_{k\ell}m_{rs}m_{tq}.$$

Proof: From the characteristic function (2.5.5), using the method of Theorem 2.3.3, the results are easily obtained. ∎

Results parallel to the ones given in Theorems 2.3.5–2.3.9 can also be derived in a similar manner using the above Theorem.

For $A\,(p \times p)$, $C\,(p \times p)$, and $D\,(p \times p)$ constant matrices we similarly have

$$E(XAX) = \frac{1}{4}[\Sigma A'\Psi + \Psi A'\Sigma + \operatorname{tr}(A\Psi)\Sigma + \operatorname{tr}(A\Sigma)\Psi] + MAM, \tag{2.5.9}$$

$$E(\operatorname{tr}(AX)X) = \frac{1}{4}[\Sigma A'\Psi + \Psi A\Sigma + \Sigma A\Psi + \Psi A'\Sigma] + \operatorname{tr}(AM)M, \tag{2.5.10}$$

and

$$E(XAXCX) = \frac{1}{4}[M A\Sigma C'\Psi + M A\Psi C'\Sigma + \Sigma C' M A'\Psi + \Psi C' M A'\Sigma + \Sigma A'\Psi CM$$
$$+ \Psi A'\Sigma CM + \operatorname{tr}(C\Sigma)M A\Psi + \operatorname{tr}(C\Psi)M A\Sigma + \operatorname{tr}(C' M A'\Sigma)\Psi$$
$$+ \operatorname{tr}(CM A\Psi)\Sigma + \operatorname{tr}(A\Sigma)\Psi CM + \operatorname{tr}(A\Psi)\Sigma CM] + MAMCM.$$

Many other higher order expectations are given by H. M. Nel (1977) and D. G. Nel (1978). It may be noted that the moments of $X = X' \sim SN_{p,p}(M, B'_p(\Sigma \otimes \Psi)B_p)$ can also be obtained from the moments of nonsymmetric $Y \sim N_{p,p}(M, \Sigma \otimes \Psi)$ by substituting $\frac{1}{2}(Y + Y')$ for $X - M$. For example, $E(XX') = E[\frac{1}{4}(Y + Y')(Y + Y')' + MM']$.

THEOREM 2.5.2. *Let* $X = X' \sim SN_{p,p}(M, B'_p(\Sigma \otimes \Psi)B_p)$, $A\,(p \times p)$ *be a symmetric matrix such that* $\Sigma A\Psi = \Psi A\Sigma$, *and* $h(X)$ *be an elementary symmetric function of* X. *Then*

$$E[\operatorname{etr}(AX)h(X)] = \operatorname{etr}\left(AM + \frac{1}{2}\Sigma A\Psi A\right)E(h(Y))$$

where $Y = Y' \sim SN_{p,p}(M + \Sigma A\Psi, B'_p(\Sigma \otimes \Psi)B_p)$.

Proof: We have

$$E[\operatorname{etr}(AX)h(X)] = (2\pi)^{-\frac{1}{4}p(p+1)}\det(B'_p(\Sigma \otimes \Psi)B_p)$$

$$\cdot \int_X h(X)\operatorname{etr}\left[AX - \frac{1}{2}\Sigma^{-1}(X - M)(X - M)\right]dX.$$

Simplifying the term within square brackets, using $\Sigma A\Psi = \Psi A\Sigma$, we get

$$E[\text{etr}(AX)h(X)] = (2\pi)^{-\frac{1}{4}p(p+1)}\det(B_p'(\Sigma \otimes \Psi)B_p)\,\text{etr}\left(AM + \frac{1}{2}\Sigma A\Psi A\right)$$

$$\cdot \int_X h(X)\,\text{etr}\left[-\frac{1}{2}\Sigma^{-1}(X - M - \Sigma A\Psi)(X - M - \Sigma A\Psi)\right]dX$$

$$= \text{etr}\left(AM + \frac{1}{2}\Sigma A\Psi A\right)\int_X h(X)f(X)\,dX$$

where $f(X)$ denotes the density $SN_{p,p}(M + \Sigma A\Psi, B_p'(\Sigma \otimes \Psi)B_p)$. This completes the proof of the theorem. ∎

THEOREM 2.5.3. *Let* $X = X' \sim SN_{p,p}(M, B_p'(\Sigma \otimes \Psi)B_p)$, *then*

$$\text{tr}(X) \sim N(\text{tr}(M), \text{tr}(\Sigma\Psi)).$$

Proof: The characteristic function of $\text{tr}(X)$ is

$$\phi_{\text{tr}(X)}(t) = E[\exp\{\iota t\,\text{tr}(X)\}]$$

$$= E[\exp\{\iota\,\text{tr}(TX)\}], \text{ where } T = tI_p$$

$$= \exp\left[\iota t\,\text{tr}(M) - \frac{1}{2}t^2\,\text{tr}(\Sigma\Psi)\right].$$

The last equality is obtained from (2.5.5). Hence, the proof is complete. ∎

THEOREM 2.5.4. *Let* $X = X' \sim SN_{p,p}(M, B_p'(\Sigma \otimes \Psi)B_p)$, *then*

$$AXA' \sim SN_{q,q}(AMA', B_q'((A\Sigma A') \otimes (A\Psi A'))B_q),$$

where $A\,(q \times p)$ *is of rank* $q \leq p$.

Proof: The characteristic function of AXA' is

$$\phi_{AXA'}(T) = E[\text{etr}(\iota TAXA')]$$

$$= E[\text{etr}(\iota(A'TA)X)]$$

$$= \text{etr}\left[\iota T(AMA') - \frac{1}{2}T(A\Sigma A')T(A\Psi A')\right],$$

from which the result follows immediately. ∎

THEOREM 2.5.5. *Let* $X = X' \sim SN_{p,p}(M, B_p'(\Sigma \otimes \Psi)B_p)$, *and partition* X, M, Σ, *and* Ψ *as*

$$X = \begin{pmatrix} X_{11} & X_{12} \\ X_{21} & X_{22} \end{pmatrix} \begin{matrix} t \\ p-t \end{matrix}, \quad M = \begin{pmatrix} M_{11} & M_{12} \\ M_{21} & M_{22} \end{pmatrix} \begin{matrix} t \\ p-t \end{matrix},$$

$$\Sigma = \begin{pmatrix} \Sigma_{11} & \Sigma_{12} \\ \Sigma_{21} & \Sigma_{22} \end{pmatrix} \begin{matrix} t \\ p-t \end{matrix}, \quad \Psi = \begin{pmatrix} \Psi_{11} & \Psi_{12} \\ \Psi_{21} & \Psi_{22} \end{pmatrix} \begin{matrix} t \\ p-t \end{matrix}.$$

Then, $X_{11} \sim N_{t,t}(M_{11}, B_t'(\Sigma_{11} \otimes \Psi_{11})B_t)$.

Proof: Let $A\,(t \times p) = (\,I_t \quad 0\,)$, then $AXA' = X_{11}$, $AMA' = M_{11}$, $A\Sigma A' = \Sigma_{11}$ and $A\Psi A' = \Psi_{11}$. Now, from Theorem 2.5.4, the result follows. $\blacksquare$

Many authors have studied the matrix variate symmetric normal distribution. H. M. Nel (1977) derived the marginal, conditional distributions and the distribution of the roots. D. G. Nel (1978) applied this distribution to derive the asymptotic expansion of a Wishart matrix. Hayakawa and Kikuchi (1979) derived moments of a function of $\mathrm{tr}(X)$ using zonal polynomials.

2.6. RESTRICTED MATRIX VARIATE NORMAL DISTRIBUTION

DEFINITION 2.6.1. *Let $X \sim N_{p,n}(M, \Sigma \otimes \Psi)$ and $C\,(n \times s)$ be a constant matrix of rank $s\,(< n)$. If the domain of definition of X is restricted to the subspace $XC = 0$ and if $MC = 0$, then the distribution of X is called restricted matrix variate normal with restriction $XC = 0$, and is denoted by $X \sim N_{p,n}(M, \Sigma \otimes \Psi|s, C)$.*

In the following theorem, we derive an explicit form of the restricted matrix variate normal density.

THEOREM 2.6.1. *Let $X \sim N_{p,n}(M, \Sigma \otimes \Psi|s, C)$, then the density of X is given by*

$$(2\pi)^{-\frac{1}{2}(n-s)p} \det(\Psi)^{-\frac{1}{2}p} \det(C'\Psi C)^{\frac{1}{2}p} \det(\Sigma)^{-\frac{1}{2}(n-s)}$$

$$\mathrm{etr}\left\{-\frac{1}{2}\Psi^{-1}(X-M)'\Sigma^{-1}(X-M)\right\},\ XC = 0.$$

Proof: The density function of unrestricted matrix X is

$$f(X) = (2\pi)^{-\frac{1}{2}np} \det(\Sigma)^{-\frac{1}{2}n} \det(\Psi)^{-\frac{1}{2}p}$$

$$\mathrm{etr}\left\{-\frac{1}{2}\Sigma^{-1}(X-M)\Psi^{-1}(X-M)'\right\},\ X \in \mathbb{R}^{p\times n}.$$

Hence, the density of the restricted matrix X is

$$\frac{f(X)}{\underset{\substack{X\in\mathbb{R}^{p\times n}\\XC=0}}{\int} f(X)\,dX} = \frac{\mathrm{etr}\{-\frac{1}{2}\Sigma^{-1}(X-M)\Psi^{-1}(X-M)'\}}{\underset{\substack{X\in\mathbb{R}^{p\times n}\\XC=0}}{\int} \mathrm{etr}\{-\frac{1}{2}\Sigma^{-1}(X-M)\Psi^{-1}(X-M)'\}\,dX}. \qquad (2.6.1)$$

From Theorem 1.4.12, the denominator on the right hand side can be evaluated as

$$\underset{\substack{X\in\mathbb{R}^{p\times n}\\XC=0}}{\int} \mathrm{etr}\left\{-\frac{1}{2}\Sigma^{-1}(X-M)\Psi^{-1}(X-M)'\right\}\,dX$$

$$= \underset{W>0}{\int}\ \underset{\substack{(X-M)\Psi^{-1}(X-M)'=W\\C'(X-M)'=0}}{\int} \mathrm{etr}\left\{-\frac{1}{2}\Sigma^{-1}(X-M)\Psi^{-1}(X-M)'\right\}\,dX\,dW$$

$$= \int\limits_{\substack{W>0 \\ Y\Psi^{-1}Y'=W \\ C'Y'=0}} \int \operatorname{etr}\left(-\frac{1}{2}\Sigma^{-1}Y\Psi^{-1}Y'\right) dY\, dW$$

$$= \frac{\pi^{\frac{1}{2}(n-s)p}}{\Gamma_p[\frac{1}{2}(n-s)]} \det(\Psi)^{\frac{1}{2}p} \det(C'\Psi C)^{-\frac{1}{2}p} \int\limits_{W>0} \det(W)^{\frac{1}{2}(n-p-s-1)} \operatorname{etr}\left(-\frac{1}{2}\Sigma^{-1}W\right) dW$$

$$= (2\pi)^{\frac{1}{2}(n-s)p} \det(\Psi)^{\frac{1}{2}p} \det(C'\Psi C)^{-\frac{1}{2}p} \det(\Sigma)^{\frac{1}{2}(n-s)}. \tag{2.6.2}$$

Now substituting (2.6.2) in (2.6.1), we get the density of the restricted matrix X as

$$(2\pi)^{-\frac{1}{2}(n-s)p} \det(\Psi)^{-\frac{1}{2}p} \det(C'\Psi C)^{\frac{1}{2}p} \det(\Sigma)^{-\frac{1}{2}(n-s)}$$

$$\operatorname{etr}\left\{-\frac{1}{2}\Psi^{-1}(X-M)'\Sigma^{-1}(X-M)\right\}, \; XC = 0. \; \blacksquare$$

THEOREM 2.6.2. *Let $X \sim N_{p,n}(M, \Sigma \otimes \Psi)$ and $B\,(r \times p)$ be a constant matrix of rank $r \leq p$. If the domain of definition of X is restricted to the subspace $BX = 0$ and if $BM = 0$, then $X' \sim N_{n,p}(M', \Psi \otimes \Sigma | r, B')$.*

Proof: From Theorem 2.3.1, $Y = X' \sim N_{n,p}(M', \Psi \otimes \Sigma)$. Also, the restriction $BX = 0$ is equivalent to $X'B' = 0$, *i.e.*, $YB' = 0$. Now, from Definition 2.6.1, it is obvious that $Y \sim N_{n,p}(M', \Psi \otimes \Sigma | r, B')$. $\blacksquare$

THEOREM 2.6.3. *Let $X \sim N_{p,n}(M, \Sigma \otimes \Psi | s, C)$. The characteristic function of X is*

$$\phi_X(Z) = \operatorname{etr}\left\{\iota M Z' - \frac{1}{2}\Sigma Z\Psi Z' + \frac{1}{2}\Sigma Z\Psi C(C'\Psi C)^{-1}C'\Psi Z'\right\}.$$

Proof: The characteristic function of X is given by

$$\phi_X(Z) = (2\pi)^{-\frac{1}{2}(n-s)p} \det(\Psi)^{-\frac{1}{2}p} \det(C'\Psi C)^{\frac{1}{2}p} \det(\Sigma)^{-\frac{1}{2}(n-s)}$$

$$\int\limits_{\substack{X\in\mathbb{R}^{p\times n} \\ XC=0}} \operatorname{etr}\left\{\iota X Z' - \frac{1}{2}\Sigma^{-1}(X-M)\Psi^{-1}(X-M)'\right\} dX$$

$$= (2\pi)^{-\frac{1}{2}(n-s)p} \det(\Psi)^{-\frac{1}{2}p} \det(C'\Psi C)^{\frac{1}{2}p} \det(\Sigma)^{-\frac{1}{2}(n-s)} \operatorname{etr}\left(\iota M Z' - \frac{1}{2}\Sigma Z\Psi Z'\right)$$

$$\int\limits_{\substack{X\in\mathbb{R}^{p\times n} \\ XC=0}} \operatorname{etr}\left\{-\frac{1}{2}\Sigma^{-1}(X-M-\iota\Sigma Z\Psi)\Psi^{-1}(X-M-\iota\Sigma Z\Psi)'\right\} dX. \tag{2.6.3}$$

Now, let $Y = X - M - \iota\Sigma Z\Psi$ so that $YC = -\iota\Sigma Z\Psi C = -\iota\Lambda$ (say). We have

$$\int\limits_{\substack{X\in\mathbb{R}^{p\times n} \\ XC=0}} \operatorname{etr}\left\{-\frac{1}{2}\Sigma^{-1}(X-M-\iota\Sigma Z\Psi)\Psi^{-1}(X-M-\iota\Sigma Z\Psi)'\right\} dX$$

$$= \int_{\substack{Y \in \mathbb{R}^{p \times n} \\ YC = -\iota \Sigma Z \Psi C}} \mathrm{etr}\left(-\frac{1}{2}\Sigma^{-1}Y\Psi^{-1}Y'\right) dY$$

$$= \int_{\substack{W + \Lambda(C'\Psi C)^{-1}\Lambda' > 0}} \int_{\substack{Y\Psi^{-1}Y' = W \\ C'Y' = -\iota\Lambda}} \mathrm{etr}\left(-\frac{1}{2}\Sigma^{-1}Y\Psi^{-1}Y'\right) dY \, dW$$

$$= \frac{\pi^{\frac{1}{2}(n-s)p}}{\Gamma_p[\frac{1}{2}(n-s)]} \det(\Psi)^{\frac{1}{2}p} \det(C'\Psi C)^{-\frac{1}{2}p}$$

$$\int_{\substack{W + \Lambda(C'\Psi C)^{-1}\Lambda' > 0}} \mathrm{etr}\left(-\frac{1}{2}\Sigma^{-1}W\right) \det(W + \Lambda(C'\Psi C)^{-1}\Lambda')^{\frac{1}{2}(n-p-s-1)} dW$$

$$= \frac{\pi^{\frac{1}{2}(n-s)p}}{\Gamma_p[\frac{1}{2}(n-s)]} \det(\Psi)^{\frac{1}{2}p} \det(C'\Psi C)^{-\frac{1}{2}p} \, \mathrm{etr}\left\{\frac{1}{2}Z\Psi C(C'\Psi C')^{-1}C'\Psi Z'\Sigma\right\}$$

$$\Gamma_p\left[\frac{1}{2}(n-s)\right] \det(2\Sigma)^{\frac{1}{2}(n-s)}$$

$$= (2\pi)^{\frac{1}{2}(n-s)p} \det(\Psi)^{\frac{1}{2}p} \det(C'\Psi C)^{-\frac{1}{2}p} \det(\Sigma)^{\frac{1}{2}(n-s)}$$

$$\mathrm{etr}\left\{\frac{1}{2}Z\Psi C(C'\Psi C')^{-1}C'\Psi Z'\Sigma\right\}. \tag{2.6.4}$$

Now by substituting (2.6.4) in (2.6.3), we get the desired result. ∎

THEOREM 2.6.4. Let $X \sim N_{p,n}(M, \Sigma \otimes \Psi | s, C)$, $B\,(m \times p)$ be of rank $m \le p$ and $D\,(n \times n)$ be a nonsingular matrix. Then $BXD \sim N_{m,n}(BMD, (B\Sigma B') \otimes (D'\Psi D) | s, D^{-1}C)$.

Proof: The characteristic function of BXD is

$$\phi_{BXD}(Z) = E[\mathrm{etr}(\iota BXDZ')]$$

$$= E[\mathrm{etr}\{\iota X(B'ZD')'\}]$$

$$= \mathrm{etr}\left[\iota BMDZ' - \frac{1}{2}(B\Sigma B')Z(D'\Psi D)Z' + \frac{1}{2}(B\Sigma B')Z(D'\Psi D)\right.$$

$$\left. \cdot D^{-1}C((D^{-1}C)'(D'\Psi D)(D^{-1}C))^{-1}(D^{-1}C)'(D'\Psi D)Z'\right]. \tag{2.6.5}$$

which is the characteristic function of a random matrix with distribution $N_{m,n}(BMD, (B\Sigma B') \otimes (D'\Psi D) | s, D^{-1}C)$. ∎

In the above theorem let $m = p$, then the p.d.f. of $Y = BXD$ becomes

$$(2\pi)^{-\frac{1}{2}(n-s)p} \det(D'\Psi D)^{-\frac{1}{2}p} \det(C'\Psi C)^{\frac{1}{2}p} \det(B\Sigma B')^{-\frac{1}{2}(n-s)}$$

$$\mathrm{etr}\left\{-\frac{1}{2}(D'\Psi D)^{-1}(Y - BMD)'(B\Sigma B')^{-1}(Y - BMD)\right\}, \quad YD^{-1}C = 0. \tag{2.6.6}$$

Also, using the transformation $Y = BXD$, with the Jacobian $J(X \rightarrow Y)$, from Theorem 2.6.1 we get the p.d.f. of Y as

$$(2\pi)^{-\frac{1}{2}(n-s)p} \det(\Psi)^{-\frac{1}{2}p} \det(C'\Psi C)^{\frac{1}{2}p} \det(\Sigma)^{-\frac{1}{2}(n-s)}$$

$$\text{etr}\left\{-\frac{1}{2}\Psi^{-1}(B^{-1}YD^{-1} - M)'\Sigma^{-1}(B^{-1}YD^{-1} - M)\right\}$$

$$J(X \rightarrow Y), \quad YD^{-1}C = 0. \tag{2.6.7}$$

Since, the density of Y is unique, comparing (2.6.6) and (2.6.7) we get the following result.

LEMMA 2.6.1. *Let the matrix X be of order $p \times n$ and transform $Y = BXD$, such that $XC = 0$ where $B\,(p \times p)$ and $D\,(n \times n)$ are nonsingular matrices and $C\,(n \times s)$ is of rank $s \leq n$. Then the Jacobian of transformation is $J(X \rightarrow Y) = \det(D)^{-p} \det(B)^{-(n-s)}$.*

2.7. MATRIX VARIATE θ-GENERALIZED NORMAL DISTRIBUTION

Another way of extending the concept of normal distribution was shown by Goodman and Kotz (1973). They introduced the multivariate θ-generalized normal distribution.

A random vector $\boldsymbol{y}\,(p \times 1)$ is said to have a vector variate θ-generalized normal distribution if it can be written as $\boldsymbol{y} = C\boldsymbol{x} + \boldsymbol{\mu}$ where $\boldsymbol{\mu}\,(p \times 1)$ is a constant vector, C is a $p \times p$ nonsingular matrix, and $\boldsymbol{x} = (x_1, \ldots, x_p)'$ is a random vector whose elements are independent and each has the probability density function

$$\frac{1}{2\Gamma\left(1 + \frac{1}{\theta}\right)} \exp\left(-|x_i|^\theta\right), \quad \theta > 0, \ x_i \in \mathbb{R}, \ i = 1, \ldots, p. \tag{2.7.1}$$

The distribution of $\boldsymbol{y}$ is denoted by $N_p(\boldsymbol{\mu}, C, \theta)$.

An extension of this concept to the matrix variate case has been given by Gupta and Varga (1995a).

DEFINITION 2.7.1. *Let $\theta > 0$. Then $X = (x_{ij})$, $i = 1, \ldots, p$, $j = 1, \ldots, n$ has a matrix variate standard θ-generalized normal distribution if x_{ij}'s are independent and identically distributed random variables with p.d.f.*

$$\frac{1}{2\Gamma\left(1 + \frac{1}{\theta}\right)} \exp\left(-|x_{ij}|^\theta\right), \quad \theta > 0, \ x_{ij} \in \mathbb{R}, \ i = 1, \ldots, p, \ j = 1, \ldots, n.$$

DEFINITION 2.7.2. *Let $\theta > 0$. Then the random matrix $Y\,(p \times n)$ is said to have a matrix variate θ-generalized normal distribution if Y can be written as $Y = AXB + M$ where $X\,(p \times n)$ is a standard θ-generalized normal random matrix, $A\,(p \times p)$, $B\,(n \times n)$, and $M\,(p \times n)$ are constant matrices, with A and B being nonsingular.*

The distribution of Y is denoted by $N_{p,n}(M, A, B, \theta)$.

For $n = 1$, we get the multivariate θ-generalized normal distribution. Furthermore, the case $n = p = 1$ reduces to the Laplace density for $\theta = 1$, and the normal density for $\theta = 2$. It approaches the uniform density as $\theta \to \infty$, and an improper uniform one over the real line as $\theta \to 0$. The probability density function of a matrix variate θ-generalized normal distribution is given in the following theorem.

THEOREM 2.7.1. *Let $Y \sim N_{p,n}(M, A, B, \theta)$. Then the probability density function of Y is*

$$\left\{ 2\Gamma\left(1 + \frac{1}{\theta}\right) \right\}^{-np} \det(A)^{-n} \det(B)^{-p}$$

$$\exp\left\{ -\sum_{i=1}^{p}\sum_{j=1}^{n} \left| \sum_{k=1}^{p}\sum_{\ell=1}^{n} a^{ik}(y_{k\ell} - m_{k\ell})b^{\ell j} \right|^{\theta} \right\} \qquad (2.7.2)$$

where $A^{-1} = (a^{ik})$, $B^{-1} = (b^{\ell j})$, $M = (m_{k\ell})$, and $Y = (y_{k\ell})$.

Proof: The p.d.f. of X is

$$\left\{ 2\Gamma\left(1 + \frac{1}{\theta}\right) \right\}^{-np} \exp\left\{ -\sum_{i=1}^{p}\sum_{j=1}^{n} |x_{ij}|^{\theta} \right\}.$$

Let $Y = AXB + M$. Substituting $x_{ij} = \sum_{k=1}^{p}\sum_{\ell=1}^{n} a^{ik}(y_{k\ell} - m_{k\ell})b^{\ell j}$ alongwith the Jacobian of the transformation $J(X \to Y) = \det(A)^{-n}\det(B)^{-p}$ in the above density we get (2.7.2). ∎

Linear transformations of matrices with matrix variate θ-generalized normal distribution also have matrix variate θ-generalized normal distribution. This is proved in the next theorem.

THEOREM 2.7.2. *Let $Y \sim N_{p,n}(M, A, B, \theta)$. Let $C\,(p \times p)$, $D\,(n \times n)$ be nonsingular matrices, L be a $p \times n$ matrix, and define $Z = CYD + L$. Then*

$$Z \sim N_{p,n}(CMD + L, CA, BD, \theta) \qquad (2.7.3)$$

Proof: Let $X \sim N_{p,n}(0, I_p, I_n, \theta)$ and $Y = AXB + M$. Then $Z = (CA)X(BD) + (CMD + L)$, where CA and BD are nonsingular. From this (2.7.3) follows. ∎

It may be remarked here that $N_{p,n}(M, A, B, 2) \equiv N_{p,n}(M, \frac{1}{2}(AA') \otimes (BB'))$. Indeed, let $Y \sim N_{p,n}(M, A, B, 2)$. Then $Y = \frac{1}{\sqrt{2}}A\sqrt{2}XB + M$, where $\sqrt{2}\,X \sim N_{p,n}(0, I_p \otimes I_n)$ from which the statement follows. Therefore the matrix variate normal distribution is a special case of the matrix variate θ-generalized normal distributions.

The relationship between matrix variate θ-generalized normal distributions and multivariate $\theta-$generalized normal distributions is pointed out in the next theorem.

THEOREM 2.7.3. *$Y \sim N_{p,n}(M, A, B, \theta)$ if and only if*

$$\text{vec}(Y') \sim N_{np}(\text{vec}(M'), A \otimes B', \theta).$$

Proof: $Y = AXB + M$ is equivalent to $\text{vec}(Y') = (A \otimes B')\text{vec}(X') + \text{vec}(M')$, from which the statement of the theorem follows. ∎

The next theorem shows that the parameters of a matrix variate θ-generalized normal distribution are not uniquely determined.

THEOREM 2.7.4. $N_{p,n}(M, A, B, \theta)$ and $N_{p,n}(M^*, A^*, B^*, \theta)$ *define the same distribution if and only if $M = M^*$ and*
(a) in the case of $\theta = 2$, there exist $G\,(p \times p)$ and $H\,(n \times n)$ orthogonal matrices and $c > 0$ such that $A^* = cAG$ and $B^* = \frac{1}{c}HB$,
(b) in the case of $\theta \neq 2$, there exist $P\,(p \times p)$ and $Q\,(n \times n)$ signed permutation matrices and $c > 0$ such that $A^* = cAP$ and $B^* = \frac{1}{c}QB$.

Proof: The sufficiency of the conditions is obvious. To prove necessity assume that $N_{pn}(\text{vec}(M'), A \otimes B', \theta)$ and $N_{pn}(\text{vec}(M^{*'}), A^* \otimes B^{*'}, \theta)$ define the same distribution. Since the first distribution is symmetric about $\text{vec}(M')$ and the second one about $\text{vec}(M^{*'})$, we must have $M = M^*$.
(a) If $\theta = 2$, we get

$$N_{pn}\left(\text{vec}(M'), \frac{1}{2}(AA') \otimes (B'B)\right) \equiv N_{pn}\left(\text{vec}(M'), \frac{1}{2}(A^*A^{*'}) \otimes (B^{*'}B^*)\right)$$

Hence there exists $c^2 > 0$ such that $A^*A^{*'} = c^2 AA'$ and $B^{*'}B^* = B'B$. But then we can find $G\,(p \times p)$ and $H\,(n \times n)$ orthogonal matrices such that $A^* = cAG$ and $B^* = \frac{1}{c}HB$.
(b) If $\theta \neq 2$ we use Theorem 3 of Goodman and Kotz (1973) which says that $A^* \otimes B^{*'}$ can differ from $A \otimes B'$ by at most a post-multiplicative signed permutation matrix R. That is $A^* \otimes B^{*'} = (A \otimes B')R$, or $A^{-1}A^* \otimes (B^{-1})'B^{*'} = R$. This last equation is equivalent to $A^{-1}A^* = cP$, $(B^{-1})'B^{*'} = \frac{1}{c}Q$, where $P\,(p \times p)$, $Q\,(n \times n)$ are signed permutation matrices, and $c > 0$. ∎

The first four moments of a matrix variate θ-generalized normal distribution are derived next. For notational ease we will write $\theta = \frac{1}{\eta}$.

THEOREM 2.7.5. *Let $X \sim N_{p,n}(0, I_p, I_n, \theta)$, then*
(i) $E(x_{ij}) = 0$,

(ii) $E(x_{i_1 j_1} x_{i_2 j_2}) = \dfrac{\Gamma(3\eta)}{\Gamma(\eta)}\delta_{i_1 i_2}\delta_{j_1 j_2}$,

(iii) $E(x_{i_1 j_1} x_{i_2 j_2} x_{i_3 j_3}) = 0$,
and
(iv) $E(x_{i_1 j_1} x_{i_2 j_2} x_{i_3 j_3} x_{i_4 j_4}) = \left[\dfrac{\Gamma(5\eta)}{\Gamma(\eta)} - 3\dfrac{\Gamma^2(3\eta)}{\Gamma^2(\eta)}\right]\delta_{i_1 i_2 i_3 i_4}\delta_{j_1 j_2 j_3 j_4}$

$$+ \frac{\Gamma^2(3\eta)}{\Gamma^2(\eta)}(\delta_{i_1 i_2}\delta_{j_1 j_2}\delta_{i_3 i_4}\delta_{j_3 j_4}$$

$$+ \delta_{i_1 i_3}\delta_{j_1 j_3}\delta_{i_2 i_4}\delta_{j_2 j_4} + \delta_{i_1 i_4}\delta_{j_1 j_4}\delta_{i_2 i_3}\delta_{j_2 j_3})$$

where $\delta_{i_1 i_2 \cdots i_k} = \begin{cases} 1 & \text{if } i_1 = i_2 = \cdots = i_k \\ 0 & \text{otherwise} \end{cases}$

Proof: If x_{ij} has the p.d.f.

$$f(x_{ij}) = \frac{1}{2\Gamma(1+\eta)} \exp\{-|x_{ij}|^{\frac{1}{\eta}}\}$$

and $k > 1$, then

$$\int_0^\infty x_{ij}^k f(x_{ij})\, dx_{ij} = \frac{\Gamma[(k+1)\eta]}{2\Gamma(\eta)}.$$

Thus if k is a non-negative integer, we have

$$E(x_{ij}^k) = \frac{1+(-1)^k}{2} \cdot \frac{\Gamma[(k+1)\eta]}{\Gamma(\eta)}. \tag{2.7.4}$$

Using (2.7.4) and the fact that the elements of X are independent of each other we obtain the results of the theorem. ∎

THEOREM 2.7.6. *Let* $Y \sim N_{p,n}(0, A, B, \theta)$, *then*

(i) $E(y_{ij}) = 0$,

(ii) $E(y_{i_1 j_1} y_{i_2 j_2}) = \dfrac{\Gamma(3\eta)}{\Gamma(\eta)} g_{i_1 i_2} h_{j_1 j_2}$,

(iii) $E(y_{i_1 j_1} y_{i_2 j_2} y_{i_3 j_3}) = 0$,

and

(iv) $E(y_{i_1 j_1} y_{i_2 j_2} y_{i_3 j_3} y_{i_4 j_4}) = \left[\dfrac{\Gamma(5\eta)}{\Gamma(\eta)} - \dfrac{\Gamma^2(3\eta)}{\Gamma^2(\eta)}\right] r_{i_1 i_2 i_3 i_4} q_{j_1 j_2 j_3 j_4}$

$$+ \frac{\Gamma^2(3\eta)}{\Gamma^2(\eta)} (g_{i_1 i_2} h_{j_1 j_2} g_{i_3 i_4} h_{j_3 j_4}$$

$$+ g_{i_1 i_3} h_{j_1 j_3} g_{i_2 i_4} h_{j_2 j_4} + g_{i_1 i_4} h_{j_1 j_4} g_{i_2 i_3} h_{j_2 j_3})$$

where $g_{uv} = \sum_{k=1}^p a_{uk} a_{vk}$, $h_{uv} = \sum_{\ell=1}^n b_{\ell u} b_{\ell v}$, $r_{uvwt} = \sum_{k=1}^p a_{uk} a_{vk} a_{wk} a_{tk}$, *and* $q_{uvwt} = \sum_{\ell=1}^p b_{\ell u} b_{\ell v} b_{\ell w} b_{\ell t}$.

Proof: The results can be obtained from Theorem 2.7.5 by expressing Y as $Y = AXB$ where $X \sim N_{p,n}(0, I_p, I_n, \theta)$. ∎

THEOREM 2.7.7. *Let* $Y \sim N_{p,n}(M, A, B, \theta)$, *then*

(i) $E(y_{ij}) = m_{ij}$,

(ii) $E(y_{i_1 j_1} y_{i_2 j_2}) = \dfrac{\Gamma(3\eta)}{\Gamma(\eta)} g_{i_1 i_2} h_{j_1 j_2} + m_{i_1 j_1} m_{i_2 j_2}$,

(iii) $E(y_{i_1 j_1} y_{i_2 j_2} y_{i_3 j_3}) = \dfrac{\Gamma(3\eta)}{\Gamma(\eta)} [g_{i_1 i_2} h_{j_1 j_2} m_{i_3 j_3} + g_{i_1 i_3} h_{j_1 j_3} m_{i_2 j_2} + g_{i_2 i_3} h_{j_2 j_3} m_{i_1 j_1}]$

$$+ m_{i_1 j_1} m_{i_2 j_2} m_{i_3 j_3},$$

and

(iv) $E(y_{i_1j_1}y_{i_2j_2}y_{i_3j_3}y_{i_4j_4}) = \left[\dfrac{\Gamma(5\eta)}{\Gamma(\eta)} - 3\dfrac{\Gamma^2(3\eta)}{\Gamma^2(\eta)}\right] r_{i_1i_2i_3i_4}q_{j_1j_2j_3j_4}$

$\qquad + \dfrac{\Gamma^2(3\eta)}{\Gamma^2(\eta)}(g_{i_1i_2}h_{j_1j_2}g_{i_3i_4}h_{j_3j_4}$

$\qquad + g_{i_1i_3}h_{j_1j_3}g_{i_2i_4}h_{j_2j_4} + g_{i_1i_4}h_{j_1j_4}g_{i_2i_3}h_{j_2j_3})$

$\qquad + \dfrac{\Gamma(3\eta)}{\Gamma(\eta)}(m_{i_1j_1}m_{i_2j_2}g_{i_3i_4}h_{j_3j_4} + m_{i_1j_1}m_{i_3j_3}g_{i_2i_4}h_{j_2j_4}$

$\qquad + m_{i_1j_1}m_{i_4j_4}g_{i_2i_3}h_{j_2j_3} + m_{i_2j_2}m_{i_3j_3}g_{i_1i_4}h_{j_1j_4}$

$\qquad + m_{i_2j_2}m_{i_4j_4}g_{i_1i_3}h_{j_1j_3} + m_{i_3j_3}m_{i_4j_4}g_{i_1i_2}h_{j_1j_2}$

$\qquad + m_{i_1j_1}m_{i_2j_2}m_{i_3j_3}m_{i_4j_4}),$

where the functions g, h, r, and q are defined in Theorem 2.7.6.

Proof: The results follow from Theorem 2.7.6 if we express Y as $Y = X + M$ where $X \sim N_{p,n}(0, A, B, \theta)$. ■

COROLLARY 2.7.7.1. Let $X \sim N_{p,n}(M, A, B, \theta)$, then $E(X) = M$ and

$$\text{cov}(\text{vec}(X')) = \frac{\Gamma(3\eta)}{\Gamma(\eta)}(AA') \otimes (B'B). \qquad (2.7.5)$$

COROLLARY 2.7.7.2. Let $X \sim N_{p,n}(M, A, B, \theta)$, then

$$\text{corr}(y_{i_1j_1}, y_{i_2j_2}) = \frac{g_{i_1i_2}h_{j_1j_2}}{\sqrt{g_{i_1i_1}g_{i_2i_2}h_{j_1j_1}h_{j_2j_2}}},$$

and hence the $\text{corr}(y_{i_1j_1}, y_{i_2j_2})$ does not depend on θ.

Using the expressions for the moments the following result can be derived.

THEOREM 2.7.8. Let $X \sim N_{p,n}(M, A, B, \theta)$ and $E(r \times p)$, $C(n \times k)$, $F(q \times p)$, and $D(n \times \ell)$ be constant matrices. Then EXC and FXD are uncorrelated if and only if either $C'B'BD = 0$ or $EAA'F' = 0$. Specially, XC and XD are uncorrelated iff $C'B'BD = 0$, and EX and FX are uncorrelated iff $EAA'F' = 0$.

Proof: Using (2.7.5) we get

$$\text{cov}(\text{vec}(EXC)', \text{vec}(FXD)') = \text{cov}((E \otimes C')\text{vec}(X'), (F \otimes D')\text{vec}(X'))$$

$$= (E \otimes C')\frac{\Gamma(3\eta)}{\Gamma(\eta)}\{(AA') \otimes (B'B)\}(F' \otimes D)$$

$$= \frac{\Gamma(3\eta)}{\Gamma(\eta)}(EAA'F') \otimes (C'B'BD),$$

and the last expression equals zero iff $EAA'F' = 0$ or $C'B'BD = 0$. ■

The next theorem shows that matrix variate θ-generalized normal distributions have maximal entropy in certain class of distributions.

THEOREM 2.7.9. *Let* $X (p \times n)$ *be a random matrix with p.d.f.* f *such that*

$$E\|AXB + M\|_\theta = c$$

where $A (p \times p)$, $B (n \times n)$ *are nonsingular matrices,* M *is* $p \times n$ *matrix,* c *is a given scalar, and for a* $p \times n$ *matrix* Y *we define* $\|Y\|_\theta$ *as*

$$\|Y\|_\theta = \sum_{i=1}^{p} \sum_{j=1}^{n} |y_{ij}|^\theta.$$

Then the entropy of X*, that is,* $E(-\ln f(X))$ *is maximized iff* $X = Y$ *a.e. where*

$$Y \sim N_{p,n}\left(-A^{-1}MB^{-1}, \left(\frac{\theta c}{pn}\right)^{\frac{1}{\theta}} A^{-1}, B^{-1}, \theta \right).$$

The maximal entropy is

$$\frac{pn}{\theta}\left[1 + \ln\left(\frac{\theta c}{pn}\right)\right] - \ln\left\{\left(2\Gamma\left(1+\frac{1}{\theta}\right)\right)^{-np} \det(A)^n \det(B)^p\right\}.$$

Proof: See Gupta and Varga (1995a). ∎

PROBLEMS

2.1.   Let the p.d.f. of $X (p \times n)$ be given by (2.2.1). Derive the characteristic function of X.

2.2.   Let $X (p \times n) \sim N_{p,n}(M_1, \Sigma_1 \otimes \Psi_1)$ and $Y (p \times n) \sim N_{p,n}(M_2, \Sigma_2 \otimes \Psi_2)$ be independently distributed. Prove that $X + Y \sim N_{p,n}(M_1 + M_2, (\Sigma_1 \otimes \Psi_1) + (\Sigma_2 \otimes \Psi_2))$.

2.3.   Let $X \sim N_{p,n}(M, \Sigma \otimes \Psi)$, and partition X, M, Σ and Ψ as

$$X = \begin{pmatrix} X_{1r} \\ X_{2r} \end{pmatrix} \begin{matrix} m \\ p-m \end{matrix} = \begin{pmatrix} X_{1c} & X_{2c} \end{pmatrix} \atop \;\;\;\; t \;\;\;\;\; n-t$$

$$M = \begin{pmatrix} M_{1r} \\ M_{2r} \end{pmatrix} \begin{matrix} m \\ p-m \end{matrix} = \begin{pmatrix} M_{1c} & M_{2c} \end{pmatrix} \atop \;\;\;\; t \;\;\;\;\; n-t$$

$$\Sigma = \begin{pmatrix} \Sigma_{11} & \Sigma_{12} \\ \Sigma_{21} & \Sigma_{22} \end{pmatrix} \begin{matrix} m \\ p-m \end{matrix} \quad \text{and} \quad \Psi = \begin{pmatrix} \Psi_{11} & \Psi_{12} \\ \Psi_{21} & \Psi_{22} \end{pmatrix} \begin{matrix} t \\ n-t \end{matrix}.$$
$$\;\;\; m \quad p-m \qquad\qquad\qquad\quad t \quad n-t$$

Then, prove that (i) X_{1r} and X_{2r} are independent if and only if $\Sigma_{12} = 0$, and (ii) X_{1c} and X_{2c} are independent if and only if $\Psi_{12} = 0$.

2.4.   Let $X = (\boldsymbol{x}_1, \ldots, \boldsymbol{x}_n) \sim N_{p,n}(M, \Sigma \otimes \Psi)$ and denote its p.d.f. by $p(X)$. Further, let $f(\boldsymbol{y}_1|\boldsymbol{y}_2)$ be the conditional density of $\boldsymbol{y}_1$ given $\boldsymbol{y}_2$. Using suitable notations for the means and covariances, write down explicitly

$$p(X) = f_1(\boldsymbol{x}_1)f_2(\boldsymbol{x}_2|\boldsymbol{x}_1)f_3(\boldsymbol{x}_3|\boldsymbol{x}_1, \boldsymbol{x}_2) \cdots f_n(\boldsymbol{x}_n|\boldsymbol{x}_1, \ldots, \boldsymbol{x}_{n-1}).$$

2.5.   Prove Theorem 2.3.3(iii).

2.6.   Let $X\,(p \times n) \sim N_{p,n}(0, \Sigma \otimes \Psi)$ and $\Sigma = (\sigma_{ij})$, $\Psi = (\psi_{ij})$. Then show that

$$
\begin{aligned}
E(x_{i_1j_1}x_{i_2j_2}x_{i_3j_3}x_{i_4j_4}x_{i_5j_5}x_{i_6j_6}) &= \sigma_{i_1i_2}\psi_{j_1j_2}\sigma_{i_3i_4}\psi_{j_3j_4}\sigma_{i_5i_6}\psi_{j_5j_6} \\
&+ \sigma_{i_1i_2}\psi_{j_1j_2}\sigma_{i_3i_5}\psi_{j_3j_5}\sigma_{i_4i_6}\psi_{j_4j_6} + \sigma_{i_1i_2}\psi_{j_1j_2}\sigma_{i_3i_6}\psi_{j_3j_6}\sigma_{i_4i_5}\psi_{j_4j_5} \\
&+ \sigma_{i_1i_3}\psi_{j_1j_3}\sigma_{i_2i_4}\psi_{j_2j_4}\sigma_{i_5i_6}\psi_{j_5j_6} + \sigma_{i_1i_3}\psi_{j_1j_3}\sigma_{i_2i_5}\psi_{j_2j_5}\sigma_{i_4i_6}\psi_{j_4j_6} \\
&+ \sigma_{i_1i_3}\psi_{j_1j_3}\sigma_{i_2i_6}\psi_{j_2j_6}\sigma_{i_4i_5}\psi_{j_4j_5} + \sigma_{i_1i_4}\psi_{j_1j_4}\sigma_{i_2i_3}\psi_{j_2j_3}\sigma_{i_5i_6}\psi_{j_5j_6} \\
&+ \sigma_{i_1i_4}\psi_{j_1j_4}\sigma_{i_2i_5}\psi_{j_2j_5}\sigma_{i_3i_6}\psi_{j_3j_6} + \sigma_{i_1i_4}\psi_{j_1j_4}\sigma_{i_2i_6}\psi_{j_2j_6}\sigma_{i_3i_5}\psi_{j_3j_5} \\
&+ \sigma_{i_1i_5}\psi_{j_1j_5}\sigma_{i_2i_3}\psi_{j_2j_3}\sigma_{i_4i_6}\psi_{j_4j_6} + \sigma_{i_1i_5}\psi_{j_1j_5}\sigma_{i_2i_4}\psi_{j_2j_4}\sigma_{i_3i_6}\psi_{j_3j_6} \\
&+ \sigma_{i_1i_5}\psi_{j_1j_5}\sigma_{i_2i_6}\psi_{j_2j_6}\sigma_{i_3i_4}\psi_{j_3j_4} + \sigma_{i_1i_6}\psi_{j_1j_6}\sigma_{i_2i_3}\psi_{j_2j_3}\sigma_{i_4i_5}\psi_{j_4j_5} \\
&+ \sigma_{i_1i_6}\psi_{j_1j_6}\sigma_{i_2i_4}\psi_{j_2j_4}\sigma_{i_3i_5}\psi_{j_3j_5} + \sigma_{i_1i_6}\psi_{j_1j_6}\sigma_{i_2i_5}\psi_{j_2j_5}\sigma_{i_3i_4}\psi_{j_3j_4}.
\end{aligned}
$$

2.7.   Prove Theorem 2.3.4(ii), (iv) and (v).

2.8.   Prove Theorem 2.3.5(v)–(vii).

2.9.   Prove Theorem 2.3.6(ii)–(vii).

2.10. Prove Theorem 2.3.8.

2.11. Prove Theorem 2.3.9.

2.12. Let $X \sim N_{p,n}(M, \Sigma \otimes \Psi)$. Then, for given matrices A, B, and C of suitable order, find
(i) $E(X'AX'BX'CX)$
(ii) $E(X'AX'BX'CX')$
(iii) $E(XAX'BXCX')$
(iv) $E(\mathrm{tr}(XAXBX')X')$
(v) $E(\mathrm{tr}(XAX'BX')X)$
(vi) $E(\mathrm{tr}(X'AX'BX)X')$
(vii) $E(\mathrm{tr}(AXBX')XCX')$
(viii) $E(\mathrm{tr}(AX)X'BX'CX)$
(ix) $E(\mathrm{tr}(AX)X'BX'CX')$
(x) $E(\mathrm{tr}(X'A)XBXCX)$
(xi) $E(\mathrm{tr}(X'A)X'BX'CX')$.

2.13. Let $X \sim N_{p,n}(M, \Sigma \otimes \Psi)$. Then, prove that

$$E(X \otimes X) = \mathrm{vec}(\Sigma)(\mathrm{vec}(\Psi))' + M \otimes M.$$

(Neudecker and Wansbeek, 1987)

2.14. Let $X \sim N_{p,n}(0, \Sigma \otimes \Psi)$. Then, for given $A\,(p \times n)$ and $a = 0, 1, 2, \ldots, \ldots$ find
  (i) $E(\operatorname{tr}(XA')^2 (AA')^a)$
  (ii) $E(\operatorname{tr}(XA'AX'(AA')^a))$
  (iii) $E(\operatorname{tr}(XX')\operatorname{tr}(XX'(AA')^a))$
  (iv) $E(\operatorname{tr}((XX')(AA')^a))$
  (v) $E(\operatorname{tr}^3(XX'))$.

2.15. Let $X\,(p \times n)$ and $Y\,(p \times n)$ be identically distributed random matrices. Suppose that $Y|X \sim N_{p,n}(aX + B, \Sigma \otimes I_n)$ with $B\,(p \times n)$ and $|a| < 1$. Then, prove that $X \sim N_{p,n}((1 - a)^{-1}B, (1 - a^2)^{-1}\Sigma \otimes I_n)$ and

$$\begin{pmatrix} X \\ Y \end{pmatrix} \sim N_{2p,n}\Big((1-a)^{-1}\begin{pmatrix} B \\ B \end{pmatrix}, (1-a^2)^{-1}\begin{pmatrix} \Sigma & a\Sigma \\ a\Sigma & \Sigma \end{pmatrix} \otimes I_n\Big).$$

(Bekker and Roux, 1990)

2.16. Let $X\,(p \times n)$ and $Y\,(p \times n)$ be identically distributed random matrices. Suppose that $Y|X \sim N_{p,n}(AX + B, I_p \otimes I_n)$ with $B\,(p \times n)$ and $A\,(p \times p)$ is symmetric. Then, prove that $X \sim N_{p,n}((I_p - A)^{-1}B, (I_p - A^2)^{-1} \otimes I_n)$ and

$$\begin{pmatrix} X \\ Y \end{pmatrix} \sim N_{2p,n}\Big(\begin{pmatrix} (I_p - A)^{-1}B \\ (I_p - A)^{-1}B \end{pmatrix}, \begin{pmatrix} (I_p - A^2)^{-1} & A(I_p - A^2)^{-1} \\ A(I_p - A^2)^{-1} & (I_p - A^2)^{-1} \end{pmatrix}\Big).$$

(Bekker and Roux, 1990)

2.17. Let $X\,(p \times n)$ and $Y\,(p \times n)$ be identically distributed random matrices. Further, let X and $V = Y - aX$ be independent with $V \sim N_{p,n}(B, \Sigma \otimes I_n)$, $B\,(p \times n)$ and $|a| < 1$. Then, prove that $X \sim N_{p,n}((1 - a)^{-1}B, \Sigma \otimes I_n)$ and $\begin{pmatrix} X \\ Y \end{pmatrix}$ also has a matrix variate normal distribution.

(Bekker and Roux, 1990)

2.18. Let X and Y be $p \times n$, identically distributed random matrices. Suppose that $Y|X \sim N_{p,n}(AX + B, \Sigma \otimes \Phi)$, where B is $p \times n$, and A is a $p \times p$ matrix which satisfies the following conditions:
  (i) A is symmetric,
  (ii) $\max_i |\operatorname{Ch}_i(A)| < 1$,
  (iii) $A\Sigma = \Sigma A$.
  Define $Z = \begin{pmatrix} X \\ Y \end{pmatrix}$. Then, prove that

$$Z \sim N_{2p,n}\Big(\begin{pmatrix} (I_p - A)^{-1}B \\ (I_p - A)^{-1}B \end{pmatrix}, \begin{pmatrix} \Sigma(I_p - A^2)^{-1} & \Sigma A(I_p - A^2)^{-1} \\ \Sigma A(I_p - A^2)^{-1} & \Sigma(I_p - A^2)^{-1} \end{pmatrix} \otimes \Phi\Big).$$

(Gupta and Varga, 1994b)

2.19. Let X and Y be $p \times n$, identically distributed random matrices. Suppose that X and $V = Y - AX$ are independent, and $V \sim N_{p,n}(B, \Sigma \otimes \Phi)$ where B is $p \times n$, and A is $p \times p$ matrix which satisfies the conditions (i)–(iii) of Problem 2.18. Define $Z = \begin{pmatrix} X \\ Y \end{pmatrix}$. Then, prove that

$$Z \sim N_{2p,n}\left(\begin{pmatrix} (I_p - A)^{-1}B \\ (I_p - A)^{-1}B \end{pmatrix}, \begin{pmatrix} \Sigma(I_p - A^2)^{-1} & \Sigma A(I_p - A^2)^{-1} \\ \Sigma A(I_p - A^2)^{-1} & \Sigma(I_p - A^2)^{-1} \end{pmatrix} \otimes \Phi \right).$$

(Gupta and Varga, 1994b)

2.20. Let X and Y be $p \times n$, identically distributed random matrices with $E(X) = E(Y) = 0$ and suppose $\text{vec}(X')$ has covariance matrix $\Sigma \otimes \Phi$. Moreover, suppose that A is nonsingular and satisfies the conditions (i)–(iii) of Problem 2.18. Let $Z = \begin{pmatrix} X \\ Y \end{pmatrix}$. Then, prove that

$$Z \sim N_{2p,n}\left(\begin{pmatrix} 0 \\ 0 \end{pmatrix}, \begin{pmatrix} \Sigma & \Sigma A \\ \Sigma A & \Sigma \end{pmatrix} \otimes \Phi \right).$$

if and only if X and $V = (I_p - A^2)^{-\frac{1}{2}}(Y - AX)$ are independent and identically distributed.

(Gupta and Varga, 1994b)

2.21. Let X ($p \times n$) and Y ($q \times n$) be random matrices. Suppose that $Y|X \sim N_{q,n}(C + DX, \Sigma_2 \otimes \Phi)$ and $X \sim N_{p,n}(F, \Sigma_1 \otimes \Phi)$. Let $Z = \begin{pmatrix} X \\ Y \end{pmatrix}$. Then prove that

$$Z \sim N_{p+q,n}\left(\begin{pmatrix} F \\ DF + C \end{pmatrix}, \begin{pmatrix} \Sigma_1 & \Sigma_1 D' \\ D\Sigma_1 & \Sigma_2 + D\Sigma_1 D' \end{pmatrix} \otimes \Phi \right).$$

(Gupta and Varga, 1994b)

2.22. Let X ($p \times n$), Y ($q \times n$) be random matrices and suppose that $Y|X \sim N_{q,n}(C + DX, \Sigma_2 \otimes \Phi)$, $X|Y = Y_0 \sim N_{p,n}(M, \Sigma_1 \otimes \Phi)$, where C ($q \times n$), D ($q \times p$), Σ_2 ($q \times q$), Φ ($n \times n$), M ($p \times n$), Σ_1 ($p \times p$), $\Sigma_1 > 0$, $\Sigma_2 > 0$, $\Phi > 0$, and Y_0 is a fixed $q \times n$ matrix. Define $B = \Sigma_1 D' \Sigma_2^{-1}$, $A = M - \Sigma_1 D' \Sigma_2^{-1} Y_0$, $N = \begin{pmatrix} (I_p - BD)^{-1}(A + BC) \\ (I_q - DB)^{-1}(C + DA) \end{pmatrix}$ and $Z = \begin{pmatrix} X \\ Y \end{pmatrix}$. Then, prove that

$$Z \sim N_{p+q,n}\left(N, \begin{pmatrix} (I_p - BD)^{-1}\Sigma_1 & (I_p - BD)^{-1}B\Sigma_2 \\ D(I_p - BD)^{-1}\Sigma_1 & (I_q - DB)^{-1}\Sigma_1 \end{pmatrix} \otimes \Phi \right).$$

(Gupta and Varga, 1992)

2.23. Let $X = X' \sim SN_{p,p}(M, B'_p(\Sigma \otimes \Psi)B_p)$. Then, for given $A\,(p \times p)$, and $C\,(p \times p)$ prove that

(i) $E(\text{tr}(CX)XAX) = \dfrac{1}{4}[M A \Sigma C' \Psi + M A \Psi C \Sigma + M A \Sigma C \Psi + M A \Psi C' \Sigma$

$\qquad\qquad\qquad\qquad + \Sigma C' \Psi A M + \Psi C \Sigma A M + \Sigma C \Psi A M + \Psi C \Sigma A M$

$\qquad\qquad\qquad\qquad + \text{tr}(CM) \Sigma A' \Psi + \text{tr}(A\Sigma)\,\text{tr}(CM) \Psi$

$\qquad\qquad\qquad\qquad + \text{tr}(A\Psi)\,\text{tr}(CM) \Sigma + \text{tr}(CM) \Psi A' \Sigma] + \text{tr}(CM) M A$

(ii) $E(\text{tr}(AXCX)X) = \dfrac{1}{4}[\text{tr}(A\Sigma C' \Psi)M + \text{tr}(A\Psi)\,\text{tr}(C\Sigma)M + \text{tr}(A\Psi C' \Sigma)M$

$\qquad\qquad\qquad\qquad + \text{tr}(A\Sigma)\,\text{tr}(\Psi C)M + \Sigma C' M A' \Psi + \Psi A M C \Sigma$

$\qquad\qquad\qquad\qquad + \Sigma A M C \Psi + \Psi C' M A' \Sigma + \Sigma A' M C' \Psi + \Psi C M A \Sigma$

$\qquad\qquad\qquad\qquad + \Sigma C M A \Psi + \Psi A' M C' \Sigma] + \text{tr}(AMM)M.$

(H. M. Nel, 1977)

2.24. Let $X \sim N_{p,n}(M, \Sigma \otimes I_n)$. Assuming *a priori* that $M \sim N_{p,n}(0, \Omega \otimes I_n)$, derive its posterior distribution.

2.25. Let $X \sim N_{p,n}(M, \Sigma \otimes \Psi | s, C)$. Partition X as $X = \begin{pmatrix} X_{1r} \\ X_{2r} \end{pmatrix} \begin{matrix} p_1 \\ p_2 \end{matrix}$, $p_1 + p_2 = p$, and derive the marginal p.d.f. of X.

CHAPTER 3

WISHART DISTRIBUTION

3.1. INTRODUCTION

Let $y_1, \ldots, y_n$ be n independent standard normal variables. Then, $w = \sum_{i=1}^{n} y_i^2 \sim \chi_n^2$ with p.d.f.

$$\left\{ 2^{\frac{1}{2}n} \Gamma\left(\frac{1}{2}n\right) \right\}^{-1} w^{\frac{1}{2}n-1} \exp\left(-\frac{1}{2}w\right), \ w > 0. \qquad (3.1.1)$$

A p-variate generalization of (3.1.1) has been given by Krishnamoorthy and Parthasarthy (1951). In this chapter, we study a matrix variate generalization of (3.1.1), known as Wishart distribution (Wishart, 1928). The discovery of this distribution has contributed enormously to the development of multivariate analysis, $e.g.$, see Roy (1957), Kshirsagar (1972), Press (1972), Giri (1977), Srivastava and Khatri (1979), Muirhead (1982), Anderson (1984), and Siotani, Hayakawa and Fujikoshi (1985).

3.2. DENSITY FUNCTION

In this section, we derive the density of a Wishart matrix using normal vectors. We begin by defining the Wishart distribution.

DEFINITION 3.2.1. *A $p \times p$ random symmetric positive definite matrix S is said to have a Wishart distribution with parameters p, n, and $\Sigma\,(p \times p) > 0$, written as $S \sim W_p(n, \Sigma)$, if its p.d.f. is given by*

$$\left\{ 2^{\frac{1}{2}np} \Gamma_p\left(\frac{1}{2}n\right) \det(\Sigma)^{\frac{1}{2}n} \right\}^{-1} \det(S)^{\frac{1}{2}(n-p-1)} \operatorname{etr}\left(-\frac{1}{2}\Sigma^{-1}S\right), \ S > 0, \ n \geq p. \qquad (3.2.1)$$

Fisher (1915) derived this distribution for $p = 2$ in order to study the distribution of correlation coefficient from a normal sample. Wishart (1928) obtained the distribution for arbitrary p as the joint distribution of sample variances and covariances from multivariate normal population. Because of its important role in multivariate statistical analysis, various authors have given different derivations, $e.g.$, see Wishart and Bartlett (1933), Ingham (1933), Mahalnobis, Bose and Roy (1937),

Madow (1938), Hsu (1939b), Elfving (1947), Sverdrup (1947), Rasch (1948), Ogawa (1953), James (1954), Mauldon (1955), Wijsman (1957), Kshirsagar (1959), and Jambunathan (1965). This distribution, for $\Sigma = I_p$, belongs to the class of orthogonally invariant and residual independent distributions discussed in Chapter 9. The orthogonal invariance and residual independence properties in this case are given in Theorems 3.3.2 and 3.3.4 respectively.

If $\boldsymbol{x}_1, \ldots, \boldsymbol{x}_n$ are independent $N_p(\boldsymbol{0}, \Sigma)$, then $X = (\boldsymbol{x}_1, \ldots, \boldsymbol{x}_n)$ has a matrix variate normal distribution. Further, if $n \geq p$, then $XX' > 0$ with probability one (Stein, 1969; Dykstra, 1970) and $XX' \sim W_p(n, \Sigma)$ as shown below.

THEOREM 3.2.1. *Let $X \sim N_{p,n}(0, \Sigma \otimes I_n)$, $n \geq p$, then $XX' > 0$ with probability one.*

Proof: Let $X = (\boldsymbol{x}_1, \ldots, \boldsymbol{x}_n)$. Then, it suffices to show that X has rank $p \leq n$, that is any p random vectors $\boldsymbol{x}_1, \ldots, \boldsymbol{x}_p$ are linearly independent with probability one. Now,

$$P\{\boldsymbol{x}_1, \ldots, \boldsymbol{x}_p \text{ are linearly independent}\}$$

$$= 1 - P\{\boldsymbol{x}_1, \ldots, \boldsymbol{x}_p \text{ are linearly dependent}\}$$

$$\geq 1 - \sum_{i=1}^{p} P\{\boldsymbol{x}_i \text{ is a linear combination of others}\}$$

$$= 1 - pP\Big\{\boldsymbol{x}_1 = \sum_{j=2}^{p} d_j \boldsymbol{x}_j, \text{ for at least one } d_j \neq 0\Big\}.$$

Since, $\boldsymbol{x}_1, \ldots, \boldsymbol{x}_p$ are independent random vectors having nondegenerate continuous distribution with covariance matrix $\Sigma > 0$,

$$P\Big\{\boldsymbol{x}_1 = \sum_{j=2}^{p} d_j \boldsymbol{x}_j, \text{ for at least one } d_j \neq 0\Big\} = 0.$$

Hence,

$$P\{\boldsymbol{x}_1, \ldots, \boldsymbol{x}_p \text{ are linearly independent}\} = 1$$

and the proof is complete. ∎

The above theorem has been proven by Eaton and Perlman (1973) without assuming normality (see also Das Gupta, 1971).

THEOREM 3.2.2. *Let $X \sim N_{p,n}(0, \Sigma \otimes I_n)$ and define $S = XX'$, $n \geq p$. Then $S \sim W_p(n, \Sigma)$.*

Proof: The density of X is

$$(2\pi)^{-\frac{1}{2}np} \det(\Sigma)^{-\frac{1}{2}n} \operatorname{etr}\Big(-\frac{1}{2}\Sigma^{-1}XX'\Big).$$

Since $XX' > 0$ with probability one, make the transformation $X = TH_1$, where $T(p \times p) = (t_{ij})$ is a lower triangular matrix with $t_{ii} > 0$, $i = 1, \ldots, p$, and $H_1(p \times n)$

is a semiorthogonal matrix, $H_1 H_1' = I_p$. The Jacobian of this transformation, $J(X \to T, H_1) = \prod_{i=1}^{p} t_{ii}^{n-i} g_{n,p}(H_1)$, is given in (1.3.25). Hence, the joint density of T and H_1 is

$$(2\pi)^{-\frac{1}{2}np} \det(\Sigma)^{-\frac{1}{2}n} \operatorname{etr}\left(-\frac{1}{2}\Sigma^{-1}TT'\right) \prod_{i=1}^{p} t_{ii}^{n-i} g_{n,p}(H_1).$$

Now, integrating out H_1 using Theorem 1.4.9, we get the marginal density of T as

$$\frac{2^{-\frac{1}{2}np+p}}{\Gamma_p(\frac{1}{2}n)} \det(\Sigma)^{-\frac{1}{2}n} \operatorname{etr}\left(-\frac{1}{2}\Sigma^{-1}TT'\right) \prod_{i=1}^{p} t_{ii}^{n-i}. \tag{3.2.2}$$

In (3.2.2), let $S = TT' (= XX')$ with the Jacobian $J(T \to S) = (2^p \prod_{i=1}^{p} t_{ii}^{p-i+1})^{-1}$, then the density of S is

$$\left\{ 2^{\frac{1}{2}np} \Gamma_p\left(\frac{1}{2}n\right) \det(\Sigma)^{\frac{1}{2}n} \right\}^{-1} \det(S)^{\frac{1}{2}(n-p-1)} \operatorname{etr}\left(-\frac{1}{2}\Sigma^{-1}S\right). \blacksquare$$

Note that in the derivation of the Wishart density given above it is assumed that $n \, (\geq p)$ is an integer, but the density (3.2.1) exists for all $n \geq p$. If $n < p$, the density of XX' is called, by some authors, a *pseudo Wishart*, e.g., Kshirsagar (1972), Siotani, Hayakawa and Fujikoshi (1985).

If Σ is of less than full rank, say p_1, then by Definition 2.4.1, $X \sim N_{p,n}(0, \Sigma \otimes I_n|p_1, n)$, and there exists a matrix $H \, (p \times p_1)$ of rank p_1 such that $X = HY$, where $Y \sim N_{p_1,n}(0, I_{p_1} \otimes I_n)$. In this case, $S = HYY'H'$, where $YY' \sim W_{p_1}(n, I_{p_1})$, and S is said to have *singular Wishart distribution*.

We now derive the c.d.f. of a Wishart matrix.

THEOREM 3.2.3. *Let* $S \sim W_p(n, \Sigma)$, *then*

$$P(S < \Lambda) = \frac{\Gamma_p[\frac{1}{2}(p+1)] \det(\Lambda)^{\frac{1}{2}n}}{2^{\frac{1}{2}np} \det(\Sigma)^{\frac{1}{2}n} \Gamma_p[\frac{1}{2}(n+p+1)]} {}_1F_1\left(\frac{1}{2}n; \frac{1}{2}(n+p+1); -\frac{1}{2}\Sigma^{-1}\Lambda\right),$$

where $\Lambda \, (p \times p) > 0$.

Proof: We have

$$P(S < \Lambda) = \frac{1}{2^{\frac{1}{2}np} \Gamma_p\left(\frac{1}{2}n\right) \det(\Sigma)^{\frac{1}{2}n}} \int_{0<S<\Lambda} \operatorname{etr}\left(-\frac{1}{2}\Sigma^{-1}S\right) \det(S)^{\frac{1}{2}(n-p-1)} \, dS. \tag{3.2.3}$$

Substituting $B = \Lambda^{-\frac{1}{2}} S \Lambda^{-\frac{1}{2}}$ with the Jacobian, $J(S \to B) = \det(\Lambda)^{\frac{1}{2}(p+1)}$, in (3.2.3) and writing $\operatorname{etr}(-\frac{1}{2}\Sigma^{-1}S) = {}_0F_0(-\frac{1}{2}\Sigma^{-1}S)$, we get

$$P(S < \Lambda) = \frac{\det(\Lambda)^{\frac{1}{2}n}}{2^{\frac{1}{2}np} \Gamma_p\left(\frac{1}{2}n\right) \det(\Sigma)^{\frac{1}{2}n}} \int_{0<B<I_p} \det(B)^{\frac{1}{2}(n-p-1)} {}_0F_0\left(-\frac{1}{2}\Lambda^{\frac{1}{2}}\Sigma^{-1}\Lambda^{\frac{1}{2}}B\right) dB.$$

The proof is completed by using the Theorem 1.6.3. $\blacksquare$

In Theorem 3.2.2, we have derived the Wishart density assuming $X \sim N_{p,n}(0, \Sigma \otimes \Psi)$ where $\Psi = I_n$. However, if $\Psi \neq I_n$, under certain conditions on Ψ, XX' is still distributed as Wishart as shown below.

THEOREM 3.2.4. *Let $X \sim N_{p,n}(0, \Sigma \otimes \Psi | p, q)$, where $\Psi \ (n \times n)$ is a symmetric idempotent matrix of rank $q \geq p$. Then $XX' \sim W_p(q, \Sigma)$.*

Proof: Since Ψ is singular, from Definition 2.4.1, we can write $X = YR$, where $Y \sim N_{p,q}(0, \Sigma \otimes I_q)$ and $R \ (q \times n)$ is a matrix of rank $q \geq p$ with $\Psi = R'R$. Note that RR' is an idempotent matrix of full rank and hence, an identity matrix. Now, $XX' = YRR'Y' = YY' \sim W_p(q, \Sigma)$ according to Theorem 3.2.2. ∎

A result closely related to the above theorem is the following.

THEOREM 3.2.5. *Let $X \sim N_{p,n}(0, \Sigma \otimes I_n)$ and $\Psi \ (n \times n)$ be a symmetric idempotent matrix of rank $q \geq p$, then $X\Psi X' \sim W_p(q, \Sigma)$.*

Proof: Since $\Psi \ (n \times n)$ is of rank $q \leq n$, one can write $\Psi = BB'$, where $B \ (n \times q)$ is of rank q. Now, from Theorem 2.3.10, $Y = XB \sim N_{p,q}(0, \Sigma \otimes B'B)$. Here, $B'B$ is an idempotent matrix of full rank and hence, $B'B = I_q$. The result follows from Theorem 3.2.2, by noting that $XBB'X' = X\Psi X' = YY'$. ∎

THEOREM 3.2.6. *Let $X \sim N_{p,m}(0, \Sigma \otimes I_m)$ and $A \ (p \times p)$ be a constant symmetric positive semidefinite matrix of rank $r \geq m$ such that $A\Sigma A = A$. Then $X'AX \sim W_m(r, I_m)$.*

Proof: Write $\Sigma = CC'$ where $C \ (p \times p)$ is a nonsingular matrix and $X = CY$. Then $Y \sim N_{p,m}(0, I_p \otimes I_m)$ and $X'AX = Y'(C'AC)Y$. Since $C'AC$ is an idempotent matrix because $A\Sigma A = A$, the result follows from Theorem 3.2.5. ∎

3.3.   PROPERTIES

3.3.1.   Invariance and Decomposition of S

THEOREM 3.3.1. *Let $S \sim W_p(n, \Sigma)$ and A be any $p \times p$ nonsingular matrix. Then, $ASA' \sim W_p(n, A\Sigma A')$.*

Proof: The result follows by making the transformation $V = ASA'$ with Jacobian $J(S \to V) = \det(A)^{-(p+1)}$ in the density of S given by (3.2.1). ∎

COROLLARY 3.3.1.1. *Let $S \sim W_p(n, \Sigma)$ and $\Sigma^{-1} = A'A$, then $ASA' \sim W_p(n, I_p)$.*

THEOREM 3.3.2. *Let $S \sim W_p(n, I_p)$ and $H \ (p \times p)$ be an orthogonal matrix, whose elements are either constants or random variables distributed independently of S. Then, the distribution of S is invariant under the transformation $S \to HSH'$ and is independent of H in the latter case.*

Proof: First, let H be a constant matrix. Then from Theorem 3.3.1, $HSH' \sim W_p(n, I_p)$. If, however, H is a random orthogonal matrix, the conditional distribution of $HSH'|H \sim W_p(n, I_p)$. Since this distribution does not depend on H, $HSH' \sim W_p(n, I_p)$. ∎

THEOREM 3.3.3. *Let $S \sim W_p(n, \Sigma)$, and n be an integer, then $S = XX'$, where $X \sim N_{p,n}(0, \Sigma \otimes I_n)$.*

Proof: Let $V = ASA'$, where $\Sigma^{-1} = A'A$. Then, according to Corollary 3.3.1.1, $V \sim W_p(n, I_p)$. Define an independent random matrix L $(p \times n)$ such that $LL' = I_p$, with the density $c^{-1}g_{n,p}(L)$, where $c = \frac{2^p \pi^{\frac{1}{2}np}}{\Gamma_p(\frac{1}{2}n)}$ and $g_{n,p}(L)$ is given in (1.3.26). Then, the joint density of L and V is

$$c^{-1}\left\{2^{\frac{1}{2}np}\Gamma_p\left(\frac{1}{2}n\right)\right\}^{-1} \operatorname{etr}\left(-\frac{1}{2}V\right) \det(V)^{\frac{1}{2}(n-p-1)} g_{n,p}(L).$$

Now, using the transformations (i) $V = TT'$, where $T = (t_{ij})$ is a lower triangular matrix with $t_{ii} > 0$ and (ii) $TL = Y$, with the Jacobians $J(V \to T) = 2^p \prod_{i=1}^p t_{ii}^{p-i+1}$ and $J(T, L \to Y) = \{g_{n,p}(L) \prod_{i=1}^p t_{ii}^{n-i}\}^{-1}$ given in (1.3.14) and (1.3.25) respectively, we get the density of Y, after some simplification as

$$(2\pi)^{-\frac{1}{2}np} \operatorname{etr}\left(-\frac{1}{2}YY'\right), \quad Y \in \mathbb{R}^{p \times n}.$$

Hence, $Y \sim N_{p,n}(0, I_p \otimes I_n)$, and $V = YY'$. It follows that $S = A^{-1}YY'(A^{-1})' = XX'$ (say), where $X \sim N_{p,n}(0, A^{-1}(A^{-1})' \otimes I_n)$. This completes the proof since $\Sigma = A^{-1}(A^{-1})'$. ∎

The following result is of importance in multivariate analysis and is known as Bartlett's decomposition, Bartlett (1933).

THEOREM 3.3.4. *Let $S \sim W_p(n, I_p)$ and $S = TT'$, where $T = (t_{ij})$ is a lower triangular matrix with $t_{ii} > 0$. Then, t_{ij}, $1 \leq j \leq i \leq p$ are independently distributed, $t_{ii}^2 \sim \chi_{n-i+1}^2$, $1 \leq i \leq p$ and $t_{ij} \sim N(0,1)$, $1 \leq j < i \leq p$.*

Proof: The density of S is

$$\left\{2^{\frac{1}{2}np}\Gamma_p\left(\frac{1}{2}n\right)\right\}^{-1} \det(S)^{\frac{1}{2}(n-p-1)} \operatorname{etr}\left(-\frac{1}{2}S\right). \tag{3.3.1}$$

Making the transformation $S = TT'$, with Jacobian $J(S \to T) = 2^p \prod_{i=1}^p t_{ii}^{p-i+1}$, in (3.3.1), we get the joint density of $t_{11}, t_{21}, \ldots, t_{p1}, t_{p2}, \ldots, t_{pp}$ as

$$\left\{2^{\frac{1}{2}np}\Gamma_p\left(\frac{1}{2}n\right)\right\}^{-1} 2^p \prod_{i=1}^p t_{ii}^{n-i} \exp\left(-\frac{1}{2}\sum_{1 \leq j \leq i \leq p} t_{ij}^2\right)$$

$$= \prod_{1 \leq j < i \leq p}\left\{\frac{1}{\sqrt{2\pi}}\exp\left(-\frac{1}{2}t_{ij}^2\right)\right\} \prod_{i=1}^p \left\{\frac{2(t_{ii}^2)^{\frac{1}{2}(n-i)}\exp(-\frac{1}{2}t_{ij}^2)}{2^{\frac{1}{2}(n-i+1)}\Gamma[\frac{1}{2}(n-i+1)]}\right\},$$

$$t_{ii} > 0, \; 1 \leq i \leq p, \; -\infty < t_{ij} < \infty, \; 1 \leq j < i \leq p. \tag{3.3.2}$$

From (3.3.2), it is easily seen that t_{ij}, $1 \leq j \leq i \leq p$, are independently distributed and $t_{ij} \sim N(0,1)$, $1 \leq j < i \leq p$. By substituting $y_{ii} = t_{ii}^2$, one can show that $t_{ii}^2 \sim \chi_{n-i+1}^2$, $1 \leq i \leq p$. ∎

A similar result can also be proved for an upper triangular factorization of S, as given in the next theorem.

THEOREM 3.3.5. Let $S \sim W_p(n, I_p)$, and $S = TT'$, where $T = (t_{ij})$ is an upper triangular matrix with $t_{ii} > 0$. Then t_{ij}, $1 \leq i \leq j \leq p$ are independently distributed, $t_{ii}^2 \sim \chi_{n-p+i}^2$, $1 \leq i \leq p$ and $t_{ij} \sim N(0,1)$, $1 \leq i < j \leq p$.

Proof: Similar to the proof of Theorem 3.3.4. ∎

3.3.2. Distribution of Sample Covariance Matrix

THEOREM 3.3.6. Let $x_1, \ldots, x_N$ be independent $N_p(\mu, \Sigma)$, $\Sigma > 0$, $N > p$. Define $\bar{x} = \frac{1}{N} \sum_{i=1}^N x_i$ and $S = \sum_{i=1}^N (x_i - \bar{x})(x_i - \bar{x})'$. Then, (i) $\bar{x}$ and S are independently distributed, (ii) $\bar{x} \sim N_p(\mu, \frac{1}{N}\Sigma)$, and (iii) $S \sim W_p(n, \Sigma)$, where $n = N - 1$.

Proof: Let $X\,(p \times N) = (x_1, \ldots, x_N)$, then $X \sim N_{p,N}(\mu e', \Sigma \otimes I_N)$, where $e'\,(1 \times N) = (1, \ldots, 1)$. The density of X is

$$(2\pi)^{-\frac{1}{2}Np} \det(\Sigma)^{-\frac{1}{2}N} \operatorname{etr}\left\{-\frac{1}{2}\Sigma^{-1}(X - \mu e')(X - \mu e')'\right\}. \tag{3.3.3}$$

Now, transform $X_1 = XH$, where $H\,(N \times N)$ is an orthogonal matrix, $H = \left(\frac{1}{\sqrt{N}}e \quad B\right)$, obtaining

$$X_1 = (\sqrt{N}\,\bar{x} \quad XB),$$
$$XX' = X_1 X_1' = N\bar{x}\bar{x}' + YY',$$

where $Y\,(p \times (N-1)) = XB$. Further,

$$(X - \mu e')(X - \mu e')' = XX' - \mu e'X' - Xe\mu' + \mu e'e\mu', \tag{3.3.4}$$

$$\mu e'X' = \mu e'HX_1'$$
$$= \mu e'\left(\frac{1}{\sqrt{N}}e \quad B\right)\begin{pmatrix}\sqrt{N}\,\bar{x}' \\ Y\end{pmatrix}$$
$$= N\mu\bar{x}' \tag{3.3.5}$$

and

$$\mu e'e\mu' = N\mu\mu'. \tag{3.3.6}$$

Hence, (3.3.4) can be written as

$$(X - \mu e')(X - \mu e')' = N\bar{x}\bar{x}' + YY' - N\mu\bar{x}' - N\bar{x}\mu' + N\mu\mu'$$
$$= N(\bar{x} - \mu)(\bar{x} - \mu)' + YY'. \tag{3.3.7}$$

Now, substituting from (3.3.5), (3.3.6), and (3.3.7) together with the Jacobian of transformation $J(X \to \sqrt{N}\,\bar{x}, Y) = 1$, in (3.3.3), we get the joint density of $\sqrt{N}\,\bar{x}$ and Y as

$$f(\sqrt{N}\,\bar{\boldsymbol{x}}, Y) = (2\pi)^{-\frac{1}{2}p} \det(\Sigma)^{-\frac{1}{2}} \operatorname{etr}\left\{-\frac{N}{2}\Sigma^{-1}(\bar{\boldsymbol{x}} - \boldsymbol{\mu})(\bar{\boldsymbol{x}} - \boldsymbol{\mu})'\right\}$$

$$(2\pi)^{-\frac{1}{2}(N-1)p} \det(\Sigma)^{-\frac{1}{2}(N-1)} \operatorname{etr}\left(-\frac{1}{2}\Sigma^{-1}YY'\right). \tag{3.3.8}$$

From (3.3.8), it is evident that $\bar{\boldsymbol{x}}$ and Y are independent, $\bar{\boldsymbol{x}} \sim N_p(\boldsymbol{\mu}, \frac{1}{N}\Sigma)$ and $Y \sim N_{p,n}(0, \Sigma \otimes I_n)$. Hence, $YY' \sim W_p(n, \Sigma)$, and since $S = YY'$, which follows from the identity (3.3.7) by substituting $\bar{\boldsymbol{x}}$ for $\boldsymbol{\mu}$. The proof of the theorem is complete. ∎

In the above theorem it has been proved that the sample covariance matrix S, while sampling from a multivariate normal population, has Wishart distribution. In this case $\frac{1}{N}S$ is the maximum likelihood estimator (MLE) of Σ under the assumption that Σ is positive definite. The distribution of S was first derived by Fisher (1915) and Wishart (1928) when Σ is positive definite. Eben (1994) derived the distribution of S, when the inverse of the covariance matrix is a band matrix. Tsai (1995) obtained the MLE of Σ under the assumption $\Sigma \geq I_p$ and has also derived its density. It may also be noted that $(\bar{\boldsymbol{x}}, S)$ form a complete sufficient set of statistics for $(\boldsymbol{\mu}, \Sigma)$ and hence Wishart matrix plays an important role in drawing inferences about the parameters of a multivariate normal distribution. There is a vast literature on this topic and the reader is referred to Roy (1957), Kshirsagar (1972), Eaton (1972), Giri (1977), Srivastava and Khatri (1979), Muirhead (1982), Anderson (1984), and Siotani, Hayakawa and Fujikoshi (1985).

It may also be noted that Ghurye and Olkin (1969) have derived the minimum variance unbiased estimate of Wishart density.

3.3.3. Characteristic Function and Additive Property of Wishart Matrices

THEOREM 3.3.7. *Let $S \sim W_p(n, \Sigma)$, then the characteristic function of S, i.e., the joint characteristic function of $s_{11}, s_{12}, \ldots, s_{pp}$ is*

$$\phi_S(Z) = \det(I_p - 2\iota Z\Sigma)^{-\frac{1}{2}n}, \tag{3.3.9}$$

where $Z = Z'\,(p \times p) = \left(\frac{1}{2}(1 + \delta_{ij})z_{ij}\right)$ and δ_{ij} is the Kronecker's delta.

Proof: The characteristic function of S is

$$\phi_S(Z) = E[\operatorname{etr}(\iota ZS)]$$

$$= \left\{2^{\frac{1}{2}np}\Gamma_p\left(\frac{1}{2}n\right)\det(\Sigma)^{\frac{1}{2}n}\right\}^{-1} \int_{S>0} \operatorname{etr}\left\{-\frac{1}{2}(I_p - 2\iota Z\Sigma)\Sigma^{-1}S\right\}\det(S)^{\frac{1}{2}(n-p-1)}\,dS$$

$$= \left\{2^{\frac{1}{2}np}\Gamma_p\left(\frac{1}{2}n\right)\det(\Sigma)^{\frac{1}{2}n}\right\}^{-1} \det\left(\frac{1}{2}(I_p - 2\iota Z\Sigma)\Sigma^{-1}\right)^{-\frac{1}{2}n}\Gamma_p\left(\frac{1}{2}n\right). \tag{3.3.10}$$

The above equality is obtained by using (1.4.6). Now, simplifying (3.3.10) we get the desired result. ∎

The above result can also be derived by assuming n to be an integer and using the decomposition given in Theorem 3.3.3, *e.g.*, see Anderson (1984).

THEOREM 3.3.8. *Let $S_1, \ldots, S_k$ be independently distributed with $S_j \sim W_p(n_j, \Sigma)$, $j = 1, \ldots, k$. Then, $\sum_{j=1}^{k} S_j \sim W_p(\sum_{j=1}^{k} n_j, \Sigma)$.*

Proof: The characteristic function of $\sum_{j=1}^{k} S_j$ is

$$E\left[\operatorname{etr}\left\{\iota\left(\sum_{j=1}^{k} S_j\right)Z\right\}\right] = \prod_{j=1}^{k} E[\operatorname{etr}(\iota S_j Z)]$$

$$= \prod_{j=1}^{k} \det(I_p - 2\iota Z\Sigma)^{-\frac{1}{2}n_j}$$

$$= \det(I_p - 2\iota Z\Sigma)^{-\frac{1}{2}\sum_{j=1}^{k} n_j}. \quad \blacksquare$$

In the above theorem when the covariance matrices are not equal, the distribution of $\sum_{j=1}^{k} S_j$ is not Wishart. For $k = 2$, the density involves $_1F_1$ function and is given in Problem 3.5. Further, let $S_j \sim W_p(n_j, \Sigma_j)$, $j = 1, \ldots, k+r$, and define $Q_1 = \sum_{j=1}^{k} \lambda_j S_j$ and $Q_2 = \sum_{j=k+1}^{k+r} \lambda_j S_j$ where λ_j's are positive constants. Then the asymptotic distributions of $-\ln\{\det(Q_1)\}$, $-\ln\{\det(Q_1 Q_2^{-1})\}$ and $-\ln\{\det(Q_1(Q_1 + Q_2)^{-1})\}$ have been derived by Gupta, Chattopadhyay, and Krishnaiah (1975).

3.3.4. Marginal and Conditional Distributions

THEOREM 3.3.9. *Let $S \sim W_p(n, \Sigma)$ and partition S and Σ as*

$$S = \begin{pmatrix} S_{11} & S_{12} \\ S_{21} & S_{22} \end{pmatrix} \begin{matrix} q \\ p-q \end{matrix}, \quad \Sigma = \begin{pmatrix} \Sigma_{11} & \Sigma_{12} \\ \Sigma_{21} & \Sigma_{22} \end{pmatrix} \begin{matrix} q \\ p-q \end{matrix}.$$
$$\quad\quad\; q \quad\; p-q \qquad\qquad\quad q \quad\; p-q$$

Let $S_{11\cdot2} = S_{11} - S_{12}S_{22}^{-1}S_{21}$, $\Sigma_{11\cdot2} = \Sigma_{11} - \Sigma_{12}\Sigma_{22}^{-1}\Sigma_{21}$, then
   (i) $S_{22} \sim W_{p-q}(n, \Sigma_{22})$,
   (ii) $S_{11\cdot2} \sim W_q(n - p + q, \Sigma_{11\cdot2})$,
   (iii) $S_{11\cdot2}$ and (S_{12}, S_{22}) are independent,
   (iv) $S_{12}|S_{22} \sim N_{q,p-q}(\Sigma_{12}\Sigma_{22}^{-1}S_{22}, \Sigma_{11\cdot2} \otimes S_{22})$.

Proof: Let $\Sigma^{-1} = \begin{pmatrix} \Sigma^{11} & \Sigma^{12} \\ \Sigma^{21} & \Sigma^{22} \end{pmatrix}$, Σ^{11} $(q \times q)$. Then $\Sigma^{11} = \Sigma_{11\cdot2}^{-1}$, $\Sigma^{22} = \Sigma_{22\cdot1}^{-1}$, $\Sigma^{12} = -\Sigma_{11}^{-1}\Sigma_{12}\Sigma_{22\cdot1}^{-1}$ and $\Sigma^{21} = -\Sigma_{22}^{-1}\Sigma_{21}\Sigma_{11\cdot2}^{-1}$. Also, note that

$$\operatorname{tr}(\Sigma^{-1}S) = \operatorname{tr}(\Sigma^{11}S_{11} + \Sigma^{12}S_{21}) + \operatorname{tr}(\Sigma^{21}S_{12} + \Sigma^{22}S_{22})$$

$$= \operatorname{tr}[\Sigma^{11}(S_{11} - S_{12}S_{22}^{-1}S_{21} + S_{12}S_{22}^{-1}S_{21})] + \operatorname{tr}(\Sigma^{12}S_{21})$$

$$\quad + \operatorname{tr}(\Sigma^{21}S_{12}) + \operatorname{tr}[(\Sigma^{22} - \Sigma^{21}(\Sigma^{11})^{-1}\Sigma^{12} + \Sigma^{21}(\Sigma^{11})^{-1}\Sigma^{12})S_{22}]$$

$$= \operatorname{tr}(\Sigma^{11}S_{11\cdot2}) + \operatorname{tr}(\Sigma^{22\cdot1}S_{22}) + \operatorname{tr}(\Sigma^{11}S_{12}S_{22}^{-1}S_{21}) + \operatorname{tr}(\Sigma^{12}S_{21})$$

$$\quad + \operatorname{tr}(\Sigma^{21}S_{12}) + \operatorname{tr}[\Sigma^{21}(\Sigma^{11})^{-1}\Sigma^{12}S_{22}]$$

$$= \text{tr}(\Sigma^{11}S_{11\cdot2}) + \text{tr}(\Sigma^{22\cdot1}S_{22})$$
$$+ \text{tr}[\Sigma^{11}(S_{12} + (\Sigma^{11})^{-1}\Sigma^{12}S_{22})S_{22}^{-1}(S_{12} + (\Sigma^{11})^{-1}\Sigma^{12}S_{22})']$$
$$= \text{tr}(\Sigma_{11\cdot2}^{-1}S_{11\cdot2}) + \text{tr}(\Sigma_{22}^{-1}S_{22})$$
$$+ \text{tr}[\Sigma_{11\cdot2}^{-1}(S_{12} - \Sigma_{12}\Sigma_{22}^{-1}S_{22})S_{22}^{-1}(S_{12} - \Sigma_{12}\Sigma_{22}^{-1}S_{22})']$$

and $\det(S) = \det(S_{11\cdot2})\det(S_{22})$, $\det(\Sigma) = \det(\Sigma_{11\cdot2})\det(\Sigma_{22})$. Now, transforming $S_{11\cdot2} = S_{11} - S_{12}S_{22}^{-1}S_{21}$ with Jacobian $J(S_{11} \to S_{11\cdot2}) = 1$, the joint density of S_{12}, S_{22}, and $S_{11\cdot2}$ obtained from the density of S, can be written as

$$f(S_{12}, S_{22}, S_{11\cdot2}) = f_1(S_{11\cdot2})f_2(S_{12}, S_{22}), \qquad (3.3.11)$$

where

$$f_1(S_{11\cdot2}) = \left\{ 2^{\frac{1}{2}(n-p+q)q}\Gamma_q\left[\frac{1}{2}(n-p+q)\right]\det(\Sigma_{11\cdot2})^{\frac{1}{2}(n-p+q)}\right\}^{-1}$$

$$\det(S_{11\cdot2})^{\frac{1}{2}(n-p+q)-\frac{1}{2}(q+1)}\,\text{etr}\left(-\frac{1}{2}\Sigma_{11\cdot2}^{-1}S_{11\cdot2}\right) \qquad (3.3.12)$$

and

$$f_2(S_{12}, S_{22}) = \left[\left\{2^{\frac{1}{2}n(p-q)}\det(\Sigma_{22})^{\frac{1}{2}n}\Gamma_{p-q}\left(\frac{1}{2}n\right)\right\}^{-1}\det(S_{22})^{\frac{1}{2}(n-p+q-1)}\,\text{etr}\left(-\frac{1}{2}\Sigma_{22}^{-1}S_{22}\right)\right]$$

$$\left[(2\pi)^{-\frac{1}{2}q(p-q)}\det(\Sigma_{11\cdot2})^{-\frac{1}{2}(p-q)}\det(S_{22})^{-\frac{1}{2}q}\right.$$

$$\left.\text{etr}\left\{-\frac{1}{2}\Sigma_{11\cdot2}^{-1}(S_{12} - \Sigma_{12}\Sigma_{22}^{-1}S_{22})S_{22}^{-1}(S_{12} - \Sigma_{12}\Sigma_{22}^{-1}S_{22})'\right\}\right]. \qquad (3.3.13)$$

From (3.3.11), it follows that $S_{11\cdot2}$ and (S_{12}, S_{22}) are independent, and from (3.3.12), we have $S_{11\cdot2} \sim W_q(n-p+q, \Sigma_{11\cdot2})$. Further, from equation (3.3.13), results (i) and (iv) are easily established. ∎

The density of S has been used to prove the above theorem. However, an alternative proof assuming n to be an integer can also be given using the decomposition given in Theorem 3.3.3, $e.g.$, see Srivastava and Khatri (1979).

THEOREM 3.3.10. Let $S = (S_{ij})$ and $\Sigma = (\Sigma_{ij})$, where $S_{ij}(p_i \times p_j)$, and $\Sigma_{ij}(p_i \times p_j)$, $i, j = 1, \ldots, k$, $p_1 + \cdots + p_k = p$. If $S \sim W_p(n, \Sigma)$, then $S_{ii} \sim W_{p_i}(n, \Sigma_{ii})$, $i = 1, \ldots, k$. Moreover, if $\Sigma_{ij} = 0$, $i \neq j$, then they are independent.

Proof: Assuming n is an integer, from Theorem 3.3.3, $S = XX'$ where $X \sim N_{p,n}(0, \Sigma \otimes I_n)$. Partition X as

$$X = \begin{pmatrix} X_1 \\ X_2 \\ \vdots \\ X_k \end{pmatrix},$$

where $X_i(p_i \times n) \sim N_{p_i,n}(0, \Sigma_{ii} \otimes I_n)$ (see Theorem 2.3.12) . Now,

$$S = XX' = \begin{pmatrix} X_1 X_1' & X_1 X_2' & \cdots & X_1 X_k' \\ \vdots & & & \\ X_k X_1' & X_k X_2' & \cdots & X_k X_k' \end{pmatrix}$$

and $S_{ii} = X_i X_i' \sim W_{p_i}(n, \Sigma_{ii})$, $i = 1, \ldots, k$. Further, if $\Sigma_{ij} = 0$, $i \neq j$, then X_j's are independent and hence, S_{ii}'s are independent. ∎

COROLLARY 3.3.10.1. *Let $S = (s_{ij}) \sim W_p(n, I_p)$. Then, $s_{ii} \sim \chi_n^2$, $i = 1, \ldots, p$, and they are independent.*

Proof: Take $p_i = 1$, $i = 1, \ldots, p$ in the above theorem. ∎

3.3.5.   Distribution of ASA' and $(AS^{-1}A')^{-1}$

THEOREM 3.3.11. *Let $S \sim W_p(n, \Sigma)$. Then, for $A\,(q \times p)$, with $\text{rank}(A) = q \leq p$, $ASA' \sim W_q(n, A\Sigma A')$.*

Proof: The characteristic function of ASA' is

$$
\begin{aligned}
\phi_{ASA'}(Z) &= E[\text{etr}(\iota ASA'Z)], \ Z\,(q \times q) = Z' \\
&= E[\text{etr}(\iota A'ZAS)] \\
&= \det(I_p - 2\iota A'ZA\Sigma)^{-\frac{1}{2}n} \\
&= \det(I_q - 2\iota ZA\Sigma A')^{-\frac{1}{2}n}.
\end{aligned}
$$

The proof is complete by observing that $\det(I_q - 2\iota ZA\Sigma A')^{-\frac{1}{2}n}$ is the characteristic function of a random matrix distributed as $W_q(n, A\Sigma A')$. ∎

COROLLARY 3.3.11.1. *Let $S \sim W_p(n, \Sigma)$. Then, $\frac{a'Sa}{a'\Sigma a} \sim \chi_n^2$ where $a\,(p \times 1) \neq 0$.*

Proof: In Theorem 3.3.11, substitute $q = 1$. ∎

Mitra (1969) has given a counter example to show that the converse of Corollary 3.3.11.1 is not true in general. However, if $\frac{a'Sa}{a'\Sigma a} \sim \chi_n^2$, for all nonnull $a \in \mathbb{R}^p$ and S can be written as $S = YAY'$, where column vectors of $Y\,(p \times n)$ are independently distributed as normal and $A\,(n \times n)$ is a symmetric nonrandom matrix of full rank, then $S \sim W_p(n, \Sigma)$, see Rao (1973), p. 535.

THEOREM 3.3.12. *Let $S \sim W_p(n, \Sigma)$ and $y\,(p \times 1)$ be a random vector distributed independently of S, and $P(y \neq 0) = 1$. Then, $\frac{y'Sy}{y'\Sigma y} \sim \chi_n^2$ and is independent of y.*

Proof: In Theorem 3.3.11, take $q = 1$ and $A = y'$. Then, the conditional distribution of $\frac{y'Sy}{y'\Sigma y}$ given y is χ_n^2, which is also the unconditional distribution. ∎

COROLLARY 3.3.12.1. *Let* $x_1, \ldots, x_N$ *be a random sample from* $N_p(\mu, \Sigma)$, $\bar{x} = \frac{1}{N} \sum_{i=1}^{N} x_i$ *and* $S = \sum_{i=1}^{N} (x_i - \bar{x})(x_i - \bar{x})'$. *Then,* $\frac{\bar{x}'S\bar{x}}{\bar{x}'\Sigma\bar{x}} \sim \chi_n^2$, $n = N - 1$, *and is independent of* $\bar{x}$.

Proof: From Theorem 3.3.6, $S \sim W_p(n, \Sigma)$ and is independent of $\bar{x}$. The result follows from above theorem. ∎

THEOREM 3.3.13. *Let* $S \sim W_p(n, \Sigma)$ *and* $A\,(q \times p)$ *be a matrix of rank* $q \leq p$. *Then,*

$$(AS^{-1}A')^{-1} \sim W_q(n - p + q, (A\Sigma^{-1}A')^{-1}).$$

Proof: Let $B = A\Sigma^{-\frac{1}{2}}$ and $\Lambda = \Sigma^{-\frac{1}{2}}S\Sigma^{-\frac{1}{2}}$ where $\Sigma^{\frac{1}{2}}$ is the symmetric positive definite square root of Σ, then $\Lambda \sim W_p(n, I_p)$ and

$$(AS^{-1}A')^{-1} = (A\Sigma^{-\frac{1}{2}}\Sigma^{\frac{1}{2}}S^{-1}\Sigma^{\frac{1}{2}}\Sigma^{-\frac{1}{2}}A')^{-1}$$
$$= (B\Lambda^{-1}B')^{-1}.$$

So, we need to prove that $(B\Lambda^{-1}B')^{-1} \sim W_q(n-p+q, (BB')^{-1})$, since $BB' = A\Sigma^{-1}A'$.

Let $B = C\,(\,I_q \quad 0\,)\,H$, where $C\,(q \times q)$ is of full rank and $HH' = H'H = I_p$. Now,

$$(B\Lambda^{-1}B')^{-1} = \left[C\,(\,I_q \quad 0\,)\,H\Lambda^{-1}H' \begin{pmatrix} I_q \\ 0 \end{pmatrix} C' \right]^{-1}$$

$$= (C^{-1})' \left[(\,I_q \quad 0\,)\,V^{-1} \begin{pmatrix} I_q \\ 0 \end{pmatrix} \right]^{-1} C^{-1}$$

$$= (C^{-1})'(V^{11})^{-1}C^{-1},$$

where $V = H\Lambda H' \sim W_p(n, I_p)$ and $V^{11}\,(q \times q) = (V_{11} - V_{12}V_{22}^{-1}V_{21})^{-1} = V_{11\cdot2}^{-1}$, where $V = \begin{pmatrix} V_{11} & V_{12} \\ V_{21} & V_{22} \end{pmatrix}$, $V_{11}\,(q \times q)$. From Theorem 3.3.9, $V_{11\cdot2} \sim W_q(n - p + q, I_q)$. Hence $(C^{-1})'(V^{11})^{-1}C^{-1} \sim W_q(n - p + q, (CC')^{-1})$. But, $CC' = BB' = A\Sigma^{-1}A'$. This completes the proof. ∎

COROLLARY 3.3.13.1. *Let* $S \sim W_p(n, \Sigma)$ *and* $a \in \mathbb{R}^p$, $a \neq 0$. *Then* $\frac{a'\Sigma^{-1}a}{a'S^{-1}a} \sim \chi_{n-p+1}^2$.

Proof: Take $q = 1$ in the above theorem. ∎

THEOREM 3.3.14. *Let* $S \sim W_p(n, \Sigma)$ *and* $y\,(p \times 1)$ *be a random vector distributed independently of* S, *and* $P(y \neq 0) = 1$. *Then,* $\frac{y'\Sigma^{-1}y}{y'S^{-1}y} \sim \chi_{n-p+1}^2$ *and is independent of* y.

Proof: In Theorem 3.3.13 take $q = 1$ and $A = y'$. Then, the conditional distribution of $\frac{y'\Sigma^{-1}y}{y'S^{-1}y}$ given y is χ_{n-p+1}^2, which is also the unconditional distribution. ∎

COROLLARY 3.3.14.1. *Let $\bar{x}$ and S be defined as in Theorem 3.3.6, then $\frac{\bar{x}'\Sigma^{-1}\bar{x}}{\bar{x}'S^{-1}\bar{x}}$ $\sim \chi^2_{n-p+1}$ and is independent of $\bar{x}$.*

Proof: From Theorem 3.3.6, $S \sim W_p(n, \Sigma)$ and is independent of $\bar{x}$. The result follows from the above theorem. ∎

It may be noted that the distribution of Hotelling's T^2 can be derived from the above corollary as in Muirhead (1982), p. 98.

3.3.6.   Expected Values

In this section, we give expected values of the elements of S and some of its scalar and matrix valued functions.

THEOREM 3.3.15. *Let $S = (s_{ij}) \sim W_p(n, \Sigma)$, then*
  (i) $E(s_{ij}) = n\sigma_{ij}$,
     $\mathrm{cov}(s_{ij}, s_{k\ell}) = n(\sigma_{ik}\sigma_{j\ell} + \sigma_{i\ell}\sigma_{jk})$,
  (ii) $E(SAS) = n\Sigma A'\Sigma + n\,\mathrm{tr}(\Sigma A)\Sigma + n^2\Sigma A\Sigma$,
  (iii) $E(\mathrm{tr}(A\Sigma)S) = n\Sigma A\Sigma + n\Sigma A'\Sigma + n^2\,\mathrm{tr}(A\Sigma)\Sigma$,
  (iv) $E(\mathrm{tr}(AS)\,\mathrm{tr}(BS)) = n\,\mathrm{tr}(A\Sigma B\Sigma) + n\,\mathrm{tr}(A'\Sigma B\Sigma) + n^2\,\mathrm{tr}(A\Sigma)\,\mathrm{tr}(B\Sigma)$,
where $\Sigma = (\sigma_{ij})$ and $A\,(p \times p)$ and $B\,(p \times p)$ are constant matrices.

Proof: (i) Assuming n to be an integer and using Theorem 3.3.3, we can write $S = YY'$, $s_{ij} = \sum_{r=1}^{n} y_{ir}y_{jr}$, where $Y\,(p \times n) = (y_{ij}) \sim N_{p,n}(0, \Sigma \otimes I_n)$. Hence, using Corollary 2.3.3.1,

$$E(s_{ij}) = \sum_{r=1}^{n} E(y_{ir}y_{jr})$$

$$= \sum_{r=1}^{n} \sigma_{ij}$$

$$= n\sigma_{ij} \qquad\qquad (3.3.14)$$

and

$$E(s_{ij}s_{k\ell}) = \sum_{r=1}^{n}\sum_{t=1}^{n} E(y_{ir}y_{jr}y_{kt}y_{\ell t})$$

$$= \sum_{r=1}^{n} E(y_{ir}y_{jr}y_{kr}y_{\ell r}) + \sum_{\substack{r=1 \\ r \neq t}}^{n}\sum_{t=1}^{n} E(y_{ir}y_{jr}y_{kt}y_{\ell t})$$

$$= \sum_{r=1}^{n}(\sigma_{ik}\sigma_{j\ell} + \sigma_{i\ell}\sigma_{jk} + \sigma_{ij}s_{k\ell}) + \sum_{\substack{r=1 \\ r \neq t}}^{n}\sum_{t=1}^{n} \sigma_{ij}\sigma_{k\ell}$$

$$= n(\sigma_{ik}\sigma_{j\ell} + \sigma_{i\ell}\sigma_{jk} + \sigma_{ij}\sigma_{k\ell}) + n(n-1)\sigma_{ij}\sigma_{k\ell}. \qquad (3.3.15)$$

From (3.3.14) and (3.3.15), we get

$$\text{cov}(s_{ij}, s_{k\ell}) = n(\sigma_{ik}\sigma_{j\ell} + \sigma_{i\ell}\sigma_{jk}).$$

(ii) The $(i,j)^{\text{th}}$ element of SAS is $\sum_{t=1}^{p}\sum_{k=1}^{p} s_{ik}a_{kt}s_{tj}$ and hence,

$$
\begin{aligned}
E(SAS) &= E\left(\left(\sum_{t=1}^{p}\sum_{k=1}^{p} s_{ik}a_{kt}s_{tj}\right)\right) \\
&= \left(\sum_{t=1}^{p}\sum_{k=1}^{p} a_{kt}E(s_{ik}s_{tj})\right) \\
&= \left(n\sum_{t=1}^{p}\sum_{k=1}^{p} a_{kt}(\sigma_{it}\sigma_{jk} + \sigma_{ij}\sigma_{kt} + n\sigma_{ik}\sigma_{tj})\right) \\
&= n\Sigma A'\Sigma + n\,\text{tr}(A\Sigma)\Sigma + n^2\Sigma A\Sigma.
\end{aligned}
$$

The proofs of the other two expected values are similar. ∎

By differentiating the moment generating function of $S \sim W_p(n, \Sigma)$, de Waal and D. G. Nel (1973) have derived the following results:

(i) $E(S^2) = n\{(n+1)\Sigma + (\text{tr}\,\Sigma)I_p\}\Sigma$

(ii) $E(S^3) = n\{(n^2 + 3n + 4)\Sigma^2 + 2(n+1)(\text{tr}\,\Sigma)\Sigma + (n+1)(\text{tr}\,\Sigma^2)I_p + (\text{tr}\,\Sigma)^2 I_p\}\Sigma$

and

(iii) $E(S^4) = n\{(n^3 + 6n^2 + 21n + 20)\Sigma^3 + (3n^2 + 10n + 12)(\text{tr}\,\Sigma)\Sigma^2$
$\qquad\qquad + (2n^2 + 5n + 5)(\text{tr}\,\Sigma^2)\Sigma + 3(n+1)(\text{tr}\,\Sigma)^2\Sigma$
$\qquad\qquad + (n^2 + 2n + 4)(\text{tr}\,\Sigma^3)I_p + 3(n+1)(\text{tr}\,\Sigma)(\text{tr}\,\Sigma^2)I_p + (\text{tr}\,\Sigma)^3 I_p\}\Sigma.$

The result (i) can also be obtained from Theorem 3.3.15(ii) by substituting $A = I_p$. For an alternative proof of Theorem 3.3.15(ii), see Styan (1979). Haff (1979), using an identity involving Wishart matrix and assuming A is positive semidefinite, has also obtained expression for $E(SAS)$. Wishart (1928) derived the central moments up to fourth order of the elements of S. Haff has also derived expected values similar to the above theorem for S^{-1} as given in the next theorem.

THEOREM 3.3.16. *Let $S \sim W_p(n, \Sigma)$, then*

(i) $E(s^{ij}) = \dfrac{\sigma^{ij}}{n - p - 1}, \; n - p - 1 > 0$

(ii) $\text{cov}(s^{ij}, s^{k\ell}) = \dfrac{2(n - p - 1)^{-1}\sigma^{ij}\sigma^{k\ell} + \sigma^{ik}\sigma^{j\ell} + \sigma^{i\ell}\sigma^{kj}}{(n - p)(n - p - 1)(n - p - 3)}, \; n - p - 3 > 0$

(iii) $E(S^{-1}AS^{-1}) = \dfrac{\text{tr}(\Sigma^{-1}A)\Sigma^{-1}}{(n - p)(n - p - 1)(n - p - 3)} + \dfrac{\Sigma^{-1}A\Sigma^{-1}}{(n - p)(n - p - 3)},$

where $S^{-1} = (s^{ij})$, $\Sigma^{-1} = (\sigma^{ij})$ and $A \, (p \times p)$ is a constant positive semidefinite matrix.

The following expected values were derived by von Rosen (1988a).

THEOREM 3.3.17. *Let* $S \sim W_p(n, \Sigma)$, *then*

(i) $E(S^{-3}) = (c_3c_1 + c_3c_2 + c_4c_1 + 5c_4c_2)\Sigma^{-3} + (2c_3c_2 + c_4c_1 + c_4c_2)(\operatorname{tr}\Sigma^{-1})\Sigma^{-2}$
$$- (c_3c_2 + c_4c_2)(\operatorname{tr}\Sigma^{-2})\Sigma^{-1} - c_4c_2(\operatorname{tr}\Sigma^{-1})^2\Sigma^{-1}, \ n - p - 5 > 0,$$

(ii) $E((\operatorname{tr}S^{-1})S^{-1}) = c_1(\operatorname{tr}\Sigma^{-1})\Sigma^{-1} + 2c_2\Sigma^{-2}, \ n - p - 3 > 0,$

and

(iii) $E((\operatorname{tr}S^{-1})S) = \dfrac{n}{(n-p-1)}(\operatorname{tr}\Sigma^{-1})\Sigma - \dfrac{2}{(n-p-1)}I_p, \ n - p - 1 > 0,$

where $c_1 = (n - p - 2)c_2$, $c_2 = \{(n - p)(n - p - 1)(n - p - 3)\}^{-1}$, $c_3 = (n - p - 3)$ $\{(n - p - 5)(n - p + 1)\}^{-1}$ *and* $c_4 = 2\{(n - p - 5)(n - p + 1)\}^{-1}$.

Marx (1981) obtained the following expected values.

THEOREM 3.3.18. *Let* $S \sim W_p(n, \Sigma)$, *then*

(i) $E(S^{-1}AS^{-1}) = c_1\Sigma^{-1}A\Sigma^{-1} + c_2[\Sigma^{-1}A'\Sigma^{-1} + \operatorname{tr}(A\Sigma^{-1})\Sigma^{-1}]$

(ii) $E(\operatorname{tr}(AS^{-1})S^{-1}) = c_1\operatorname{tr}(A\Sigma^{-1})\Sigma^{-1} + c_2[\Sigma^{-1}A'\Sigma^{-1} + \Sigma^{-1}A\Sigma^{-1}]$

(iii) $E(\operatorname{tr}(AS^{-1})\operatorname{tr}(BS^{-1})) = c_1\operatorname{tr}(A\Sigma^{-1})\operatorname{tr}(B\Sigma^{-1}) + c_2[\operatorname{tr}(B\Sigma^{-1}A\Sigma^{-1})$
$$+ \operatorname{tr}(B\Sigma^{-1}A'\Sigma^{-1})],$$

where c_1 *and* c_2 *are defined in Theorem 3.3.17 and* $A\,(p \times p)$, $B\,(p \times p)$ *are constant matrices.*

Proof: (i) Let $S^{-1} = (s^{ij})$ and $A = (a_{ij})$. Then the $(i,j)^{\text{th}}$ element of $S^{-1}AS^{-1}$ is $\sum_{t=1}^{p}\sum_{k=1}^{p}s^{it}a_{tk}s^{kj}$, and

$$E(S^{-1}AS^{-1}) = E\left(\left(\sum_{t=1}^{p}\sum_{k=1}^{p}s^{it}a_{tk}s^{kj}\right)\right)$$

$$= \left(\sum_{t=1}^{p}\sum_{k=1}^{p}a_{tk}(s^{it}s^{kj})\right)$$

$$= \left(\sum_{t=1}^{p}\sum_{k=1}^{p}a_{tk}\{\operatorname{cov}(s^{it}, s^{kj}) + E(s^{it})E(s^{kj})\}\right).$$

By substituting for $\operatorname{cov}(s^{it}, s^{kj})$ and $E(s^{ij})$ from Theorem 3.3.16, we obtain

$E(S^{-1}AS^{-1})$

$$= \left(\sum_{t=1}^{p}\sum_{k=1}^{p}a_{tk}\{c_2(2(n-p-1)^{-1}\sigma^{it}\sigma^{kj} + \sigma^{ik}s^{tj} + \sigma^{ij}\sigma^{kt}) + (n-p-1)^{-2}\sigma^{it}\sigma^{kj}\}\right)$$

$$= c_2[2(n-p-1)^{-1}\Sigma^{-1}A\Sigma^{-1} + \Sigma^{-1}A'\Sigma^{-1} + \operatorname{tr}(A\Sigma^{-1})\Sigma^{-1}] + (n-p-1)^{-2}\Sigma^{-1}A\Sigma^{-1}$$

$$= c_1\Sigma^{-1}A\Sigma^{-1} + c_2[\Sigma^{-1}A'\Sigma^{-1} + \operatorname{tr}(A\Sigma^{-1})\Sigma^{-1}].$$

(ii) The $(i,j)^{\text{th}}$ element of $\operatorname{tr}(AS^{-1})S^{-1}$ is $s^{ij}\sum_{\ell=1}^{p}\sum_{k=1}^{p}a_{k\ell}s^{\ell k}$. Thus,

$E(\mathrm{tr}(AS^{-1})S^{-1})$

$$= E\left(\left(s^{ij}\sum_{\ell=1}^{p}\sum_{k=1}^{p}a_{k\ell}s^{\ell k}\right)\right)$$

$$= \left(\sum_{\ell=1}^{p}\sum_{k=1}^{p}a_{k\ell}E(s^{ij}s^{\ell k})\right)$$

$$= \left(\sum_{\ell=1}^{p}\sum_{k=1}^{p}a_{k\ell}\left\{c_2(2(n-p-1)^{-1}\sigma^{ij}\sigma^{k\ell}+\sigma^{ik}\sigma^{j\ell}+\sigma^{i\ell}\sigma^{kj})+(n-p-1)^{-2}\sigma^{ij}\sigma^{\ell k}\right\}\right)$$

$$= c_2[2(n-p-1)^{-1}\mathrm{tr}(A\Sigma^{-1})\Sigma^{-1}+\Sigma^{-1}A\Sigma^{-1}+\Sigma^{-1}A'\Sigma^{-1}]$$

$$+(n-p-1)^{-2}\mathrm{tr}(A\Sigma^{-1})\Sigma^{-1}$$

$$= c_1\mathrm{tr}(A\Sigma^{-1})\Sigma^{-1}+c_2[\Sigma^{-1}A\Sigma^{-1}+\Sigma^{-1}A'\Sigma^{-1}].$$

(iii) Pre-multiplying the result (ii) by the constant matrix B and taking the trace, the desired result follows. ∎

Styan (1989), using a result from Olkin and Rubin (1962), has also proved Theorem 3.3.18(i). He has also derived expression for $E(SAS^{-1})$, where $S \sim W_p(n,\Sigma)$ and A is a square nonrandom matrix not necessarily symmetric, as

$$E(SAS^{-1}) = \frac{1}{n-p-1}\left[n\Sigma A\Sigma^{-1}-A'-\mathrm{tr}(A)I_p\right].$$

THEOREM 3.3.19. *Let $S \sim W_p(n,\Sigma)$, then*

(i) $E(C_\kappa(S)) = 2^k\left(\dfrac{1}{2}n\right)_\kappa C_\kappa(\Sigma)$

and

(ii) $E(C_\kappa(S^{-1})) = 2^{-k}\dfrac{\Gamma_p\left(\frac{1}{2}n,-\kappa\right)}{\Gamma_p\left(\frac{1}{2}n\right)}C_\kappa(\Sigma^{-1}),\ \dfrac{1}{2}n > \dfrac{1}{2}(p-1)+k_1.$

Proof: (i) We have

$$E(C_\kappa(S)) = \left\{2^{\frac{1}{2}np}\Gamma_p\left(\frac{1}{2}n\right)\det(\Sigma)^{\frac{1}{2}n}\right\}^{-1}\int_{S>0}C_\kappa(S)\det(S)^{\frac{1}{2}(n-p-1)}\mathrm{etr}\left(-\frac{1}{2}\Sigma^{-1}S\right)dS$$

$$= \left\{2^{\frac{1}{2}np}\Gamma_p\left(\frac{1}{2}n\right)\det(\Sigma)^{\frac{1}{2}n}\right\}^{-1}\Gamma_p\left(\frac{1}{2}n,\kappa\right)2^{\frac{1}{2}np}\det(\Sigma^{-1})^{-\frac{1}{2}n}C_\kappa(2\Sigma)$$

$$= 2^k\left(\frac{1}{2}n\right)_\kappa C_\kappa(\Sigma),$$

where we have used the Lemma 1.5.2.

(ii) The proof is similar to the above, using Lemma 1.5.2. ∎

From (i) above we have

$$E[(\operatorname{tr} S)^k] = \sum_\kappa E(C_\kappa(S))$$

$$= 2^k \sum_\kappa \left(\frac{1}{2}n\right)_\kappa C_\kappa(\Sigma).$$

Since $\Sigma^{-\frac{1}{2}} S \Sigma^{-\frac{1}{2}} \sim W_p(n, I_p)$, it follows that

$$E[\{\operatorname{tr}(\Sigma^{-1}S)\}^k] = 2^k \sum_\kappa \left(\frac{1}{2}n\right)_\kappa C_\kappa(I_p)$$

$$= 2^k \left(\frac{1}{2}np\right)_k$$

where the last step has been obtained by using (see Subrahmaniam, 1976),

$$\sum_\kappa (n)_\kappa C_\kappa(I_p) = (np)_k.$$

This result has also been obtained by Muirhead (1986) who also derived the following results.

$$E[\{\operatorname{tr}(\Sigma^{-1}S)\}^{-k}] = \left(-\frac{1}{2}\right)^k \left(-\frac{1}{2}np + 1\right)_k, \ 2k < np,$$

$$E[\{\operatorname{tr}(\Sigma^{-1}S)\}^k \operatorname{tr}(S)] = n2^k \left(\frac{1}{2}np + 1\right)_k (\operatorname{tr} \Sigma),$$

$$E[\{\operatorname{tr}(\Sigma^{-1}S)\}^k \operatorname{tr}(S^{-1})] = (n - p - 1)^{-1} 2^k \left(\frac{1}{2}np - 1\right)_k (\operatorname{tr} \Sigma^{-1}), \ n > p + 1,$$

$$E[\{\operatorname{tr}(\Sigma^{-1}S)\}^k \operatorname{tr}(\Sigma S^{-1})] = (n - p - 1)^{-1} p2^k \left(\frac{1}{2}np - 1\right)_k, \ n > p + 1,$$

$$E[\{\operatorname{tr}(\Sigma^{-1}S)\}^k \det(S)^h] = 2^{ph+k} \left(\frac{1}{2}np + ph\right)_k \frac{\Gamma_p(\frac{1}{2}n + h)}{\Gamma_p(\frac{1}{2}n)} \det(\Sigma)^h,$$

$$E[\{\operatorname{tr}(\Sigma^{-1}S)\}^r C_\kappa(SB)] = 2^{k+r} \left(\frac{1}{2}np + k\right)_r \left(\frac{1}{2}n\right)_\kappa C_\kappa(B\Sigma),$$

$$E[\{\operatorname{tr}(\Sigma^{-1}S)\}^{-r} C_\kappa(SB)] = \frac{(-1)^r 2^{k-r}}{(-\frac{1}{2}np - k + 1)_r} \left(\frac{1}{2}n\right)_\kappa C_\kappa(B\Sigma), \ r < \frac{1}{2}np + k,$$

where $B \ (p \times p)$ is a constant matrix.

THEOREM 3.3.20. *Let $S = TT' \sim W_p(n, I_p)$, where $T = (t_{ij})$ is a lower triangular matrix with positive diagonal elements, then*

$$E(T'T)^{-1} = B$$

where $B = \operatorname{diag}(b_1, \ldots, b_p)$ with

$$b_1 = \frac{1}{n - 2}$$

and

$$b_j = \frac{(n-1)}{(n-j-1)(n-j)}, \quad j = 2, \ldots, p.$$

Proof: From Theorem 3.3.4, it is known that t_{ij}'s $(1 \le j \le i \le p)$ are independent, with $t_{ij} \sim N(0,1)$, $1 \le j < i \le p$ and $t_{ii}^2 \sim \chi_{n-i+1}^2$, $i = 1, \ldots, p$. Let

$$a_i = E\left(\frac{1}{t_{ii}^2}\right) = \frac{1}{n-i-1}, \quad i = 1, \ldots, p. \tag{3.3.16}$$

For any diagonal matrix D, with diagonal elements ± 1, DTD and T have the same distribution and therefore,

$$B = E(T'T)^{-1} = E[(DTD)'(DTD)]^{-1}$$
$$= DBD,$$

which implies that B is a diagonal matrix. Writing

$$T = \begin{pmatrix} T_{11} & 0 \\ T_{21} & T_{22} \end{pmatrix}$$

we get

$$T^{-1} = \begin{pmatrix} T_{11}^{-1} & 0 \\ -T_{22}^{-1}T_{21}T_{11}^{-1} & T_{22}^{-1} \end{pmatrix}$$

and

$$(T'T)^{-1} = \begin{pmatrix} T_{11}^{-1}(T_{11}')^{-1} & T_{11}^{-1}R_{21}' \\ R_{21}(T_{11}')^{-1} & R_{21}R_{21}' + T_{22}^{-1}(T_{22}')^{-1} \end{pmatrix},$$

where $R_{21} = -T_{22}^{-1}T_{21}T_{11}^{-1}$. Now, taking expectation we get

$$E(T'T)^{-1} = B = \begin{pmatrix} E(T_{11}'T_{11})^{-1} & 0 \\ 0 & E(R_{21}R_{21}' + (T_{22}'T_{22})^{-1}) \end{pmatrix}. \tag{3.3.17}$$

Letting T_{11} be $(p-1) \times (p-1)$, $T_{22} = t_{pp}$, $T_{21}(1 \times (p-1)) = t_{21}'$ and using the independence of T_{11}, t_{pp} and t_{21}', we get

$$b_p = E\left[\frac{1}{t_{pp}^2}\{1 + t_{21}'(T_{11}'T_{11})^{-1}t_{21}\}\right]$$

$$= E\left[\frac{1}{t_{pp}^2}\right]E\{1 + t_{21}'(T_{11}'T_{11})^{-1}t_{21}\}. \tag{3.3.18}$$

Since $t_{21} \sim N_{p-1}(0, I_{p-1})$, and from (3.3.17), $E(T_{11}'T_{11})^{-1} = \text{diag}(b_1, \ldots, b_{p-1})$, we have

$$E[t'_{21}(T'_{11}T_{11})^{-1}t_{21}] = E[\text{tr}\{t_{21}t'_{21}(T'_{11}T_{11})^{-1}\}]$$
$$= \text{tr}[E(t_{21}t'_{21})E(T'_{11}T_{11})^{-1}]$$
$$= \sum_{j=1}^{p-1} b_j. \tag{3.3.19}$$

Now, using (3.3.16) and (3.3.19) in (3.3.18), one obtains

$$b_p = a_p\left(1 + \sum_{j=1}^{p-1} b_j\right).$$

By an inductive process, it is straightforward to show that

$$b_j = a_j\left(1 + \sum_{i=1}^{j-1} b_i\right), j = 2, \ldots, p \tag{3.3.20}$$

and

$$b_1 = a_1.$$

Solving equations (3.3.20), in terms of a_j's, we get

$$b_1 = a_1,$$

$$b_j = a_j\prod_{i=1}^{j-1}(1 + a_i), j = 2, \ldots, p,$$

and using (3.3.16), we finally get

$$b_1 = \frac{1}{n-2},$$

and

$$b_j = \frac{(n-1)}{(n-j-1)(n-j)}, j = 2, \ldots, p. \quad \blacksquare$$

The above result has been derived by Eaton and Olkin (1987). Using this procedure one can derive a similar result for an upper triangular factorization of S as given in the next theorem.

THEOREM 3.3.21. Let $S = TT' \sim W_p(n, I_p)$, where $T = (t_{ij})$ is an upper triangular matrix with positive diagonal elements. Then

$$E(T'T)^{-1} = B$$

where $B = \text{diag}(b_1, \ldots, b_p)$ with

$$b_j = \frac{n-1}{(n-p+j-1)(n-p+j-2)}, j = 1, 2, \ldots, p-1,$$

and

$$b_p = \frac{1}{n-2}.$$

Proof: Similar to the proof of Theorem 3.3.20. ∎

THEOREM 3.3.22. *If $S \sim W_p(n, \Sigma)$, then*

(i) $\dfrac{\det(S)}{\det(\Sigma)} \sim \prod\limits_{i=1}^{p} u_i$, *where u_i's are independent and $u_i \sim \chi^2_{n-i+1}$, $i = 1, \ldots, p$,*

and

(ii) $E[\det(S)^h] = 2^{ph} \det(\Sigma)^h \prod\limits_{i=1}^{p} \dfrac{\Gamma\left[\frac{1}{2}(n-i+1)+h\right]}{\Gamma\left[\frac{1}{2}(n-i+1)\right]}$, $\operatorname{Re}(h) > -\dfrac{1}{2}n + \dfrac{1}{2}(p-1)$.

$$(3.3.21)$$

Proof: (i) Let $V = \Sigma^{-\frac{1}{2}} S \Sigma^{-\frac{1}{2}}$, then from Corollary 3.3.1.1, $V \sim W_p(n, I_p)$. Now, from Theorem 3.3.4, V can be written as TT' and

$$\det(V) = \det(S)\det(\Sigma)^{-1} = \prod_{i=1}^{p} u_i, \qquad (3.3.22)$$

where $u_i = t_{ii}^2$ are independently distributed as χ^2_{n-i+1}, $i = 1, \ldots, p$.

(ii) From (3.3.22), we have

$$E[\det(S)^h] = \det(\Sigma)^h E\left(\prod_{i=1}^{p} u_i\right)^h$$

$$= \det(\Sigma)^h \prod_{i=1}^{p} \left\{ 2^h \frac{\Gamma[\frac{1}{2}(n-i+1)+h]}{\Gamma[\frac{1}{2}(n-i+1)]} \right\}. \quad ∎$$

Alternately, $E[\det(S)^h]$ can be evaluated using the density of S as follows.

$$E[\det(S)^h] = \int_{S>0} \det(S)^h \frac{\det(S)^{\frac{1}{2}(n-p-1)} \operatorname{etr}(-\frac{1}{2}\Sigma^{-1}S)}{2^{\frac{1}{2}np} \det(\Sigma)^{\frac{1}{2}n} \Gamma_p(\frac{1}{2}n)} \, dS$$

$$= 2^{ph} \det(\Sigma)^h \frac{\Gamma_p(\frac{1}{2}n+h)}{\Gamma_p(\frac{1}{2}n)}.$$

Substituting $\Gamma_p(\cdot)$ from Theorem 1.4.1 we get (3.3.21).

The statistic $n^{-p} \det(S)$ is known as the *sample generalized variance* (Wilks, 1932). Many test statistics in multivariate statistical analysis are functions of sample generalized variance (*e.g.*, see Anderson, 1984; Gupta and Tang, 1984, 1986a, 1986b, 1987, 1988; Sen Gupta, 1987).

THEOREM 3.3.23. *Let $S \sim W_p(n, \Sigma)$, then the characteristic function of $\mathrm{tr}(S)$ is*

$$\phi_{\mathrm{tr}(S)}(z) = \det(I_p - 2\iota z \Sigma)^{-\frac{1}{2}n}$$

and the kth moment of $\mathrm{tr}(S)$ is

$$E[(\mathrm{tr}\, S)^k] = 2^k \sum_{\kappa} \left(\tfrac{1}{2}n\right)_{\kappa} C_{\kappa}(\Sigma), \; k = 0, 1, 2, \ldots, \ldots .$$

Proof: The characteristic function of $\mathrm{tr}(S)$ is

$$\begin{aligned}
\phi_{\mathrm{tr}(S)}(z) &= E[\exp\{\iota z\, \mathrm{tr}(S)\}] \\
&= E[\exp\{\iota\, \mathrm{tr}(ZS)\}], \; Z = zI_p, \\
&= \det(I_p - 2\iota z \Sigma)^{-\frac{1}{2}n}.
\end{aligned} \tag{3.3.23}$$

The last equality is obtained from the characteristic function of S. Now, expanding (3.3.23), for $\|2z\Sigma\| < 1$,

$$\phi_{\mathrm{tr}(S)}(z) = \sum_{k=0}^{\infty} \sum_{\kappa} \frac{(2\iota z)^k}{k!} \left(\tfrac{1}{2}n\right)_{\kappa} C_{\kappa}(\Sigma)$$

from which the coefficient of $\frac{(\iota z)^k}{k!}$ gives the kth moment of $\mathrm{tr}(S)$. ∎

A number of results has also been obtained on the expected values of the elementary symmetric functions of $S \sim W_p(n, \Sigma)$. The following results have been derived by de Waal and D. G. Nel (1973).

$E(\mathrm{tr}_j S) = n(n-1) \cdots (n-j+1)(\mathrm{tr}_j \Sigma)$

$E[(\mathrm{tr}_1 S)(\mathrm{tr}_2 S)] = n(n-1)(n+2)(\mathrm{tr}_1 \Sigma)(\mathrm{tr}_2 \Sigma) - 6n(n-1)(\mathrm{tr}_3 \Sigma)$

$E[(\mathrm{tr}_1 S)(\mathrm{tr}_3 S)] = n(n-1)(n-2)(n+2)(\mathrm{tr}_1 \Sigma)(\mathrm{tr}_3 \Sigma) - 8n(n-1)(n-2)(\mathrm{tr}_4 \Sigma)$

$E(\mathrm{tr}_2 S)^2 = n(n+2)(n-1)(n+1)(\mathrm{tr}_2 \Sigma)^2 - 4n(n+2)(n-1)(\mathrm{tr}_1 \Sigma)(\mathrm{tr}_3 \Sigma)$
$\qquad - 4n(n-1)(2n-5)(\mathrm{tr}_4 \Sigma)$

$E[(\mathrm{tr}_1 S)^2 (\mathrm{tr}_2 S)] = n(n+2)(n+4)(n-1)(\mathrm{tr}_1 \Sigma)^2(\mathrm{tr}_2 \Sigma)$
$\qquad - 4n(n-1)(n+2)(\mathrm{tr}_2 \Sigma)^2$
$\qquad - 12n(n-1)(n+2)(\mathrm{tr}_1 \Sigma)(\mathrm{tr}_3 \Sigma) + 48n(n-1)(\mathrm{tr}_4 \Sigma)$

$E[(\mathrm{tr}_1 S)^2] = n(n+2)(\mathrm{tr}_1 \Sigma)^2 - 4n(\mathrm{tr}_2 \Sigma)$

$E[(\mathrm{tr}_1 S)^3] = n(n+2)(n+4)(\mathrm{tr}_1 \Sigma)^3 - 12n(n+2)(\mathrm{tr}_1 \Sigma)(\mathrm{tr}_2 \Sigma) + 24n(\mathrm{tr}_3 \Sigma)$

$E[(\mathrm{tr}_1 S)^4] = n(n+2)(n+4)(n+6)(\mathrm{tr}_1 \Sigma)^4 - 24n(n+2)(n+4)(\mathrm{tr}_2 \Sigma)(\mathrm{tr}_1 \Sigma)^2$
$\qquad + 48n(n+2)(\mathrm{tr}_2 \Sigma)^2 + 96n(n+2)(\mathrm{tr}_1 \Sigma)(\mathrm{tr}_3 \Sigma) - 192n(\mathrm{tr}_4 \Sigma),$

where $\mathrm{tr}_j S$ is the jth elementary symmetric function of the matrix S. For further work in this direction, the reader is referred to Pillai and Gupta (1967, 1968), de Waal (1972a, 1978), de Waal and D. G. Nel (1973), Saw (1973), and Shah and Khatri (1974).

3.3.7. Distributions of Correlation, Regression Matrices and S^{-1}

THEOREM 3.3.24. *Let $R = (r_{ij})$ be the correlation matrix of a random sample of size $N = n+1$ from $N_p(\boldsymbol{\mu}, \Sigma)$. Then, the density of R when $\Sigma = \mathrm{diag}(\sigma_{11}, \ldots, \sigma_{pp})$ is*

$$\frac{\Gamma^p(\frac{1}{2}n)}{\Gamma_p(\frac{1}{2}n)} \det(R)^{\frac{1}{2}(n-p-1)}, \quad -1 < r_{ij} < 1, \; i < j. \tag{3.3.24}$$

Proof: Note that $r_{ij} = \frac{s_{ij}}{\sqrt{s_{ii} s_{jj}}}$, where $S = (s_{ij})$ is defined in Theorem 3.3.6. The density of S is

$$\left\{ 2^{\frac{1}{2}np} \Gamma_p\left(\frac{1}{2}n\right) \det(\Sigma)^{\frac{1}{2}n} \right\}^{-1} \det(S)^{\frac{1}{2}(n-p-1)} \, \mathrm{etr}\left(-\frac{1}{2}\Sigma^{-1} S \right).$$

Now, making the transformation

$$S = \mathrm{diag}(s_{11}^{\frac{1}{2}}, \ldots, s_{pp}^{\frac{1}{2}}) R \, \mathrm{diag}(s_{11}^{\frac{1}{2}}, \ldots, s_{pp}^{\frac{1}{2}}),$$

with the Jacobian $J(S \to s_{11}, \ldots, s_{pp}, R) = \prod_{i=1}^{p} s_{ii}^{\frac{1}{2}(p-1)}$, we get the joint density of $s_{11}, \ldots, s_{pp}$ and R as

$$\prod_{i=1}^{p} \left\{ \frac{s_{ii}^{\frac{1}{2}n-1} \exp(-\frac{s_{ii}}{2\sigma_{ii}})}{2^{\frac{1}{2}n} \sigma_{ii}^{\frac{1}{2}n} \Gamma(\frac{1}{2}n)} \right\} \frac{\Gamma^p(\frac{1}{2}n) \det(R)^{\frac{1}{2}(n-p-1)}}{\Gamma_p(\frac{1}{2}n)}. \tag{3.3.25}$$

From (3.3.25), it is seen that $s_{11}, \ldots, s_{pp}$ and R are independently distributed and $s_{ii} \sim \sigma_{ii} \chi_n^2$, $i = 1, \ldots, p$. The density of R is obtained from (3.3.25) and is given by (3.3.24). ∎

From the above theorem it is clear that if $S = (s_{ij}) \sim W_p(n, \Sigma)$, $\Sigma = (\sigma_{ij})$, then for $\sigma_{ij} = 0$, $i \neq j$, $x_i = \frac{s_{ii}}{\sigma_{ii}}$, $i = 1, \ldots, p$ are independently distributed as chi-square with n degrees of freedom. In the general case when $\sigma_{ij} \neq 0$, the joint distribution of $x_1, \ldots, x_p$ is given by the following theorem (Mathai and Tan, 1977).

THEOREM 3.3.25. *Let $S = (s_{ij}) \sim W_p(n, \Sigma)$, where $\Sigma = (\sigma_{ij})$. Then the joint distribution of $u_i = \frac{s_{ii}}{2\sigma_{ii}}$, $i = 1, \ldots, p$ is*

$$\sum_{m=0}^{\infty} \frac{\Gamma(\frac{1}{2}n+m)}{\Gamma(\frac{1}{2}n) m!} \sum_{a_1=0}^{m} \sum_{a_2=0}^{m} \cdots \sum_{a_p=0}^{m} A_a f_a\left(u_1, \ldots, u_p; \frac{1}{2}n, \ldots, \frac{1}{2}n\right), \tag{3.3.26}$$

where A_a is the coefficient of $z_1^{a_1} z_2^{a_2} \cdots z_p^{a_p}$ in the expansion of $[b(z)]^m = \sum_{a_1=0}^{m} \sum_{a_2=0}^{m} \cdots \sum_{a_p=0}^{m} A_a z_1^{a_1} z_2^{a_2} \cdots z_p^{a_p}$ with $\det(I_p - A(z)) = 1 - b(z)$,

$$A(z) = \begin{pmatrix} 0 & z_1 \rho_{12} & \cdots & z_1 \rho_{1p} \\ z_2 \rho_{21} & 0 & \cdots & z_2 \rho_{2p} \\ \vdots & \vdots & & \vdots \\ z_p \rho_{p1} & z_p \rho_{p2} & \cdots & 0 \end{pmatrix}, \quad \rho_{ij} = \frac{\sigma_{ij}}{\sqrt{\sigma_{ii} \sigma_{jj}}},$$

and

$$f_a\left(u_1,\ldots,u_p;\frac{1}{2}n,\ldots,\frac{1}{2}n\right) = \prod_{i=1}^{p}\left\{\sum_{j_i=0}^{a_i}\binom{a_i}{j_i}(-1)^{j_i}\frac{\exp(-u_i)}{\Gamma(a_i+\frac{1}{2}n-j_i)}u^{a_i+\frac{1}{2}n-j_i-1}\right\}.$$

Theorem 3.3.5 is a special case of a general result derived by Jensen (1970). He obtained the joint distribution of $u_i = \frac{1}{2}\operatorname{tr}(\Sigma_{ii}^{-1}S_{ii})$, $i = 1,\ldots,q$, where $S = (S_{ij})$, $S_{ij}\,(p_i \times p_j)$, $i,j = 1,\ldots,q$, $p_1 + p_2 + \cdots + p_q = p$.

THEOREM 3.3.26. *Let $S \sim W_p(n,\Sigma)$, and partition S as*

$$S = \begin{pmatrix} S_{11} & S_{12} \\ S_{21} & S_{22} \end{pmatrix}\begin{matrix} q \\ p-q \end{matrix}.$$
$$\begin{matrix} q & p-q \end{matrix}$$

Then the distribution of the regression coefficient matrix $B = S_{11}^{-1}S_{12}$ is

$$\frac{\Gamma_q[\frac{1}{2}(n-q+p)]}{\pi^{\frac{1}{2}q(p-q)}\Gamma_q(\frac{1}{2}n)}\det(\Sigma_{11})^{-\frac{1}{2}n}\det(\Sigma_{22\cdot1})^{-\frac{1}{2}q}\det(\Sigma_{11}^{-1}+(B-\beta)\Sigma_{22\cdot1}^{-1}(B-\beta)')^{-\frac{1}{2}(n-q+p)},$$

(3.3.27)

where $\beta = \Sigma_{11}^{-1}\Sigma_{12}$.

Proof: From Theorem 3.3.9, it is known that $S_{11} \sim W_q(n,\Sigma_{11})$ and $S_{21}|S_{11} \sim N_{p-q,q}(\Sigma_{21}\Sigma_{11}^{-1}S_{11},\Sigma_{22\cdot1}\otimes S_{11})$. Now, using Theorem 2.3.1 and 2.3.10, we get

$$S_{11}^{-1}S_{12}|S_{11} \sim N_{q,p-q}(\Sigma_{11}^{-1}\Sigma_{12}, S_{11}^{-1}\otimes\Sigma_{22\cdot1}).$$

The distribution of $B = S_{11}^{-1}S_{12}$ is then derived by integrating out S_{11} from the joint density of $S_{11}^{-1}S_{12}$ and S_{11}. Thus, we have the density of B as

$$(2\pi)^{-\frac{1}{2}q(p-q)}\det(\Sigma_{22\cdot1})^{-\frac{1}{2}q}\left\{2^{\frac{1}{2}nq}\Gamma_q\left(\frac{1}{2}n\right)\det(\Sigma_{11})^{\frac{1}{2}n}\right\}^{-1}$$

$$\int_{S_{11}>0}\det(S_{11})^{\frac{1}{2}(n+p-2q-1)}\operatorname{etr}\left[-\frac{1}{2}S_{11}\{(B-\beta)\Sigma_{22\cdot1}^{-1}(B-\beta)'+\Sigma_{11}^{-1}\}\right]dS_{11}.$$

The above integral is evaluated using (1.4.6), finally giving the density of B as (3.3.27). ∎

The above density was derived by Kshirsagar (1961a) and is known as matrix variate t-density. This density is studied in the next chapter. It may be remarked here that the distribution of B is not known when S has a noncentral Wishart distribution. However, in the linear case the result has been derived by Juritz and Troskie (1976).

THEOREM 3.3.27. *Let $S \sim W_p(n,\Sigma)$, then the density of $V = S^{-1}$ is*

$$\left\{2^{\frac{1}{2}np}\Gamma_p\left(\frac{1}{2}n\right)\det(\Sigma)^{\frac{1}{2}n}\right\}^{-1}\det(V)^{-\frac{1}{2}(n+p+1)}\operatorname{etr}\left(-\frac{1}{2}\Sigma^{-1}V^{-1}\right),\ V>0. \quad (3.3.28)$$

Proof: In the density of S given by

$$\left\{2^{\frac{1}{2}np}\Gamma_p\left(\frac{1}{2}n\right)\det(\Sigma)^{\frac{1}{2}n}\right\}^{-1}\det(S)^{\frac{1}{2}(n-p-1)}\operatorname{etr}\left(-\frac{1}{2}\Sigma^{-1}S\right).$$

making the transformation $V = S^{-1}$, with the Jacobian, $J(S \to V) = \det(V)^{-(p+1)}$, we get the density of V as given in (3.3.28). ∎

THEOREM 3.3.28. *Let* $S \sim W_p(n, I_p)$, *and* $\boldsymbol{x} \sim N_p(\boldsymbol{0}, I_p)$ *be independent. Then,*

$$\boldsymbol{x}'(C'C)^{-1}\boldsymbol{x} \sim \frac{p}{n-p+1}F_{p,n-p+1},$$

where $S = CC'$, *the matrix* C *being either triangular or nonsingular, and* $F_{p,q}$ *is the* F-*distribution with* p *and* q *degrees of freedom.*

Proof: Let $\boldsymbol{y} = (C^{-1})'\boldsymbol{x}$. Then $\boldsymbol{y}|C \sim N_p(\boldsymbol{0}, (C^{-1})'C^{-1})$. Denote the conditional and the unconditional densities of $\boldsymbol{y}$ by $f(\boldsymbol{y}|C)$ and $f(y)$ respectively. Then,

$$
\begin{aligned}
f(\boldsymbol{y}) &= E_C[f(\boldsymbol{y}|C)]\\
&= E_C[f(\boldsymbol{y}|CC')]\\
&= E_{CC'}[f(\boldsymbol{y}|CC')]\\
&= \int_{S>0} f(\boldsymbol{y}|S)g(S)\,dS,
\end{aligned}
$$

where $g(S)$ is the p.d.f. of S. Now,

$$
\begin{aligned}
f(\boldsymbol{y}) &= (2\pi)^{-\frac{1}{2}p}\left\{2^{\frac{1}{2}np}\Gamma_p\left(\frac{1}{2}n\right)\right\}^{-1}\int_{S>0}\det(S)^{\frac{1}{2}(n-p)}\operatorname{etr}\left(-\frac{1}{2}S - \frac{1}{2}S\boldsymbol{y}\boldsymbol{y}'\right)dS\\
&= \frac{\Gamma_p[\frac{1}{2}(n+1)]}{2^{\frac{1}{2}(n+1)p}\pi^{\frac{1}{2}p}\Gamma_p(\frac{1}{2}n)}\det\left(\frac{1}{2}(I_p+\boldsymbol{y}\boldsymbol{y}')\right)^{-\frac{1}{2}(n+1)}\\
&= \frac{\Gamma_p[\frac{1}{2}(n+1)]}{\pi^{\frac{1}{2}p}\Gamma_p(\frac{1}{2}n)}\det(I_p+\boldsymbol{y}\boldsymbol{y}')^{-\frac{1}{2}(n+1)}\\
&= \frac{\Gamma[\frac{1}{2}(n+1)]}{\pi^{\frac{1}{2}p}\Gamma[\frac{1}{2}(n-p+1)]}\det(I_p+\boldsymbol{y}'\boldsymbol{y})^{-\frac{1}{2}(n+1)}.
\end{aligned}
$$

Finally using Theorem 1.4.10, we get the density of $\boldsymbol{y}'\boldsymbol{y} = \boldsymbol{x}'(C'C)^{-1}\boldsymbol{x} = v$ (say) as

$$\left\{\beta\left(\frac{1}{2}p, \frac{1}{2}(n-p+1)\right)\right\}^{-1}v^{\frac{1}{2}(p-2)}(1+v)^{-\frac{1}{2}(n+1)}, \; v>0,$$

which is the desired result. ∎

THEOREM 3.3.29. *Let* $S \sim W_p(n, I_p)$, *and* $\boldsymbol{a} \in \mathbb{R}^p$, $\boldsymbol{a} \neq \boldsymbol{0}$. *Then* $\frac{\boldsymbol{a}'S^{-1}\boldsymbol{a}}{\boldsymbol{a}'S^{-2}\boldsymbol{a}}$ *is distributed as* xy, *where* x *and* y *are independent,*

$$x \sim B^I\left(\frac{1}{2}(n-p+2), \frac{1}{2}(p-1)\right) \text{ and } y \sim \chi^2_{n-p+1}.$$

Proof: Let

$$w = \frac{a'S^{-1}a}{a'S^{-2}a}. \tag{3.3.29}$$

From Theorem 3.3.2, it is known that for any orthogonal matrix Γ $(p \times p)$, the distribution of $\Gamma S \Gamma'$ is $W_p(n, I_p)$. Now, let $V = (v_{ij}) = \Gamma S \Gamma'$ and choose the orthogonal matrix Γ as

$$\Gamma' = ((a'a)^{-\frac{1}{2}} a \quad C).$$

Then,

$$S^{-1} = \Gamma' V^{-1} \Gamma, \; S^{-2} = \Gamma' V^{-2} \Gamma,$$

$$a'S^{-1}a = (a'a) v^{11}, \tag{3.3.30}$$

and

$$a'S^{-2}a = (a'a) \sum_{j=1}^{p} (v^{1j})^2, \tag{3.3.31}$$

where $V^{-1} = (v^{ij})$. By substituting from (3.3.30) and (3.3.31) in (3.3.29), we get

$$w = \frac{v^{11}}{\sum_{j=1}^{p} (v^{1j})^2}.$$

Now, let $V = TT'$, where T is an upper triangular matrix with positive diagonal elements and partition T as

$$T = \begin{pmatrix} t_{11} & t' \\ 0 & T_{22} \end{pmatrix}, \; T_{22} \, ((p-1) \times (p-1)).$$

Then,

$$T^{-1} = \begin{pmatrix} t_{11}^{-1} & -t_{11}^{-1} t' T_{22}^{-1} \\ 0 & T_{22}^{-1} \end{pmatrix},$$

and

$$
\begin{aligned}
V^{-1} &= (TT')^{-1} \\
&= (T')^{-1} T^{-1} \\
&= \begin{pmatrix} t_{11}^{-2} & -t_{11}^{-2} t' T_{22}^{-1} \\ -t_{11}^{-2} (T_{22}')^{-1} t & (T_{22} T_{22}')^{-1} + t_{11}^{-2} (T_{22}')^{-1} t t' T_{22}^{-1} \end{pmatrix}.
\end{aligned} \tag{3.3.32}
$$

From (3.3.32) it follows that

$$
\begin{aligned}
w &= \frac{t_{11}^{-2}}{t_{11}^{-4} + t_{11}^{-4} t' (T_{22}' T_{22})^{-1} t} \\
&= \frac{t_{11}^{2}}{1 + t' (T_{22}' T_{22})^{-1} t}.
\end{aligned}
$$

From Theorems 3.3.5 and 3.3.28, it is known that t_{11}^2 and $t'(T_{22}'T_{22})^{-1}t$ are independent, with $t_{11}^2 \sim \chi_{n-p+1}^2$ and $t'(T_{22}'T_{22})^{-1}t \sim \frac{p-1}{n-p\mid 2} F_{p-1,n-p\mid 2}$. Since, $\frac{1}{1+t'(T_{22}'T_{22})^{-1}t}$
$\sim B^I(\frac{1}{2}(n-p+2), \frac{1}{2}(p-1))$, the theorem follows. ∎

The above result has been derived by Gupta and Nagar (1994). They have also derived the distribution of w in terms of the Whittaker function.

THEOREM 3.3.30. *Let* $S \sim W_p(n, I_p)$. *Then* $\lambda = \frac{p^p \det(S)}{\{\operatorname{tr}(S)\}^p}$ *and* $\operatorname{tr}(S)$ *are independent.*

Proof: Let

$$R = \operatorname{diag}(s_{11}^{-\frac{1}{2}}, \ldots, s_{pp}^{-\frac{1}{2}}) S \operatorname{diag}(s_{11}^{-\frac{1}{2}}, \ldots, s_{pp}^{-\frac{1}{2}}).$$

Then from Theorem 3.3.24, $s_{11}, \ldots, s_{pp}$ and R are independent, and $s_{ii} \sim \chi_n^2$, $i = 1, \ldots, p$. Further, let $y_i = \frac{s_{ii}}{z}$, $i = 1, \ldots, p-1$ and $z = \sum_{i=1}^p s_{ii}$. Then $(y_1, \ldots, y_{p-1})$ and z are independent. Now, since

$$\lambda = p^p \prod_{i=1}^p y_i \left(1 - \sum_{i=1}^{p-1} y_i\right) \det(R)$$

is a function of $y_1, \ldots, y_{p-1}$ and $\det(R)$ only, it is independent of $\operatorname{tr}(S) = z$. ∎

The statistics λ given above is the likelihood ratio test statistic for sphericity hypothesis first studied by Mauchly (1940) (also see Gupta, 1977; Muirhead, 1982; Anderson, 1984; Amey and Gupta, 1992).

3.4. INVERTED WISHART DISTRIBUTION

DEFINITION 3.4.1. *A random matrix* V $(p \times p)$ *is said to be distributed as inverted Wishart, with* m *degrees of freedom and parameter matrix* Ψ $(p \times p)$, *denoted by* $V \sim IW_p(m, \Psi)$, *if its density is given by*

$$\frac{2^{-\frac{1}{2}(m-p-1)p} \det(\Psi)^{\frac{1}{2}(m-p-1)}}{\Gamma_p[\frac{1}{2}(m-p-1)] \det(V)^{\frac{1}{2}m}} \operatorname{etr}\left(-\frac{1}{2}V^{-1}\Psi\right), \quad V > 0, \ \Psi > 0, \ m > 2p.$$

The inverted Wishart distribution is the matrix variate generalization of the inverted gamma distribution. This distribution has been used as conjugate prior for the covariance matrix in a normal distribution. The relation between the Wishart and inverted Wishart distributions is given in the following theorem.

THEOREM 3.4.1. *Let* $V \sim IW_p(m, \Psi)$, *then* $V^{-1} \sim W_p(m-p-1, \Psi^{-1})$.

Proof: The density of V is

$$\frac{2^{-\frac{1}{2}(m-p-1)p} \det(\Psi)^{\frac{1}{2}(m-p-1)}}{\Gamma_p[\frac{1}{2}(m-p-1)] \det(V)^{\frac{1}{2}m}} \operatorname{etr}\left(-\frac{1}{2}V^{-1}\Psi\right).$$

Transforming $S = V^{-1}$ with Jacobian $J(V \to S) = \det(S)^{-(p+1)}$, we get the density of S as

$$\frac{2^{-\frac{1}{2}(m-p-1)p} \det(\Psi)^{\frac{1}{2}(m-p-1)}}{\Gamma_p[\frac{1}{2}(m-p-1)]} \det(S)^{\frac{1}{2}m-p-1} \operatorname{etr}\left(-\frac{1}{2}S\Psi\right).$$

which is the Wishart density with parameters $m - p - 1$ and Ψ^{-1}. ∎

The marginal distribution of any square submatrix on the main diagonal of an inverted Wishart matrix is also an inverted Wishart.

THEOREM 3.4.2. *Let $V \sim IW_p(m, \Psi)$ and partition V and Ψ as*

$$V = \begin{pmatrix} V_{11} & V_{12} \\ V_{21} & V_{22} \end{pmatrix} \begin{matrix} q \\ p-q \end{matrix}, \quad \Psi = \begin{pmatrix} \Psi_{11} & \Psi_{12} \\ \Psi_{21} & \Psi_{22} \end{pmatrix} \begin{matrix} q \\ p-q \end{matrix}.$$
$$\quad\quad q \quad p-q \quad\quad\quad\quad\quad q \quad p-q$$

Then, $V_{11} \sim IW_q(m - 2p + 2q, \Psi_{11})$.

Proof: From Theorem 3.4.1, $V^{-1} \sim W_p(m - p - 1, \Psi^{-1})$. Let

$$V^{-1} = \begin{pmatrix} V^{11} & V^{12} \\ V^{21} & V^{22} \end{pmatrix} \begin{matrix} q \\ p-q \end{matrix}.$$
$$\quad\quad q \quad p-q$$

Then from Theorem 3.3.9, $V^{11 \cdot 2} = V^{11} - V^{12}(V^{22})^{-1}V^{21} \sim W_q(m - 2p + q - 1, \Psi^{11 \cdot 2})$, where $\Psi^{11 \cdot 2} = \Psi^{11} - \Psi^{12}(\Psi^{22})^{-1}\Psi^{21}$ and $\Psi^{-1} = \begin{pmatrix} \Psi^{11} & \Psi^{12} \\ \Psi^{21} & \Psi^{22} \end{pmatrix}$. Now, since $V^{11 \cdot 2} = V_{11}^{-1}$ and $\Psi^{11 \cdot 2} = \Psi_{11}^{-1}$, we have $V_{11}^{-1} \sim W_q(m - 2p + q - 1, \Psi_{11}^{-1})$ and hence, $V_{11} \sim IW_q(m - 2p + 2q, \Psi_{11})$. ∎

COROLLARY 3.4.2.1. *Any diagonal element of an inverted Wishart matrix is distributed as inverted gamma.*

Proof: Take $q = 1$, in Theorem 3.4.2, and write $V_{11} = v_{11}$, $\Psi_{11} = \psi_{11}$, then $v_{11} \sim IW_1(m - 2p + 2, \psi_{11})$. The density of v_{11} from Definition 3.4.1 is

$$\left\{2^{\frac{1}{2}(m-2p)}\Gamma\left[\frac{1}{2}(m-2p)\right]\right\}^{-1} \psi_{11}^{\frac{1}{2}(m-2p)} v_{11}^{-\frac{1}{2}(m-2p+2)} \exp\left(-\frac{1}{2}\frac{\psi_{11}}{v_{11}}\right), \ v_{11} > 0, \ m > 2p,$$

which is an inverted gamma density. ∎

Different techniques have been used to derive the first and second order moments of inverted Wishart matrix. Kaufman (1967) derived the moments using a factorization theorem. Das Gupta (1968) employed the invariance arguments. Haff (1979) established an identity by applying Stokes' theorem and derived the first two moments. von Rosen (1988a) gave a general method to obtain the rth order moment and obtained explicit expressions up to fourth order. Some of these results are given in the next two theorems.

THEOREM 3.4.3. *Let $V \sim IW_p(m, \Psi)$, then*

(i) $E(v_{ij}) = \dfrac{\psi_{ij}}{m - 2p - 2}, \ m - 2p - 2 > 0,$

(ii) $\mathrm{cov}(v_{ij}, v_{k\ell}) = \dfrac{2(m - 2p - 2)^{-1}\psi_{ij}\psi_{k\ell} + \psi_{ik}\psi_{j\ell} + \psi_{i\ell}\psi_{kj}}{(m - 2p - 1)(m - 2p - 2)(m - 2p - 4)}, \ m - 2p - 4 > 0,$

(iii) $E(VAV) = \dfrac{\mathrm{tr}(A\Psi)\Psi + (m - 2p - 2)\Psi A\Psi}{(m - 2p - 1)(m - 2p - 2)(m - 2p - 4)}, \ m - 2p - 4 > 0,$

where $V = (v_{ij})$, $\Psi = (\psi_{ij})$, and A $(p \times p)$ is a constant positive semidefinite matrix.

Proof: See Haff (1979). ∎

THEOREM 3.4.4. *Let $V \sim IW_p(m, \Psi)$, then*

(i) $E(V^3) = (c_1 c_3 + c_2 c_3 + c_1 c_4 + 5c_2 c_4)\Psi^3 + (2c_2 c_3 + c_1 c_4 + c_2 c_4)(\mathrm{tr}\ \Psi)\Psi^2$

$\qquad - (c_2 c_3 + c_2 c_4)(\mathrm{tr}\ \Psi^2)\Psi - c_2 c_4(\mathrm{tr}^2\ \Psi)\Psi, \ m - 2p - 6 > 0,$

(ii) $E(\mathrm{tr}(V)V) = c_1(\mathrm{tr}\ \Psi)\Psi + 2c_2\Psi^2, \ m - 2p - 4 > 0,$

(iii) $E(\mathrm{tr}(V)V^{-1}) = \dfrac{(m - p - 1)(\mathrm{tr}\ \Psi)\Psi^{-1} - 2I_p}{m - 2p - 2}, \ m - 2p - 2 > 0,$

where $c_1 = (m - 2p - 3)c_2$, $c_2 = \{(m - 2p - 1)(m - 2p - 2)(m - 2p - 4)\}^{-1}$, $c_3 = (m - 2p - 4)\{(m - 2p - 6)(m - 2p)\}^{-1}$, and $c_4 = 2\{(m - 2p - 6)(m - 2p)\}^{-1}$.

Proof: See von Rosen (1988a). ∎

THEOREM 3.4.5. *Let $V \sim IW_p(m, \Psi)$, then*
(i) $E(VAV) = c_1 \Psi A\Psi + c_2[\Psi A'\Psi + \mathrm{tr}(A\Psi)\Psi],$
(ii) $E(\mathrm{tr}(AV)V) = c_1\mathrm{tr}(A\Psi)\Psi + c_2[\Psi A'\Psi + \Psi A\Psi],$
(iii) $E(\mathrm{tr}(AV)\mathrm{tr}(BV)) = c_1\mathrm{tr}(A\Psi)\mathrm{tr}(B\Psi) + c_2[\mathrm{tr}(B\Psi A\Psi) + \mathrm{tr}(B\Psi A'\Psi)],$
where c_1, c_2 are defined in Theorem 3.4.4 and A $(p \times p)$, B $(p \times p)$ are constant matrices.

Proof: From Theorem 3.4.1 we know that $V^{-1} \sim W_p(m - p - 1, \Psi^{-1})$. The results then follow by using Theorem 3.3.18. ∎

3.5. NONCENTRAL WISHART DISTRIBUTION

Noncentral Wishart distribution is the matrix variate generalization of noncentral chi-square distribution. It is useful in studying robustness and power of most of the multivariate tests.

DEFINITION 3.5.1. *A $p \times p$ random symmetric positive definite matrix S is said to have a noncentral Wishart distribution with parameters p, n, $\Sigma > 0$ and Θ, written* as $S \sim W_p(n, \Sigma, \Theta)$, if its p.d.f. is given by

$$\left\{2^{\frac{1}{2}np}\Gamma_p\left(\frac{1}{2}n\right)\det(\Sigma)^{\frac{1}{2}n}\right\}^{-1} \operatorname{etr}\left(-\frac{1}{2}\Theta\right)\operatorname{etr}\left(-\frac{1}{2}\Sigma^{-1}S\right)\det(S)^{\frac{1}{2}(n-p-1)}$$

$$_0F_1\left(\frac{1}{2}n; \frac{1}{4}\Theta\Sigma^{-1}S\right), \quad S > 0, \; n \geq p. \quad (3.5.1)$$

where $_0F_1$ is the hypergeometric function (Bessel function).

The matrix Θ is called the noncentrality parameter matrix. When $\Theta = 0$, the noncentral Wishart distribution reduces to the Wishart distribution defined in Section 3.2. This distribution, like Wishart distribution, can also be derived from normal distribution.

THEOREM 3.5.1. Let $X \sim N_{p,n}(M, \Sigma \otimes I_n)$, $n \geq p$, then $S = XX' \sim W_p(n, \Sigma, \Theta)$, where $\Theta = \Sigma^{-1}MM'$.

Proof: The Laplace transform of $f(S)$, the density of $S = XX'$, is

$$g(Z) = E[\operatorname{etr}(-ZS)], \; Z\,(p \times p) = Z'$$
$$= E[\operatorname{etr}(-ZXX')]$$
$$= (2\pi)^{-\frac{1}{2}np}\det(\Sigma)^{-\frac{1}{2}n}$$
$$\int_{X \in \mathbb{R}^{p \times n}} \operatorname{etr}\left\{-ZXX' - \frac{1}{2}\Sigma^{-1}(X-M)(X-M)'\right\}dX. \quad (3.5.2)$$

Now, write trace of the quadratic form in the exponent as

$$\operatorname{tr}\left\{-ZXX' - \frac{1}{2}\Sigma^{-1}(X-M)(X-M)'\right\}$$
$$= \operatorname{tr}\left\{-\frac{1}{2}(2Z+\Sigma^{-1})(X-(2Z+\Sigma^{-1})^{-1}\Sigma^{-1}M)(X-(2Z+\Sigma^{-1})^{-1}\Sigma^{-1}M)'\right.$$
$$\left.+\frac{1}{2}\Sigma^{-1}MM'\Sigma^{-1}(2Z+\Sigma^{-1})^{-1} - \frac{1}{2}\Sigma^{-1}MM'\right\}. \quad (3.5.3)$$

Substituting from (3.5.3) in (3.5.2) and evaluating the integral we get

$$g(Z) = \det(\Sigma)^{-\frac{1}{2}n}\det(2Z+\Sigma^{-1})^{-\frac{1}{2}n}$$
$$\operatorname{etr}\left\{-\frac{1}{2}\Sigma^{-1}MM' + \frac{1}{2}\Sigma^{-1}MM'\Sigma^{-1}(2Z+\Sigma^{-1})^{-1}\right\}$$
$$= 2^{-\frac{1}{2}np}\det(\Sigma)^{-\frac{1}{2}n}\det\left(Z+\frac{1}{2}\Sigma^{-1}\right)^{-\frac{1}{2}n}\operatorname{etr}\left(-\frac{1}{2}\Theta\right)$$
$$_0F_0\left(\frac{1}{4}\Theta\Sigma^{-1}\left(Z+\frac{1}{2}\Sigma^{-1}\right)^{-1}\right), \quad \operatorname{Re}\left(Z+\frac{1}{2}\Sigma^{-1}\right) > 0. \quad (3.5.4)$$

The density $f(S)$ of S is obtained by finding the inverse Laplace transform of (3.5.4) as

$$f(S) = \frac{2^{\frac{1}{2}p(p-1)}}{(2\pi\iota)^{\frac{1}{2}p(p+1)}} \int_{\mathrm{Re}(Z)>0} \mathrm{etr}(SZ) g(Z)\, dZ$$

$$= 2^{-\frac{1}{2}np} \det(\Sigma)^{-\frac{1}{2}n} \mathrm{etr}\left(-\frac{1}{2}\Theta\right) \frac{2^{\frac{1}{2}p(p-1)}}{(2\pi\iota)^{\frac{1}{2}p(p+1)}} \int_{\mathrm{Re}(Z)>0} \mathrm{etr}(SZ)$$

$$\det\left(Z + \frac{1}{2}\Sigma^{-1}\right)^{-\frac{1}{2}n} {}_0F_0\left(\frac{1}{4}\Theta\Sigma^{-1}\left(Z + \frac{1}{2}\Sigma^{-1}\right)^{-1}\right) dZ$$

$$= \frac{2^{-\frac{1}{2}np} \det(\Sigma)^{-\frac{1}{2}n} \mathrm{etr}(-\frac{1}{2}\Theta)}{\Gamma_p(\frac{1}{2}n)}$$

$$\det(S)^{\frac{1}{2}(n-p-1)} \mathrm{etr}\left(-\frac{1}{2}\Sigma^{-1}S\right) {}_0F_1\left(\frac{1}{2}n; \frac{1}{4}\Theta\Sigma^{-1}S\right). \qquad (3.5.5)$$

The last equality is obtained by applying the result (1.5.14). ∎

The noncentral Wishart density (3.5.5) was derived by Herz (1955) and James (1954, 1955, 1964). In the case rank$(\Theta) = 1, 2$, the results were obtained by Anderson and Girshick (1944), Anderson (1946) and Herz (1955) whereas Weibull (1953) and James (1955) gave the results for rank$(\Theta) = 3$. When $\Sigma = I_p$ and the only nonzero element of Θ is θ_{11}, then the p.d.f. of $S = (s_{ij})$ simplifies to

$$\left\{2^{\frac{1}{2}np}\Gamma_p\left(\frac{1}{2}n\right)\right\}^{-1} \det(S)^{\frac{1}{2}(n-p-1)} \exp\left(-\frac{1}{2}\,\mathrm{tr}\,S - \frac{1}{2}\theta_{11}\right) {}_0F_1\left(\frac{1}{2}n; \frac{1}{4}\theta_{11}s_{11}\right), \qquad (3.5.6)$$

where now ${}_0F_1(\cdot)$ is the Bessel function of a scalar argument.

In the rest of this section, we study some basic properties of noncentral Wishart distribution.

THEOREM 3.5.2. *Let* $X \sim N_{p,n}(M, \Sigma \otimes I_n)$, $n \geq p$, *and* $A\,(q \times p)$ *be any matrix of rank* $q \leq p$. *Then,*

$$AXX'A' \sim W_q(n, A\Sigma A', (A\Sigma A')^{-1}AMM'A').$$

Proof: Let $Y = AX$, then from Theorem 2.3.10, $Y \sim N_{q,n}(AM, (A\Sigma A') \otimes I_n)$. From Theorem 3.5.1, we get

$$YY' = AXX'A' \sim W_q(n, A\Sigma A', (A\Sigma A')^{-1}AMM'A'). \quad ∎$$

THEOREM 3.5.3. *Let* $S \sim W_p(n, \Sigma, \Theta)$. *Then, the characteristic function of* S *is*

$$\det(I_p - 2\iota\Sigma Z)^{-\frac{1}{2}n} \mathrm{etr}\left\{-\frac{1}{2}\Theta + \frac{1}{2}(I_p - 2\iota\Sigma Z)^{-1}\Theta\right\},$$

where $Z = Z'\,(p \times p) = (\frac{1}{2}(1 + \delta_{ij})z_{ij})$ *and* δ_{ij} *is the Kronecker's delta.*

Proof: By definition, the characteristic function of S is

$$\phi_S(Z) = \left\{2^{\frac{1}{2}np} \det(\Sigma)^{\frac{1}{2}n}\Gamma_p\left(\frac{1}{2}n\right)\right\}^{-1} \mathrm{etr}\left(-\frac{1}{2}\Theta\right)$$

$$\int_{S>0} \det(S)^{\frac{1}{2}(n-p-1)} \mathrm{etr}\left(\iota ZS - \frac{1}{2}\Sigma^{-1}S\right) {}_0F_1\left(\frac{1}{2}n; \frac{1}{4}\Theta\Sigma^{-1}S\right) dS. \qquad (3.5.7)$$

Now using (1.6.4), we get

$$\int_{S>0} \det(S)^{\frac{1}{2}(n-p-1)} \, \mathrm{etr} \left(\iota Z S - \frac{1}{2}\Sigma^{-1}S \right) {}_0F_1 \left(\frac{1}{2}n; \frac{1}{4}\Theta\Sigma^{-1}S \right) dS$$

$$= \Gamma_p \left(\frac{1}{2}n \right) \det\left(\frac{1}{2}\Sigma^{-1} - \iota Z \right)^{-\frac{1}{2}n} {}_1F_1 \left(\frac{1}{2}n; \frac{1}{2}n; \frac{1}{4}\left(\frac{1}{2}\Sigma^{-1} - \iota Z \right)^{-1}\Theta\Sigma^{-1} \right)$$

$$= 2^{\frac{1}{2}np} \Gamma_p \left(\frac{1}{2}n \right) \det(\Sigma)^{\frac{1}{2}n} \det(I_p - 2\iota\Sigma Z)^{-\frac{1}{2}n} \, \mathrm{etr} \left\{ \frac{1}{2}(I_p - 2\iota\Sigma Z)^{-1}\Theta \right\}. \quad (3.5.8)$$

Substituting from (3.5.8) in (3.5.7), we get

$$\phi_S(Z) = \det(I_p - 2\iota\Sigma Z)^{-\frac{1}{2}n} \, \mathrm{etr} \left\{ -\frac{1}{2}\Theta + \frac{1}{2}(I_p - 2\iota\Sigma Z)^{-1}\Theta \right\}. \quad \blacksquare$$

Next we derive a differential equation for the characteristic function of the non-central Wishart matrix which is useful in the study of approximation of noncentral distribution by a central distribution (Steyn and Roux, 1972).

THEOREM 3.5.4. *Let* $X \sim N_{p,n}(M, \Sigma \otimes I_n)$, $n \geq p$, $S = XX'$ *and* $\Gamma = (\gamma_{ij})$, *where* $\gamma_{ij} = \frac{1}{2}(1 + \delta_{ij})z_{ij}$, $z_{ij} = z_{ji}$, $i, j = 1, \ldots, p$ *and* δ_{ij} *is the Kronecker's delta. Then the characteristic function* ϕ *of* S *satisfies the differential equation*

$$\frac{\partial\phi}{\partial Z} = \iota[n(\Psi - 2\iota\Gamma)^{-1} + (\Psi - 2\iota\Gamma)^{-1}\Psi MM'\Psi(\Psi - 2\iota\Gamma)^{-1}]\phi$$

where $\Psi = \Sigma^{-1}$ *and* $\frac{\partial\phi}{\partial Z} = \left(\frac{\partial\phi}{\partial z_{ij}} \right)$.

Proof: Let $X = (\boldsymbol{x}_1, \ldots, \boldsymbol{x}_n)$, $\boldsymbol{x}_\alpha = (x_{1\alpha}, \ldots, x_{p\alpha})'$, $M = (\boldsymbol{m}_1, \ldots, \boldsymbol{m}_n)$, $\boldsymbol{m}_\alpha = (m_{1\alpha}, \ldots, m_{p\alpha})'$, $\alpha = 1, \ldots, n$ and $\Psi = (\psi_{ij})$. Then $S = \sum_{\alpha=1}^n \boldsymbol{x}_\alpha \boldsymbol{x}_\alpha' = (s_{rt})$, $s_{rt} = \sum_{\alpha=1}^n x_{r\alpha}x_{t\alpha}$, and $\boldsymbol{x}_\alpha$, $\alpha = 1, \ldots, n$ are independently distributed as

$$f_\alpha = c\exp\left[-\frac{1}{2}\sum_{i,j=1}^p \psi_{ij}(x_{i\alpha} - m_{i\alpha})(x_{j\alpha} - m_{j\alpha}) \right]$$

where $c = (2\pi)^{-\frac{1}{2}p} \det(\Psi)^{\frac{1}{2}}$. The characteristic function of S is

$$\phi = E[\mathrm{etr}(\iota\Gamma S)]$$

$$= \int_{-\infty}^\infty \cdots \int_{-\infty}^\infty \exp\left[\iota\sum_{r\leq t} z_{rt}s_{rt} \right]\left(\prod_{\beta=1}^n f_\beta \right)\prod_{i=1}^p \prod_{\beta=1}^n dx_{i\beta}.$$

Differentiating ϕ w.r.t. z_{1j} we get

$$\frac{\partial\phi}{\partial z_{1j}} = \int_{-\infty}^\infty \cdots \int_{-\infty}^\infty \iota s_{1j} \, \mathrm{etr}(\iota\Gamma S)\left(\prod_{\beta=1}^n f_\beta \right)\prod_{i=1}^p \prod_{\beta=1}^n dx_{i\beta}. \quad (3.5.9)$$

Multiplying (3.5.9) by ψ_{ij} and summing over j we get

$$\sum_{j=1}^{p} \psi_{ij} \frac{\partial \phi}{\partial z_{1j}} = \int_{-\infty}^{\infty} \cdots \int_{-\infty}^{\infty} \left(\iota \sum_{j=1}^{p} \sum_{\alpha=1}^{n} \psi_{ij} x_{1\alpha} x_{j\alpha} \right) \text{etr}(\iota \Gamma S) \left(\prod_{\beta=1}^{n} f_{\beta} \right) \prod_{i=1}^{p} \prod_{\beta=1}^{n} dx_{i\beta}.$$

Now using the result $\sum_{j=1}^{p} \psi_{ij} x_{j\alpha} = \sum_{j=1}^{p} \psi_{ij}(x_{j\alpha} - m_{j\alpha}) + \sum_{j=1}^{p} \psi_{ij} m_{j\alpha}$, the above expression can be rewritten as

$$\sum_{j=1}^{p} \psi_{ij} \frac{\partial \phi}{\partial z_{1j}} = w_1 + \iota \sum_{j=1}^{p} \sum_{\alpha=1}^{n} \psi_{ij} m_{j\alpha} y_{1\alpha} \qquad (3.5.10)$$

where

$$w_1 = \iota \sum_{\alpha=1}^{n} \int_{-\infty}^{\infty} \cdots \int_{-\infty}^{\infty} x_{1\alpha} \sum_{j=1}^{p} \psi_{ij}(x_{j\alpha} - m_{j\alpha}) \, \text{etr}(\iota \Gamma S) \left(\prod_{\beta=1}^{n} f_{\beta} \right) \prod_{i=1}^{p} \prod_{\beta=1}^{n} dx_{i\beta}$$

and

$$y_{1\alpha} = \int_{-\infty}^{\infty} \cdots \int_{-\infty}^{\infty} x_{1\alpha} \, \text{etr}(\iota \Gamma S) \left(\prod_{\beta=1}^{n} f_{\beta} \right) \prod_{i=1}^{p} \prod_{\beta=1}^{n} dx_{i\beta}.$$

Further, w_1 can be written as

$$w_1 = \iota \sum_{\alpha=1}^{n} \int_{-\infty}^{\infty} \cdots \int_{-\infty}^{\infty} \left\{ \int_{-\infty}^{\infty} x_{1\alpha} \left(\sum_{j=1}^{p} \psi_{ij}(x_{j\alpha} - m_{j\alpha}) \right) \text{etr}(\iota \Gamma S) f_{\alpha} \, dx_{i\alpha} \right\}$$
$$\left(\prod_{\substack{\beta=1 \\ \neq \alpha}}^{n} f_{\beta} \right) \prod_{\substack{g=1 \\ (g,\beta) \neq (i,\alpha)}}^{p} \prod_{\beta=1}^{n} dx_{g\beta}. \qquad (3.5.11)$$

Now integrating out $x_{i\alpha}$ using the result

$$\sum_{j=1}^{p} \psi_{ij}(x_{j\alpha} - m_{j\alpha}) f_{\alpha} = -\frac{\partial}{\partial x_{i\alpha}} f_{\alpha},$$

and

$$\frac{\partial}{\partial x_{i\alpha}} \left(x_{1\alpha} \, \text{etr}(\iota \Gamma S) \right) = \left(\iota x_{1\alpha} \sum_{j=1}^{p} 2\gamma_{ij} x_{j\alpha} + \delta_{i1} \right) \text{etr}(\iota \Gamma S)$$

the expression (3.5.11) is simplified as

$$w_1 = \iota \int_{-\infty}^{\infty} \cdots \int_{-\infty}^{\infty} \left(2\iota \sum_{j=1}^{p} \gamma_{ij} \sum_{\alpha=1}^{n} x_{1\alpha} x_{j\alpha} + n\delta_{i1} \right) \text{etr}(\iota \Gamma S) \left(\prod_{\beta=1}^{n} f_{\beta} \right) \prod_{g=1}^{p} \prod_{\beta=1}^{n} dx_{g\beta}$$
$$= \iota \left\{ n\delta_{i1} \phi + 2 \sum_{j=1}^{p} \gamma_{ij} \frac{\partial \phi}{\partial z_{1j}} \right\} \qquad (3.5.12)$$

Substituting w_1 from (3.5.12) in (3.5.10) we finally get

$$\sum_{j=1}^{p} \psi_{ij} \frac{\partial \phi}{\partial z_{1j}} = \iota \left\{ n\delta_{i1} \phi + 2 \sum_{j=1}^{p} \gamma_{ij} \frac{\partial \phi}{\partial z_{1j}} + \sum_{\alpha=1}^{n} \sum_{j=1}^{p} \psi_{ij} m_{j\alpha} y_{1\alpha} \right\}. \qquad (3.5.13)$$

Similarly by differentiating ϕ w.r.t. $z_{2j}, \ldots, z_{pj}$, we can derive $(p-1)$ differential equations which together with (3.5.13) can, in general, be written as

$$\sum_{j=1}^{p} \psi_{ij}\frac{\partial\phi}{\partial z_{\ell j}} = \iota\Big\{ n\delta_{i\ell}\phi + 2\sum_{j=1}^{p}\gamma_{ij}\frac{\partial\phi}{\partial z_{\ell j}} + \sum_{\alpha=1}^{n}\sum_{j=1}^{p}\psi_{ij}m_{j\alpha}y_{\ell\alpha}\Big\} \qquad (3.5.14)$$

where

$$y_{\ell\alpha} = \int_{-\infty}^{\infty}\cdots\int_{-\infty}^{\infty} x_{\ell\alpha}\,\mathrm{etr}(\iota\Gamma S)\Big(\prod_{\beta=1}^{n} f_\beta\Big)\prod_{i=1}^{p}\prod_{\beta=1}^{n} dx_{i\beta}, \;\ell=1,\ldots,p. \qquad (3.5.15)$$

Further using the result

$$\sum_{j=1}^{p}\psi_{ij}x_{j\alpha}m_{\ell\alpha} = \sum_{j=1}^{p}\psi_{ij}(x_{j\alpha}-m_{j\alpha})m_{\ell\alpha} + \sum_{j=1}^{p}\psi_{ij}m_{j\alpha}m_{\ell\alpha}$$

together with (3.5.15) we have

$$\sum_{j=1}^{p}\psi_{ij}\sum_{\alpha=1}^{n} y_{j\alpha}m_{\ell\alpha} = \sum_{\alpha=1}^{n} m_{\ell\alpha}\int_{-\infty}^{\infty}\cdots\int_{-\infty}^{\infty}\Big\{\int_{-\infty}^{\infty}\Big(\sum_{j=1}^{p}\psi_{ij}(x_{j\alpha}-m_{j\alpha})\Big)\mathrm{etr}(\iota\Gamma S)f_\alpha\,dx_{i\alpha}\Big\}$$

$$\Big(\prod_{\substack{\beta=1\\\neq\alpha}}^{n} f_\beta\Big)\prod_{\substack{g=1\\(g,\beta)\neq(i,\alpha)}}^{p}\prod_{\beta=1}^{n} dx_{g\beta} + \phi\sum_{j=1}^{p}\psi_{ij}\sum_{\alpha=1}^{n} m_{j\alpha}m_{\ell\alpha}.$$

Now solving the integral inside the curly brackets, as before, we have

$$\sum_{j=1}^{p}\psi_{ij}\sum_{\alpha=1}^{n} y_{j\alpha}m_{\ell\alpha} = \iota\sum_{j=1}^{p} 2\gamma_{ij}\sum_{\alpha=1}^{n} y_{j\alpha}m_{\ell\alpha} + \phi\sum_{j=1}^{p}\psi_{ij}\sum_{\alpha=1}^{n} m_{j\alpha}m_{\ell\alpha}. \qquad (3.5.16)$$

Further let $\boldsymbol{y}_\alpha = (y_{1\alpha}, \ldots, y_{p\alpha})'$ then equations (3.5.14) and (3.5.16) can be written as

$$(\Psi - 2\iota\Gamma)\frac{\partial\phi}{\partial Z} = \iota\Big\{ nI_p\phi + \Psi\sum_{\alpha=1}^{n}\boldsymbol{m}_\alpha\boldsymbol{y}'_\alpha\Big\} \qquad (3.5.17)$$

and

$$(\Psi - 2\iota\Gamma)\sum_{\alpha=1}^{n}\boldsymbol{y}_\alpha\boldsymbol{m}'_\alpha = \phi\Psi\sum_{\alpha=1}^{n}\boldsymbol{m}_\alpha\boldsymbol{m}'_\alpha = \phi\Psi MM' \qquad (3.5.18)$$

respectively. Finally substituting for $\sum_{\alpha=1}^{n}\boldsymbol{m}_\alpha\boldsymbol{y}'_\alpha$ from (3.5.18) in (3.5.17), we get

$$(\Psi - 2\iota\Gamma)\frac{\partial\phi}{\partial Z} = \iota\Big\{ nI_p + \Psi MM'\Psi(\Psi - 2\iota\Gamma)^{-1}\Big\}\phi$$

i.e.

$$\frac{\partial\phi}{\partial Z} = \iota\Big\{ n(\Psi - 2\iota\Gamma)^{-1} + (\Psi - 2\iota\Gamma)^{-1}\Psi MM'\Psi(\Psi - 2\iota\Gamma)^{-1}\Big\}\phi. \quad \blacksquare$$

THEOREM 3.5.5. *Let $S_j \sim W_p(n_j, \Sigma, \Theta_j)$, $j = 1, \ldots, k$ be independently distributed, then $\sum_{j=1}^{k} S_j \sim W_p(n, \Sigma, \Theta)$ where $n = \sum_{j=1}^{k} n_j$ and $\Theta = \sum_{j=1}^{k}\Theta_j$.*

Proof: The characteristic function of $S = \sum_{j=1}^{k} S_j$ is

$$\phi_S(Z) = E[\text{etr}(\iota Z S)]$$

$$= \prod_{j=1}^{k} E[\text{etr}(\iota Z S_j)]$$

$$= \prod_{j=1}^{k} \left[\det(I_p - 2\iota \Sigma Z)^{-\frac{1}{2}n_j} \, \text{etr} \left\{ -\frac{1}{2}\Theta_j + \frac{1}{2}(I_p - 2\iota \Sigma Z)^{-1}\Theta_j \right\} \right]$$

$$= \det(I_p - 2\iota \Sigma Z)^{-\frac{1}{2}n} \, \text{etr} \left\{ -\frac{1}{2}\Theta + \frac{1}{2}(I_p - 2\iota \Sigma Z)^{-1}\Theta \right\},$$

which is the characteristic function of a noncentral Wishart matrix with parameters n, Σ and Θ. ∎

When $S_j \sim W_p(n_j, \Sigma_j, \Theta_j)$, $j = 1, \ldots, k$ are independently distributed, Chikuse and Davis (1986) derived the distribution of $\sum_{j=1}^{k} S_j$ in series involving generalized Laguerre polynomials.

THEOREM 3.5.6. *Let* $S \sim W_p(n, \Sigma, \Theta)$, *then*

$$E[\det(S)^h] = \frac{2^{ph}\Gamma_p(\frac{1}{2}n + h)\det(\Sigma)^h}{\Gamma_p(\frac{1}{2}n)} \, \text{etr} \left(-\frac{1}{2}\Theta \right) {}_1F_1\left(\frac{1}{2}n + h; \frac{1}{2}n; \frac{1}{2}\Theta \right),$$

$$\text{Re}(h) > -\frac{1}{2}n + \frac{1}{2}(p-1). \quad (3.5.19)$$

Proof: From the density (3.5.1), we get

$$E[\det(S)^h] = \left\{ 2^{\frac{1}{2}np}\Gamma_p\left(\frac{1}{2}n\right)\det(\Sigma)^{\frac{1}{2}n} \right\}^{-1} \text{etr}\left(-\frac{1}{2}\Theta \right)$$

$$\int_{S>0} \text{etr}\left(-\frac{1}{2}\Sigma^{-1}S \right) \det(S)^{\frac{1}{2}(n-p-1)+h} \, {}_0F_1\left(\frac{1}{2}n; \frac{1}{4}\Theta\Sigma^{-1}S \right) dS.$$

Now, using (1.6.4) and simplifying, the result follows. ∎

THEOREM 3.5.7. *Let* $Y \sim N_{p,n}(M, \Sigma \otimes I_n)$, $n \geq p$, $S = YY' = (s_{ij})$ *and* $MM' = (\omega_{ij})$. *Then*
  (i) $E(s_{ij}) = n\sigma_{ij} + \omega_{ij}$ *and*
  (ii) $E(s_{ij}s_{k\ell}) = (n\sigma_{ij} + \omega_{ij})(n\sigma_{k\ell} + \omega_{k\ell}) + n(\sigma_{ik}\sigma_{j\ell} + \sigma_{i\ell}\sigma_{jk})$
      $+ \sigma_{j\ell}\omega_{ik} + \sigma_{i\ell}\omega_{jk} + \sigma_{jk}\omega_{i\ell} + \sigma_{ik}\omega_{j\ell}.$

Proof: (i) From Theorem 2.3.5(ii) we get

$$E(YY') = n\Sigma + MM', \quad (3.5.20)$$

and hence

$$E(s_{ij}) = n\sigma_{ij} + \omega_{ij}.$$

(ii) From Theorem 2.3.8(v), we get

$$
\begin{aligned}
E(YY'BYY') = {} & n\operatorname{tr}(B\Sigma)\Sigma + n^2\Sigma B\Sigma + n\Sigma B'\Sigma + nMM'B\Sigma + MM'B'\Sigma \\
& + \operatorname{tr}(BMM')\Sigma + \operatorname{tr}(B\Sigma)MM' + \Sigma B'MM' + n\Sigma BMM' \\
& + MM'BMM'.
\end{aligned}
\tag{3.5.21}
$$

Now

$$
E(YY'BYY') = E(SBS)
$$

$$
= E\left(\left(\sum_{k=1}^{p}\sum_{j=1}^{p} s_{ij}b_{jk}s_{k\ell}\right)\right),
$$

and hence

$$
\begin{aligned}
E\left(\sum_{k=1}^{p}\sum_{j=1}^{p} s_{ij}b_{jk}s_{k\ell}\right) = {} & \sum_{k=1}^{p}\sum_{j=1}^{p}[n\sigma_{i\ell}b_{jk}\sigma_{kj} + n^2\sigma_{ij}b_{jk}\sigma_{k\ell} + n\sigma_{ij}b_{kj}\sigma_{k\ell} + n\omega_{ij}b_{jk}\sigma_{k\ell} \\
& + \omega_{ij}b_{kj}\sigma_{k\ell} + \sigma_{i\ell}b_{jk}\omega_{kj} + \omega_{i\ell}b_{jk}\sigma_{kj} + \sigma_{ij}b_{kj}\omega_{k\ell} \\
& + n\sigma_{ij}b_{jk}\omega_{k\ell} + \omega_{ij}b_{jk}\omega_{k\ell}].
\end{aligned}
$$

Next, substituting $b_{jk} = 1$ and $= 0$ otherwise, we get

$$
\begin{aligned}
E(s_{ij}s_{k\ell}) = {} & (n\sigma_{ij} + \omega_{ij})(n\sigma_{k\ell} + \omega_{k\ell}) + n(\sigma_{i\ell}\sigma_{jk} + \sigma_{ik}\sigma_{j\ell}) \\
& + \sigma_{j\ell}\omega_{ik} + \sigma_{i\ell}\omega_{kj} + \sigma_{jk}\omega_{i\ell} + \sigma_{ik}\omega_{j\ell}. \quad\blacksquare
\end{aligned}
$$

In the case $M = 0$, the above theorem gives the first two moments of S, where $S \sim W_p(n, \Sigma)$. Premultiplying (3.5.20) and (3.5.21) by Σ^{-1}, setting $B = \Sigma^{-1}$ in (3.5.21), and taking the trace of the resulting equation, for $n \geq p$, we get

$$
E[\operatorname{tr}(\Sigma^{-1}S)] = np + \operatorname{tr}(\Theta)
$$

and

$$
E[\operatorname{tr}(\Sigma^{-1}S\Sigma^{-1}S)] = np(n + p + 1) + 2(n + p + 1)\operatorname{tr}(\Theta) + \operatorname{tr}(\Theta^2)
$$

where $S \sim W_p(n, \Sigma, \Theta)$. For an identity involving expectation of noncentral Wishart matrices, the reader is referred to Leung (1994).

Shah and Khatri (1974) have proved that if $S \sim W_p(n, \Sigma, \Theta)$ with $\Theta = \Sigma^{-1}W$, $W = MM'$ and $\operatorname{tr}_i S$ is the ith elementary symmetric function of S, then

(i) $E(\operatorname{tr}_p S) = E[\det(A)] = \det(\Sigma)\left[n^{(p)} + \displaystyle\sum_{i=1}^{p}(n - i)^{(p-i)}\operatorname{tr}_i \Theta\right]$

and

(ii) $E(\operatorname{tr}_j S) = n^{(j)}\operatorname{tr}_j \Sigma + \displaystyle\sum_{k=1}^{j}(n - k)^{(j-k)}\sum_{i(j)}\det(\Sigma(i(j)))$
$\qquad\qquad \operatorname{tr}_k\{[\Sigma(i(j))]^{-1}W(i(j))\},\ j = 1, 2, \ldots, p - 1,$

where $n^{(j)} = n(n-1)\cdots(n-j+1)$, $\Sigma(i(j))$ and $W(i(j))$ are submatrices obtained by considering $i_1, i_2, \ldots, i_j$ rows and $i_1, i_2, \ldots, i_j$ columns of matrices Σ and W respectively and $\Sigma_{i(j)} = \sum_{i_1=1}^{p} \cdots \sum_{i_j=1}^{p}$. Saw (1973) has shown that
$$\underset{i_1 > i_2 > \cdots > i_j}{}$$

$$E[\mathrm{tr}_j(\Sigma^{-1}S)] = \sum_{i=0}^{j}(n-i)^{(j-i)}\binom{p-i}{j-i}\mathrm{tr}_i(\Theta), \ i \le j \le p \le n.$$

THEOREM 3.5.8. *Let $S \sim W_p(n, I_p, \Theta)$, $\Theta = \mathrm{diag}(\theta, 0, \ldots, 0)$ and $S = TT'$ where $T\,(p \times p) = (t_{ij})$ is a lower triangular matrix with diagonal elements $t_{ii} > 0$. Then, $t_{ij}, \ 1 \le j \le i \le p$ are independently distributed $t_{11}^2 \sim \chi_n'^2(\theta)$, $t_{ii}^2 \sim \chi_{n-i+1}^2$, $i = 2, \ldots, p$, and $t_{ij} \sim N(0,1)$, $1 \le j < i \le p$.*

Proof: The density of S for $\Theta = \mathrm{diag}(\theta, 0, ..., 0)$ and $\Sigma = I_p$ from (3.5.1) is

$$\left\{ 2^{\frac{1}{2}np}\Gamma_p\left(\frac{1}{2}n\right)\right\}^{-1}\exp\left(-\frac{1}{2}\theta\right)\mathrm{etr}\left(-\frac{1}{2}S\right)\det(S)^{\frac{1}{2}(n-p-1)}\,_0F_1\left(\frac{1}{2}n; \frac{1}{4}\theta s_{11}\right). \quad (3.5.22)$$

Let $S = TT'$ so that

$$\mathrm{tr}(S) = \sum_{i=1}^{p}\sum_{j=1}^{i}t_{ij}^2$$

$$\det(S) = \prod_{i=1}^{p}t_{ii}^2$$

and from (1.3.14),

$$J(S \to T) = 2^p \prod_{i=1}^{p}t_{ii}^{p-i+1}.$$

The joint density of t_{ij}, $1 \le j \le i \le p$, obtained from (3.5.22) is

$$\prod_{1 \le j < i \le p}\left\{\frac{1}{\sqrt{2\pi}}\exp\left(-\frac{1}{2}t_{ij}^2\right)\right\}\prod_{i=1}^{p}\left\{\frac{2t_{ii}^{n-i}\exp(-\frac{1}{2}t_{ii}^2)}{2^{\frac{1}{2}(n-i+1)}\Gamma[\frac{1}{2}(n-i+1)]}\right\}\exp\left(-\frac{1}{2}\theta\right)\,_0F_1\left(\frac{1}{2}n; \frac{1}{4}\theta t_{11}^2\right),$$

$$t_{ii} > 0, \ 1 \le i \le p, \ -\infty < t_{ij} < \infty, \ 1 \le j < i \le p. \quad (3.5.23)$$

From (3.5.23) it is easily seen that t_{ij}, $1 \le j \le i \le p$ are independently distributed and $t_{ij} \sim N(0,1)$, $1 \le j < i \le p$. Substituting $y_{ii} = t_{ii}^2$, one can show that $t_{11}^2 \sim \chi_n'^2(\theta)$ and $t_{ii}^2 \sim \chi_{n-i+1}^2$, $i = 2, \ldots, p$. ∎

There is also the noncentral inverted Wishart distribution defined by Roux and Becker (1984).

DEFINITION 3.5.2. *A random matrix $V\,(p \times p)$ is said to be distributed as noncentral inverted Wishart with m degrees of freedom and parameter matrices $\Psi\,(p \times p)$ and $\Theta\,(p \times p)$, denoted by $V \sim IW_p(m, \Psi, \Theta)$, if its density is given by*

$$\frac{2^{-\frac{1}{2}(m-p-1)p}\det(\Psi)^{\frac{1}{2}(m-p-1)}}{\Gamma_p[\frac{1}{2}(m-p-1)]}\,\mathrm{etr}\left(-\frac{1}{2}\Theta\right)\mathrm{etr}\left(-\frac{1}{2}V^{-1}\Psi\right)\det(V)^{-\frac{1}{2}m}$$

$$_0F_1\left(\frac{1}{2}(m-p-1); \frac{1}{4}\Theta\Psi V^{-1}\right), \ V > 0, \ \Psi > 0, \ m > 2p.$$

This distribution is a matrix variate generalization of inverted noncentral gamma distribution. It may be noted that if $V \sim IW_p(m, \Psi, \Theta)$, then $V^{-1} \sim W_p(m - p - 1, \Psi^{-1}, \Theta)$.

3.6. MATRIX VARIATE GAMMA DISTRIBUTION

Asoo (1969) defined the matrix variate gamma distribution as follows.

DEFINITION 3.6.1. *A random positive definite matrix W $(p \times p)$ is said to follow a matrix variate gamma distribution, denoted as $W \sim G_p(a, C)$, if its p.d.f. is*

$$\left\{\Gamma_p(a) \det(C)^{-a}\right\}^{-1} \mathrm{etr}(-CW) \det(W)^{a-\frac{1}{2}(p+1)}, \ W > 0,$$

where C $(p \times p) > 0$ and $a > \frac{1}{2}(p-1)$.

Note that if $S \sim W_p(n, \Sigma)$, then $S \sim G_p\left(\frac{1}{2}n, \frac{1}{2}\Sigma^{-1}\right)$. Similarly the random positive definite matrix W $(p \times p)$ has the noncentral matrix variate gamma distribution, $G_p(a, C, \Theta)$, if its p.d.f. is

$$\left\{\Gamma_p(a) \det(C)^{-a}\right\}^{-1} \mathrm{etr}(-\Theta - CW) \det(W)^{a-\frac{1}{2}(p+1)} \, {}_0F_1(a; \Theta CW), \ W > 0,$$

where C $(p \times p) > 0$, $a > \frac{1}{2}(p-1)$ and the symmetric matrix Θ is the noncentrality parameter. In this case if $S \sim W_p(n, \Sigma, \Theta)$, then $S \sim G_p(\frac{1}{2}n, \frac{1}{2}\Sigma^{-1}, \frac{1}{2}\Theta)$.

From Definitions 3.4.1 and 3.6.1, we define the matrix variate inverted gamma distribution with the notation, $W \sim IG_p(m, B)$, if its p.d.f. is

$$\frac{\det(B)^{m-\frac{1}{2}(p+1)}}{\Gamma_p[m - \frac{1}{2}(p+1)]} \det(W)^{-m} \mathrm{etr}(-BW^{-1}), \ W > 0,$$

where B $(p \times p) > 0$ and $m > p$. If $W \sim IG_p(m, B)$, then $W^{-1} \sim G_p(m - \frac{1}{2}(p+1), B)$. Conversely if $W \sim G_p(a, C)$, then $W^{-1} \sim IG_p(a + \frac{1}{2}(p+1), C)$.

Using Bellman's (1956) integral identities, one can also give the following generalizations of matrix variate gamma distribution (see Olkin, 1959).

DEFINITION 3.6.2. *A random positive definite matrix W $(p \times p)$ is said to follow Bellman gamma type I distribution, denoted by $W \sim BG_p^I(a_1, \ldots, a_p; C)$, if its p.d.f. is given by*

$$\left\{\Gamma_p^*(a_1, \ldots, a_p) \prod_{\alpha=1}^{p} \det(C_{(\alpha)})^{-m_\alpha}\right\}^{-1} \mathrm{etr}(-CW) \det(W)^{a_p - \frac{1}{2}(p+1)} \prod_{\alpha=1}^{p-1} \det(W^{[\alpha]})^{-m_{\alpha+1}},$$

where C $(p \times p) > 0$ is a constant matrix, $a_j = m_1 + \cdots + m_j$, and $a_j > \frac{1}{2}(j-1)$, $j = 1, \ldots, p$.

The generalized multivariate gamma function $\Gamma_p^*(a_1, \ldots, a_p)$ is defined in Theorem 1.4.6, and the matrices $A^{[\alpha]}$ and $A_{(\alpha)}$ are given in Definition 1.2.4.

DEFINITION 3.6.3. *A random positive definite matrix, W $(p \times p)$, is said to follow Bellman gamma type II distribution, denoted by $W \sim BG_p^{II}(b_1, \ldots, b_p; B)$, if its p.d.f. is given by*

$$\left\{ \Gamma_p^*(b_1, \ldots, b_p) \prod_{\alpha=1}^{p} \det(B^{[\alpha]})^{-k_\alpha} \right\}^{-1} \text{etr}(-BW) \det(W)^{b_p - \frac{1}{2}(p+1)} \prod_{\alpha=2}^{p} \det(W_{(\alpha)})^{-k_\alpha - 1}$$

where B $(p \times p) > 0$ is a constant matrix, $b_j = k_{p-j+1} + \cdots + k_p$, and $b_j > \frac{1}{2}(j-1)$, $j = 1, \ldots, p$.

THEOREM 3.6.1. *Let $S = TT' \sim W_p(n, \Sigma)$, where T $(p \times p)$ is a lower triangular matrix with positive diagonal elements, then the distribution of the matrix $R = T'\Sigma^{-1}T$ is*

$$\left\{ 2^{\frac{1}{2}np} \Gamma_p\left(\frac{1}{2}n\right) \right\}^{-1} \det(R)^{\frac{1}{2}(n-2)} \prod_{i=2}^{p} \det(R_{(i)})^{-1} \text{etr}\left(-\frac{1}{2}R\right), \ R > 0.$$

Proof: The density S is

$$\left\{ 2^{\frac{1}{2}np} \Gamma_p\left(\frac{1}{2}n\right) \det(\Sigma)^{\frac{1}{2}n} \right\}^{-1} \det(S)^{\frac{1}{2}(n-p-1)} \text{etr}\left(-\frac{1}{2}\Sigma^{-1}S\right).$$

Let $S = TT'$, then the Jacobian of transformation is $J(S \to T) = 2^p \prod_{i=1}^{p} t_{ii}^{p-i+1}$, and the density of $T = (t_{ij})$ is

$$2^p \left\{ 2^{\frac{1}{2}np} \Gamma_p\left(\frac{1}{2}n\right) \det(\Sigma)^{\frac{1}{2}n} \right\}^{-1} \det(TT')^{\frac{1}{2}(n-p-1)} \text{etr}\left(-\frac{1}{2}\Sigma^{-1}TT'\right) \prod_{i=1}^{p} t_{ii}^{p-i+1}.$$

Write $\Sigma^{-1} = A'A$ where $A = (a_{ij})$ is a lower triangular matrix and transform $R_1 = (r_{ij(1)}) = AT$, which is a lower triangular matrix and $r_{ii(1)} = a_{ii}t_{ii}$. The Jacobian of transformation from (1.3.7) is $J(T \to R_1) = \prod_{i=1}^{p} a_{ii}^{-i}$, and the density of R_1 is given by

$$2^p \left\{ 2^{\frac{1}{2}np} \Gamma_p\left(\frac{1}{2}n\right) \right\}^{-1} \det(R_1'R_1)^{\frac{1}{2}(n-p-1)} \text{etr}\left(-\frac{1}{2}R_1'R_1\right) \prod_{i=1}^{p} r_{ii(1)}^{p+1-i}.$$

Now, let $R = R_1'R_1 = T'\Sigma^{-1}T$ and get

$$r_{ii(1)} = \begin{cases} \left\{ \dfrac{\det(R_{(i)})}{\det(R_{(i+1)})} \right\}^{\frac{1}{2}}, & i = 1, \ldots, p-1 \\ r_{pp}^{\frac{1}{2}}, & i = p \end{cases}.$$

The Jacobian of this transformation is $J(R_1 \to R) = 2^{-p} \prod_{i=1}^{p} r_{ii(1)}^{-i}$, and from the density of R_1, we get the density of R as

$$\left\{ 2^{\frac{1}{2}np} \Gamma_p\left(\frac{1}{2}n\right) \right\}^{-1} \det(R)^{\frac{1}{2}(n-2)} \prod_{i=2}^{p} \det(R_{(i)})^{-1} \text{etr}\left(-\frac{1}{2}R\right). \ \blacksquare$$

Tan and Guttman (1971) derived the above density in a slightly different form and called it the disguised Wishart distribution. However, this distribution is a special case of Bellman gamma distribution type II given above. The disguised inverted Wishart distribution has been studied by Gupta and Ofori-Nyarko (1995).

3.7. APPROXIMATIONS

In this section we derive approximations to the distributions of a linear combination of Wishart matrices and a noncentral Wishart matrix. The linear combination of independent Wishart matrices arise in matrix quadratic forms, MANOVA random effects models, and robustness studies involving mixtures of multivariate normal distributions.

Let $S_j \sim W_p(n_j, \Sigma_j)$, $j = 1, \ldots, k$ be mutually independent. Consider a linear combination

$$S = \sum_{j=1}^{k} a_j S_j, \, a_j > 0.$$

In the univariate case, the distribution of a linear combination of chi-square variables has been approximated by a chi-square distribution by equating the first two moments. In the present case Tan and R. P. Gupta (1983) have approximated the distribution of S by the distribution of W where $W \sim W_p(n, \Sigma)$ and n and Σ have been obtained by comparing their expected values and the generalized variances.

Write $S = (s_{uv})$, vecp$(S) = (s_{11}, s_{12}, s_{22}, \ldots, s_{1p}, \ldots, s_{pp})'$, $A_1 = \text{cov}(\text{vecp}(S))$, and $A_2 = \text{cov}(\text{vecp}(W))$. Then

$$E(S) = \sum_{j=1}^{k} a_j n_j \Sigma_j, \tag{3.7.1}$$

$$E(W) = n\Sigma, \tag{3.7.2}$$

$$A_1 = 2\sum_{j=1}^{k} a_j^2 n_j B_p'(\Sigma_j \otimes \Sigma_j) B_p, \tag{3.7.3}$$

and

$$A_2 = 2n B_p'(\Sigma \otimes \Sigma) B_p, \tag{3.7.4}$$

where the expressions (3.7.3) and (3.7.4) have been obtained by using a result given in Problem 3.19, and the matrix $B_p\left(p^2 \times \frac{1}{2}p(p+1)\right)$ has been defined in Section 1.2. Now equating the expected values from (3.7.1) and (3.7.2) and the generalized variances from (3.7.3) and (3.7.4) we get

$$\Sigma = \frac{1}{n}\sum_{j=1}^{k} a_j n_j \Sigma_j \tag{3.7.5}$$

and

$$n = \left\{\frac{n^{\frac{1}{2}p(p+1)}\det(A_1)}{\det(A_2)}\right\}^{\frac{2}{p(p+1)}} \tag{3.7.6}$$

It may be noted that $n^{\frac{1}{2}p(p+1)}\det(A_1)$ does not depend on n. Using (1.2.18), we get

$$
\begin{aligned}
\det(A_2) &= (2n)^{\frac{1}{2}p(p+1)}\det(B'_p(\Sigma\otimes\Sigma)B_p)\\
&= 2^p n^{\frac{1}{2}p(p+1)}\det(\Sigma)^{p+1}\\
&= 2^p n^{\frac{1}{2}p(p+1)}\det\Big(\frac{1}{n}\sum_{j=1}^k a_j n_j \Sigma_j\Big)^{p+1}\\
&= 2^p n^{-\frac{1}{2}p(p+1)}\det\Big(\sum_{j=1}^k a_j n_j \Sigma_j\Big)^{p+1}
\end{aligned}
$$

and therefore

$$
\begin{aligned}
n^{\frac{1}{2}p(p+1)}\det(A_1) &= n^{\frac{1}{2}p(p+1)}\det(A_2)\\
&= 2^p \det\Big(\sum_{j=1}^k a_j n_j \Sigma_j\Big)^{p+1}
\end{aligned}
$$

Another approximation to the distribution of S has been obtained by Khatri (1989), by comparing the expected values and the *total variance*.

Yet another approximation can be given by generalized Gram-Charlier series expansion, which becomes quite complicated if higher order derivatives are included (Tan, 1980 and Tan and R. P. Gupta, 1982).

The noncentral Wishart distribution has been approximated by a Wishart distribution (Steyn and Roux, 1972) by using the representation of noncentral Wishart matrix in normal vectors. Let $X \sim N_{p,n}(M,\Sigma\otimes I_n)$, $n \geq p$. Then $S = XX' = (s_{ij})$ has a noncentral Wishart distribution. From Theorem 3.5.8 the first two moments of S are given by

$$E(s_{ij}) = n\sigma_{ij} + \omega_{ij} \tag{3.7.7}$$

and

$$
\begin{aligned}
E(s_{ij}s_{k\ell}) &= (n\sigma_{ij} + \omega_{ij})(n\sigma_{k\ell} + \omega_{k\ell}) + n(\sigma_{ik}\sigma_{j\ell} + \sigma_{i\ell}\sigma_{jk})\\
&\quad + \sigma_{j\ell}\omega_{ik} + \sigma_{i\ell}\omega_{jk} + \sigma_{jk}\omega_{i\ell} + \sigma_{ik}\omega_{j\ell},
\end{aligned}\tag{3.7.8}
$$

where $MM' = (\omega_{ij})$. When $M = 0$, i.e., $\omega_{ij} = 0$, the above moments reduce to the moments of Wishart distribution given by

$$E(s_{ij}) = n\sigma_{ij} \tag{3.7.9}$$

and

$$E(s_{ij}s_{k\ell}) = n^2\sigma_{ij}\sigma_{k\ell} + n(\sigma_{ik}\sigma_{j\ell} + \sigma_{i\ell}\sigma_{jk}). \tag{3.7.10}$$

Now consider a Wishart matrix $B = (b_{ij})$, $B \sim W_p(n,\Sigma^*)$, where $\Sigma^* = \Sigma + \frac{1}{n}MM'$. Then from (3.7.9) and (3.7.10) we have

$$E(b_{ij}) = n\sigma_{ij} + \omega_{ij} \tag{3.7.11}$$

and

$$E(b_{ij}b_{k\ell}) = (n\sigma_{ij} + \omega_{ij})(n\sigma_{k\ell} + \omega_{k\ell}) + n(\sigma_{ik}\sigma_{j\ell} + \sigma_{i\ell}\sigma_{jk}) + \sigma_{j\ell}\omega_{ik}$$

$$+ \sigma_{i\ell}\omega_{jk} + \sigma_{jk}\omega_{i\ell} + \sigma_{ik}\omega_{j\ell} + \frac{1}{n}(\omega_{ik}\omega_{j\ell} + \omega_{i\ell}\omega_{jk}). \qquad (3.7.12)$$

Comparing (3.7.7) with (3.7.11) and (3.7.8) with (3.7.12) it is seen that the first order moments of S and B are identical, where as the second order moments differ in terms of order $O(n^{-1})$, i.e.,

$$E(b_{ij}) = E(s_{ij})$$

and

$$E(b_{ij}b_{k\ell}) = E(s_{ij}s_{k\ell}) + O(n^{-1}).$$

This suggests that we can approximate the distribution of S by a Wishart distribution with parameters n and $\Sigma + \frac{1}{n}MM'$. Note that the characteristic function ϕ of S satisfies the differential equation given in Theorem 3.5.5, viz.

$$\frac{\partial \phi}{\partial Z} = \iota \left\{ n(\Psi - 2\iota\Gamma)^{-1} + (\Psi - 2\iota\Gamma)^{-1}\Psi MM'\Psi(\Psi - 2\iota\Gamma)^{-1} \right\}\phi \qquad (3.7.13)$$

where $\Psi = \Sigma^{-1}$, and $\Gamma = \left(\frac{1}{2}(1 + \delta_{ij})z_{ij} \right)$. When $M = 0$, this differential equation reduces to

$$\frac{\partial \phi}{\partial Z} = n\iota(\Psi - 2\iota\Gamma)^{-1}\phi$$

$$= n\iota(I_p - 2\iota\Gamma\Sigma)^{-1}\Sigma\phi. \qquad (3.7.14)$$

From (3.7.14), the characteristic function, ϕ^*, of B satisfies the following differential equation

$$\frac{\partial \phi^*}{\partial Z} = n\iota \left\{ I_p - 2\iota\Gamma\left(\Sigma + \frac{1}{n}MM'\right) \right\}^{-1}\left(\Sigma + \frac{1}{n}MM'\right)\phi^*. \qquad (3.7.15)$$

Now, by taking Γ such that the conditions for convergence of matrix series are satisfied, from (3.7.15) it follows that

$$\frac{\partial \phi^*}{\partial Z} = n\iota \left[(I_p - 2\iota\Gamma\Sigma)^{-1} + (I_p - 2\iota\Gamma\Sigma)^{-1}\frac{2\iota\Gamma MM'}{n}(I_p - 2\iota\Gamma\Sigma)^{-1} \right.$$

$$\left. + O(n^{-2}) \right]\left(\Sigma + \frac{1}{n}MM'\right)\phi^*.$$

Thus

$$\frac{\partial \phi^*}{\partial Z} = n\iota \left[(\Psi - 2\iota\Gamma)^{-1} + (\Psi - 2\iota\Gamma)^{-1}\frac{\Psi MM'\Psi}{n}(\Psi - 2\iota\Gamma)^{-1} + O(n^{-2}) \right]\phi^*. \qquad (3.7.16)$$

The expressions in (3.7.13) and (3.7.16) differ only in terms of order $O(n^{-2})$, which indicates the closeness of approximation of the noncentral Wishart distribution by a central Wishart distribution. For further results on approximation of noncentral

Wishart distribution by a Wishart distribution see Tan (1979), Tan and R. P. Gupta (1982), and Kollo and von Rosen (1995). For results on the asymptotic expansion of the Wishart density, the reader is referred to Sugiura (1973), D. G. Nel (1978), and D. G. Nel and Groenewald (1979).

PROBLEMS

3.1. Let $X = (x_1, \ldots, x_N)$, where $x_i \sim N_p(\mu, \Sigma)$, $i = 1, \ldots, N$ are independently distributed. Further, let $A(N \times N)$ be a constant matrix of rank $(N - r)$. Then, prove that XAX' is positive definite with probability one if $N \geq p + r$.

3.2. Let $S \sim W_p(n, \Sigma)$ and $X \sim N_{p,m}(0, \Sigma \otimes I_m)$ are independently distributed. Assuming $m < p$, prove that $S + XX' \sim W_p(m + n, \Sigma)$.

3.3. Let $X \sim N_{p,m}(M, \Sigma \otimes \Psi | s, C)$, $n \geq p + s$. Show that $(X - M)\Psi^{-1}(X - M)' \sim W_p(n - s, \Sigma)$.

3.4. Prove Theorem 3.3.7, when n is an integer by expressing the matrix S in normal variables.

3.5. Let $S_1 \sim W_p(n_1, \Sigma_1)$ and $S_2 \sim W_p(n_2, \Sigma_2)$ be independent. Show that the p.d.f. of $S = S_1 + S_2$ is given by

$$\left\{ 2^{\frac{1}{2}(n_1+n_2)p} \Gamma_p \left[\frac{1}{2}(n_1 + n_2) \right] \det(\Sigma_1)^{\frac{1}{2}n_1} \det(\Sigma_2)^{\frac{1}{2}n_2} \right\}^{-1} \mathrm{etr} \left(-\frac{1}{2}\Sigma_1^{-1} S \right)$$

$$\det(S)^{\frac{1}{2}(n_1+n_2-p-1)} {}_1F_1\left(\frac{1}{2}n_2; \frac{1}{2}(n_1 + n_2); \frac{1}{2}(\Sigma_1^{-1} - \Sigma_2^{-1})S \right), \ S > 0.$$

3.6. Prove Theorem 3.3.9, when n is an integer by expressing the matrix S in normal variables.

3.7. Let $S \sim W_p(n, \Sigma)$ and partition S and Σ as in Theorem 3.3.9. Then, prove that S_{11} and S_{22} are independent if and only if $\Sigma_{12} = 0$.

3.8. Let $S \sim W_p(n, \Sigma)$ and for $A(p \times p) = (a_{ij})$ define $A^{[r]} = (a_{ij})$, $i, j = 1, \ldots, r$. Then prove that

$$\frac{\det(S^{[r]})}{\det(S^{[r-1]})} \frac{\det(\Sigma^{[r-1]})}{\det(\Sigma^{[r]})}, \ r = 1, \ldots, p,$$

where $\det(S^{[0]}) = \det(\Sigma^{[0]}) = 1$, are independently distributed as χ^2_{n-r+1}, $r = 1, \ldots, p$.

3.9. Let $S \sim W_p\left(n, \frac{1}{n}\Sigma\right)$. Prove that the asymptotic distribution, as $n \to \infty$, of $\left(\frac{n}{2p}\right)^{\frac{1}{2}} \ln \frac{\det(S)}{\det(\Sigma)}$ is $N(0, 1)$.

3.10. Let $S \sim W_p(n, \Sigma)$. Prove that the asymptotic distribution of $\sqrt{n} \left(\frac{\det(S)}{n^p \det(\Sigma)} - 1 \right)$ is normal with mean 0 and variance $2p$.

3.11. Let $S_1 \sim W_p(n_1, \Sigma)$ and $S_2 \sim W_p(n_2, \Sigma)$ be independent. Show that $S_1 + S_2$ and $(S_1 + S_2)^{-\frac{1}{2}} S_1 S_2^{-1} (S_1 + S_2)^{\frac{1}{2}}$ are independent, where $(S_1 + S_2)^{\frac{1}{2}}$ is any square root depending only on $S_1 + S_2$ and not on the individual values of S_1 and S_2.

(Perlman, 1977)

3.12. Let $S_i \sim W_p(n_i, \Sigma)$, $i = 1, \ldots, d$ be independently distributed.
(i) If $S = \sum_{j=1}^{d} S_j$ and $g(S_1, \ldots, S_d)$ are independently distributed, then show that the random variable $g(AS_1 A', \ldots, AS_d A')$ has the same distribution as $g(S_1, \ldots, S_d)$ for any nonsingular matrix A ($p \times p$).
(ii) If for each $B > 0$, there is an M with $B = MM'$ and $g(MS_1 M', \ldots, MS_d M')$ and $g(S_1, \ldots, S_d)$ have identical distribution, then prove that $S = \sum_{j=1}^{d} S_j$ and $g(S_1, \ldots, S_d)$ are independent.

(Olkin and Rubin, 1964)

3.13. Let $S_i \sim W_p(n_i, \Sigma)$, $i = 1, \ldots, d$ be independently distributed. Then, prove that the random matrices
(a) $W_j = (S_1 + \cdots + S_j)^{-\frac{1}{2}} S_{j+1} (S_1 + \cdots + S_j)^{-\frac{1}{2}}$, $j = 1, \ldots, d - 1$, where $(S_1 + \cdots + S_j)^{\frac{1}{2}}$ is the triangular root of $S_1 + \cdots + S_j$, are independently distributed, and
(b) $Z_j = (S_1 + \cdots + S_{j+1})^{-\frac{1}{2}} S_{j+1} (S_1 + \cdots + S_{j+1})^{-\frac{1}{2}}$, $j = 1, \ldots, d - 1$ where $(S_1 + \cdots + S_{j+1})^{\frac{1}{2}}$ is any nonsingular square root depending only on $S_1 + \cdots + S_{j+1}$, are independently distributed.

3.14. Let $S_i \sim W_p(n_i, \Sigma)$, $i = 1, \ldots, d$ be independently distributed, and $(S_1 + \cdots + S_j)^{\frac{1}{2}}$ be any square root depending only on $S_1 + \cdots + S_j$. Then, show that the random matrices

$$W_j = (S_1 + \cdots + S_j)^{-\frac{1}{2}} S_{j+1} (S_1 + \cdots + S_j)^{-\frac{1}{2}}, \ j = 1, \ldots, d - 1$$

are not independent. However, $\tilde{W}_1, \ldots, \tilde{W}_{d-1}$ are independent where

$$\tilde{W}_j = (S_1 + \cdots + S_{j+1})^{-\frac{1}{2}} S_{j+1} (S_1 + \cdots + S_{j+1})^{-1} (S_1 + \cdots + S_{j+1})^{\frac{1}{2}}.$$

(Olkin and Rubin, 1964; Perlman, 1977)

3.15. Let $S \sim W_p(n, \Sigma)$. Then, show that

(i) $E[\ln\{\det(S)\}] = \ln\{\det(\Sigma)\} + p \ln 2 + \sum_{i=1}^{p} \psi\left[\frac{1}{2}(n - i + 1)\right]$,

where $\psi(\cdot)$ is the psi-function.
(ii) When $\Sigma = I_p$, $a \in \mathbb{R}^p$, $a \neq 0$,

$$E\left[\frac{(a'S^{-1}a)(a'S^{-2}a)}{(a'a)^2}\right] = \frac{(n-1)}{(n-p)(n-p-1)(n-p-3)(n-p-5)}, \ n > p + 5.$$

3.16. Let $nS \sim W_p(n, I_p)$ and $S = I_p + n^{-\frac{1}{2}}W$, where $W = (w_{ij})$. Furthermore, let $\boldsymbol{a}$ be a fixed vector. Then prove that
(i) $E(w_{11}^2) = 2$
(ii) $E(w_{12}^2) = 1$
(iii) $E(\boldsymbol{a}'W\boldsymbol{a})^2 = 2(\boldsymbol{a}'\boldsymbol{a})^2$
(iv) $E(\boldsymbol{a}'W^2\boldsymbol{a}) = (p+1)\boldsymbol{a}'\boldsymbol{a}$.

3.17. Let $S \sim W_p(n, \Sigma)$ and put $\alpha = \frac{1}{\delta}E\left(\frac{pn-2}{\operatorname{tr} S}\right)$, where $\delta = \frac{1}{p}\operatorname{tr}(\Sigma^{-1})$. Prove that

(i) $\alpha = \frac{1}{\delta}E\left(\dfrac{\operatorname{tr}(\Sigma^{-1}S)}{\operatorname{tr} S}\right)$

(ii) $0 < \alpha \le 1$ for all $\Sigma > 0$.

3.18. Let $S \sim W_p(n, \Sigma)$ and u be distributed as beta with parameters $(\frac{1}{2}m, \frac{1}{2}(n-m))$ independent of S, $n > m$. Further, let $A = uS$ and $\boldsymbol{a}$ be any $p \times 1$ vector of constants. Then prove that

(i) $\dfrac{\boldsymbol{a}'A\boldsymbol{a}}{\boldsymbol{a}'\Sigma\boldsymbol{a}} \sim \chi_m^2$, $\boldsymbol{a} \ne 0$

(ii) $E(A) = m\Sigma$.

3.19. Let $S \sim W_p(n, \Sigma)$. Prove that

$$\operatorname{cov}(\operatorname{vec}(S)) = n(I_{p_2} + K_{pp})(\Sigma \otimes \Sigma),$$

and

$$\operatorname{cov}(\operatorname{vecp}(S)) = 2nB_p'(\Sigma \otimes \Sigma)B_p$$

where the matrices K_{pp} and B_p are defined in Section 1.2.

3.20. Let $S = TT' \sim W_3(n, I_3)$, where $T = (t_{ij})$ is a lower triangular matrix with positive diagonal elements. Prove that

(i) $E\left(\dfrac{1}{t_{11}^4}\right) = \dfrac{1}{(n-2)(n-4)}$, $n > 4$

(ii) $E\left(\dfrac{t_{21}^2}{t_{11}^4 t_{22}^2}\right) = \dfrac{1}{(n-2)(n-3)(n-4)}$, $n > 4$

(iii) $E\left(\dfrac{t_{31} - t_{32}t_{22}^{-1}t_{21}}{t_{11}^2 t_{33}}\right)^2 = \dfrac{1}{(n-3)(n-4)^2}$, $n > 4$

(iv) $E\left(\dfrac{t_{21}^2}{t_{11}^2 t_{22}^2} + \dfrac{1}{t_{22}^2}\right)^2 = \dfrac{(n-1)}{(n-2)(n-4)(n-5)}$, $n > 5$

(v) $E\left(\dfrac{t_{21}(t_{31} - t_{32}t_{22}^{-1}t_{21})}{t_{11}^2 t_{22}t_{33}} - \dfrac{t_{32}}{t_{22}^2 t_{33}}\right)^2 = \dfrac{n^2 - 3n - 2}{(n-2)(n-3)(n-4)^2(n-5)}$, $n > 5$

and hence, show that for $n > 6$,

$$
E(T'T)^{-2} = \begin{pmatrix} \frac{1}{(n-4)^2} & 0 & 0 \\ 0 & \frac{n^2-3n-2}{(n-2)(n-4)^2(n-5)} & 0 \\ 0 & 0 & \frac{(n-1)(n^2-3n-6)}{\prod_{i=2}^{6}(n-i)} \end{pmatrix}.
$$

3.21. Let $S \sim W_p(n, I_p)$ and $S = TT'$, where T is a lower triangular matrix with positive diagonal elements. Further, let $Q = E(T'DT)^{-2}$, where D is a diagonal matrix with elements ± 1. Then
(i) show that Q is a diagonal matrix, and
(ii) find a recurrence relation between the diagonal elements of Q.

(Krishnamoorthy and Gupta, 1989)

3.22. Let $S \sim W_p(n, \Sigma)$ and $S = TT'$, where T is a lower triangular matrix with positive diagonal elements. Show that

(i) $t_{pp}^2 = \dfrac{1}{s^{pp}}$, where $S^{-1} = (s^{ij})$,

(ii) $t_{pp}^2 \sim \dfrac{1}{\sigma^{pp}}\chi_{n-p+1}^2$, where $\Sigma^{-1} = (\sigma^{ij})$,

(iii) when $\Sigma = I_p$, $E(T'AT)^{-1} = (w_{ij})$, where $w_{ij} = \beta_i b_{ij} \beta_j$, $i \neq j$; $w_{11} = \dfrac{b_{11}}{(n-2)}$, $w_{ii} = \dfrac{1}{n-i-1}\left[\dfrac{1}{n-1}\sum_{j=1}^{i-1} b_{ij} + b_{ii}\right]$, $i = 2, \ldots, p$, $A^{-1} = B = (b_{ij})$, and $\beta_i = \sqrt{2}\,\dfrac{\Gamma[\frac{1}{2}(n-i)]}{\Gamma[\frac{1}{2}(n-i+1)]}$, $i = 1, \ldots, p$.

3.23. Prove the results (i) and (ii) in Problem 3.21, where T is an upper triangular matrix.

3.24. Let $S \sim W_p(n, \Sigma)$ and $S = TT'$, where T is an upper triangular matrix with positive diagonal elements. Show that

(i) $t_{11}^2 = \dfrac{1}{s_{11}}$, where $S^{-1} = (s^{ij})$.

(ii) $t_{11}^2 \sim \dfrac{1}{\sigma^{11}}\chi_{n-p+1}^2$, where $\Sigma^{-1} = (\sigma^{ij})$.

3.25. Let r be the sample correlation coefficient from a sample of size $n + 1$ from a bivariate normal population. Assuming that the population correlation coefficient r is different from zero, show that the p.d.f. of r is given by

$$
\frac{(n-1)\Gamma(n)}{(2\pi)^{\frac{1}{2}}\Gamma(n+\frac{1}{2})}(1-\rho^2)^{\frac{1}{2}n}(1-r^2)^{\frac{1}{2}(n-3)}(1-\rho r)^{-n+\frac{1}{2}}\,{}_2F_1\left(\frac{1}{2}, \frac{1}{2}; n+\frac{1}{2}; \frac{1}{2}(1+\rho r)\right).
$$

3.26. Let R be the correlation matrix of a random sample of size $n+1$ from $N_p(\boldsymbol{\mu}, I_p)$. Then, prove that

$$
E[\det(R)^h] = \prod_{i=2}^{p}\left\{\frac{\Gamma[\frac{1}{2}(n-i+1)+h]\Gamma(\frac{1}{2}n)}{\Gamma[\frac{1}{2}(n-i+1)]\Gamma(\frac{1}{2}n+h)}\right\}, \quad \mathrm{Re}(h) > -\frac{1}{2}(n-p+1).
$$

3.27. Let $S \sim W_p(n, \Sigma)$ and a priori $\Sigma \sim IW_p(m, \Psi)$. Show that given S, the posterior distribution of Σ is $IW_p(n + m, S + \Psi)$.

3.28. Let $x_i \sim N_p(\mu_i, \Sigma)$, $i = 1, \ldots, N$ be independently distributed. Prove that under suitable transformation $S = \sum_{i=1}^{N}(x_i - \bar{x})(x_i - \bar{x})'$, where $\bar{x} = \frac{1}{N}\sum_{i=1}^{N} x_i$ can be represented as $S = \sum_{i=1}^{N-1} y_i y_i'$ with $y_i \sim N_p(\nu_i, \Sigma)$, $i = 1, \ldots, N-1$ independent.

3.29. Let $X \sim N_{p,n}(M, \Sigma \otimes I_n)$, $M = \begin{pmatrix} m_1 & \cdots & m_n \\ & 0 & \end{pmatrix}$ where $m_1, \ldots, m_n$ are scalars. Derive the p.d.f. of XX'.

3.30. Let $S \sim W_p(n, \Sigma, \Theta)$ and a $(p \times 1)$ be a vector of constants. Then prove that $a'Sa \sim (a'\Sigma a)\chi_m'^2(\lambda)$, where $\lambda = \frac{2a'\Sigma \Theta a}{a'\Sigma a}$.

3.31. Let $S \sim W_p(n, \Sigma, \Theta)$. Prove that the characteristic function of $\mathrm{tr}(S)$ is

$$\phi_{\mathrm{tr}(S)}(t) = \det(I_p - 2\iota t \Sigma)^{-\frac{1}{2}n} \exp[\iota t \, \mathrm{tr}\{\Theta\Sigma(I_p - 2\iota t\Sigma)^{-1}\}].$$

3.32. Let $S \sim W_p(n, \Sigma)$. Then show that

$$E(S^{-1} \otimes S^{-1}) = c_1(\Sigma^{-1} \otimes \Sigma^{-1}) + c_2 \, \mathrm{vec}(\Sigma^{-1})(\mathrm{vec}(\Sigma^{-1}))' + c_2 K_{pp}(\Sigma^{-1} \otimes \Sigma^{-1})$$

and

$$\begin{aligned}
E(S^{-1} \otimes S^{-1} \otimes S^{-1}) &= c_3 c_1(\Sigma^{-1} \otimes \Sigma^{-1} \otimes \Sigma^{-1}) \\
&\quad + (c_4 c_1 + c_3 c_2) \, \mathrm{vec}(\Sigma^{-1})(\mathrm{vec}(\Sigma^{-1}))' \otimes \Sigma^{-1} \\
&\quad + (c_3 c_2 + c_4 c_2) P_1(\Sigma^{-1} \otimes \Sigma^{-1} \otimes \Sigma^{-1}) \\
&\quad + c_3 c_2 P_2 \, \mathrm{vec}(\Sigma^{-1})(\mathrm{vec}(\Sigma^{-1}))' \otimes \Sigma^{-1} P_2 \\
&\quad + c_3 c_2 P_2 P_1(\Sigma^{-1} \otimes \Sigma^{-1} \otimes \Sigma^{-1}) P_2 \\
&\quad + (c_4 c_1 - c_3 c_2) P_2(\Sigma^{-1} \otimes \Sigma^{-1} \otimes \Sigma^{-1}) \\
&\quad + c_4 c_2 P_2 P_1(\Sigma^{-1} \otimes \Sigma^{-1} \otimes \Sigma^{-1}) \\
&\quad + 2 c_4 c_2 \, \mathrm{vec}(\Sigma^{-1})(\mathrm{vec}(\Sigma^{-1}))' \otimes \Sigma^{-1} P_2 P_1 \\
&\quad - (c_3 c_2 + c_4 c_2)\Sigma^{-1} \otimes (\mathrm{vec}(\Sigma^{-1})(\mathrm{vec}(\Sigma^{-1}))') \\
&\quad - c_4 c_2 P_1 P_2 \, \mathrm{vec}(\Sigma^{-1})(\mathrm{vec}(\Sigma^{-1}))' \otimes \Sigma^{-1},
\end{aligned}$$

where $P_1 = K_{pp} \otimes I_p$, $P_2 = I_p \otimes K_{pp}$, and c_1, c_2, c_3 and c_4 are defined in Theorem 3.3.17.

3.33. Let $S \sim W_p(n, \Sigma)$. Then show that

$$E(S^{-1} \otimes S) = \frac{n}{n - p - 1}\Sigma^{-1} \otimes \Sigma$$

$$- \frac{1}{n - p - 1}(\mathrm{vec}(I_p)(\mathrm{vec}(I_p))' + K_{pp}), \quad n - p - 1 > 0.$$

CHAPTER 4

MATRIX VARIATE
t-DISTRIBUTION

4.1. INTRODUCTION

Let x and v be independent random variables distributed as standard normal and chi-square with n degrees of freedom respectively. Then, the random variable

$$t = \left(\frac{v}{n}\right)^{-\frac{1}{2}} x$$

is said to have t-distribution with n degrees of freedom. In the multivariate case, x is replaced by the vector $\boldsymbol{x}$, which is distributed as $N_p(\boldsymbol{0}, \Sigma)$ and define

$$\boldsymbol{t} = \left(\frac{v}{n}\right)^{-\frac{1}{2}} \boldsymbol{x}, \tag{4.1.1}$$

which is distributed as multivariate t with parameters n and Σ. The density of $\boldsymbol{t}$ is given by

$$\frac{\Gamma[\frac{1}{2}(n+p)]}{(n\pi)^{\frac{1}{2}p}\Gamma(\frac{1}{2}n)} \det(\Sigma)^{-\frac{1}{2}} \left(1 + \frac{1}{n}\boldsymbol{t}'\Sigma^{-1}\boldsymbol{t}\right)^{-\frac{1}{2}(n+p)}, \ \boldsymbol{t} \in \mathbb{R}^p. \tag{4.1.2}$$

It is also known that $\boldsymbol{t}$ has the representation

$$\boldsymbol{t} = (S^{-\frac{1}{2}})'\boldsymbol{y} \tag{4.1.3}$$

where $S = S^{\frac{1}{2}}(S^{\frac{1}{2}})' \sim W_p(n+p-1, \Sigma^{-1})$ and $\boldsymbol{y} \sim N_p(\boldsymbol{0}, nI_p)$ are independent. In this chapter matrix variate generalization of (4.1.2) is studied. Because of its applications in Bayesian inference, many researchers have studied this distribution, *e.g.*, Khatri (1959a), Kshirsagar (1961a), Tan (1964), Tiao and Zellner (1964), Geisser (1965), Dickey (1967, 1976), Juritz (1973), Rinco (1973), Haq and Rinco (1976), Marx (1981), Marx and Nel (1982), Javier (1982), Javier and Gupta (1985b), and Phillips (1985).

4.2.   DENSITY FUNCTION

The matrix variate t-distribution is defined as follows.

DEFINITION 4.2.1. *The random matrix $T\,(p \times m)$ is said to have a matrix variate t-distribution with parameters M, Σ, Ω, and n if its p.d.f. is given by*

$$\frac{\Gamma_p[\frac{1}{2}(n+m+p-1)]}{\pi^{\frac{1}{2}mp}\Gamma_p[\frac{1}{2}(n+p-1)]}\det(\Sigma)^{-\frac{1}{2}m}\det(\Omega)^{-\frac{1}{2}p}$$

$$\det(I_p + \Sigma^{-1}(T-M)\Omega^{-1}(T-M)')^{-\frac{1}{2}(n+m+p-1)}, \qquad (4.2.1)$$

where $T \in \mathbb{R}^{p\times m}$, $M \in \mathbb{R}^{p\times m}$, $\Omega\,(m \times m) > 0$, $\Sigma\,(p \times p) > 0$ and $n > 0$.

We shall denote this by $T \sim T_{p,m}(n, M, \Sigma, \Omega)$. Dickey, Dawid and Kadane (1986) call the matrices Ω and Σ, the spread matrices and n the degrees of freedom. This distribution belongs to the class of matrix variate elliptically contoured distributions studied in Chapter 9. In particular, for $M = 0$, this distribution belongs to (i) the class of right spherical distributions if $\Omega = I_m$, (ii) the class of left spherical distributions if $\Sigma = I_p$, and (iii) the class of spherical distributions if $\Omega = I_m$ and $\Sigma = I_p$. When $n = 1$, this distribution may be called the matrix variate Cauchy. When $m = 1$ or $p = 1$ it reduces to a multivariate t-distribution (Cornish, 1954, 1955, 1962; Dunnett and Sobel, 1954; Lin, 1972). More specifically when $m = 1$, $T = \boldsymbol{t}\,(p \times 1)$, $M = \boldsymbol{\mu}(p \times 1)$, $\Omega = \omega$ and (4.2.1) becomes

$$\pi^{-\frac{1}{2}p}\frac{\Gamma[\frac{1}{2}(n+p)]}{\Gamma(\frac{1}{2}n)}\det(\Sigma)^{-\frac{1}{2}}\omega^{-\frac{1}{2}p}\Big(1 + \frac{1}{\omega}(\boldsymbol{t}-\boldsymbol{\mu})'\Sigma^{-1}(\boldsymbol{t}-\boldsymbol{\mu})\Big)^{-\frac{1}{2}(n+p)}, \; \boldsymbol{t} \in \mathbb{R}^p,$$

which will be denoted by $\boldsymbol{t} \sim t_p(n, \omega, \boldsymbol{\mu}, \Sigma)$. For $p = 1$, by taking $M = \boldsymbol{\nu}'$ and $\Sigma = \sigma$, it is easily seen that $T' = \boldsymbol{t} \sim t_m(n, \sigma, \boldsymbol{\nu}, \Omega)$.

This distribution can be derived in a manner similar to the univariate theory as shown in the following theorem.

THEOREM 4.2.1. *Let $S \sim W_p(n+p-1, \Sigma^{-1})$, independent of $X \sim N_{p,m}(0, I_p \otimes \Omega)$. Define*

$$T = (S^{-\frac{1}{2}})'X + M, \qquad (4.2.2)$$

where $M\,(p \times m)$ is a constant matrix, and $S^{\frac{1}{2}}(S^{\frac{1}{2}})' = S$. Then, $T \sim T_{p,m}(n, M, \Sigma, \Omega)$.

Proof: The joint density of S and X is given by

$$\frac{\pi^{-\frac{1}{2}mp}\det(\Sigma)^{\frac{1}{2}(n+p-1)}\det(\Omega)^{-\frac{1}{2}p}}{2^{\frac{1}{2}(n+m+p-1)p}\Gamma_p[\frac{1}{2}(n+p-1)]}\det(S)^{\frac{1}{2}(n-2)}\operatorname{etr}\Big(-\frac{1}{2}\Sigma S - \frac{1}{2}X\Omega^{-1}X'\Big),$$

$$S > 0, \; X \in \mathbb{R}^{p\times n}.$$

Now, let $T = (S^{-\frac{1}{2}})'X + M$. The Jacobian of the transformation is $J(X \to T) = \det(S)^{\frac{1}{2}m}$. Substituting for X in terms of T in the joint density of X and S, and multiplying the resulting expression by $J(X \to T)$, we get the joint p.d.f. of T and S as

$$\frac{\pi^{-\frac{1}{2}mp} \det(\Sigma)^{\frac{1}{2}(n+p-1)} \det(\Omega)^{-\frac{1}{2}p}}{2^{\frac{1}{2}(n+m+p-1)p} \Gamma_p[\frac{1}{2}(n+p-1)]} \det(S)^{\frac{1}{2}(n+m-2)}$$

$$\text{etr}\left[-\frac{1}{2}S\{\Sigma + (T-M)\Omega^{-1}(T-M)'\}\right], \ S > 0, \ T \in \mathbb{R}^{p \times m}.$$

Now, integrating out S using multivariate gamma integral (1.4.6) the density of T is obtained as

$$\frac{\Gamma_p[\frac{1}{2}(n+m+p-1)]}{\pi^{\frac{1}{2}mp} \Gamma_p[\frac{1}{2}(n+p-1)]} \det(\Sigma)^{-\frac{1}{2}m} \det(\Omega)^{-\frac{1}{2}p}$$

$$\det\left(I_p + \Sigma^{-1}(T-M)\Omega^{-1}(T-M)'\right), \ T \in \mathbb{R}^{p \times m}. \ \blacksquare$$

The above result was proved by Dickey (1967). Another representation of T when Σ and Ω are symmetric nonnegative definite matrices is given by Dickey, Dawid and Kadane (1986).

4.3.   PROPERTIES

In this section, various properties of the random matrix T are studied using its p.d.f. and the representation (4.2.2). First, we derive expected values of the random matrix T and some of its functions.

THEOREM 4.3.1. Let $T \sim T_{p,m}(n, M, \Sigma, \Omega)$, then

$$E(T) = M$$

and

$$\text{cov}(\text{vec}(T')) = \frac{1}{(n-2)}\Sigma \otimes \Omega, \ n > 2.$$

Proof: According to Theorem 4.2.1 the random matrix T can be represented as

$$T = (S^{-\frac{1}{2}})'X + M, \tag{4.3.1}$$

where $S^{\frac{1}{2}}(S^{\frac{1}{2}})' \sim W_p(n+p-1, \Sigma^{-1})$ and $X \sim N_{p,m}(0, I_p \otimes \Omega)$ are independent. From (4.3.1), it is seen that $T|S \sim N_{p,m}(M, S^{-1} \otimes \Omega)$ and therefore

$$E(T|S) = M \tag{4.3.2}$$

and

$$\text{cov}(\text{vec}(T')|S) = S^{-1} \otimes \Omega. \tag{4.3.3}$$

Now, from (4.3.2), we have $E(T) = M$ since the conditional expectation does not depend on S. Also, from (4.3.3) we have

$$\begin{aligned}
\text{cov}(\text{vec}(T')) &= E_S\{\text{cov}(\text{vec}(T')|S)\} \\
&= E_S(S^{-1} \otimes \Omega) \\
&= (n-2)^{-1}\Sigma \otimes \Omega.
\end{aligned}$$

The last step follows from Theorem 3.3.16. ■

THEOREM 4.3.2. *Let* $T \sim T_{p,m}(n, M, \Sigma, \Omega)$, *then*
  (i) $E(TCT) = (n-2)^{-1}\Sigma C'\Omega + MCM$, $C\,(m \times p)$,
  (ii) $E(TCT') = (n-2)^{-1}\text{tr}(C'\Omega)\Sigma + MCM'$, $C\,(m \times m)$,
  (iii) $E(TCT'DT) = (n-2)^{-1}\Sigma D'MC'\Omega + (n-2)^{-1}\text{tr}(D\Sigma)MC\Omega$
                       $+(n-2)^{-1}\text{tr}(C'\Omega)\Sigma DM + MCM'DM$,
                       $C\,(m \times m),\ D\,(p \times p)$,
  (iv) $E(TCTDT) = (n-2)^{-1}\Sigma D'M'C'\Omega + (n-2)^{-1}MC\Sigma D'\Omega$
                      $+(n-2)^{-1}\Sigma C'\Omega DM + MCMDM$, $C\,(m \times p),\ D\,(m \times p)$,
where $n > 2$.

Proof: The representation (4.3.1) yields

$$\begin{aligned}
E(TCT) &= E_S[E_X\{((S^{-\frac{1}{2}})'X + M)C((S^{-\frac{1}{2}})'X + M)|S\}] \\
&= E_S[E_X\{((S^{-\frac{1}{2}})'XC(S^{-\frac{1}{2}})'X + (S^{-\frac{1}{2}})'XCM \\
&\quad + MC(S^{-\frac{1}{2}})'X + MCM)|S\}] \\
&= E_S[(S^{-\frac{1}{2}})'S^{-\frac{1}{2}}C'\Omega + MCM]
\end{aligned}$$

where the last equality follows from Theorem 2.3.5, since $X \sim N_{p,m}(0, I_p \otimes \Omega)$. Now, using Theorem 3.3.16, the result is easily obtained. The derivation of $E(TCT')$ is similar. Also,

$$\begin{aligned}
E(TCTDT) &= E_S[E_X\{((S^{-\frac{1}{2}})'X + M)C((S^{-\frac{1}{2}})'X + M)D((S^{-\frac{1}{2}})'X + M)|S\}] \\
&= E_S[E_X\{((S^{-\frac{1}{2}})'XC(S^{-\frac{1}{2}})'XD(S^{-\frac{1}{2}})'X + (S^{-\frac{1}{2}})'XCMD(S^{-\frac{1}{2}})'X \\
&\quad + MC(S^{-\frac{1}{2}})'XD(S^{-\frac{1}{2}})'X + MCMD(S^{-\frac{1}{2}})'X \\
&\quad + (S^{-\frac{1}{2}})'XC(S^{-\frac{1}{2}})'XDM + (S^{-\frac{1}{2}})'XCMDM \\
&\quad + MC(S^{-\frac{1}{2}})'XDM + MCMDM)|S\}] \\
&= E_S[(S^{-\frac{1}{2}})'S^{-\frac{1}{2}}D'M'C'\Omega + MC(S^{-\frac{1}{2}})'S^{-\frac{1}{2}}D'\Omega \\
&\quad + (S^{-\frac{1}{2}})'S^{-\frac{1}{2}}C'\Omega DM + MCMDM],
\end{aligned}$$

where the last equality has been obtained from Theorems 2.3.5 and 2.3.6. The desired result now follows from Theorem 3.3.16. The derivation of $E(TCT'DT)$ is similar. ■

THEOREM 4.3.3. *If $T \sim T_{p,m}(n, M, \Sigma, \Omega)$, then $T' \sim T_{m,p}(n, M', \Omega, \Sigma)$.*

Proof: The result follows by noting that

$$f(T) \propto \det(I_p + \Sigma^{-1}(T - M)\Omega^{-1}(T - M)')^{-\frac{1}{2}(n+m+p-1)}$$

$$= \det(I_m + \Omega^{-1}(T' - M')\Sigma^{-1}(T' - M')')^{-\frac{1}{2}(n+p+m-1)}. \quad \blacksquare$$

It may be noted that the matrix variate t-distribution is a mixture of matrix variate normal distributions and matrix variate normal distribution is, itself, a limiting case of the matrix variate t-distribution as shown below.

THEOREM 4.3.4. *Let $T \sim T_{p,m}(n, M, n\Sigma, \Omega)$, then $T \xrightarrow{\mathcal{D}} X$ as $n \to \infty$ where $X \sim N_{p,m}(M, \Sigma \otimes \Omega)$ and "$\xrightarrow{\mathcal{D}}$" denotes convergence in distribution.*

Proof: The p.d.f. of T is

$$f(T) = \frac{\Gamma_p[\frac{1}{2}(n + m + p - 1)]}{\pi^{\frac{1}{2}mp}\Gamma_p[\frac{1}{2}(n + p - 1)]} \det(n\Sigma)^{-\frac{1}{2}m} \det(\Omega)^{-\frac{1}{2}p}$$

$$\det\left(I_p + \frac{1}{n}\Sigma^{-1}(T - M)\Omega^{-1}(T - M)'\right)^{-\frac{1}{2}(n+m+p-1)}.$$

Now, since $\lim_{n\to\infty} \det(I_p + \frac{1}{n}A)^{-\frac{1}{2}(n+m+p-1)} = \text{etr}(-\frac{1}{2}A)$, where $A = \Sigma^{-1}(T-M)\Omega^{-1}(T - M)'$, and $\lim_{n\to\infty} n^{-\frac{1}{2}mp} \frac{\Gamma_p[\frac{1}{2}(n + m + p - 1)]}{\Gamma_p[\frac{1}{2}(n + p - 1)]} = (\frac{1}{2})^{\frac{1}{2}mp}$, we have

$$\lim_{n\to\infty} f(T) = (2\pi)^{-\frac{1}{2}mp} \det(\Sigma)^{-\frac{1}{2}m} \det(\Omega)^{-\frac{1}{2}p}$$

$$\text{etr}\left\{-\frac{1}{2}\Sigma^{-1}(T - M)\Omega^{-1}(T - M)'\right\}, \ T \in \mathbb{R}^{p \times m}. \quad \blacksquare$$

In the next three theorems, we will derive distributions of certain linear transformations of the matrix T. Some of these results were derived by Tan (1969a).

THEOREM 4.3.5. *Let $T \sim T_{p,m}(n, M, \Sigma, \Omega)$, and $A\,(p \times p)$ and $B\,(m \times m)$ be nonsingular matrices, then $ATB \sim T_{p,m}(n, AMB, A\Sigma A', B'\Omega B)$.*

Proof: Transforming $W = ATB$, with the Jacobian of transformation $J(T \to W) = \det(A)^{-m} \det(B)^{-p}$, from the density (4.2.1) of T we get the density of W as

$$\frac{\Gamma_p[\frac{1}{2}(n + m + p - 1)]}{\pi^{\frac{1}{2}mp}\Gamma_p[\frac{1}{2}(n + p - 1)]} \det(A\Sigma A')^{-\frac{1}{2}m} \det(B'\Omega B)^{-\frac{1}{2}p}$$

$$\det(I_p + (A\Sigma A')^{-1}(W - AMB)(B'\Omega B)^{-1}(W - AMB)')^{-\frac{1}{2}(n+m+p-1)}, \ W \in \mathbb{R}^{p \times m}.$$

and, hence, the result. $\quad \blacksquare$

COROLLARY 4.3.5.1. *In the above theorem,*
   (i) if $A = \Sigma^{-\frac{1}{2}}$, then

$$\Sigma^{-\frac{1}{2}}TB \sim T_{p,m}(n, \Sigma^{-\frac{1}{2}}MB, I_p, B'\Omega B),$$

and
   (ii) if $B = \Omega^{-\frac{1}{2}}$, then

$$AT\Omega^{-\frac{1}{2}} \sim T_{p,m}(n, AM\Omega^{-\frac{1}{2}}, A\Sigma A', I_m).$$

THEOREM 4.3.6. *Let $T \sim T_{p,m}(n, M, \Sigma, \Omega)$ and $B\,(m \times r)$ be a matrix of rank $r \leq m$. Then, $TB \sim T_{p,r}(n, MB, \Sigma, B'\Omega B)$.*

Proof: According to the Theorem 4.2.1, T can be represented as

$$T = (S^{-\frac{1}{2}})'X + M \qquad (4.3.4)$$

where $S^{\frac{1}{2}}(S^{\frac{1}{2}})' \sim W_p(n + p - 1, \Sigma^{-1})$ and $X \sim N_{p,m}(0, I_p \otimes \Omega)$ are independently distributed. Post multiplying (4.3.4) by the matrix B, we get the representation for TB as

$$TB = (S^{-\frac{1}{2}})'(XB) + MB$$

where, from Theorem 2.3.10, $XB \sim N_{p,r}(0, I_p \otimes (B'\Omega B))$. Hence it follows, from Theorem 4.2.1, that $TB \sim T_{p,r}(n, MB, \Sigma, B'\Omega B)$. ∎

THEOREM 4.3.7. *Let $T \sim T_{p,m}(n, M, \Sigma, \Omega)$ and $A\,(s \times p)$ be a matrix of rank $s \leq p$. Then, $AT \sim T_{s,m}(n, AM, A\Sigma A', \Omega)$.*

Proof: Let $Y = AT$ then $Y' = T'A'$. From Theorem 4.3.3, we have $T' \sim T_{m,p}(n, M', \Omega, \Sigma)$ and from Theorem 4.3.6, we get $Y' = T'A' \sim T_{m,s}(n, M'A', \Omega, A\Sigma A')$. Now, using Theorem 4.3.3, again we get $Y = AT \sim T_{s,m}(n, AM, A\Sigma A', \Omega)$. ∎

Combining the above two results, we get the following theorem.

THEOREM 4.3.8. *Let $T \sim T_{p,m}(n, M, \Sigma, \Omega)$ and $A\,(s \times p)$, $B\,(m \times r)$ be constant matrices of ranks $s\,(\leq p)$ and $r\,(\leq m)$, respectively. Then $ATB \sim T_{s,r}(n, AMB, A\Sigma A', B'\Omega B)$.*

Proof: Let $W = AY$ and $Y = TB$. From Theorem 4.3.6, $Y \sim T_{p,r}(n, MB, \Sigma, B'\Omega B)$ and from Theorem 4.3.7, $W = ATB \sim T_{s,r}(n, AMB, A\Sigma A', B'\Omega B)$. ∎

The marginal and conditional distributions for column (row) partitions of T were derived by Dickey (1967) and are presented below (see also Box and Tiao, 1973).

THEOREM 4.3.9. *Let $T \sim T_{p,m}(n, M, \Sigma, \Omega)$ and partition T, M, Σ, and Ω as*

$$T = \begin{pmatrix} T_{1r} \\ T_{2r} \end{pmatrix} \begin{matrix} p_1 \\ p_2 \end{matrix} = (\, T_{1c} \quad T_{2c}\,), \quad M = \begin{pmatrix} M_{1r} \\ M_{2r} \end{pmatrix} \begin{matrix} p_1 \\ p_2 \end{matrix} = (\, M_{1c} \quad M_{2c}\,),$$
$$\qquad\qquad\qquad\quad m_1 \quad m_2 \qquad\qquad\qquad\qquad\quad m_1 \quad m_2$$

$$\Sigma = \begin{pmatrix} \Sigma_{11} & \Sigma_{12} \\ \Sigma_{21} & \Sigma_{22} \end{pmatrix} \begin{matrix} p_1 \\ p_2 \end{matrix}, \ and \ \Omega = \begin{pmatrix} \Omega_{11} & \Omega_{12} \\ \Omega_{21} & \Omega_{22} \end{pmatrix} \begin{matrix} m_1 \\ m_2 \end{matrix}.$$
$$\qquad\quad p_1 \quad p_2 \qquad\qquad\qquad m_1 \quad m_2$$

Then, (i) $T_{2r} \sim T_{p_2,m}(n, M_{2r}, \Sigma_{22}, \Omega)$,

$$T_{1r}|T_{2r} \sim T_{p_1,m}(n + p_2, M_{1r} + \Sigma_{12}\Sigma_{22}^{-1}(T_{2r} - M_{2r}), \Sigma_{11\cdot2},$$
$$\Omega(I_m + \Omega^{-1}(T_{2r} - M_{2r})'\Sigma_{22}^{-1}(T_{2r} - M_{2r})))$$

and (ii) $T_{2c} \sim T_{p,m_2}(n, M_{2c}, \Sigma, \Omega_{22})$,

$$T_{1c}|T_{2c} \sim T_{p,m_1}(n + m_2, M_{1c} + (T_{2c} - M_{2c})\Omega_{22}^{-1}\Omega_{21},$$
$$\Sigma(I_p + \Sigma^{-1}(T_{2c} - M_{2c})\Omega_{22}^{-1}(T_{2c} - M_{2c})'), \Omega_{11\cdot2}).$$

Proof: (i) From (4.2.1), the density of T is

$$f(T) = \frac{\Gamma_p[\frac{1}{2}(n + m + p - 1)]}{\pi^{\frac{1}{2}mp}\Gamma_p[\frac{1}{2}(n + p - 1)]} \det(\Sigma)^{-\frac{1}{2}m} \det(\Omega)^{-\frac{1}{2}p}$$

$$\det(I_p + \Sigma^{-1}(T - M)\Omega^{-1}(T - M)')^{-\frac{1}{2}(n+m+p-1)},$$

$$= \frac{\Gamma_p[\frac{1}{2}(n + m + p - 1)]}{\pi^{\frac{1}{2}mp}\Gamma_p[\frac{1}{2}(n + p - 1)]} \det(\Sigma)^{-\frac{1}{2}m} \det(\Omega)^{-\frac{1}{2}p}$$

$$\det(I_m + \Omega^{-1}(T - M)'\Sigma^{-1}(T - M))^{-\frac{1}{2}(n+m+p-1)}. \qquad (4.3.5)$$

Now, the quadratic form $(T - M)'\Sigma^{-1}(T - M)$ can be written as

$$(T - M)'\Sigma^{-1}(T - M)$$
$$= (T_{1r} - M_{1r} - \Sigma_{12}\Sigma_{22}^{-1}(T_{2r} - M_{2r}))'\Sigma_{11\cdot2}^{-1}(T_{1r} - M_{1r} - \Sigma_{12}\Sigma_{22}^{-1}(T_{2r} - M_{2r}))$$
$$+ (T_{2r} - M_{2r})'\Sigma_{22}^{-1}(T_{2r} - M_{2r}). \qquad (4.3.6)$$

Substituting (4.3.6) in (4.3.5) and noting that $\det(\Sigma) = \det(\Sigma_{22})\det(\Sigma_{11\cdot2})$, we can factorize the density of T as

$$f(T) = f_1(T_{2r})f_2(T_{1r}|T_{2r}),$$

where

$$f_1(T_{2r}) = \frac{\Gamma_{p_2}[\frac{1}{2}(n + m + p_2 - 1)]}{\pi^{\frac{1}{2}mp_2}\Gamma_{p_2}[\frac{1}{2}(n + p_2 - 1)]} \det(\Sigma_{22})^{-\frac{1}{2}m} \det(\Omega)^{-\frac{1}{2}p_2}$$

$$\det(I_m + \Omega^{-1}(T_{2r} - M_{2r})'\Sigma_{22}^{-1}(T_{2r} - M_{2r}))^{-\frac{1}{2}(n+m+p_2-1)} \qquad (4.3.7)$$

and

$$f_2(T_{1r}|T_{2r}) = \frac{\Gamma_{p_1}[\frac{1}{2}(n+m+p-1)]}{\pi^{\frac{1}{2}mp_1}\Gamma_{p_1}[\frac{1}{2}(n+p-1)]}$$

$$\det(I_m + \Omega^{-1}(T_{2r} - M_{2r})'\Sigma_{22}^{-1}(T_{2r} - M_{2r}))^{-\frac{1}{2}p_1}\det(\Omega)^{-\frac{1}{2}p_1}$$

$$\det(\Sigma_{11\cdot2})^{-\frac{1}{2}m}\det(I_m + (I_m + \Omega^{-1}(T_{2r} - M_{2r})'\Sigma_{22}^{-1}(T_{2r} - M_{2r}))^{-1}$$

$$\Omega^{-1}(T_{1r} - M_{1r} - \Sigma_{12}\,\Sigma_{22}^{-1}(T_{2r} - M_{2r}))'$$

$$\Sigma_{11\cdot2}^{-1}(T_{1r} - M_{1r} - \Sigma_{12}\Sigma_{22}^{-1}(T_{2r} - M_{2r})))^{-\frac{1}{2}(n+m+p-1)}. \tag{4.3.8}$$

From (4.3.7) and (4.3.8), it follows that the marginal distribution of T_{2r} is $T_{p_2,m}(n, M_{2r}, \Sigma_{22}, \Omega)$ and the conditional distribution of T_{1r} given T_{2r} is $T_{p_1,m}(n+p_2, M_{1r} + \Sigma_{12}\Sigma_{22}^{-1}(T_{2r} - M_{2r}), \Sigma_{11\cdot2}, \Omega(I_m + \Omega^{-1}(T_{2r} - M_{2r})'\Sigma_{22}^{-1}(T_{2r} - M_{2r})))$ respectively.

(ii) From Theorem 4.3.3, $T' = \begin{pmatrix} T'_{1c} \\ T'_{2c} \end{pmatrix} \sim T_{m,p}(n, M', \Omega, \Sigma)$, and from part (i) $T'_{2c} \sim T_{m_2,p}(n, M'_{2c}, \Omega_{22}, \Sigma)$ and $T'_{1c}|T'_{2c} \sim T_{m_1,p}(n+m_2, M'_{1c}+\Omega_{12}\Omega_{22}^{-1}(T'_{2c}-M'_{2c}), \Omega_{11\cdot2}, \Sigma(I_p + \Sigma^{-1}(T_{2c}-M_{2c})\Omega_{22}^{-1}(T_{2c}-M_{2c})'))$. Now, the distributions of T_{2c} and $T_{1c}|T_{2c}$ are obtained using Theorem 4.3.3. ∎

From the above theorem, matrix variate t-density can be written as the product of multivariate t-densities. Setting $m_1 = 1$ and $m_2 = m - 1$, $T_{1c} = t_1$, and $T_{2c} = (t_2, \ldots, t_m)$ in (ii), we get

$$t_1|T_{2c} \sim T_{p,1}(n+m-1, M_{1c} + (T_{2c} - M_{2c})\Omega_{22}^{-1}\Omega_{21},$$

$$\Sigma(I_p + \Sigma^{-1}(T_{2c} - M_{2c})\Omega_{22}^{-1}(T_{2c} - M_{2c})'), \Omega_{11\cdot2}).$$

which is the p-dimensional multivariate t-density. Next, from the marginal distribution of T_{2c}, one can see that $t_2|t_3, \ldots, t_m$ is also multivariate t. Repeating this procedure $(m-1)$ times, it is easy to see that the density of T can be expressed as

$$f(T) = f_1(t_1|t_2, \ldots, t_m)f_2(t_2|t_3, \ldots, t_m) \cdots f_m(t_m)$$

where every density on the right hand side is a p-dimensional multivariate t.

Similarly, using part (i) of Theorem 4.3.9, it can be proved that the density of T is the product of p m-dimensional multivariate t-densities.

It may be noted that while in Chapter 2, the normal matrix X is merely an arrangement of multivariate normal vector $\text{vec}(X')$, but this is not the case with the matrix T, as pointed out by Dickey, Dawid, and Kadane (1986). For consider the matrix $T (2 \times m) = (t_1^* \quad t_2^*)' \sim T_{2,m}(n, 0, \Sigma, \Omega)$. Then, according to Theorem 4.3.9, the marginal distribution of t_1^* is multivariate t, with n degrees of freedom and the conditional distribution of $t_2^*|t_1^*$, will have $n+1$ degrees of freedom. If T is merely an arrangement of the elements of the vector $\text{vec}(T')$, then the distribution of $\text{vec}(T') = \begin{pmatrix} t_1^* \\ t_2^* \end{pmatrix}$ would be $2m$-variate t-distribution with n degrees of freedom. Now, contrary to the above, the distribution of $t_2^*|t_1^*$ will have $n + m$ degrees of freedom.

In (4.2.1) if $\Omega = I_m$, then the columns of the matrix T are uncorrelated. Further if $M = \mu e'$, where μ $(p \times 1)$ is a constant vector and e $(m \times 1) = (1, \ldots, 1)'$, then the p.d.f. of $T = (t_1, \ldots, t_m)$ is

$$f(t_1, \ldots, t_m) = \frac{\Gamma_p[\frac{1}{2}(n+m+p-1)]}{\pi^{\frac{1}{2}mp}\Gamma_p[\frac{1}{2}(n+p-1)]} \det(\Sigma)^{-\frac{1}{2}m}$$

$$\det(I_p + \Sigma^{-1}(T - \mu e')(T - \mu e')')^{-\frac{1}{2}(n+m+p-1)}. \qquad (4.3.9)$$

The distribution of $A = \sum_{j=1}^{m}(t_j - \bar{t})(t_j - \bar{t})'$, where $\bar{t} = \frac{1}{m}\sum_{j=1}^{m} t_j$, is given in the following theorem.

THEOREM 4.3.10. *Let* $T = (t_1, \ldots, t_m)$ *be distributed as (4.3.9). Then, the distribution of* $A = \sum_{j=1}^{m}(t_j - \bar{t})(t_j - \bar{t})'$ *is*

$$\left\{ B_p\left(\frac{1}{2}(m-1), \frac{1}{2}(n+p-1)\right) \right\}^{-1} \det(\Sigma)^{-\frac{1}{2}(m-1)}$$

$$\det(A)^{\frac{1}{2}(m-p-2)} \det(I_p + \Sigma^{-1}A)^{-\frac{1}{2}(n+m+p-2)}, \quad A > 0$$

and $\sqrt{m}\,\bar{t}|A \sim T_{p,1}(n+m-1, \mu, \Sigma + A, 1)$.

Proof: Let H $(m \times m) = \left(\frac{1}{\sqrt{m}}e \quad B \right)$ be an orthogonal matrix. Transform $Y = TH = (\, y_1 \quad Y_2 \,)$, where y_1 $(p \times 1)$ and Y_2 $(p \times (m-1))$. Then, from (3.3.7)

$$(T - \mu e')(T - \mu e')' = m(\bar{t} - \mu)(\bar{t} - \mu)' + Y_2 Y_2'. \qquad (4.3.10)$$

Note that if μ is replaced by $\bar{t}$ in (4.3.10), then we get $A = Y_2 Y_2'$. Substitute from (4.3.10) in (4.3.9) together with the Jacobian of transformation $J(T \to \sqrt{m}\,\bar{t}, Y_2) = 1$, to get the joint density of $\sqrt{m}\,\bar{t}$ and Y_2 as

$$\frac{\Gamma_p[\frac{1}{2}(n+m+p-1)]}{\pi^{\frac{1}{2}mp}\Gamma_p[\frac{1}{2}(n+p-1)]} \det(\Sigma)^{-\frac{1}{2}m}$$

$$\det(I_p + m\Sigma^{-1}(\bar{t} - \mu)(\bar{t} - \mu)' + \Sigma^{-1}Y_2 Y_2')^{-\frac{1}{2}(n+m+p-1)}, \quad \bar{t} \in \mathbb{R}^p, \ Y_2 \in \mathbb{R}^{p \times (m-1)}.$$

Making use of the Theorem 1.4.10, the joint density of $\sqrt{m}\,\bar{t}$ and A is given by

$$\frac{\Gamma_p[\frac{1}{2}(n+m+p-1)]}{\pi^{\frac{1}{2}mp}\Gamma_p[\frac{1}{2}(n+p-1)]} \det(\Sigma)^{-\frac{1}{2}m}$$

$$\int_{Y_2 Y_2' = A} \det(I_p + m\Sigma^{-1}(\bar{t} - \mu)(\bar{t} - \mu)' + \Sigma^{-1}Y_2 Y_2')^{-\frac{1}{2}(n+m+p-1)} \, dY_2$$

$$= \pi^{-\frac{1}{2}p} \frac{\Gamma_p[\frac{1}{2}(n+m+p-1)]}{\Gamma_p[\frac{1}{2}(n+p-1)]\Gamma_p[\frac{1}{2}(m-1)]} \det(\Sigma)^{-\frac{1}{2}m} \det(A)^{\frac{1}{2}(m-p-2)}$$

$$\det(I_p + m\Sigma^{-1}(\bar{t} - \mu)(\bar{t} - \mu)' + \Sigma^{-1}A)^{-\frac{1}{2}(n+m+p-1)}$$

$$= f_1(A)f_2(\sqrt{m}\,\bar{t}|A), \ \bar{t} \in \mathbb{R}^p, \ A > 0,$$

where

$$f_1(A) = \left\{ \beta_p\!\left(\tfrac{1}{2}(m-1), \tfrac{1}{2}(n+p-1)\right) \right\}^{-1} \det(\Sigma)^{-\frac{1}{2}(m-1)}$$

$$\det(A)^{\frac{1}{2}(m-p-2)} \det(I_p + \Sigma^{-1}A)^{-\frac{1}{2}(n+m+p-2)}, \; A > 0$$

and

$$f_2(\sqrt{m}\,\bar{t}|A) = \pi^{-\frac{1}{2}p} \frac{\Gamma[\frac{1}{2}(n+m+p-1)]}{\Gamma[\frac{1}{2}(n+m-1)]} \det(\Sigma + A)^{-\frac{1}{2}}$$

$$\det(I_p + m(\Sigma + A)^{-1}(\bar{t} - \boldsymbol{\mu})(\bar{t} - \boldsymbol{\mu})')^{-\frac{1}{2}(n+m+p-1)}, \; \bar{t} \in \mathbb{R}^p. \; \blacksquare$$

4.4. INVERTED MATRIX VARIATE t-DISTRIBUTION

In this section, we define the inverted matrix variate t-distribution.

DEFINITION 4.4.1. *The random matrix $T\,(p \times m)$ is said to have an inverted matrix variate t-distribution with parameters $M \in \mathbb{R}^{p \times m}$, $\Sigma\,(p \times p) > 0$, and $\Omega\,(m \times m) > 0$, if its p.d.f. is given by*

$$\frac{\Gamma_p[\frac{1}{2}(n+m+p-1)]}{\pi^{\frac{1}{2}mp} \Gamma_p[\frac{1}{2}(n+p-1)]} \det(\Sigma)^{-\frac{1}{2}m} \det(\Omega)^{-\frac{1}{2}p}$$

$$\det(I_p - \Sigma^{-1}(T-M)\Omega^{-1}(T-M)')^{\frac{1}{2}(n-2)}, \; T \in \mathbb{R}^{p \times m} \qquad (4.4.1)$$

where $I_p - \Sigma^{-1}(T-M)\Omega^{-1}(T-M)' > 0$.

We shall denote this by $T \sim IT_{p,m}(n, M, \Sigma, \Omega)$. When $m = 1$, $T = \boldsymbol{t}\,(p \times 1)$, $M = \boldsymbol{\mu}\,(p \times 1)$, $\Omega = \omega$ and the density (4.4.1) reduces to

$$\frac{\Gamma[\frac{1}{2}(n+p)]}{\pi^{\frac{1}{2}p}\Gamma(\frac{1}{2}n)} \det(\Sigma)^{-\frac{1}{2}} \omega^{-\frac{1}{2}p} \left(1 - \frac{1}{\omega}(\boldsymbol{t} - \boldsymbol{\mu})'\Sigma^{-1}(\boldsymbol{t} - \boldsymbol{\mu})\right)^{\frac{1}{2}n-1}, \; \boldsymbol{t} \in \mathbb{R}^p,$$

which is the inverted multivariate t-density and will be denoted by $\boldsymbol{t} \sim It_p(n, \omega, \boldsymbol{\mu}, \Sigma)$. When $p = 1$, by taking $M = \boldsymbol{\nu}'$, $\Sigma = \sigma$, it is easily seen that $T' = \boldsymbol{t} \sim It_m(n, \sigma, \boldsymbol{\nu}, \Omega)$.

Khatri (1959a) derived the above density, but the following derivation of the inverted matrix variate t-distribution is due to Dickey (1967).

THEOREM 4.4.1. *Let $S \sim W_p(n+p-1, I_p)$ and $X \sim N_{p,m}(0, I_p \otimes I_m)$ be independently distributed. For $M \in \mathbb{R}^{p \times m}$, define*

$$T = \Sigma^{\frac{1}{2}}(S + XX')^{-\frac{1}{2}} X \Omega^{\frac{1}{2}} + M, \qquad (4.4.2)$$

where $S + XX' = (S+XX')^{\frac{1}{2}}((S+XX')^{\frac{1}{2}})'$ and $\Sigma^{\frac{1}{2}}$ and $\Omega^{\frac{1}{2}}$ are the symmetric square roots of the positive definite matrices $\Sigma\,(p \times p)$ and $\Omega\,(m \times m)$, respectively. Then, $T \sim IT_{p,m}(n, M, \Sigma, \Omega)$.

Proof: The joint density of S and X is

$$\frac{\pi^{-\frac{1}{2}mp}}{2^{\frac{1}{2}(n+m+p-1)p}\Gamma_p[\frac{1}{2}(n+p-1)]}\det(S)^{\frac{1}{2}(n-2)}\,\text{etr}\left\{-\frac{1}{2}(S+XX')\right\},\ S>0,\ X\in\mathbb{R}^{p\times m}.$$

Transforming $U=S+XX'$, with Jacobian $J(S\to U)=1$, we get the joint density of U and X as

$$\frac{\pi^{-\frac{1}{2}mp}}{2^{\frac{1}{2}(n+m+p-1)p}\Gamma_p[\frac{1}{2}(n+p-1)]}\det(U-XX')^{\frac{1}{2}(n-2)}\,\text{etr}\left(-\frac{1}{2}U\right),$$

$$U-XX'>0,\ X\in\mathbb{R}^{p\times m}. \qquad (4.4.3)$$

Now, let $T=\Sigma^{\frac{1}{2}}U^{-\frac{1}{2}}X\Omega^{\frac{1}{2}}+M$. Then, $J(X\to T)=\det(U)^{\frac{1}{2}m}\det(\Sigma)^{-\frac{1}{2}m}\det(\Omega)^{-\frac{1}{2}p}$ and the joint density of T and U, from (4.4.3), is

$$\left\{2^{\frac{1}{2}(n+m+p-1)p}\Gamma_p\left[\frac{1}{2}(n+m+p-1)\right]\right\}^{-1}\det(U)^{\frac{1}{2}(n+m-2)}\,\text{etr}\left(-\frac{1}{2}U\right)$$

$$\frac{\Gamma_p[\frac{1}{2}(n+m+p-1)]}{\pi^{\frac{1}{2}mp}\Gamma_p[\frac{1}{2}(n+p-1)]}\det(\Sigma)^{-\frac{1}{2}m}\det(\Omega)^{-\frac{1}{2}p}$$

$$\det(I_p-\Sigma^{-1}(T-M)\Omega^{-1}(T-M)')^{\frac{1}{2}(n-2)},\ U>0,\ T\in\mathbb{R}^{p\times m}. \qquad (4.4.4)$$

From (4.4.4), it is easily seen that $T\sim IT_{p,m}(n,M,\Sigma,\Omega)$ and is independent of U. ∎
It may also be noted that
(i) if $T\sim IT_{p,m}(n,M,\Sigma,\Omega)$ then $T'\sim IT_{p,m}(n,M',\Omega,\Sigma)$,
(ii) if $T\sim IT_{p,m}(n,M,n\Sigma,\Omega)$ then $T\overset{\mathcal{D}}{\to}X$ as $n\to\infty$, where $X\sim N_{p,m}(M,\Sigma\otimes\Omega)$ and
(iii) if $T\sim IT_{p,m}(n,M,\Sigma,\Omega)$, and $A\,(p\times p)$, $B\,(m\times m)$ are nonsingular matrices then $ATB\sim IT_{p,m}(n,AMB,A\Sigma A',B'\Omega B)$.
The corresponding results on marginal and conditional distributions can also be derived (see Problems 4.11–4.15).

4.5. DISGUISED MATRIX VARIATE t-DISTRIBUTION

The type of distributions introduced in this section were derived by Olkin and Rubin (1964). Tan (1973) studied their properties and called them disguised matrix variate t-distributions because of their similarities with matrix variate t-distribution and relationship with matrix variate beta distribution.
First, we derive the lower and upper disguised matrix variate t-densities.

THEOREM 4.5.1. *Let* $S\sim W_m(n,\Psi)$ *and* $X\sim N_{p,m}(0,\Sigma\otimes\Psi)$ *be independent and*

$$T=XU^{-1}, \qquad (4.5.1)$$

where $S = U'U$ and U is a lower triangular matrix with positive diagonal elements. Then, T is said to have a lower disguised matrix variate t-distribution given by

$$\{K(m, p, n + p)\}^{-1} \det(\Sigma)^{-\frac{1}{2}m} \det(I_m + T'\Sigma^{-1}T)^{-\frac{1}{2}(n+p-m-1)}$$

$$\prod_{i=1}^{m} \det((I_m + T'\Sigma^{-1}T)_{[i]})^{-1}, \ T \in \mathbb{R}^{p \times m}, \qquad (4.5.2)$$

where

$$K(m, p, n) = \frac{\pi^{\frac{1}{2}mp}\Gamma_m[\frac{1}{2}(n - p)]}{\Gamma_m(\frac{1}{2}n)}$$

$$= K(p, m, n). \qquad (4.5.3)$$

Proof: Since $T = XU^{-1} = X\Psi^{-\frac{1}{2}}(U\Psi^{-\frac{1}{2}})^{-1}$, where $\Psi^{\frac{1}{2}}$ is the lower triangular matrix with positive diagonal elements such that $\Psi = (\Psi^{\frac{1}{2}})'\Psi^{\frac{1}{2}}$, the distribution of T remains invariant under the transformation $X \to X\Psi^{-\frac{1}{2}}$ and $U \to U\Psi^{-\frac{1}{2}}$. Hence, without loss of generality take $\Psi = I_m$. Now, the joint density of X and S is

$$\frac{\pi^{-\frac{1}{2}mp}}{2^{\frac{1}{2}m(n+p)}\Gamma_m(\frac{1}{2}n)} \det(\Sigma)^{-\frac{1}{2}m} \operatorname{etr}\left\{-\frac{1}{2}(X'\Sigma^{-1}X + S)\right\} \det(S)^{\frac{1}{2}(n-m-1)}.$$

Transforming $S = U'U$, $T = XU^{-1}$, with Jacobian $J(S, X \to U, T) = 2^m \det(U)^p \prod_{i=1}^{m} u_{ii}^i$, where $U = (u_{ij})$, the joint density of U and T is given by

$$\frac{\pi^{-\frac{1}{2}mp}\det(\Sigma)^{-\frac{1}{2}m}}{2^{\frac{1}{2}m(n+p-2)}\Gamma_m(\frac{1}{2}n)} \det(U'U)^{\frac{1}{2}(n+p-m-1)} \operatorname{etr}\left\{-\frac{1}{2}(U'T'\Sigma^{-1}TU + U'U)\right\} \prod_{i=1}^{m} u_{ii}^i.$$

Now, let $W = U'(I_m + T'\Sigma^{-1}T)U$. Then, the Jacobian of transformation is $J(U \to W) = 2^{-m} \prod_{i=1}^{m} \left\{u_{ii}^{-i} \det((I_m + T'\Sigma^{-1}T)_{[i]})^{-1}\right\}$ and the joint density of W and T is

$$\frac{\pi^{-\frac{1}{2}mp}\det(\Sigma)^{-\frac{1}{2}m}}{2^{\frac{1}{2}m(n+p)}\Gamma_m(\frac{1}{2}n)} \prod_{i=1}^{m} \det((I_m + T'\Sigma^{-1}T)_{[i]})^{-1} \det(I_m + T'\Sigma^{-1}T)^{-\frac{1}{2}(n+p-m-1)}$$

$$\det(W)^{\frac{1}{2}(n+p-m-1)} \operatorname{etr}\left(-\frac{1}{2}W\right).$$

Integrating out W in the above density using multivariate gamma integral (1.4.6) one obtains the p.d.f. of T as

$$\frac{\pi^{-\frac{1}{2}mp}\Gamma_m[\frac{1}{2}(n + p)]}{\Gamma_m(\frac{1}{2}n)} \det(\Sigma)^{-\frac{1}{2}m} \prod_{i=1}^{m} \det((I_m + T'\Sigma^{-1}T)_{[i]})^{-1}$$

$$\det(I_m + T'\Sigma^{-1}T)^{-\frac{1}{2}(n+p-m-1)}, \ T \in \mathbb{R}^{p \times m}. \ \blacksquare$$

It may be noted that if $T = (\boldsymbol{t}_1, \ldots, \boldsymbol{t}_m)$, then

$$I_m + T'\Sigma^{-1}T = I_m + \begin{pmatrix} t_1'\Sigma^{-1}t_1 & \cdots & t_1'\Sigma^{-1}t_m \\ & \vdots & \\ t_m'\Sigma^{-1}t_1 & \cdots & t_m'\Sigma^{-1}t_m \end{pmatrix} \tag{4.5.4}$$

from which it is seen that

$$(I_m + T'\Sigma^{-1}T)_{[i]} = I_i + \begin{pmatrix} t_{m-i+1}'\Sigma^{-1}t_{m-i+1} & \cdots & t_{m-i+1}'\Sigma^{-1}t_m \\ \vdots & & \\ t_m'\Sigma^{-1}t_{m-i+1} & \cdots & t_m'\Sigma^{-1}t_m \end{pmatrix}$$

and

$$\det((I_m + T'\Sigma^{-1}T)_{[i]}) = \det(\Sigma)^{-1}\det\left(\Sigma + \sum_{j=m-i+1}^{m} t_j t_j'\right). \tag{4.5.5}$$

Substituting (4.5.5) in (4.5.2) the density of T can be equivalently written as

$$\{K(m,p,n+p)\}^{-1}\det(\Sigma)^{\frac{1}{2}(n+p-1)} \prod_{i=1}^{m} \det\left(\Sigma + \sum_{j=m-i+1}^{m} t_j t_j'\right)^{-1}$$

$$\det(\Sigma + TT')^{-\frac{1}{2}(n+p-m-1)}, \ T \in \mathbb{R}^{p \times m}. \tag{4.5.6}$$

If, in (4.5.1), we take U as an upper triangular matrix, we obtain what is known as the upper disguised matrix variate t-distribution.

THEOREM 4.5.2. *Let $S \sim W_m(n, \Psi)$ and $X \sim N_{p,m}(0, \Sigma \otimes \Psi)$ be independent and*

$$T = XU^{-1},$$

where $S = U'U$ and U is the upper triangular matrix with positive diagonal elements. Then, T is said to have an upper disguised matrix variate t-distribution given by

$$\{K(m,p,n+p)\}^{-1}\det(\Sigma)^{-\frac{1}{2}m}\det(I_m + T'\Sigma^{-1}T)^{-\frac{1}{2}(n+p-m-1)}$$

$$\prod_{i=1}^{m} \det((I_m + T'\Sigma^{-1}T)^{[i]})^{-1}, \ T \in \mathbb{R}^{p \times m}, \tag{4.5.7}$$

where $K(m,p,n)$ is defined by (4.5.3).

Proof: The proof is similar to the one given for Theorem 4.5.1. ∎

From (4.5.4), we get

$$(I_m + T'\Sigma^{-1}T)^{[i]} = I_i + \begin{pmatrix} t_1'\Sigma^{-1}t_1 & \cdots & t_1'\Sigma^{-1}t_i \\ & \vdots & \\ t_i'\Sigma^{-1}t_1 & \cdots & t_i'\Sigma^{-1}t_i \end{pmatrix} \tag{4.5.8}$$

and

$$\det((I_m + T'\Sigma^{-1}T)^{[i]}) = \det(\Sigma)^{-1} \det\left(\Sigma + \sum_{j=1}^{i} t_j t_j'\right). \qquad (4.5.9)$$

Substituting (4.5.9) in (4.5.7) the upper disguised matrix variate t-density can be equivalently written as

$$\{K(m,p,n+p)\}^{-1} \det(\Sigma)^{\frac{1}{2}(n+p-1)} \prod_{i=1}^{m} \det\left(\Sigma + \sum_{j=1}^{i} t_j t_j'\right)^{-1}$$

$$\det(\Sigma + TT')^{-\frac{1}{2}(n+p-m-1)}, \; T \in \mathbb{R}^{p \times m}. \qquad (4.5.10)$$

The above two results were derived by Olkin and Rubin (1964, p. 465). Tan (1973) called these distributions as lower and upper disguised matrix variate t. If the density of $W = T - M$ is (4.5.2), we shall write $T \sim \underline{D}T_{p,m}(n+p, M, \Sigma)$ and if it is (4.5.7), then $T \sim \overline{D}T_{p,m}(n+p, M, \Sigma)$.

Next, we derive expected values of the matrix T and some of its functions.

THEOREM 4.5.3. *(i) If* $T \sim \underline{D}T_{p,m}(n, M, \Sigma)$, *then* $E(T) = M$ *and* $\mathrm{cov}(\mathrm{vec}(T'))$ $= \Sigma \otimes B$ *where* $B = \mathrm{diag}(b_1, \ldots, b_m)$ *with*

$$b_j = \frac{n-p-1}{(n-p-m+j-2)(n-p-m+j-1)}, \; j = 1, 2, \ldots, m-1,$$

and

$$b_m = \frac{1}{n-p-2}.$$

(ii) If $T \sim \overline{D}T_{p,m}(n, M, \Sigma)$, *then* $E(T) = M$ *and* $\mathrm{cov}(\mathrm{vec}(T')) = \Sigma \otimes B$, *where* $B = \mathrm{diag}(b_1, \ldots, b_m)$, *with*

$$b_1 = \frac{1}{n-p-2}$$

and

$$b_j = \frac{n-p-1}{(n-p-j-1)(n-p-j)}, \; j = 2, \ldots, m.$$

Proof: (i) Notice that the random matrix T can be represented as

$$T = XU^{-1} + M, \qquad (4.5.11)$$

where X and U are independently distributed, $X \sim N_{p,m}(0, \Sigma \otimes I_m)$, $U'U \sim W_m(n-p, I_m)$, and U is a lower triangular matrix with positive diagonal elements. From (4.5.11), it is clear that $T|U \sim N_{p,m}(M, \Sigma \otimes (UU')^{-1})$, *i.e.*, the conditional mean of T given U is M, which is independent of U and hence the unconditional mean of T is also M.

Further, the conditional variance of $\mathrm{vec}(T')$ given U is $\Sigma \otimes (UU')^{-1}$. Therefore, the unconditional variance of $\mathrm{vec}(T')$ is $E_U(\Sigma \otimes (UU')^{-1})$. Now, from Theorem 3.3.21, we get the desired result.

(ii) The proof is similar to part (i). ∎

THEOREM 4.5.4. *If* $T \sim \underline{D}T_{p,m}(n, M, \Sigma)$*, then*
  (i) $E(TCT) = \Sigma C'B + MCM$, $C\,(m \times p)$,
  (ii) $E(TCT') = \mathrm{tr}(CB)\Sigma + MCM'$, $C\,(m \times m)$,
  (iii) $E(TCT'DT) = \mathrm{tr}(\Sigma D')MCB + \Sigma D'MC'B + \mathrm{tr}(CB)\Sigma DM$
                    $+ MCM'DM$, $C\,(m \times m)$, $D\,(p \times p)$,
  (iv) $E(TCTDT) = \Sigma D'M'C'B + MC\Sigma D'B + \Sigma C'BDM$
                    $+ MCMDM$, $C\,(m \times p)$, $D\,(m \times p)$,
where the matrix B *is defined in Theorem 4.5.3(i).*

Proof: (i) Using the representation (4.5.11), we get

$$E(TCT) = E[(XU^{-1} + M)C(XU^{-1} + M)]$$
$$= E_U E_X[(XU^{-1}CXU^{-1} + MCXU^{-1} + XU^{-1}CM + MCM)|U]$$
$$= E_U[E_X\{(XU^{-1}CXU^{-1} + MCM)|U\}]$$
$$= E_U[\Sigma C'(U^{-1})'U^{-1} + MCM], \;\; \text{from Theorem 2.3.5}$$
$$= \Sigma C' E_U(UU')^{-1} + MCM$$
$$= \Sigma C'B + MCM.$$

The last step follows from Theorem 3.3.21.
  (ii) Derivation of E(TCT') is similar to (i).
  (iii) As in part (i), we have

$$E(TCT'DT) = E[(XU^{-1} + M)C(XU^{-1} + M)'D(XU^{-1} + M)]$$
$$= E_U E_X[(XU^{-1}C(U^{-1})'X'DXU^{-1} + MC(U^{-1})'X'DXU^{-1}$$
$$+ XU^{-1}CM'DXU^{-1} + MCM'DXU^{-1} + XU^{-1}C(U^{-1})'X'DM$$
$$+ MC(U^{-1})'X'DM + XU^{-1}CM'DM + MCM'DM)|U]$$
$$= E_U[\mathrm{tr}(\Sigma D')MC(UU')^{-1} + \Sigma D'MC'(UU')^{-1} + \mathrm{tr}(C(UU')^{-1})\Sigma DM]$$
$$+ MCM'DM$$
$$= \mathrm{tr}(\Sigma D')MCB + \Sigma D'MC'B + \mathrm{tr}(CB)\Sigma DM + MCM'DM.$$

  (iv) This result can be derived in the same manner as (iii). ∎
  It may be noted that if $T \sim \overline{D}T_{p,m}(n, M, \Sigma)$, then the results (i)–(iv) given above still hold, but the matrix B is now given by Theorem 4.5.3(ii).
  We now derive the distribution of certain functions of lower (upper) disguised matrix T.

THEOREM 4.5.5. *Let* $A\,(p \times p)$ *be a constant nonsingular matrix.*
  (i) *If* $T \sim \underline{D}T_{p,m}(n, M, \Sigma)$*, then*

$$AT \sim \underline{D}T_{p,m}(n, AM, A\Sigma A'),$$

and

(ii) if $T \sim \overline{D}T_{p,m}(n, M, \Sigma)$, then

$$AT \sim \overline{D}T_{p,m}(n, AM, A\Sigma A').$$

Proof: (i) The density of T is

$$\{K(m,p,n)\}^{-1} \det(\Sigma)^{-\frac{1}{2}m} \prod_{i=1}^{m} \det((I_m + (T - M)'\Sigma^{-1}(T - M))_{[i]})^{-1}$$

$$\det(I_m + (T - M)'\Sigma^{-1}(T - M))^{-\frac{1}{2}(n-m-1)}, \; T \in \mathbb{R}^{p \times m}. \quad (4.5.12)$$

Substituting $W = AT$, with Jacobian $J(T \to W) = \det(A)^{-m}$, in (4.5.12) we get the density of W as

$$\{K(m,p,n)\}^{-1} \det(A\Sigma A')^{-\frac{1}{2}m} \prod_{i=1}^{m} \det((I_m + (W - AM)'(A\Sigma A')^{-1}(W - AM))_{[i]})^{-1}$$

$$\det(I_m + (W - AM)'(A\Sigma A')^{-1}(W - AM))^{-\frac{1}{2}(n-m-1)}, \; T \in \mathbb{R}^{p \times m}.$$

Hence, the result.

(ii) This can be proved in the same way as part (i). ■

COROLLARY 4.5.5.1. *(i) If $T \sim \underline{D}T_{p,m}(n, M, \Sigma)$, then*

$$\Sigma^{-\frac{1}{2}}T \sim \underline{D}T_{p,m}(n, \Sigma^{-\frac{1}{2}}M, I_p), \; and$$

(ii) if $T \sim \overline{D}T_{p,m}(n, M, \Sigma)$, then

$$\Sigma^{-\frac{1}{2}}T \sim \overline{D}T_{p,m}(n, \Sigma^{-\frac{1}{2}}M, I_p).$$

Proof: Put $A = \Sigma^{-\frac{1}{2}}$ in Theorem 4.5.5. ■

THEOREM 4.5.6. *Let $A\,(r \times p)$ be a constant matrix of rank $r \leq p$.*
(i) If $T \sim \underline{D}T_{p,m}(n, M, \Sigma)$, then $AT \sim \underline{D}T_{r,m}(n - p + r, AM, A\Sigma A')$, and
(ii) if $T \sim \overline{D}T_{p,m}(n, M, \Sigma)$, then $AT \sim \overline{D}T_{r,m}(n - p + r, AM, A\Sigma A')$.

Proof: Here we give the proof for (i) only since the proof of (ii) is similar. Let $W = T - M$, then from (4.5.11) W can be represented as

$$W = T - M = XU^{-1}$$

and

$$AW = A(T - M) = AXU^{-1}$$

where $AX \sim N_{r,m}(0, A\Sigma A' \otimes I_m)$. Therefore, by definition, $A(T - M) \sim \underline{D}T_{r,m}(n - p + r, 0, A\Sigma A')$ and the result follows immediately. ■

THEOREM 4.5.7. *Let* $T(p \times m) = (T_{1c} \quad T_{2c})$, $T_{1c}(p \times m_1)$ *and* $M(p \times m) = (M_{1c} \quad M_{1c})$, $M_{1c}(p \times m_1)$.

(i) If $T \sim \underline{D}T_{p,m}(n, M, \Sigma)$, *then* $T_{2c} \sim \underline{D}T_{p,m_2}(n, M_{2c}, \Sigma)$, *and* $W_1^{-1}(T_{1c} - M_{1c}) \sim \underline{D}T_{p,m_1}(n - m_2, 0, I_p)$, *where* $W_1 = \{\Sigma + (T_{2c} - M_{2c})(T_{2c} - M_{2c})'\}^{\frac{1}{2}}$, *and*

(ii) if $T \sim \overline{D}T_{p,m}(n, M, \Sigma)$, *then* $T_{1c} \sim \overline{D}T_{p,m_1}(n, M_{1c}, \Sigma)$, *and* $W_2^{-1}(T_{2c} - M_{2c}) \sim \overline{D}T_{p,m_2}(n - m_1, 0, I_p)$, *where* $W_2 = \{\Sigma + (T_{1c} - M_{1c})(T_{1c} - M_{1c})'\}^{\frac{1}{2}}$.

Proof: We shall give the proof of (i) only since proof of (ii) follows similar steps. Without loss of generality, it can be assumed that $M = 0$. The lower disguised matrix variate t-density given by (4.5.6) is

$$\{K(p,m,n)\}^{-1} \det(\Sigma)^{\frac{1}{2}(n-1)} \prod_{i=1}^{m} \det\left(\Sigma + \sum_{j=m-i+1}^{m} t_j t_j'\right)^{-1}$$

$$\det(\Sigma + TT')^{-\frac{1}{2}(n-m-1)}, \quad T \in \mathbb{R}^{p \times m}. \qquad (4.5.13)$$

where $K(p,m,n)$ is defined by (4.5.3) and $T = (t_1, \ldots, t_m)$. Integrating (4.5.13), with respect to t_1, we get the joint density of $t_2, t_3, \ldots, t_m$ as

$$\{K(p,m,n)\}^{-1} \det(\Sigma)^{\frac{1}{2}(n-1)} \prod_{i=1}^{m-1} \det\left(\Sigma + \sum_{j=m-i+1}^{m} t_j t_j'\right)^{-1}$$

$$\int_{t_1 \in \mathbb{R}^p} \det\left(\Sigma + \sum_{j=1}^{m} t_j t_j'\right)^{-\frac{1}{2}(n-m+1)} dt_1. \qquad (4.5.14)$$

Substituting $y_1 = (\Sigma + \sum_{j=2}^{m} t_j t_j')^{-\frac{1}{2}} t_1$, with the Jacobian $J(t_1 \to y_1) = \det(\Sigma + \sum_{j=2}^{m} t_j t_j')^{\frac{1}{2}}$, in (4.5.14), we get

$$\{K(p,m,n)\}^{-1} \det(\Sigma)^{\frac{1}{2}(n-1)} \prod_{i=1}^{m-1} \det\left(\Sigma + \sum_{j=m-i+1}^{m} t_j t_j'\right)^{-1}$$

$$\det\left(\Sigma + \sum_{j=2}^{m} t_j t_j'\right)^{-\frac{1}{2}(n-m)} \int_{y_1 \in \mathbb{R}^p} \det(I_p + y_1 y_1')^{-\frac{1}{2}(n-m+1)} dy_1. \qquad (4.5.15)$$

Evaluating the integral in (4.5.15), using Theorem 1.4.11 and simplifying, we get the joint density of $t_2, \ldots, t_m$ as

$$\{K(p,m-1,n)\}^{-1} \det(\Sigma)^{\frac{1}{2}(n-1)} \prod_{i=1}^{m-1} \det\left(\Sigma + \sum_{j=m-i+1}^{m} t_j t_j'\right)^{-1}$$

$$\det\left(\Sigma + \sum_{j=2}^{m} t_j t_j'\right)^{-\frac{1}{2}(n-m)}. \qquad (4.5.16)$$

Now, integrate (4.5.16), with respect to t_2, using the same procedure, to get the joint density of $t_3, \ldots, t_m$ as

$$\{K(p, m-2, n)\}^{-1} \det(\Sigma)^{\frac{1}{2}(n-1)} \prod_{i=1}^{m-2} \det \left(\Sigma + \sum_{j=m-i+1}^{m} t_j t_j' \right)^{-1}$$

$$\det \left(\Sigma + \sum_{j=3}^{m} t_j t_j' \right)^{-\frac{1}{2}(n-m+1)}.$$

Repeating this procedure m_1 times, we get the joint density of $t_{m_1+1}, \ldots, t_m$ as

$$\{K(p, m_2, n)\}^{-1} \det(\Sigma)^{\frac{1}{2}(n-1)} \prod_{i=1}^{m_2} \det \left(\Sigma + \sum_{j=m-i+1}^{m} t_j t_j' \right)^{-1}$$

$$\det \left(\Sigma + \sum_{j=m_1+1}^{m} t_j t_j' \right)^{-\frac{1}{2}(n-m_2-1)}.$$

i.e., $(t_{m_1+1}, \ldots, t_m) \sim \underline{D}T_{p,m_2}(n, M_{2c}, \Sigma)$.

To prove the second part, note that

$$\det(\Sigma + TT') = \det(\Sigma + T_{2c}T_{2c}') \det(I_p + W_1^{-1} T_{1c} T_{1c}' W_1^{-1}), \qquad (4.5.17)$$

$$\prod_{i=1}^{m} \det \left(\Sigma + \sum_{j=m-i+1}^{m} t_j t_j' \right)$$

$$= \prod_{i=1}^{m_2} \det \left(\Sigma + \sum_{j=m-i+1}^{m} t_j t_j' \right) \prod_{i=m_2+1}^{m} \det \left(\Sigma + \sum_{j=m-i+1}^{m} t_j t_j' \right) \qquad (4.5.18)$$

and

$$\prod_{i=m_2+1}^{m} \det \left(\Sigma + \sum_{j=m-i+1}^{m} t_j t_j' \right) = \prod_{i=1}^{m_1} \det \left(\Sigma + \sum_{j=m_1-i+1}^{m} t_j t_j' \right)$$

$$= \det(\Sigma + T_{2c}T_{2c}')^{m_1}$$

$$\prod_{i=1}^{m_1} \det \left(I_p + W_1^{-1} \sum_{j=m_1-i+1}^{m_1} t_j t_j' W_1^{-1} \right). \qquad (4.5.19)$$

Now, substituting (4.5.19) in (4.5.18) and (4.5.17) and (4.5.18) in (4.5.13), we get the joint density of T_{1c} and T_{2c} as

$$\{K(p, m, n)\}^{-1} \det(\Sigma)^{\frac{1}{2}(n-1)} \prod_{i=1}^{m_2} \det \left(\Sigma + \sum_{j=m-i+1}^{m} t_j t_j' \right)^{-1} \det(\Sigma + T_{2c}T_{2c}')^{-\frac{1}{2}(n-m_2-1+m_1)}$$

$$\prod_{i=1}^{m_1} \det \left(I_p + W_1^{-1} \sum_{j=m_1-i+1}^{m_1} t_j t_j' W_1^{-1} \right) \det \left(I_p + W_1^{-1} T_{1c} T_{1c}' W_1^{-1} \right)^{-\frac{1}{2}(n-m-1)}.$$

Now, transforming $y_j = W_1^{-1} t_j$, $j = 1, \ldots, m_1$, so that $Y = (y_1, \ldots, y_{m_1}) = W_1^{-1} T_{1c}$, with Jacobian $J(T_{1c} \to Y) = \det(W_1)^{m_1}$, the joint density of Y and T_{2c} is

$$\{K(p, m, n)\}^{-1} \det(\Sigma)^{\frac{1}{2}(n-1)} \prod_{i=1}^{m_2} \det \left(\Sigma + \sum_{j=m-i+1}^{m} t_j t_j' \right)^{-1} \det(\Sigma + T_{2c}T_{2c}')^{-\frac{1}{2}(n-m_2-1)}$$

$$\prod_{i=1}^{m_1} \det \left(I_p + \sum_{j=m_1-i+1}^{m_1} y_j y_j' \right)^{-1} \det(I_p + YY')^{-\frac{1}{2}(n-m-1)}. \qquad (4.5.20)$$

4.6. RESTRICTED MATRIX VARIATE t-DISTRIBUTION

From (4.5.20), it is easily seen that Y and T_{2c} are independently distributed, $T_{2c} \sim \underline{D}T_{p,m_2}(n, 0, \Sigma)$ and $Y = W_1^{-1}T_{1c} \sim \underline{D}T_{p,m_1}(n - m_2, 0, I_p)$. ∎

THEOREM 4.5.8. *Let* $T \sim \overline{D}T_{p,m}(n, M, \Sigma)$ *and partition* T, M *and* Σ *as*

$$T = \begin{pmatrix} T_{1r} \\ T_{2r} \end{pmatrix} \begin{matrix} p_1 \\ p_2 \end{matrix}, \quad M = \begin{pmatrix} M_{1r} \\ M_{2r} \end{pmatrix} \begin{matrix} p_1 \\ p_2 \end{matrix}, \quad and \; \Sigma = \begin{pmatrix} \Sigma_{11} & \Sigma_{12} \\ \Sigma_{21} & \Sigma_{22} \end{pmatrix} \begin{matrix} p_1 \\ p_2 \end{matrix}.$$
$$ \begin{matrix} p_1 & p_2 \end{matrix}$$

Then,

   (i) $T_{1r} \sim \overline{D}T_{p_1,m}(n - p_2, M_{1r}, \Sigma_{11})$, $T_{2r} \sim \overline{D}T_{p_2,m}(n - p_1, M_{2r}, \Sigma_{22})$,
and

   (ii) $(T_{1r} - M_{1r}) - \Sigma_{12}\Sigma_{22}^{-1}(T_{2r} - M_{2r}) \sim \overline{D}T_{p_1,m}(n - p_2, 0, \Sigma_{11\cdot2})$, $(T_{2r} - M_{2r}) - \Sigma_{21}\Sigma_{11}^{-1}(T_{1r} - M_{1r}) \sim \overline{D}T_{p_2,m}(n - p_1, 0, \Sigma_{22\cdot1})$.

Proof: (i) According to Theorem 4.5.2, we can write

$$T - M = XU^{-1},$$

where $X \sim N_{p,m}(0, \Sigma \otimes I_m)$ and $U'U \sim W_m(n - p, I_m)$ are independent and U is an upper triangular matrix with positive diagonal elements. Partitioning X as $X = \begin{pmatrix} X_{1r} \\ X_{2r} \end{pmatrix} \begin{matrix} p_1 \\ p_2 \end{matrix}$, we have $T_{1r} - M_{1r} = X_{1r}U^{-1}$ and $T_{2r} - M_{2r} = X_{2r}U^{-1}$, where $X_{1r} \sim N_{p_1,m}(0, \Sigma_{11} \otimes I_m)$ and $X_{2r} \sim N_{p_2,m}(0, \Sigma_{22} \otimes I_m)$. Hence, the results follow from Theorem 4.5.2.

   (ii) Here, we have

$$(T_{1r} - M_{1r}) - \Sigma_{12}\Sigma_{22}^{-1}(T_{2r} - M_{2r}) = (X_{1r} - \Sigma_{12}\Sigma_{22}^{-1}X_{2r})U^{-1},$$

where $X_{1r} - \Sigma_{12}\Sigma_{22}^{-1}X_{2r} \sim N_{p_1,m}(0, \Sigma_{11\cdot2} \otimes I_m)$ and hence,

$$(T_{1r} - M_{1r}) - \Sigma_{12}\Sigma_{22}^{-1}(T_{2r} - M_{2r}) \sim \overline{D}T_{p_1,m}(n - p_2, 0, \Sigma_{11\cdot2}).$$

The proof of the second part is similar. ∎

   When T has lower disguised matrix variate t-distribution, results similar to Theorem 4.5.8 can also be derived.

4.6. RESTRICTED MATRIX VARIATE t-DISTRIBUTION

Tan (1969b) defined a restricted form of the matrix variate t-distribution which occurs in the derivation of the posterior distribution of a parameter of a generalized multivariate normal process. In this section, we study this restricted matrix variate t-distribution.

DEFINITION 4.6.1. *A random matrix $T\,(p \times m)$, such that $TC = 0$, where $C\,(m \times s)$ is constant matrix of rank $s\,(\leq m)$, is said to have restricted matrix variate t-distribution if its p.d.f. is given by*

$$\frac{\Gamma_p[\frac{1}{2}(n+m+p-s-1)]}{\pi^{\frac{1}{2}(m-s)p}\Gamma_p[\frac{1}{2}(n+p-1)]} \det(\Omega)^{-\frac{1}{2}p} \det(\Sigma)^{-\frac{1}{2}(m-s)} \det(C'\Omega C)^{\frac{1}{2}p}$$

$$\det(I_p + \Sigma^{-1}T\Omega^{-1}T')^{-\frac{1}{2}(n+m+p-s-1)}.$$

This density will be denoted by $T_{p,m}(n,0,\Sigma,\Omega|s,C)$. Further, if for $M\,(p \times m)$, $MC = 0$, and $T - M \sim T_{p,m}(n,0,\Sigma,\Omega|s,C)$, then $T \sim T_{p,m}(n,M,\Sigma,\Omega|s,C)$.

This density can be derived by using the representation of T given in Theorem 4.2.1, where now $X \sim N_{p,m}(0,\Sigma \otimes \Omega|s,C)$.

THEOREM 4.6.1. *Let $S \sim W_p(n+p-1,\Sigma^{-1})$ independent of $X \sim N_{p,m}(0,I_p \otimes \Omega|s,C)$. Define $T = (S^{-\frac{1}{2}})'X + M$, where $M\,(p \times m)$ is a constant matrix, such that $MC = 0$, and $S^{\frac{1}{2}}(S^{\frac{1}{2}})' = S$. Then, $T \sim T_{p,m}(n,M,\Sigma,\Omega|s,C)$.*

Proof: Given S, T is distributed as $N_{p,m}(M,S^{-1} \otimes \Omega|s,C)$. Hence, the unconditional density of T is given by

$$\frac{2^{-\frac{1}{2}(n+m+p-s-1)p}\pi^{-\frac{1}{2}(m-s)p}}{\Gamma_p[\frac{1}{2}(n+p-1)]} \det(\Sigma)^{\frac{1}{2}(n+p-1)} \det(\Omega)^{-\frac{1}{2}p} \det(C'\Omega C)^{\frac{1}{2}p}$$

$$\int_{S>0} \det(S)^{\frac{1}{2}(n+m-s-2)} \operatorname{etr}\left\{-\frac{1}{2}S(T-M)\Omega^{-1}(T-M)' - \frac{1}{2}\Sigma S\right\} dS$$

$$= \frac{\pi^{-\frac{1}{2}(m-s)p}\Gamma_p[\frac{1}{2}(n+m+p-s-1)]}{\Gamma_p[\frac{1}{2}(n+p-1)]\det(\Sigma)^{\frac{1}{2}(m-s)}} \det(\Omega)^{-\frac{1}{2}p} \det(C'\Omega C)^{\frac{1}{2}p}$$

$$\det(I_p + \Sigma^{-1}(T-M)\Omega^{-1}(T-M)')^{-\frac{1}{2}(n+m+p-s-1)}, \quad TC = 0,$$

which is the required result. ∎

The above theorem can also be proved along the lines as Theorem 4.2.1.

THEOREM 4.6.2. *If $T \sim T_{p,m}(n,M,\Sigma,\Omega|s,C)$, and $B\,(p \times p)$ and $D\,(m \times m)$ are constant nonsingular matrices, then $BTD \sim T_{p,m}(n,BMD,B\Sigma B',D'\Omega D|s,D^{-1}C)$.*

Proof: See Problem 4.24. ∎

4.7. NONCENTRAL MATRIX VARIATE t-DISTRIBUTION

In Section 4.2, we defined the matrix variate t-distribution. In the subsequent section, it was represented as $T = (S^{-\frac{1}{2}})'X + M$, where $X \sim N_{p,m}(0,I_p \otimes \Omega)$, and $S \sim W_p(n+p-1,\Sigma^{-1})$. When $S \sim W_p(n+p-1,\Sigma^{-1},\Theta)$ or $X \sim N_{p,m}(M,I_p \otimes \Omega)$, the distribution of T, so obtained is called the noncentral matrix variate t-distribution.

THEOREM 4.7.1. *Let* $X \sim N_{p,m}(0, I_p \otimes \Omega)$ *and* $S \sim W_p(n + p - 1, \Sigma^{-1}, \Theta)$ *be independent. Then the distribution of*

$$T = (S^{-\frac{1}{2}})'X,$$

where $S = S^{\frac{1}{2}}(S^{\frac{1}{2}})'$, *is the lower noncentral matrix variate t and the density of* T *is given by*

$$\frac{\pi^{-\frac{1}{2}mp}\Gamma_p[\frac{1}{2}(n + m + p - 1)]}{\Gamma_p[\frac{1}{2}(n + p - 1)]} \det(\Omega)^{-\frac{1}{2}p} \det(\Sigma)^{-\frac{1}{2}m} \operatorname{etr}\left(-\frac{1}{2}\Theta\right)$$

$$\det(I_p + \Sigma^{-1}T\Omega^{-1}T')^{-\frac{1}{2}(n+m+p-1)} {}_1F_1\left(\frac{1}{2}(n + m + p - 1); \frac{1}{2}(n + p - 1);\right.$$

$$\left.\frac{1}{2}(I_p + \Sigma^{-1}T\Omega^{-1}T')^{-1}\Theta\right), \ T \in \mathbb{R}^{p\times m}. \qquad (4.7.1)$$

Proof: The joint density of X and S is

$$(2\pi)^{-\frac{1}{2}mp} \det(\Omega)^{-\frac{1}{2}p} \operatorname{etr}\left(-\frac{1}{2}\Omega^{-1}X'X\right)\left\{2^{\frac{1}{2}(n+p-1)p}\Gamma_p\left[\frac{1}{2}(n + p - 1)\right]\right\}^{-1}$$

$$\det(\Sigma)^{\frac{1}{2}(n+p-1)} \det(S)^{\frac{1}{2}(n-2)} \operatorname{etr}\left(-\frac{1}{2}\Sigma S - \frac{1}{2}\Theta\right) {}_0F_1\left(\frac{1}{2}(n + p - 1); \frac{1}{2}\Theta\Sigma S\right).$$

Transforming $T = (S^{-\frac{1}{2}})'X$, with the Jacobian $J(X \to T) = \det(S)^{\frac{1}{2}m}$, we get the joint density of T and S as

$$\left\{2^{\frac{1}{2}(n+m+p-1)p}\pi^{\frac{1}{2}mp}\Gamma_p\left[\frac{1}{2}(n + p - 1)\right]\right\}^{-1} \det(\Sigma)^{\frac{1}{2}(n+p-1)} \det(\Omega)^{-\frac{1}{2}p} \det(S)^{\frac{1}{2}(n+m-2)}$$

$$\operatorname{etr}\left\{-\frac{1}{2}(T\Omega^{-1}T' + \Sigma)S - \frac{1}{2}\Theta\right\} {}_0F_1\left(\frac{1}{2}(n + p - 1); \frac{1}{4}\Theta\Sigma S\right). \qquad (4.7.2)$$

To find the marginal density of T we integrate out (4.7.2), with respect to S. Let $U = \frac{1}{2}(T\Omega^{-1}T' + \Sigma)^{\frac{1}{2}}S(T\Omega^{-1}T' + \Sigma)^{\frac{1}{2}}$, then $J(S \to U) = \det(\frac{1}{2}(T\Omega^{-1}T' + \Sigma))^{-\frac{1}{2}(p+1)}$ and we can write

$$\int_{S>0} \det(S)^{\frac{1}{2}(n+m-2)} \operatorname{etr}\left\{-\frac{1}{2}(T\Omega^{-1}T' + \Sigma)S\right\} {}_0F_1\left(\frac{1}{2}(n + p - 1); \frac{1}{4}\Theta\Sigma S\right) dS$$

$$= 2^{\frac{1}{2}(n+m+p-1)p} \det(T\Omega^{-1}T' + \Sigma)^{-\frac{1}{2}(n+m+p-1)} \int_{U>0} \operatorname{etr}(-U) \det(U)^{\frac{1}{2}(n+m-2)}$$

$$\quad {}_0F_1\left(\frac{1}{2}(n + p - 1); \frac{1}{2}(T\Omega^{-1}T' + \Sigma)^{-\frac{1}{2}}\Theta\Sigma(T\Omega^{-1}T' + \Sigma)^{-\frac{1}{2}}U\right) dU$$

$$= 2^{\frac{1}{2}(n+m+p-1)p}\Gamma_p\left[\frac{1}{2}(n + m + p - 1)\right] \det(T\Omega^{-1}T' + \Sigma)^{-\frac{1}{2}(n+m+p-1)}$$

$$\quad {}_1F_1\left(\frac{1}{2}(n + m + p - 1); \frac{1}{2}(n + p - 1); \frac{1}{2}(I_p + \Sigma^{-1}T\Omega^{-1}T')^{-1}\Theta\right).$$

The last equality is obtained by using Theorem 1.6.2. Hence, the density of T is given by (4.7.1). ∎

In the above theorem, if $\Omega = I_m$, $\Sigma = I_p$, and $\Theta = \text{diag}(\theta_{11}, 0, \ldots, 0)$, then the p.d.f. of T simplifies to

$$\frac{\Gamma_p[\frac{1}{2}(n+m+p-1)]}{\pi^{\frac{1}{2}mp}\Gamma_p[\frac{1}{2}(n+p-1)]} \text{etr}\left(-\frac{1}{2}\theta_{11}\right) \det(I_p + TT')^{-\frac{1}{2}(n+m+p-1)}$$

$${}_1F_1\left(\frac{1}{2}(n+m+p-1); \frac{1}{2}(n+p-1); \frac{1}{2}\tau^{11}\theta_{11}\right),$$

where $(I_p + TT')^{-1} = (\tau^{ij})$. These distributions were studied by Juritz and Troskie (1976), Marx (1981), and Hayakawa (1985). Marx (1981) and Hayakawa (1985) also derived asymptotic expansions of the p.d.f. of T given in the next theorem.

THEOREM 4.7.2. *The asymptotic expansion for the lower noncentral matrix variate t-density (4.7.1),*
(i) if $\Sigma = N\Phi$, $\Theta = 0(1)$, $N = n + p - 1$, is

$$(2\pi)^{-\frac{1}{2}mp} \det(\Phi)^{-\frac{1}{2}m} \det(\Omega)^{-\frac{1}{2}p} \text{etr}\left(-\frac{1}{2}\Phi^{-1}T\Omega^{-1}T'\right)\left[1 + \frac{c}{N} + O(N^{-2})\right],$$

where

$$c = \frac{1}{4}mp(m-p-1) + \frac{1}{4}\{\text{tr}(\Omega^{-1}T'\Phi^{-1}T)^2 - 2m\,\text{tr}(\Omega^{-1}T'\Phi^{-1}T)\}$$

$$+ \frac{1}{2}m\,\text{tr}(\Theta) - \frac{1}{2}\text{tr}(T\Omega^{-1}T'\Phi^{-1}\Theta),$$

and (ii) if $\Sigma = N\Phi$, $\Theta = N\Theta_1$, $\Theta_1 = 0(1)$, $N = n + p - 1$, is

$$(2\pi)^{-\frac{1}{2}mp} \det(\Phi)^{-\frac{1}{2}m} \det(\Omega)^{-\frac{1}{2}p} \det(I_p + \Theta_1)^{\frac{1}{2}m} \text{etr}\left\{-\frac{1}{2}(I_p + \Theta_1)\Phi^{-1}T\Omega^{-1}T'\right\}$$

$$\left[1 + \frac{d}{N} + O(N^{-2})\right]$$

where

$$d = \frac{1}{2}mp(m-p-1) - \frac{1}{2}m\,\text{tr}(\Phi^{-1}T\Omega^{-1}T') - \frac{1}{2}m\,\text{tr}\{\Phi^{-1}T\Omega^{-1}T'\Theta_1(I_p+\Theta_1)^{-1}\}$$

$$- \frac{1}{4}\text{tr}\{(I_p - 2\Theta_1)(\Phi^{-1}T\Omega^{-1}T')^2\} + \frac{1}{4}m[\{\text{tr}\,\Theta_1(I_p+\Theta_1)^{-1}\}^2$$

$$- (m-1)\,\text{tr}\{\Theta_1(I_p+\Theta_1)^{-1}\}^2].$$

Proof: See Hayakawa (1985). ∎

THEOREM 4.7.3. *Let $X \sim N_{p,m}(M, I_p \otimes \Omega)$ and $S \sim W_p(n+p-1, \Sigma^{-1})$ be independent. Then, the distribution of*

$$T = (S^{-\frac{1}{2}})'X,$$

where $S = S^{\frac{1}{2}}(S^{\frac{1}{2}})'$, is the upper noncentral matrix variate t and the density of T is given by

$$\frac{\Gamma_p[\frac{1}{2}(n+m+p-1)]}{\pi^{\frac{1}{2}mp}\Gamma_p[\frac{1}{2}(n+p-1)]}\det(\Omega)^{-\frac{1}{2}p}\det(\Sigma)^{-\frac{1}{2}m}\det(I_p+\Sigma^{-1}T\Omega^{-1}T')^{-\frac{1}{2}(n+m+p-1)}$$

$$I(M,T,\Sigma,\Omega). \qquad (4.7.3)$$

where

$$I(M,T,\Sigma,\Omega) = \left\{\Gamma_p\!\left[\frac{1}{2}(n+m+p-1)\right]\right\}^{-1}\operatorname{etr}\left(-\frac{1}{2}\Omega^{-1}M'M\right)\int_{S>0}\det(S)^{\frac{1}{2}(n+m-2)}$$

$$\operatorname{etr}\left\{-S+\sqrt{2}\,M\Omega^{-1}T'(T\Omega^{-1}T'+\Sigma)^{-\frac{1}{2}}S^{\frac{1}{2}}\right\}dS. \qquad (4.7.4)$$

Proof: The joint density of X and S is

$$(2\pi)^{-\frac{1}{2}mp}\det(\Omega)^{-\frac{1}{2}p}\operatorname{etr}\left\{-\frac{1}{2}\Omega^{-1}(X-M)'(X-M)\right\}$$

$$\left\{2^{\frac{1}{2}(n+p-1)p}\Gamma_p\!\left[\frac{1}{2}(n+p-1)\right]\right\}^{-1}\det(\Sigma)^{\frac{1}{2}(n+p-1)}\det(S)^{\frac{1}{2}(n-2)}\operatorname{etr}\left(-\frac{1}{2}\Sigma S\right).$$

Transforming $T = (S^{-\frac{1}{2}})'X$, with $J(X \to T) = \det(S)^{\frac{1}{2}m}$, we get the joint density of T and S as

$$\frac{2^{-\frac{1}{2}(n+m+p-1)p}\pi^{-\frac{1}{2}mp}}{\Gamma_p[\frac{1}{2}(n+p-1)]}\det(\Omega)^{-\frac{1}{2}p}\det(\Sigma)^{\frac{1}{2}(n+p-1)}\operatorname{etr}\left(-\frac{1}{2}\Omega^{-1}M'M\right)$$

$$\det(S)^{\frac{1}{2}(n+m-2)}\operatorname{etr}\left\{-\frac{1}{2}(T\Omega^{-1}T'+\Sigma)S+T\Omega^{-1}M'(S^{\frac{1}{2}})'\right\}. \qquad (4.7.5)$$

To find the marginal density of T, we integrate (4.7.5) with respect to S. Let $U = \frac{1}{2}(T\Omega^{-1}T'+\Sigma)^{\frac{1}{2}}S(T\Omega^{-1}T'+\Sigma)^{\frac{1}{2}}$, then $J(S \to U) = \det(\frac{1}{2}(T\Omega^{-1}T'+\Sigma))^{-\frac{1}{2}(p+1)}$ and we can write

$$\int_{S>0}\operatorname{etr}\left\{-\frac{1}{2}(T\Omega^{-1}T'+\Sigma)S+T\Omega^{-1}M'(S^{\frac{1}{2}})'\right\}\det(S)^{\frac{1}{2}(n+m-2)}\,dS$$

$$= 2^{\frac{1}{2}(n+m+p-1)p}\det(T\Omega^{-1}T'+\Sigma)^{-\frac{1}{2}(n+m+p-1)}\int_{U>0}\det(U)^{\frac{1}{2}(n+m-2)}$$

$$\operatorname{etr}\left\{-U+\sqrt{2}\,M\Omega^{-1}T'(T\Omega^{-1}T'+\Sigma)^{-\frac{1}{2}}U^{\frac{1}{2}}\right\}dU. \qquad (4.7.6)$$

Substituting (4.7.6) in (4.7.5), we get the result (4.7.3). ∎

The integral (4.7.4) has not been evaluated so far. However, when $M = 0$, $I(0,T,\Sigma,\Omega) = 1$ and we get the central case.

When $X \sim N_{p,m}(M, I_p \otimes \Omega)$ and $S \sim W_p(n+p-1, \Sigma^{-1}, \Theta)$, the distribution of $T = (S^{-\frac{1}{2}})'X$ is called doubly noncentral matrix variate t. However, its density has not been evaluated due to complexity of certain integrals involved in the derivation. Marx (1981) has given an asymptotic expansion for the density of T in this case.

4.8. DISTRIBUTION OF QUADRATIC FORMS

In this section we study the distribution of quadratic forms of the type TAT', where $A\,(m \times m)$ is symmetric positive definite and the random matrix $T\,(p \times m)$, $p \leq m$ has t-distribution. It may be recalled here that for $p \leq m$, $TAT' > 0$ with probability 1. The next result is due to Javier and Gupta (1985b).

THEOREM 4.8.1. *Let $T \sim T_{p,m}(n, 0, \Sigma, \Omega)$ and $A\,(m \times m)$ be a symmetric positive definite matrix. Then the p.d.f. of $W = TAT'$ for $p \leq m$ is given by*

$$\left\{ \beta_p \left(\frac{1}{2}(n+m+p-1), \frac{1}{2}m \right) \right\}^{-1} \det(\Sigma)^{-\frac{1}{2}m} \det(\Omega A)^{-\frac{1}{2}p}$$

$$\det(W)^{\frac{1}{2}(m-p-1)} \det(I_p + \Sigma^{-1}W)^{-\frac{1}{2}(n+m+p-1)}$$

$$_1F_0^{(m)}\left(\frac{1}{2}(n+m+p-1); (I_p + \Sigma^{-1}W)^{-1}\Sigma^{-1}W, B \right), \quad W > 0, \qquad (4.8.1)$$

where $B = I_m - A^{-\frac{1}{2}}\Omega^{-1}A^{-\frac{1}{2}}$.

Proof: The density of T is

$$\frac{\Gamma_p[\frac{1}{2}(n+m+p-1)]}{\pi^{\frac{1}{2}mp}\Gamma_p[\frac{1}{2}(n+p-1)]} \det(\Sigma)^{-\frac{1}{2}m} \det(\Omega)^{-\frac{1}{2}p}$$

$$\det(I_p + \Sigma^{-1}T\Omega^{-1}T')^{-\frac{1}{2}(n+m+p-1)}, \quad \mathbb{R}^{p \times m}.$$

Therefore the density of $W = TAT'$, is given by

$$f(W) = \frac{\Gamma_p[\frac{1}{2}(n+m+p-1)]}{\pi^{\frac{1}{2}mp}\Gamma_p[\frac{1}{2}(n+p-1)]} \det(\Sigma)^{-\frac{1}{2}m} \det(\Omega)^{-\frac{1}{2}p}$$

$$\int_{TAT'=W} \det(I_p + \Sigma^{-1}T\Omega^{-1}T')^{-\frac{1}{2}(n+m+p-1)} \, dT.$$

Now let $Y = TA^{\frac{1}{2}}$. Then $J(T \to Y) = \det(A)^{-\frac{1}{2}p}$, and hence

$$f(W) = \frac{\Gamma_p[\frac{1}{2}(n+m+p-1)]}{\pi^{\frac{1}{2}mp}\Gamma_p[\frac{1}{2}(n+p-1)]} \det(\Sigma)^{-\frac{1}{2}m} \det(A\Omega)^{-\frac{1}{2}p}$$

$$\int_{YY'=W} \det(I_p + \Sigma^{-1}YA^{-\frac{1}{2}}\Omega^{-1}A^{-\frac{1}{2}}Y')^{-\frac{1}{2}(n+m+p-1)} \, dY. \qquad (4.8.2)$$

Next write

$$\det(I_p + \Sigma^{-1}YA^{-\frac{1}{2}}\Omega^{-1}A^{-\frac{1}{2}}Y')^{-\frac{1}{2}(n+m+p-1)}$$

$$= \det(I_p + \Sigma^{-1}YY')^{-\frac{1}{2}(n+m+p-1)}$$

$$\det(I_p - (I_p + \Sigma^{-1}YY')^{-1}\Sigma^{-1}Y(I_m - A^{-\frac{1}{2}}\Omega^{-1}A^{-\frac{1}{2}})Y')^{-\frac{1}{2}(n+m+p-1)}$$

$$= \det(I_p + \Sigma^{-1}YY')^{-\frac{1}{2}(n+m+p-1)}$$

$$\det(I_m - Y'(I_p + \Sigma^{-1}YY')^{-1}\Sigma^{-1}YB)^{-\frac{1}{2}(n+m+p-1)}$$

$$= \det(I_p + \Sigma^{-1}YY')^{-\frac{1}{2}(n+m+p-1)}$$

$$\quad {}_1F_0^{(m)}\left(\frac{1}{2}(n+m+p-1); Y'(I_p + \Sigma^{-1}YY')^{-1}\Sigma^{-1}YB\right). \tag{4.8.3}$$

Substituting from (4.8.3) in (4.8.2), we get

$$f(W) = \frac{\Gamma_p[\frac{1}{2}(n+m+p-1)]}{\pi^{\frac{1}{2}mp}\Gamma_p[\frac{1}{2}(n+p-1)]} \det(\Sigma)^{-\frac{1}{2}m}\det(A\Omega)^{-\frac{1}{2}p}\, g(B), \tag{4.8.4}$$

where

$$g(B) = \int_{YY'=W} \det(I_p + \Sigma^{-1}YY')^{-\frac{1}{2}(n+m+p-1)}$$

$$\quad {}_1F_0^{(m)}\left(\frac{1}{2}(n+m+p-1); Y'(I_p + \Sigma^{-1}YY')^{-1}\Sigma^{-1}YB\right) dY. \tag{4.8.5}$$

Since B is a symmetric matrix, the integral (4.8.5) is invariant under the transformation $B \to HBH'$, $H \in O(m)$. Hence, from (4.8.5), using Theorem 1.6.1, we obtain

$$f(W) = \frac{\Gamma_p[\frac{1}{2}(n+m+p-1)]}{\pi^{\frac{1}{2}mp}\Gamma_p[\frac{1}{2}(n+p-1)]} \det(\Sigma)^{-\frac{1}{2}m}\det(A\Omega)^{-\frac{1}{2}p}$$

$$\int_{YY'=W} \det(I_p + \Sigma^{-1}YY')^{-\frac{1}{2}(n+m+p-1)}$$

$$\quad {}_1F_0^{(m)}\left(\frac{1}{2}(n+m+p-1); (I_p + \Sigma^{-1}YY')^{-1}\Sigma^{-1}YY', B\right) dY. \tag{4.8.6}$$

Finally using Theorem 1.4.10 we get the desired result. ∎

If $m < p$, then $T' \sim T_{m,p}(n, 0, \Omega, \Sigma)$ and for $A\,(p \times p) > 0$, the density of $R = T'AT$ is obtained from the above theorem as

$$\left\{\beta_m\left(\frac{1}{2}(n+p+m-1), \frac{1}{2}p\right)\right\}^{-1} \det(\Omega)^{-\frac{1}{2}p}\det(\Sigma A)^{-\frac{1}{2}m}$$

$$\det(R)^{\frac{1}{2}(p-m-1)}\det(I_m + \Omega^{-1}R)^{-\frac{1}{2}(n+p+m-1)}$$

$$\quad {}_1F_0^{(p)}\left(\frac{1}{2}(n+p+m-1); (I_m + \Omega^{-1}R)^{-1}\Omega^{-1}R, B\right), \ R > 0,$$

where $B = I_p - A^{-\frac{1}{2}}\Sigma^{-1}A^{-\frac{1}{2}}$.

The synthetic representation (4.2.2) can also be used to obtain the density of $W = TAT'$, where $T \sim T_{p,m}(n, M, \Sigma, \Omega)$, $p \leq m$. Briefly the approach is as follows. The matrix T can be represented as

$$T = (S^{-\frac{1}{2}})'X + M,$$

where independently $S \sim W_p(n + p - 1, \Sigma^{-1})$, and $X \sim N_{p,m}(0, I_p \otimes \Omega)$. Then $T|S \sim N_{p,m}(M, S^{-1} \otimes \Omega)$, and $W|S$ is distributed as YAY', where $Y \sim N_{p,m}(M, S^{-1} \otimes \Omega)$. Such quadratic forms have been studied in Chapter 7. From there, the conditional density of W given S, $f_{W|S}(W|S)$ can be obtained. Then the unconditional density of W, $f_W(W)$, is

$$f_W(W) = E_S[f_{W|S}(W|S)],$$

where E_S denotes the expectation with respect to $S \sim W_p(n + p - 1, \Sigma^{-1})$. When $M = 0$, the null density of W can be obtained by using either (7.2.1) or (7.2.5) or (7.2.7). For $M \neq 0$, Marx (1983), using the density function in Theorem 7.6.2, has obtained the nonnull density of W in terms of an integral.

The density of $T'BT$, where the matrix T $(p \times m)$, $(p \geq m)$, has a lower noncentral t-distribution, has been derived by Hayakawa (1985) in terms of invariant polynomials (Davis, 1979, 1980).

Quadratic forms in disguised matrix variate t have been found useful in the study of simultaneous equations in econometric problems, e.g., see Tiao and Zellner (1964), Goldberger (1970), Chang (1972), Tiao, Tan and Chang (1970), and Tan (1973). The next two theorems give the distribution of quadratic form in T.

THEOREM 4.8.2. Let T $(p \times m)$ be distributed as a lower or upper disguised matrix variate t with parameters n, M, and Σ. Then $(T - M)(T - M)'$ is distributed as $Y'Y$ where $Y \sim T_{m,p}(n - p - m + 1, 0, I_m, \Sigma)$.

Proof: According to Theorems 4.5.1 and 4.5.2, $T - M$ can be represented as

$$T - M = XU^{-1}$$

where X and U are independent, $X \sim N_{p,m}(0, \Sigma \otimes I_m)$, $U'U = S \sim W_m(n - p, I_m)$ and U is a lower or upper triangular matrix with positive diagonal elements. Then $(T - M)(T - M)' = X(U'U)^{-1}X' = XS^{-1}X' = (S^{-\frac{1}{2}}X')'(S^{-\frac{1}{2}}X') = Y'Y$, where $S^{\frac{1}{2}}$ is the symmetric positive definite square root of S. Since $X' \sim N_{m,p}(0, I_m \otimes \Sigma)$ and $S \sim W_m(n - p, I_m)$ are independent, $S^{-\frac{1}{2}}X' = Y \sim T_{m,p}(n - p - m + 1, 0, I_m, \Sigma)$. ∎

The density of $Y'Y$ is given in (4.8.1).

THEOREM 4.8.3. Let A $(p \times p)$ be a positive semidefinite matrix of rank r $(\leq p)$.
(i) If $T \sim \underline{D}T_{p,m}(n, M, \Sigma)$, and $A\Sigma A = A$, then the p.d.f. of $W_1 = (T - M)'A(T - M)$ is

$$\left\{\beta_m\left(\frac{1}{2}r, \frac{1}{2}(n - p)\right)\right\}^{-1} \det(W_1)^{\frac{1}{2}(r-m-1)} \det(I_m + W_1)^{-\frac{1}{2}(n-p+r-m-1)}$$

$$\prod_{i=1}^m \det((I_m + W_1)_{[i]})^{-1}, \, r \geq m, \, n \geq m + p,$$

and (ii) if $T \sim \overline{D}T_{p,m}(n, M, \Sigma)$, and $A\Sigma A = A$, then the p.d.f. of $W_2 = (T-M)'A(T-M)$ is

$$\left\{\beta_m\left(\frac{1}{2}r, \frac{1}{2}(n-p)\right)\right\}^{-1} \det(W_2)^{\frac{1}{2}(r-m-1)} \det(I_m + W_2)^{-\frac{1}{2}(n-p+r-m-1)}$$

$$\prod_{i=1}^{m} \det((I_m + W_2)^{[i]})^{-1}, \ r \geq m, \ n \geq m+p,$$

Proof: From Theorem 4.5.1, we can write

$$(T - M)'A(T - M) = (U^{-1})'QU^{-1}$$

where $Q = X'AX$, $X \sim N_{p,m}(0, \Sigma \otimes I_m)$, $U'U \sim W_m(n - p, I_m)$ and U is a lower triangular matrix with positive diagonal elements. Using Theorem 3.2.6 we have $Q \sim W_m(r, I_m)$.

Now the joint density of U and Q is

$$\left\{2^{\frac{1}{2}m(n-p+r-2)} \Gamma_m\left(\frac{1}{2}r\right)\Gamma_m\left[\frac{1}{2}(n-p)\right]\right\}^{-1} \det(U'U)^{\frac{1}{2}(n-p-m-1)} \prod_{i=1}^{m} u_{ii}^{i}$$

$$\det(Q)^{\frac{1}{2}(r-m-1)} \operatorname{etr}\left\{-\frac{1}{2}(Q + UU')\right\}.$$

Transforming $W_1 = (U^{-1})'QU^{-1}$, $Z = U'(I_m + W_1)U$ with the Jacobian $J(Q, U \to W_1, Z) = \det(U'U)^{\frac{1}{2}(m+1)} 2^{-m} \prod_{i=1}^{m}\{u_{ii}^{i} \det((I_m + W_1)_{[i]})\}^{-1}$, the joint density of W_1 and Z is given by

$$\left\{2^{\frac{1}{2}m(n-p+r)} \Gamma_m\left[\frac{1}{2}(n+r-p)\right]\right\}^{-1} \det(Z)^{\frac{1}{2}(n+r-p-m-1)} \operatorname{etr}\left(-\frac{1}{2}Z\right)$$

$$\left\{\beta_m\left(\frac{1}{2}r, \frac{1}{2}(n-p)\right)\right\}^{-1} \det(W_1)^{\frac{1}{2}(r-m-1)} \det(I_m + W_1)^{-\frac{1}{2}(n-p+r-m-1)}$$

$$\prod_{i=1}^{m} \det((I_m + W_1)_{[i]})^{-1}. \tag{4.8.7}$$

From (4.8.7) it is easy to see that W_1 and Z are independent and the distribution of W_1 is given in (i) above.

Similarly one can prove the result (ii). ∎

PROBLEMS

4.1. Let $S \sim W_p(n + p - 1, \Sigma^{-1})$, independent of $X \sim N_{p,m}(0, \Psi \otimes \Omega)$. Define $T = \Psi^{\frac{1}{2}}(S^{-\frac{1}{2}})'\Psi^{-\frac{1}{2}}X + M$, where M $(p \times m)$ is a constant matrix, $S^{\frac{1}{2}}(S^{\frac{1}{2}})' = S$ and $\Psi^{\frac{1}{2}}(\Psi^{\frac{1}{2}})' = \Psi$. Prove that $T \sim T_{p,m}(n, M, \Psi^{\frac{1}{2}}\Sigma(\Psi^{\frac{1}{2}})', \Omega)$.

(Marx, 1981)

4.2. Let $T \sim T_{p,m}(n, M, \Sigma, \Omega)$. Then show that the characteristic function of T is

$$\{\Gamma_p(\delta)\}^{-1} \operatorname{etr}(\iota ZM') B_{-\delta}\left(\frac{1}{4}Z\Omega Z'\Sigma\right), \ Z \ (p \times m),$$

where $\delta = \frac{1}{2}(n + p - 1)$ and $B_\lambda(\cdot)$ is the type two Bessel function of Herz defined in Section 1.6.

4.3. Let $T \sim T_{p,m}(n, M, \Sigma, \Omega)$, then prove that

(i) $E(TCTDTFT) = c_1 \Sigma F' \Omega D \Sigma C' \Omega + c_2 \{ \Sigma D' \Omega F \Sigma C' \Omega$

$\qquad\qquad\qquad + \text{tr}(F' \Omega D \Sigma) \Sigma C' \Omega \} + c_1 \Sigma C' \Omega D \Sigma F' \Omega$

$\qquad\qquad\qquad + c_2 \{ \Sigma D' \Omega C \Sigma F' \Omega + \text{tr}(C' \Omega D \Sigma) \Sigma F' \Omega \}$

$\qquad\qquad\qquad + c_1 \text{tr}(F' \Omega C \Sigma) \Sigma D' \Omega + c_2 \{ \Sigma C' \Omega F \Sigma D' \Omega$

$\qquad\qquad\qquad + \Sigma F' \Omega C \Sigma D' \Omega \} + (n-2)^{-1} \{ \Sigma D' M' C' \Omega F M$

$\qquad\qquad\qquad + \Sigma F' M' D' M' C' \Omega + \Sigma C' \Omega D M F M + M C \Sigma D' \Omega F M$

$\qquad\qquad\qquad + M C \Sigma F' M' D' \Omega + M C M D \Sigma F' \Omega \} + M C M D M F M.$

(ii) $E(TCT'DTFT') = \text{tr}(\Omega F' \Omega C') \{ c_1 \text{tr}(D \Sigma) \Sigma + c_2 \Sigma D' \Sigma + c_2 \Sigma D \Sigma \}$

$\qquad\qquad\qquad + \text{tr}(C \Omega) \text{tr}(F \Omega) \{ c_1 \Sigma D \Sigma + c_2 \Sigma D' \Sigma + c_2 \text{tr}(D \Sigma) \Sigma \}$

$\qquad\qquad\qquad + \text{tr}(C \Omega F' \Omega) \{ c_1 \Sigma D' \Sigma + c_2 \Sigma D \Sigma + c_2 \text{tr}(D' \Sigma) \Sigma \}$

$\qquad\qquad\qquad + (n-2)^{-1} \{ \text{tr}(C' \Omega) \Sigma D M F M' + \Sigma D' M C' \Omega F M'$

$\qquad\qquad\qquad + \text{tr}(F' M' D' M C' \Omega) \Sigma + \text{tr}(D' \Sigma) M C \Omega F M'$

$\qquad\qquad\qquad + M C \Omega F' M' D' \Sigma + \text{tr}(F' \Omega) M C M' D \Sigma \}$

$\qquad\qquad\qquad + M C M' D M F M',$

where the matrices C, D, and F are of appropriate order, $c_1 = (n-3)c_2$, $c_2 = \{(n-1)(n-2)(n-4)\}^{-1}$, and $n > 4$.

4.4. Let the joint density of X $(p \times m)$ and Ω $(m \times m) > 0$ be

$$\frac{2^{-\frac{1}{2}m(p+n+m-1)} \pi^{-\frac{1}{2}mp}}{\Gamma_m[\frac{1}{2}(n+m-1)]} \det(\Sigma)^{-\frac{1}{2}m} \det(\Psi)^{-\frac{1}{2}(n+m-1)} \det(\Omega)^{-\frac{1}{2}(n+p+2m)}$$

$$\text{etr} \left[-\frac{1}{2} \{ (X-M)' \Sigma^{-1} (X-M) + \Psi \} \Omega^{-1} \right], \quad X \in \mathbb{R}^{p \times m}.$$

Then, prove that (i) given Ω, $X \sim N_{p,m}(M, \Sigma \otimes \Omega)$, (ii) $\Omega \sim IW_m(n+2m, \Psi)$, and (iii) $X \sim T_{p,m}(n, M, \Sigma, \Psi)$.

4.5. In Problem 4.4, let $\Omega = \begin{pmatrix} \Omega_{11} & \Omega_{12} \\ \Omega_{21} & \Omega_{22} \end{pmatrix}$, Ω_{11} $(m_1 \times m_1)$, $\Psi = \begin{pmatrix} \Psi_{11} & \Psi_{12} \\ \Psi_{21} & \Psi_{22} \end{pmatrix}$, Ψ_{11} $(m_1 \times m_1)$, $m_1 + m_2 = m$, $\Omega_{22 \cdot 1} = \Omega_{22} - \Omega_{21} \Omega_{11}^{-1} \Omega_{12}$, $\Psi_{22 \cdot 1} = \Psi_{22} - \Psi_{21} \Psi_{11}^{-1} \Psi_{12}$ and $T = \Omega_{11}^{-1} \Omega_{12}$. Prove that
(i) Ω_{11} and $(T, \Omega_{22 \cdot 1})$ are independent,
(ii) $\Omega_{11} \sim IW_{m_1}(n+2m, \Psi_{11})$,
(iii) $\Omega_{22 \cdot 1} \sim IW_{m_2}(n+2m+m_1, \Psi_{22 \cdot 1})$, and
(iv) $T \sim T_{m_1, m_2}(n+2m+m_1, \Psi_{11}^{-1} \Psi_{12}, \Psi_{11}^{-1}, \Psi_{22 \cdot 1})$.

4.6.   Let $X \sim N_{p,m}(0, \Sigma \otimes \Omega)$ and $v \sim \chi_n^2$ be independent. Prove that the p.d.f. of
$T = \left(\frac{v}{n}\right)^{-\frac{1}{2}} X$ is

$$\frac{\Gamma[\frac{1}{2}(n+mp)]}{(n\pi)^{\frac{1}{2}mp}\Gamma(\frac{1}{2}n)} \det(\Sigma)^{-\frac{1}{2}m} \det(\Omega)^{-\frac{1}{2}p}$$

$$\left(1 + \frac{1}{n}\operatorname{tr}(\Sigma^{-1}T\Omega^{-1}T')\right)^{-\frac{1}{2}(n+mp)}, \quad T \in \mathbb{R}^{p \times m}.$$

4.7.   In Problem 4.6, prove that the distribution of $S = T\Omega^{-1}T'$, for $m \geq p$, is

$$\frac{\Gamma[\frac{1}{2}(n+mp)]}{n^{\frac{1}{2}mp}\Gamma(\frac{1}{2}n)\Gamma_p(\frac{1}{2}m)} \det(\Sigma)^{-\frac{1}{2}m} \det(S)^{\frac{1}{2}(m-p-1)}\left(1 + \frac{1}{n}\operatorname{tr}(\Sigma^{-1}S)\right)^{-\frac{1}{2}(n+mp)}, \quad S > 0.$$

4.8.   Let $S \sim W_p(n + p - 1, I_p)$ and $X \sim N_{p,m}(0, \Sigma^{-1} \otimes \Omega)$ be independent. Prove
that $T = (S^{-\frac{1}{2}})'X + M \sim T_{p,m}(n, M, \Sigma, \Omega)$, where $M \ (p \times m)$ is a constant
matrix and $S^{\frac{1}{2}}(S^{\frac{1}{2}})' = S$.

4.9.   Let $T \sim T_{p,m}(n, M, \Sigma, \Omega)$, and $A \ (m \times t)$, and $C \ (m \times s)$ be constant matrices.
Prove that
$$\operatorname{cov}(TA, TC) = (n - 2)^{-1}\Sigma \otimes (A'\Omega C), \quad n > 2$$

and hence show that
$$\operatorname{cov}(\boldsymbol{t}_i, \boldsymbol{t}_j) = (n - 2)^{-1}\omega_{ij}\Sigma,$$

where $\Omega = (\omega_{ij})$ and $T = (\boldsymbol{t}_1, \ldots, \boldsymbol{t}_m)$.

4.10. Let $T \sim T_{p,m}(n, M, \Sigma, \Omega)$, and $B \ (r \times p)$ and $D \ (s \times p)$ be constant matrices.
Prove that
$$\operatorname{cov}(BT, DT) = (n - 2)^{-1}\Omega \otimes B\Sigma D', \quad n > 2$$

and hence show that
$$\operatorname{cov}(\boldsymbol{t}_i^*, \boldsymbol{t}_j^*) = (n - 2)^{-1}\sigma_{ij}\Omega,$$

where $\Sigma = (\sigma_{ij})$ and $\boldsymbol{t}_i^{*\prime}$ is the ith row of the matrix T.

4.11. Let $T \sim IT_{p,m}(n, M, \Sigma, \Omega)$. Show that the m.g.f. of T is

$$\operatorname{etr}(ZM')\, _0F_1\left(\frac{1}{2}(n + m + p - 1); \frac{1}{4}Z\Omega Z'\Sigma\right), \quad Z \ (p \times m),$$

where $_0F_1(\cdot)$ is defined in Section 1.6.

4.12. Let $T \sim IT_{p,m}(n, M, \Sigma, \Omega)$ and $B \ (m \times r)$ be a matrix of rank $r \leq m$. Then,
prove that $TB \sim IT_{p,m}(n + m - r, MB, \Sigma, B'\Omega B)$.
(HINT: Let $B_0 = (\ B \quad B_1\ )$ where $B_1 \ (m \times (m-r))$ is such that B is nonsingular
and then find the marginal distribution of TB from the distribution of TB_0.)

4.13. Let $X_i \sim N_{p,m_i}(0, \Sigma \otimes I_{m_i})$, $i = 1, \ldots, k$ and $S \sim W_p(n, \Sigma)$ be independent.
Further, let $S_k = S + \sum_{j=1}^{k} X_j X_j'$ and $T_j = (S + \sum_{i=1}^{j} X_i X_i')^{-\frac{1}{2}}X_j$, $j = 1, \ldots, k$.
Then, show that $T_1, \ldots, T_k$ are independent and derive the distribution of T_j.

4.14. Let $T \sim IT_{p,m}(n, M, \Sigma, \Omega)$, and partition T, M, Σ and Ω as

$$T = \begin{pmatrix} T_{1r} \\ T_{2r} \end{pmatrix} \begin{matrix} p_1 \\ p_2 \end{matrix} = (T_{1c} \ \ T_{2c}), \quad M = \begin{pmatrix} M_{1r} \\ M_{2r} \end{pmatrix} \begin{matrix} p_1 \\ p_2 \end{matrix} = (M_{1c} \ \ M_{2c}),$$
$$\begin{matrix} m_1 \ \ m_2 \end{matrix} \qquad\qquad\qquad\qquad\qquad \begin{matrix} m_1 \ \ m_2 \end{matrix}$$

$$\Sigma = \begin{pmatrix} \Sigma_{11} & \Sigma_{12} \\ \Sigma_{21} & \Sigma_{22} \end{pmatrix} \begin{matrix} p_1 \\ p_2 \end{matrix}, \text{ and } \Omega = \begin{pmatrix} \Omega_{11} & \Omega_{12} \\ \Omega_{21} & \Omega_{22} \end{pmatrix} \begin{matrix} m_1 \\ m_2 \end{matrix}.$$
$$\begin{matrix} p_1 \ \ \ p_2 \end{matrix} \qquad\qquad\qquad \begin{matrix} m_1 \ \ \ m_2 \end{matrix}$$

Then, prove that (i) $T_{2r} \sim IT_{p_2,m}(n + p_2, M_{2r}, \Sigma_{22}, \Omega)$,

$$T_{1r}|T_{2r} \sim IT_{p_1,m}(n, M_{1r} + \Sigma_{12}\Sigma_{22}^{-1}(T_{2r} - M_{2r}), \Sigma_{11 \cdot 2},$$
$$\Omega(I_m - (T_{2r} - M_{2r})'\Sigma_{22}^{-1}(T_{2r} - M_{2r})))$$

and (ii) $T_{2c} \sim IT_{p,m_2}(n + m_2, M_{2c}, \Sigma, \Omega_{22})$,

$$T_{1c}|T_{2c} \sim IT_{p,m_1}(n, M_{1c} + (T_{2c} - M_{2c})\Omega_{22}^{-1}\Omega_{21},$$
$$(I_p - (T_{2c} - M_{2c})\Omega_{22}^{-1}(T_{2c} - M_{2c})')\Sigma, \Omega_{11 \cdot 2}).$$

4.15. Let $T = (t_1, t_2, \ldots, t_m) \sim IT_{p,m}(n, M, \Sigma, \Omega)$ and denote its p.d.f. by $p(T)$. Further let $f(y_1|y_2)$ be the conditional density of y_1 given y_2. Using suitable notation, write down explicitly

$$p(T) = f_1(t_1)f_2(t_2|t_1)f_3(t_3|t_1, t_2) \cdots f_m(t_m|t_1, \ldots, t_{m-1}).$$

4.16. Prove Theorem 4.5.2.

4.17. Prove Theorem 4.5.3(ii).

4.18. Prove Theorem 4.5.4(iv).

4.19. Let $T \sim \underline{D}T_{p,m}(n, M, \Sigma)$, and $A(t \times p)$ and $C(s \times p)$ be constant matrices. Prove that

$$\text{cov}(AT, CT) = B \otimes (A\Sigma C'),$$

where B is a diagonal matrix given in Theorem 4.5.3(i). Also, show that

$$\text{cov}(t_i^*, t_j^*) = \sigma_{ij}B,$$

where $\Sigma = (\sigma_{ij})$ and $t_i^{*\prime}$ is the ith row of the matrix T.

4.20. Let $T \sim \overline{D}T_{p,m}(n, M, \Sigma)$, and $A(t \times p)$ and $C(s \times p)$ be constant matrices. Derive the results stated in Problem 4.19, where the matrix B is now given in Theorem 4.5.3(ii).

4.21. Let $S \sim W_p(n, I_p, \Theta)$, where $\Theta = \text{diag}(\theta_{11}, 0, \ldots, 0)$ and partition S as $S = \begin{pmatrix} S_{11} & S_{12} \\ S_{21} & S_{22} \end{pmatrix}$, S_{11} $(q \times q)$. Prove that the distribution of $B = S_{11}^{-1}S_{12}$ is given by

$$\frac{\Gamma_q[\frac{1}{2}(n + p - q)]}{\pi^{\frac{1}{2}q(p-q)}\Gamma_q(\frac{1}{2}n)}e^{-\frac{1}{2}\theta_{11}}\det(I_q + BB')^{-\frac{1}{2}(n+p-q)}\,_1F_1\Big(\frac{1}{2}(n + p - q); \frac{1}{2}n; \frac{1}{2}\theta_{11}w^{11}\Big),$$

where $(w^{ij}) = (I_q + BB')^{-1}$.

(Juritz and Troskie, 1976)

4.22. Let the joint density of X $(p \times m)$ and Σ $(p \times p) > 0$ be

$$\frac{2^{-\frac{1}{2}(n+m+p-s-1)p} \pi^{-\frac{1}{2}(m-s)p}}{\Gamma_p[\frac{1}{2}(n+p-1)]} \det(\Omega)^{-\frac{1}{2}p} \det(C'\Omega C)^{\frac{1}{2}p} \det(\Psi)^{\frac{1}{2}(n+p-1)}$$

$$\det(\Sigma)^{\frac{1}{2}(m+n-s-2)} \operatorname{etr}\left\{-\frac{1}{2}\Sigma(X-M)\Omega^{-1}(X-M)' - \frac{1}{2}\Sigma\Psi\right\},$$

where the domain of definition of X is restricted to all $X \in \mathbb{R}^{p \times m}$ such that $XC = 0$, and $MC = 0$, for a fixed matrix C $(m \times s)$ of rank $s \leq m$. Then, prove that
(i) given Σ, $X \sim N_{p,m}(M, \Sigma^{-1} \otimes \Omega | s, C)$
(ii) $\Sigma \sim W_p(n+p-1, \Psi^{-1})$, and
(iii) $X \sim T_{p,m}(n, M, \Psi, \Omega | s, C)$.

4.23. If $T \sim T_{p,m}(n, M, n\Sigma, \Omega | s, C)$, then prove that $T \xrightarrow{D} X$ as $n \to \infty$, where $X \sim N_{p,m}(M, \Sigma \otimes \Omega | s, C)$.

4.24. Prove Theorem 4.6.2.
(HINT: Use the transformation $W = BTD$, with the Jacobian $J(T \to W)$ given in Lemma 2.6.1.)

4.25. Let $T \sim T_{p,m}(n, M, \Sigma \otimes \Omega | s, C)$ and partition T, M and Ω as

$$T = \begin{pmatrix} T_{1c} & T_{2c} \\ m_1 & m_2 \end{pmatrix}, \ M = \begin{pmatrix} M_{1c} & M_{2c} \\ m_1 & m_2 \end{pmatrix}, \ \Omega = \begin{pmatrix} \Omega_{11} & \Omega_{12} \\ \Omega_{21} & \Omega_{22} \end{pmatrix} \begin{matrix} m_1 \\ m_2 \end{matrix}$$
$$\begin{matrix} & m_1 & m_2 \end{matrix}$$

where $m_1 + m_2 = m$. Then, prove that
(i) $T_{1c} \sim T_{p,m_1}(n, M_{1c}, \Sigma, \Omega_{11} | s, C)$ and
(ii) $T_{2c}|T_{1c} \sim T_{p,m_2}(n+m_1, M_{2c} + \Omega_{21}\Omega_{11}^{-1}(T_{1c} - M_{1c})',$
$(\Sigma + (T_{1c} - M_{1c})\Omega_{11}^{-1}(T_{1c} - M_{1c})'), \Omega_{22\cdot1} | s, C)$.

4.26. Let $S \sim W_p(n+p-1, \alpha I_p)$, $\alpha > 0$, and $X \sim N_{p,m}(0, I_p \otimes I_m)$ be independently distributed. Derive the distribution of $T = (S + XX')^{-\frac{1}{2}}X$.

4.27. Let $p \times 1$ random vectors $\boldsymbol{x}_1, \ldots, \boldsymbol{x}_N$ have the joint density

$$\frac{\nu^{\frac{1}{2}\nu}\Gamma[\frac{1}{2}(\nu + Np)]}{\pi^{\frac{1}{2}Np}\Gamma(\frac{1}{2}\nu)}\det(\Lambda)^{-\frac{1}{2}N}\left\{\nu + \sum_{j=1}^{N}(\boldsymbol{x}_j - \boldsymbol{\theta})'\Lambda^{-1}(\boldsymbol{x}_j - \boldsymbol{\theta})\right\}^{-\frac{1}{2}(\nu + Np)}, \ \boldsymbol{x}_j \in \mathbb{R}^p,$$

where $\nu > 0$ and Λ $(p \times p) > 0$. Define $A = \sum_{j=1}^{N}(\boldsymbol{x}_j - \bar{\boldsymbol{x}})(\boldsymbol{x}_j - \bar{\boldsymbol{x}})'$, $N\bar{\boldsymbol{x}} = \sum_{j=1}^{N}\boldsymbol{x}_j$ and $n = N - 1 \geq p$. Show that
(i) the p.d.f. of $\bar{\boldsymbol{x}}$ is

$$\frac{n^{\frac{1}{2}p}\nu^{\frac{1}{2}\nu}\Gamma[\frac{1}{2}(\nu + p)]}{\pi^{\frac{1}{2}p}\Gamma(\frac{1}{2}\nu)}\det(\Lambda)^{-\frac{1}{2}}\left\{\nu + n(\bar{\boldsymbol{x}} - \boldsymbol{\theta})'\Lambda^{-1}(\bar{\boldsymbol{x}} - \boldsymbol{\theta})\right\}^{-\frac{1}{2}(\nu + p)}, \ \bar{\boldsymbol{x}} \in \mathbb{R}^p,$$

(ii) the p.d.f. of A is

$$\frac{\nu^{\frac{1}{2}\nu}\Gamma[\frac{1}{2}(\nu+np)]}{\Gamma(\frac{1}{2}\nu)\Gamma_p(\frac{1}{2}n)}\det(\Lambda)^{-\frac{1}{2}n}\det(A)^{\frac{1}{2}(n-p-1)}\{\nu+\mathrm{tr}(\Lambda^{-1}A)\}^{-\frac{1}{2}(\nu+np)}, \ A>0,$$

(iii) $E[\det(A)^r] = \dfrac{\nu^{rp}\Gamma(\frac{1}{2}\nu-rp)}{\Gamma(\frac{1}{2}\nu)}\dfrac{\Gamma_p[\frac{1}{2}(n+2r)]}{\Gamma_p(\frac{1}{2}n)}\det(\Lambda)^r, \ \nu>2rp,$

(iv) $E[\det(A)^r A] = \nu^{rp+1}\left(\dfrac{1}{2}n+r\right)\dfrac{\Gamma(\frac{1}{2}\nu-rp-1)}{\Gamma(\frac{1}{2}\nu)}\dfrac{\Gamma_p(\frac{1}{2}n+r)}{\Gamma_p(\frac{1}{2}n)}\det(\Lambda)^r\Lambda,$

$$\nu>2(rp+1)$$

and

(v) $E[C_\kappa(A)] = \nu^k\dfrac{\Gamma(\frac{1}{2}\nu-k)}{\Gamma(\frac{1}{2}\nu)}\left(\dfrac{1}{2}n\right)_\kappa C_\kappa(\Lambda), \ \nu>2k.$

(Sutradhar and Ali, 1989; Joarder and Ali, 1992)

CHAPTER 5

MATRIX VARIATE BETA DISTRIBUTIONS

5.1. INTRODUCTION

The random variable u with the p.d.f.

$$\{\beta(a,b)\}^{-1}u^{a-1}(1-u)^{b-1}, \; 0 < u < 1, \tag{5.1.1}$$

where $a > 0$ and $b > 0$, is said to have a beta type I distribution with parameters (a, b). The random variable v with p.d.f.

$$\{\beta(a,b)\}^{-1}v^{a-1}(1+v)^{-(a+b)}, \; v > 0, \tag{5.1.2}$$

where $a > 0$ and $b > 0$, is said to have beta type II distribution with parameters (a, b). Since (5.1.2) can be obtained from (5.1.1) by the transformation $v = \frac{u}{1-u}$, some authors call the distribution of v an *inverted beta distribution*.

In this chapter, several generalizations which lead to matrix variate analogs of beta type I and type II distributions have been studied.

5.2. DENSITY FUNCTIONS

First we shall define the matrix variate beta distributions of type I and type II.

DEFINITION 5.2.1. *A $p \times p$ random symmetric positive definite matrix U is said to have a matrix variate beta type I distribution with parameters (a, b), denoted as $U \sim B_p^I(a, b)$, if its p.d.f. is given by*

$$\{\beta_p(a,b)\}^{-1}\det(U)^{a-\frac{1}{2}(p+1)}\det(I_p - U)^{b-\frac{1}{2}(p+1)}, \; 0 < U < I_p, \tag{5.2.1}$$

where $a > \frac{1}{2}(p-1)$, $b > \frac{1}{2}(p-1)$, and $\beta_p(a,b)$ is the multivariate beta function given by (1.4.8).

Using Theorem 1.6.8, the c.d.f. of U is obtained as

$$P(U < \Lambda) = \frac{\Gamma_p(a+b)\Gamma_p[\frac{1}{2}(p+1)]}{\Gamma_p(b)\Gamma_p[a+\frac{1}{2}(p+1)]} \det(\Lambda)^a$$

$$_2F_1\left(a, -b+\frac{1}{2}(p+1); a+\frac{1}{2}(p+1); \Lambda\right), \ 0 < \Lambda < I_p.$$

DEFINITION 5.2.2. *A $p \times p$ random symmetric positive definite matrix V is said to have a matrix variate beta type II distribution with parameters (a,b), denoted as $V \sim B_p^{II}(a,b)$, if its p.d.f. is given by*

$$\{\beta_p(a,b)\}^{-1} \det(V)^{a-\frac{1}{2}(p+1)} \det(I_p + V)^{-(a+b)}, \ V > 0, \tag{5.2.2}$$

where $a > \frac{1}{2}(p-1)$, and $b > \frac{1}{2}(p-1)$.

As in the univariate case, the density (5.2.2) can be obtained from (5.2.1) by transforming $U = (I_p + V)^{-1}V$, together with the Jacobian $J(U \to V) = \det(I_p + V)^{-(p+1)}$. The matrix variate beta type II distribution is also known as matrix variate F-distribution. These distributions belong to the class of orthogonally invariant and residual independent distributions discussed in Chapter 9.

The c.d.f. of V, using the transformation $U = (I_p + V)^{-1}V$ and Theorem 1.6.8, is given by

$$P(V < \Lambda) = \frac{\Gamma_p(a+b)\Gamma_p[\frac{1}{2}(p+1)]}{\Gamma_p(b)\Gamma_p[a+\frac{1}{2}(p+1)]} \det(\Lambda)^a \det(I_p + \Lambda)^{-a}$$

$$_2F_1\left(a, -b+\frac{1}{2}(p+1); a+\frac{1}{2}(p+1); (I_p+\Lambda)^{-1}\Lambda\right), \ \Lambda > 0.$$

By means of a bilinear transformation of the random matrix U, a generalized matrix variate beta type I distribution is generated as given in the following theorem.

THEOREM 5.2.1. *Let $U \sim B_p^I(a,b)$. Then for given $p \times p$ symmetric matrices $\Psi (\geq 0)$ and $\Omega (> \Psi)$, the random matrix X $(p \times p)$ defined by*

$$X = (\Omega - \Psi)^{\frac{1}{2}} U (\Omega - \Psi)^{\frac{1}{2}} + \Psi \tag{5.2.3}$$

has the p.d.f.

$$\frac{\det(X - \Psi)^{a-\frac{1}{2}(p+1)} \det(\Omega - X)^{b-\frac{1}{2}(p+1)}}{\beta_p(a,b) \det(\Omega - \Psi)^{(a+b)-\frac{1}{2}(p+1)}}, \ \Psi < X < \Omega. \tag{5.2.4}$$

Proof: The Jacobian of the transformation (5.2.3) is $J(U \to X) = \det(\Omega - \Psi)^{-\frac{1}{2}(p+1)}$. Hence, the p.d.f. of U is transformed to the p.d.f. of X given by (5.2.4). ∎

DEFINITION 5.2.3. *A $p \times p$ random symmetric positive definite matrix X is said to have a generalized matrix variate beta type I distribution with parameters a, b; Ω, Ψ denoted by $X \sim GB_p^I(a,b;\Omega,\Psi)$ if its p.d.f. is given by (5.2.4).*

When $\Psi = 0$ and $\Omega = I_p$, the above definition yields the standard beta type I distribution (5.2.1). Further if $X \sim GB_p^I(a, b; \Omega, \Psi)$, then $(\Omega - \Psi)^{-\frac{1}{2}}(X - \Psi)(\Omega - \Psi)^{-\frac{1}{2}} \sim B_p^I(a, b)$.

THEOREM 5.2.2. *Let $V \sim B_p^{II}(a, b)$. For given $p \times p$ symmetric matrices $\Psi (\geq 0)$ and $\Omega (> \Psi)$, the random matrix Y defined by*

$$Y = (\Omega + \Psi)^{\frac{1}{2}} V (\Omega + \Psi)^{\frac{1}{2}} + \Psi \qquad (5.2.5)$$

has the p.d.f.

$$\frac{\det(Y - \Psi)^{a - \frac{1}{2}(p+1)} \det(\Omega + Y)^{-(a+b)}}{\beta_p(a, b) \det(\Omega + \Psi)^{-b}}, \quad Y > \Psi. \qquad (5.2.6)$$

Proof: The Jacobian of transformation (5.2.5) is $J(V \to Y) = \det(\Omega + \Psi)^{-\frac{1}{2}(p+1)}$, from which the p.d.f. of Y follows. ∎

DEFINITION 5.2.4. *A $p \times p$ random symmetric positive definite matrix Y is said to have a generalized matrix variate beta type II distribution with parameters a, b; Ω and Ψ if its p.d.f. is given by (5.2.6).*

In this case we write $Y \sim GB_p^{II}(a, b; \Omega, \Psi)$. When $\Psi = 0$ and $\Omega = I_p$, the above definition yields the standard beta type II distribution. Further, if $Y \sim GB_p^{II}(a, b; \Omega, \Psi)$, then $(\Omega + \Psi)^{-\frac{1}{2}}(Y - \Psi)(\Omega + \Psi)^{-\frac{1}{2}} \sim B_p^{II}(a, b)$.

In univariate statistical analysis if x and y are independent chi-square random variables with degrees of freedom n_1 and n_2 respectively, then $\frac{x}{x+y}$ is distributed as beta type I, and $\frac{x}{y}$ is distributed as beta type II. In the multivariate case, the Wishart distribution plays the role of the chi-square distribution, and these ratios have been generalized in many ways. As is often the case, these generalizations can take a number of different forms. Many of these generalizations have been studied extensively in the literature, *e.g.*, see Hsu (1939a), Khatri (1959a, 1970a), Olkin (1959), Olkin and Rubin (1964), Tan (1969c), Mitra (1970), Javier (1982), Javier and Gupta (1985a), and Uhlig (1994).

In the next two theorems, we give derivations of beta distributions of type I and II, generalizing the ratios $\frac{x}{y}$ and $\frac{x}{x+y}$ to the matrix case. The other generalizations, which do not lead to the beta distributions (5.2.1) or (5.2.2), will be studied in Section 5.

To derive the beta density from Wishart density, the following result is needed.

THEOREM 5.2.3. *Let $X \sim N_{p,n_1}(0, \Sigma \otimes I_{n_1})$, and $S_2 \sim W_p(n_2, \Sigma)$ be independent. Further let $S = S_2 + XX'$ and $Z = S^{-\frac{1}{2}}X$, where $S^{-\frac{1}{2}}$ is a nonsingular square root of S. Then S and Z are independent, $S \sim W_p(n_1 + n_2, \Sigma)$ and $Z \sim IT_{p,n_1}(n_2 - p + 1, 0, I_p, I_{n_1})$.*

Proof: The joint density of X and S_2 is

$$\frac{\pi^{-\frac{1}{2}n_1 p} \det(\Sigma)^{-\frac{1}{2}(n_1+n_2)}}{2^{\frac{1}{2}(n_1+n_2)p} \Gamma_p(\frac{1}{2}n_2)} \det(S_2)^{\frac{1}{2}(n_2-p-1)} \text{etr}\left(-\frac{1}{2}\Sigma^{-1}S_2 - \frac{1}{2}\Sigma^{-1}XX'\right),$$

$$X \in \mathbb{R}^{p \times n_1}, \ S_2 > 0. \qquad (5.2.7)$$

Making the transformation $S_2 = S - XX'$, $Z = S^{-\frac{1}{2}}X$, with Jacobian $J(S_2, X \to S, Z) = \det(S)^{\frac{1}{2}n_1}$, from (5.2.7), we get the joint density of S and Z as

$$\left\{ 2^{\frac{1}{2}(n_1+n_2)p} \Gamma_p\left[\frac{1}{2}(n_1+n_2)\right] \det(\Sigma)^{\frac{1}{2}(n_1+n_2)} \right\}^{-1} \det(S)^{\frac{1}{2}(n_1+n_2-p-1)} \operatorname{etr}\left(-\frac{1}{2}\Sigma^{-1}S\right)$$

$$\frac{\Gamma_p[\frac{1}{2}(n_1+n_2)]}{\pi^{\frac{1}{2}n_1 p}\Gamma_p(\frac{1}{2}n_2)} \det(I_p - ZZ')^{\frac{1}{2}(n_2-p-1)}, \quad Z \in \mathbb{R}^{p \times n_1}, \ S > 0. \qquad (5.2.8)$$

From (5.2.8), it is easily seen that S and Z are independent, $S \sim W_p(n_1+n_2, \Sigma)$ and $Z \sim IT_{p,n_1}(n_2 - p + 1, 0, I_p, I_{n_1})$ as defined in Section 4.4. ∎

Now, we give the derivation of the matrix variate beta type I density.

THEOREM 5.2.4. *Let Z be defined as in Theorem 5.2.3.*
   (i) *If $n_1 \geq p$, then $ZZ' \sim B_p^I\left(\frac{1}{2}n_1, \frac{1}{2}n_2\right)$.*
   (ii) *If $n_1 < p$, then $Z'Z \sim B_{n_1}^I\left(\frac{1}{2}p, \frac{1}{2}(n_1+n_2-p)\right)$.*

Proof: (i) From Theorems 5.2.3 and 1.4.10, the density of $U = ZZ'$ is obtained as

$$\frac{\Gamma_p[\frac{1}{2}(n_1+n_2)]}{\pi^{\frac{1}{2}n_1 p}\Gamma_p(\frac{1}{2}n_2)} \int_{ZZ'=U} \det(I_p - ZZ')^{\frac{1}{2}(n_2-p-1)} \, dZ$$

$$= \frac{\Gamma_p[\frac{1}{2}(n_1+n_2)]}{\pi^{\frac{1}{2}n_1 p}\Gamma_p(\frac{1}{2}n_2)} \cdot \frac{\pi^{\frac{1}{2}n_1 p}}{\Gamma_p(\frac{1}{2}n_2)} \det(U)^{\frac{1}{2}(n_1-p-1)} \det(I_p - U)^{\frac{1}{2}(n_2-p-1)}$$

$$= \left\{ \beta_p\left(\frac{1}{2}n_1, \frac{1}{2}n_2\right) \right\}^{-1} \det(U)^{\frac{1}{2}(n_1-p-1)} \det(I_p - U)^{\frac{1}{2}(n_2-p-1)},$$

which proves that $U \sim B_p^I\left(\frac{1}{2}n_1, \frac{1}{2}n_2\right)$.

   (ii) Again from Theorem 1.4.10 if $n_1 < p$, we get

$$\int_{Z'Z=W} \det(I_{n_1} - Z'Z)^{\frac{1}{2}(n_2-p-1)} \, dZ$$

$$= \frac{\pi^{\frac{1}{2}n_1 p}}{\Gamma_{n_1}(\frac{1}{2}p)} \det(W)^{\frac{1}{2}(p-n_1-1)} \det(I_{n_1} - W)^{\frac{1}{2}(n_2-p-1)},$$

and hence $W \sim B_{n_1}^I\left(\frac{1}{2}p, \frac{1}{2}(n_1+n_2-p)\right)$. ∎

It may be noted that if in (i) above, $n_1 < p$ or in (ii), $n_1 > p$, then the density functions of U and W do not exist and are called singular beta distributions (Mitra, 1970).

The above results were derived by Hsu (1939a), and by Khatri (1959a) for a triangular root of S. For $n_1 \geq p$, from Theorems 5.2.3 and 5.2.4(i), it is observed that $XX' = S_1 \sim W_p(n_1, \Sigma)$ and therefore, $U = ZZ' = (S_1 + S_2)^{-\frac{1}{2}}S_1((S_1 + S_2)^{-\frac{1}{2}})' \sim B_p^I(\frac{1}{2}n_1, \frac{1}{2}n_2)$. This, therefore gives a natural generalization of the ratio $\frac{x}{x+y}$ in the univariate case. It may be noted here that $(S_1 + S_2)^{\frac{1}{2}}$ can be taken any reasonable square root depending on $S_1 + S_2$. Mitra (1970) took this square root to be a lower triangular matrix and assumed $n_1 + n_2 > p$ only. He then studied certain properties of the random matrix U using the density free approach. Khatri (1970a) further relaxed the restrictions on n_1 and n_2 and derived Mitra's results.

THEOREM 5.2.5. *Let $S_1 \sim W_p(n_1, I_p)$ and $S_2 \sim W_p(n_2, I_p)$ be independent. Define*

$$V = S_2^{-\frac{1}{2}} S_1 S_2^{-\frac{1}{2}},$$

where $S_2^{\frac{1}{2}}$ is a symmetric square root of S_2. Then, $V \sim B_p^{II}(\frac{1}{2}n_1, \frac{1}{2}n_2)$.

Proof: The joint density of S_1 and S_2 is

$$\left\{ 2^{\frac{1}{2}(n_1+n_2)p} \Gamma_p\left(\frac{1}{2}n_1\right) \Gamma_p\left(\frac{1}{2}n_2\right) \right\}^{-1} \operatorname{etr}\left\{ -\frac{1}{2}(S_1 + S_2) \right\}$$

$$\det(S_1)^{\frac{1}{2}(n_1-p-1)} \det(S_2)^{\frac{1}{2}(n_2-p-1)}, \ S_1 > 0, \ S_2 > 0.$$

Transforming $V = S_2^{-\frac{1}{2}} S_1 S_2^{-\frac{1}{2}}$, with Jacobian $J(S_1, S_2 \to V, S_2) = \det(S_2)^{\frac{1}{2}(p+1)}$, we get the joint density of S_2 and V as

$$\left\{ 2^{\frac{1}{2}(n_1+n_2)p} \Gamma_p\left(\frac{1}{2}n_1\right) \Gamma_p\left(\frac{1}{2}n_2\right) \right\}^{-1} \operatorname{etr}\left\{ -\frac{1}{2}(I_p + V)S_2 \right\}$$

$$\det(V)^{\frac{1}{2}(n_1-p-1)} \det(S_2)^{\frac{1}{2}(n_1+n_2-p-1)}, \ V > 0, \ S_2 > 0. \qquad (5.2.9)$$

Now, integrating out S_2 from (5.2.9), using the multivariate gamma integral, completes the proof. ∎

The above result has been derived by Olkin and Rubin (1964). The converse of this result is not true. That is if X_1 $(p \times p)$ and X_2 $(p \times p)$ are independent random matrices and $X_2^{-\frac{1}{2}} X_1 X_2^{-\frac{1}{2}}$ has beta type II distribution, then it does not necessarily follow that X_1 and X_2 have Wishart density. Hence this property does not characterize Wishart distribution. Roux (1975) has given the following result.

THEOREM 5.2.6. *Let X_i $(p \times p)$, $i = 1, 2$ be independent random matrices with density*

$$\frac{\Gamma_p[a_i + \frac{1}{2}\nu + \frac{1}{4}(p+1)]}{\Gamma_p(a_i)\Gamma_p[\nu + \frac{1}{2}(p+1)]} \det(X_i)^{\frac{1}{4}(2\nu-p-1)}$$

$$_1F_1\left(a_i + \frac{1}{2}\nu + \frac{1}{4}(p+1); \nu + \frac{1}{2}(p+1); -X_i\right), \ X_i > 0,$$

where $\operatorname{Re}(\frac{1}{2}\nu + \frac{1}{4}(p+1) - a_i) > \frac{1}{2}(p-1)$, and $\operatorname{Re}(a_i) > \frac{1}{2}(p-1)$, $i = 1, 2$. Then $X_2^{-\frac{1}{2}} X_1 X_2^{-\frac{1}{2}} \sim B_p^{II}(a_2, a_1)$.

Proof: Making the transformation $Y = X_2^{-\frac{1}{2}} X_1 X_2^{-\frac{1}{2}}$ with Jacobian $J(X_1, X_2 \to Y, X_2) = \det(X_2)^{\frac{1}{2}(p+1)}$, in the joint density of X_1 and X_2 we get the joint density of Y and X_2 as

$$\prod_{i=1}^{2} \left\{ \frac{\Gamma_p[a_i + \frac{1}{2}\nu + \frac{1}{4}(p+1)]}{\Gamma_p(a_i)\Gamma_p[\nu + \frac{1}{2}(p+1)]} \right\} \det(Y)^{\frac{1}{4}(2\nu-p-1)} \det(X_2)^{\nu}$$

$$_1F_1\left(a_1 + \frac{1}{2}\nu + \frac{1}{4}(p+1); \nu + \frac{1}{2}(p+1); -X_2 Y\right)$$

$$_1F_1\left(a_2 + \frac{1}{2}\nu + \frac{1}{4}(p+1); \nu + \frac{1}{2}(p+1); -X_2\right). \qquad (5.2.10)$$

Integrating (5.2.10) with respect to X_2, the p.d.f. of Y is given by

$$\prod_{i=1}^{2} \left\{ \frac{\Gamma_p[a_i + \frac{1}{2}\nu + \frac{1}{4}(p+1)]}{\Gamma_p(a_i)\Gamma_p[\nu + \frac{1}{2}(p+1)]} \right\} \det(Y)^{\frac{1}{4}(2\nu-p-1)} \int_{X_2>0} \det(X_2)^{\nu}$$

$$_1F_1\left(a_1 + \frac{1}{2}\nu + \frac{1}{4}(p+1); \nu + \frac{1}{2}(p+1); -X_2 Y\right)$$

$$_1F_1\left(a_2 + \frac{1}{2}\nu + \frac{1}{4}(p+1); \nu + \frac{1}{2}(p+1); -X_2\right) dX_2. \qquad (5.2.11)$$

Using (1.6.7), we can write

$$_1F_1\left(a_1 + \frac{1}{2}\nu + \frac{1}{4}(p+1); \nu + \frac{1}{2}(p+1); -X_2 Y\right)$$

$$= \frac{\Gamma_p[\nu + \frac{1}{2}(p+1)]}{\Gamma_p[\frac{1}{2}\nu + \frac{1}{4}(p+1) + a_1]\Gamma_p[\frac{1}{2}\nu + \frac{1}{4}(p+1) - a_1]} \int_{0<S<I_p} \text{etr}\left(-Y^{\frac{1}{2}} X_2 Y^{\frac{1}{2}} S\right)$$

$$\det(S)^{\frac{1}{4}(2\nu+p+1)+a_1-\frac{1}{2}(p+1)} \det(I_p - S)^{\frac{1}{4}(2\nu+p+1)-a_1-\frac{1}{2}(p+1)} dS. \qquad (5.2.12)$$

Substituting from (5.2.12) in (5.2.11), and using (1.6.4) we get

$$\frac{\Gamma_p[a_2 + \frac{1}{2}\nu + \frac{1}{4}(p+1)]}{\Gamma_p(a_1)\Gamma_p(a_2)\Gamma_p[\frac{1}{2}\nu + \frac{1}{4}(p+1) - a_1]} \det(Y)^{\frac{1}{4}(2\nu-p-1)}$$

$$\int_{0<S<I_p} \det(S)^{\frac{1}{4}(2\nu+p+1)+a_1-\frac{1}{2}(p+1)} \det(I_p - S)^{\frac{1}{4}(2\nu+p+1)-a_1-\frac{1}{2}(p+1)} \det(YS)^{-\nu-\frac{1}{2}(p+1)}$$

$$_2F_1\left(\nu + \frac{1}{2}(p+1); a_2 + \frac{1}{2}\nu + \frac{1}{4}(p+1); \nu + \frac{1}{2}(p+1); -(YS)^{-1}\right) dS$$

$$= \frac{\Gamma_p[a_2 + \frac{1}{2}\nu + \frac{1}{4}(p+1)]}{\Gamma_p(a_1)\Gamma_p(a_2)\Gamma_p[\frac{1}{2}\nu + \frac{1}{4}(p+1) - a_1]} \det(Y)^{a_2-\frac{1}{2}(p+1)} \int_{0<S<I_p} \det(S)^{a_1+a_2-\frac{1}{2}(p+1)}$$

$$\det(I_p - S)^{\frac{1}{4}(2\nu+p+1)-a_1-\frac{1}{2}(p+1)} \det(I_p + YS)^{-a_2-\frac{1}{4}(2\nu+p+1)} dS$$

$$= \frac{\Gamma_p(a_1 + a_2)}{\Gamma_p(a_1)\Gamma_p(a_2)} \det(Y)^{a_2-\frac{1}{2}(p+1)}$$

$$_2F_1\left(a_1 + a_2; a_2 + \frac{1}{2}\nu + \frac{1}{4}(p+1); a_2 + \frac{1}{2}\nu + \frac{1}{4}(p+1); -Y\right),$$

$$[\text{ from (1.6.8)}],$$

$$= \frac{\Gamma_p(a_1 + a_2)}{\Gamma_p(a_1)\Gamma_p(a_2)} \det(Y)^{a_2-\frac{1}{2}(p+1)} \det(I_p + Y)^{-(a_1+a_2)}, \quad Y > 0. \quad \blacksquare$$

From Theorems 5.2.3 and 5.2.4, it is clear that

$$U = (S_1 + S_2)^{-\frac{1}{2}} S_1 ((S_1 + S_2)^{-\frac{1}{2}})' \sim B_p^I \left(\frac{1}{2} n_1, \frac{1}{2} n_2 \right)$$

and is independent of $S_1 + S_2$, which holds for all $\Sigma > 0$ and all square roots of $S_1 + S_2$ depending on $S_1 + S_2$ only. However analogous results do not hold for $V = S_2^{-\frac{1}{2}} S_1 S_2^{-\frac{1}{2}}$, since the distribution of V and its independence from $S_1 + S_2$ depend on Σ and the choice of the square root of S_2. If $S_2^{\frac{1}{2}}$ is a symmetric square root and $\Sigma = \alpha I_p$, then from Theorem 5.2.5, we have $V \sim B_p^{II}(\frac{1}{2} n_1, \frac{1}{2} n_2)$, but is not independent of $S_1 + S_2$. If $\Sigma \neq \alpha I_p$, then the distribution of V is not beta type II. If $S_2^{\frac{1}{2}}$ is taken as a triangular matrix, then V and $S_1 + S_2$ are independent for all $\Sigma > 0$, but again distribution of V is not beta type II (see Section 4). Considering these facts, Perlman (1977) defined

$$\begin{aligned} \tilde{V} &= (S_1 + S_2)^{-\frac{1}{2}} S_1 S_2^{-1} (S_1 + S_2)^{\frac{1}{2}} \\ &= ((S_1 + S_2)^{\frac{1}{2}})' S_2^{-1} S_1 ((S_1 + S_2)^{-\frac{1}{2}})' \\ &= \tilde{V}' \end{aligned}$$

and proved that $\tilde{V} \sim B_p^{II} \left(\frac{1}{2} n_1, \frac{1}{2} n_2 \right)$ and is independent of $S_1 + S_2$ for all $\Sigma > 0$ and all choices of square root $(S_1 + S_2)^{\frac{1}{2}}$, provided that the choice is made in a measurable way depending only on $S_1 + S_2$.

5.3. PROPERTIES

In this section, we study some properties of the random matrices distributed as matrix variate beta type I and II.

THEOREM 5.3.1. Let $U \sim B_p^I(a, b)$ and $A\,(p \times p)$ be a constant nonsingular matrix. Then $AUA' \sim GB_p^I(a, b; AA', 0)$.

Proof: The density of U is

$$\{\beta_p(a, b)\}^{-1} \det(U)^{a - \frac{1}{2}(p+1)} \det(I_p - U)^{b - \frac{1}{2}(p+1)}, \ 0 < U < I_p. \tag{5.3.1}$$

Making the transformation $X = AUA'$, with the Jacobian $J(U \to X) = \det(A)^{-(p+1)}$, the density of X, obtained from (5.3.1), is

$$\{\beta_p(a, b)\}^{-1} \det(AA')^{-(a+b)+\frac{1}{2}(p+1)} \det(X)^{a-\frac{1}{2}(p+1)} \det(AA' - X)^{b-\frac{1}{2}(p+1)}, \ 0 < X < AA',$$

which is the desired result. ∎

Similar result for beta type II distribution is given in the next theorem.

THEOREM 5.3.2. Let $V \sim B_p^{II}(a, b)$ and $A\,(p \times p)$ be a constant nonsingular matrix. Then $AVA' \sim GB_p^{II}(a, b; AA', 0)$.

Proof: The density of V is

$$\{\beta_p(a,b)\}^{-1} \det(V)^{a-\frac{1}{2}(p+1)} \det(I_p + V)^{-(a+b)}, \ V > 0. \qquad (5.3.2)$$

Making the transformation $Y = AVA'$, with the Jacobian $J(V \to Y) = \det(A)^{-(p+1)}$, the density of Y, obtained from (5.3.2), is

$$\{\beta_p(a,b)\}^{-1} \det(AA')^b \det(Y)^{a-\frac{1}{2}(p+1)} \det(AA' + Y)^{-(a+b)}, \ Y > 0,$$

which completes the proof of the theorem. ∎

   In the next two theorems, it is shown that matrix variate beta distributions are orthogonally invariant.

THEOREM 5.3.3. *Let $U \sim B_p^I(a,b)$ and H $(p \times p)$ be an orthogonal matrix, whose elements are either constants or random variables distributed independently of U. Then, the distribution of U is invariant under the transformation $U \to HUH'$, and is independent of H in the latter case.*

Proof: First, let H be a constant matrix. Then, from Theorem 5.3.1, $HUH' \sim B_p^I(a,b)$ since $HH' = I_p$. If, however, H is a random orthogonal matrix, then $HUH'|H \sim B_p^I(a,b)$. Since this distribution does not depend on H, $HUH' \sim B_p^I(a,b)$. ∎

THEOREM 5.3.4. *Let $V \sim B_p^{II}(a,b)$ and H $(p \times p)$ be an orthogonal matrix whose elements are either constants or random variables distributed independently of V. Then, the distribution of V is invariant under the transformation $V \to HVH'$, and is independent of H in the latter case.*

Proof: Similar to the proof of Theorem 5.3.3. ∎

   The relationship between beta type I and type II matrices is now exhibited. First, we derive densities of U^{-1} and V^{-1}.

THEOREM 5.3.5. *Let $U \sim B_p^I(a,b)$, then the density of $X = U^{-1}$ is*

$$\{\beta_p(a,b)\}^{-1} \det(X)^{-(a+b)} \det(X - I_p)^{b-\frac{1}{2}(p+1)}, \ X > I_p, \qquad (5.3.3)$$

where $a > \frac{1}{2}(p-1)$, and $b > \frac{1}{2}(p-1)$.

Proof: Making the transformation $X = U^{-1}$, with the Jacobian $J(U \to X) = \det(X)^{-(p+1)}$, in the density of U the result follows. ∎

   Now (5.3.3) may be called the inverse beta type I density and denoted by $IB_p(a,b)$. From Theorem 5.3.5, it is clear that if $U \sim B_p^I(a,b)$, then U^{-1} does not follow the beta I distribution. However it is easily seen that $I_p - U \sim B_p^I(b,a)$, and $U^{-1} - I_p \sim B_p^{II}(b,a)$. For beta type II random matrix V, the distribution of V^{-1} is also beta type II as shown in the following theorem.

THEOREM 5.3.6. *Let $V \sim B_p^{II}(a,b)$, then $Y = V^{-1} \sim B_p^{II}(b,a)$.*

Proof: Making the transformation $Y = V^{-1}$, with the Jacobian $J(V \to Y) = \det(Y)^{-(p+1)}$, in the density of V the result follows. ∎

THEOREM 5.3.7. *(i) Let $U \sim B_p^I(a,b)$ and $V = (I_p - U)^{-\frac{1}{2}}U(I_p - U)^{-\frac{1}{2}}$, then $V \sim B_p^{II}(a,b)$.*

(ii) Similarly, if $V \sim B_p^{II}(a,b)$ and $U = (I_p + V)^{-\frac{1}{2}}V(I_p + V)^{-\frac{1}{2}}$, then $U \sim B_p^I(a,b)$.

Proof: (i) Since U commutes with any rational function of U, $V = (I_p - U)^{-\frac{1}{2}}U(I_p - U)^{-\frac{1}{2}} = (I_p - U)^{-1}U$, and the Jacobian of this transformation is $J(U \to V) = \det(I_p + V)^{-(p+1)}$. Now, making the substitution in the density of U given by (5.2.1) the result follows.

(ii) The proof is similar to part (i). ∎

The characteristic functions of U and V are now obtained in the following theorems.

THEOREM 5.3.8. *Let $U \sim B_p^I(a,b)$. Then the characteristic function of $U = (u_{ij})$, i.e., the joint characteristic function of $u_{11}, u_{12}, \ldots, u_{pp}$ is*

$$\phi_U(Z) = {}_1F_1(a; a + b; \iota Z),$$

where $Z = Z'\ (p \times p) = \left(\frac{1}{2}(1 + \delta_{ij})z_{ij}\right)$ and $\iota = \sqrt{-1}$.

Proof: By definition,

$$\phi_U(Z) = E[\text{etr}(\iota Z U)]$$

$$= \{\beta_p(a,b)\}^{-1} \int_{0 < U < I_p} \text{etr}(\iota Z U) \det(U)^{a - \frac{1}{2}(p+1)} \det(I_p - U)^{b - \frac{1}{2}(p+1)}\, dU$$

$$= {}_1F_1(a; a + b; \iota Z).$$

The last equality follows from Corollary 1.6.3.1. ∎

It may be noted here that if $X \sim GB_p^I(a, b; \Omega, \Psi)$, then the characteristic function of X can be obtained from the above theorem. Since $X = (\Omega - \Psi)^{\frac{1}{2}}U(\Omega - \Psi)^{\frac{1}{2}} + \Psi$, where $U \sim B_p^I(a,b)$, we have

$$\phi_X(Z) = E[\text{etr}(\iota Z X)]$$

$$= E[\text{etr}\{\iota Z((\Omega - \Psi)^{\frac{1}{2}}U(\Omega - \Psi)^{\frac{1}{2}} + \Psi)\}]$$

$$= \text{etr}(\iota Z \Psi)E[\text{etr}\{\iota(\Omega - \Psi)^{\frac{1}{2}}Z(\Omega - \Psi)^{\frac{1}{2}}U\}]$$

$$= \text{etr}(\iota Z \Psi)\phi_U((\Omega - \Psi)^{\frac{1}{2}}Z(\Omega - \Psi)^{\frac{1}{2}})$$

$$= \text{etr}(\iota Z \Psi)\ {}_1F_1(a; a + b; \iota Z(\Omega - \Psi)). \tag{5.3.4}$$

THEOREM 5.3.9. *Let $V \sim B_p^{II}(a, b)$. Then, the characteristic function of $V = (v_{ij})$, i.e., the joint characteristic function of $v_{11}, v_{12}, \ldots, v_{pp}$ is*

$$\phi_V(Z) = \frac{\Gamma_p(a+b)}{\Gamma_p(b)} \Psi\left(a; -b + \frac{1}{2}(p+1); -\iota Z\right),$$

where $Z = Z' (p \times p) = (\frac{1}{2}(1 + \delta_{ij})z_{ij})$, $\iota = \sqrt{-1}$, and $\mathrm{Re}(-\iota Z) > 0$.

Proof: Here, the characteristic function of V is given by

$$\phi_V(Z) = \{\beta_p(a, b)\}^{-1} \int_{V>0} \mathrm{etr}(\iota Z V) \det(V)^{a - \frac{1}{2}(p+1)} \det(I_p + V)^{-(a+b)} dV.$$

Now, using the Definition 1.6.3 of the confluent hypergeometric function Ψ, the result follows. ∎

In this case as well, if $Y \sim GB_p^{II}(a, b; \Omega, \Psi)$, then

$$\phi_Y(Z) = \frac{\Gamma_p(a+b)}{\Gamma_p(b)} \mathrm{etr}(\iota Z \Psi) \, \Psi\left(a; -b + \frac{1}{2}(p+1); -\iota Z(\Omega + \Psi)\right), \tag{5.3.5}$$

where $\mathrm{Re}(-\iota Z(\Omega + \Psi)) > 0$.

The marginal and conditional distributions of U are given next.

THEOREM 5.3.10. *Let $U = \begin{pmatrix} U_{11} & U_{12} \\ U_{21} & U_{22} \end{pmatrix}$, U_{ij} $(p_i \times p_j)$, $p_1 + p_2 = p$, and $U_{22 \cdot 1} = U_{22} - U_{21}U_{11}^{-1}U_{12}$. If $U \sim B_p^I(a, b)$, then U_{11} and $U_{22 \cdot 1}$ are independently distributed, $U_{11} \sim B_{p_1}^I(a, b)$ and $U_{22 \cdot 1} \sim B_{p_2}^I(a - \frac{1}{2}p_1, b)$. Further, $U_{21}|U_{11}, U_{22 \cdot 1} \sim IT_{p_2, p_1}(2b - p + 1, 0, I_{p_2} - U_{22 \cdot 1}, U_{11}(I_{p_1} - U_{11}))$.*

Proof: The density of U is

$$f(U) = \{\beta_p(a, b)\}^{-1} \det(U)^{a - \frac{1}{2}(p+1)} \det(I_p - U)^{b - \frac{1}{2}(p+1)}, \ 0 < U < I_p. \tag{5.3.6}$$

From the partition of U, we have

$$\det(U) = \det(U_{11}) \det(U_{22 \cdot 1}) \tag{5.3.7}$$

and

$$\begin{aligned}
\det(I_p - U) &= \det(I_{p_1} - U_{11}) \det(I_{p_2} - U_{22} - U_{21}(I_{p_1} - U_{11})^{-1}U_{12}) \\
&= \det(I_{p_1} - U_{11}) \det(I_{p_2} - U_{22 \cdot 1} - U_{21}(U_{11}^{-1} + (I_{p_1} - U_{11})^{-1})U_{12}) \\
&= \det(I_{p_1} - U_{11}) \det(I_{p_2} - U_{22 \cdot 1} - U_{21}U_{11}^{-1}(I_{p_1} - U_{11})^{-1}U_{12}). \tag{5.3.8}
\end{aligned}$$

Now making the transformation $U_{11} = U_{11}$, $U_{21} = U_{21}$ and $U_{22 \cdot 1} = U_{22} - U_{21}U_{11}^{-1}U_{12}$ with Jacobian $J(U_{11}, U_{22}, U_{21} \to U_{11}, U_{22 \cdot 1}, U_{21}) = 1$ and substituting (5.3.7) and (5.3.8)in (5.3.6), we get the joint density of U_{11}, $U_{22 \cdot 1}$, and U_{21} as

$f(U_{11}, U_{22 \cdot 1}, U_{21})$

$$= \{\beta_p(a,b)\}^{-1} \det(U_{11})^{a-\frac{1}{2}(p+1)} \det(U_{22 \cdot 1})^{a-\frac{1}{2}(p+1)} \det(I_{p_1} - U_{11})^{b-\frac{1}{2}(p+1)}$$

$$\det(I_{p_2} - U_{22 \cdot 1} - U_{21} U_{11}^{-1}(I_{p_1} - U_{11})^{-1} U_{12})^{b-\frac{1}{2}(p+1)}$$

$$= \{\beta_{p_1}(a,b)\}^{-1} \det(U_{11})^{a-\frac{1}{2}(p_1+1)} \det(I_{p_1} - U_{11})^{b-\frac{1}{2}(p_1+1)}$$

$$\left\{\beta_{p_2}\left(a - \frac{1}{2}p_1, b\right)\right\}^{-1} \det(U_{22 \cdot 1})^{a-\frac{1}{2}p_1-\frac{1}{2}(p_2+1)} \det(I_{p_2} - U_{22 \cdot 1})^{b-\frac{1}{2}(p_2+1)}$$

$$\frac{\pi^{-\frac{1}{2}p_1 p_2} \Gamma_{p_2}(b)}{\Gamma_{p_2}(b - \frac{1}{2}p_1)} \det(U_{11}(I_{p_1} - U_{11}))^{-\frac{1}{2}p_2} \det(I_{p_2} - U_{22 \cdot 1})^{-b+\frac{1}{2}(p_2+1)}$$

$$\det(I_{p_2} - U_{22 \cdot 1} - U_{21} U_{11}^{-1}(I_{p_1} - U_{11})^{-1} U_{12})^{b-\frac{1}{2}(p+1)}, \qquad (5.3.9)$$

where $0 < U_{11} < I_{p_1}$, $0 < U_{22 \cdot 1} < I_{p_2}$ and $I_{p_2} - U_{22 \cdot 1} - U_{21} U_{11}^{-1}(I_{p_1} - U_{11})^{-1} U_{12} > 0$. From the factorization (5.3.9), the result follows. ∎

THEOREM 5.3.11. Let $V = \begin{pmatrix} V_{11} & V_{12} \\ V_{21} & V_{22} \end{pmatrix}$, V_{ij} $(p_i \times p_j)$, $p_1 + p_2 = p$, and $V_{22 \cdot 1} = V_{22} - V_{21} V_{11}^{-1} V_{12}$. If $V \sim B_p^{II}(a,b)$, then V_{11} and $V_{22 \cdot 1}$ are independently distributed, $V_{11} \sim B_{p_1}^{II}(a, b - \frac{1}{2}p_2)$, $V_{22 \cdot 1} \sim B_{p_2}^{II}(a - \frac{1}{2}p_1, b)$, and $V_{21}|V_{11}, V_{22 \cdot 1} \sim T_{p_2,p_1}(2a + 2b - p + 1, 0, I_{p_2} + V_{22 \cdot 1}, V_{11}(I_{p_1} + V_{11}))$.

Proof: Similar to the proof of Theorem 5.3.10. ∎

The distributions of certain matrix valued functions, *viz* AUA', AVA', $(AU^{-1}A')^{-1}$, $(AV^{-1}A')^{-1}$ where $A\,(q \times p)$ is a constant matrix of rank $q\,(< p)$, are now derived.

THEOREM 5.3.12. Let $U \sim B_p^I(a,b)$. Then, for a constant matrix $A\,(q \times p)$ of rank $q\,(\le p)$, $AUA' \sim GB_q^I(a,b;AA',0)$.

Proof: Write $A = M\,(\,I_q\ \ 0\,)\Gamma$, where $M\,(q \times q)$ is nonsingular and $\Gamma\,(p \times p)$ is orthogonal. Now $AUA' = M\,(\,I_q\ \ 0\,)\Gamma U\Gamma'\,(\,I_q\ \ 0\,)'\,M' = MX_{11}M'$, where $X = \Gamma U\Gamma'$ and $X_{11}\,(q \times q)$ is the first principal diagonal block of X. From Theorems 5.3.3 and 5.3.10, we know that $X \sim B_p^I(a,b)$ and $X_{11} \sim B_q^I(a,b)$. Hence, using Theorem 5.3.1, $MX_{11}M' \sim GB_q^I(a,b;MM',0)$ and the result follows by noting that $MM' = AA'$. ∎

COROLLARY 5.3.12.1. Let $U \sim B_p^I(a,b)$ and $\mathbf{a} \in \mathbb{R}^p$, $\mathbf{a} \ne \mathbf{0}$, then $\frac{\mathbf{a}'U\mathbf{a}}{\mathbf{a}'\mathbf{a}} \sim B^I(a,b)$.

Proof: Take $q = 1$ in Theorem 5.3.12. ∎

In Corollary 5.3.12.1 the distribution of $\frac{\mathbf{a}'U\mathbf{a}}{\mathbf{a}'\mathbf{a}}$ does not depend on $\mathbf{a}$. Thus if $\mathbf{y}\,(p \times 1)$ is a random vector, independent of U, and $P(\mathbf{y} \ne \mathbf{0}) = 1$, then it follows that $\frac{\mathbf{y}'U\mathbf{y}}{\mathbf{y}'\mathbf{y}} \sim B^I(a,b)$.

THEOREM 5.3.13. Let $V \sim B_p^{II}(a,b)$. Then, for a constant matrix $A\,(q \times p)$ of rank $q\,(\le p)$, $AVA' \sim GB_q^{II}\left(a, b - \frac{1}{2}(p - q); AA', 0\right)$.

Proof: Similar to the proof of Theorem 5.3.12. ∎

COROLLARY 5.3.13.1. Let $V \sim B_p^{II}(a,b)$ and $a \in \mathbb{R}^p$, $a \neq 0$, then $\frac{a'Va}{a'a} \sim B^{II}(a, b - \frac{1}{2}(p-1))$.

Proof: Take $q = 1$ in Theorem 5.3.13. ∎

In Corollary 5.3.13.1, the distribution of $\frac{a'Va}{a'a}$ does not depend on a. Thus, if $y\,(p \times 1)$ is a random vector, independent of V, and $P(y \neq 0) = 1$, then also $\frac{y'Vy}{y'y} \sim B^{II}(a, b - \frac{1}{2}(p-1))$.

THEOREM 5.3.14. Let $A\,(q \times p)$ be a constant matrix of rank $q\,(\leq p)$.
  (i) If $U \sim B_p^I(a,b)$, then $(AU^{-1}A')^{-1} \sim GB_q^I\left(a - \frac{1}{2}(p-q), b; (AA')^{-1}, 0\right)$.
  (ii) If $V \sim B_p^{II}(a,b)$, then $(AV^{-1}A')^{-1} \sim GB_q^{II}\left(a - \frac{1}{2}(p-q), b; (AA')^{-1}, 0\right)$.

Proof: Write $A = M\,(\,I_q \quad 0\,)\,\Gamma$, where $M\,(q \times q)$ is nonsingular and $\Gamma\,(p \times p)$ is orthogonal. Now,

$$(AV^{-1}A')^{-1} = [M\,(\,I_q \quad 0\,)\,\Gamma V^{-1}\Gamma'\,(\,I_q \quad 0\,)'\,M']^{-1}$$

$$= (M')^{-1}\left[(\,I_q \quad 0\,)\,Y^{-1}\begin{pmatrix} I_q \\ 0 \end{pmatrix}\right]^{-1} M^{-1}$$

$$= (M')^{-1}(Y^{11})^{-1}M^{-1},$$

where $Y = \begin{pmatrix} Y_{11} & Y_{12} \\ Y_{21} & Y_{22} \end{pmatrix} = \Gamma V\Gamma' \sim B_p^{II}(a,b)$, $Y_{11}\,(q \times q)$, and $Y^{11} = (Y_{11} - Y_{12}Y_{22}^{-1}Y_{21})^{-1} = Y_{11\cdot2}^{-1}$. From Theorem 5.3.11, $Y_{11\cdot2} \sim B_q^{II}\left(a - \frac{1}{2}(p-q), b\right)$ and from Theorem 5.3.2, $(M')^{-1}Y_{11\cdot2}M^{-1} \sim GB_q^{II}\left(a - \frac{1}{2}(p-q), b; (MM')^{-1}, 0\right)$. The proof of (ii) is now completed by observing that $MM' = AA'$.

Similarly, one can prove part (i). ∎

From the above theorem, when $a \in \mathbb{R}^p$, $a \neq 0$, it follows that

$$\frac{a'a}{a'U^{-1}a} \sim B^I\left(a - \frac{1}{2}(p-1), b\right)$$

and

$$\frac{a'a}{a'V^{-1}a} \sim B^{II}\left(a - \frac{1}{2}(p-1), b\right).$$

In the next six theorems we give expected values of the elements of beta type I and type II matrices and some of their scalar and matrix valued functions.

THEOREM 5.3.15. Let $U \sim B_p^I(a,b)$. Then,

  (i) $E[\det(U)^h] = \dfrac{\Gamma_p(a+h)\Gamma_p(a+b)}{\Gamma_p(a+b+h)\Gamma_p(a)}$, $\operatorname{Re}(a+h) > \dfrac{1}{2}(p-1)$,

and

  (ii) $E[\det(I_p - U)^h] = \dfrac{\Gamma_p(b+h)\Gamma_p(a+b)}{\Gamma_p(a+b+h)\Gamma_p(b)}$, $\operatorname{Re}(b+h) > \dfrac{1}{2}(p-1)$.

Proof: From the density of U, we have

$$E[\det(U)^h] = \{\beta_p(a,b)\}^{-1} \int_{0<U<I_p} \det(U)^{h+a-\frac{1}{2}(p+1)} \det(I_p - U)^{b-\frac{1}{2}(p+1)}\, dU$$

$$= \frac{\beta_p(a+h,b)}{\beta_p(a,b)}, \ \mathrm{Re}(a+h) > \frac{1}{2}(p-1)$$

$$= \frac{\Gamma_p(a+h)\Gamma_p(a+b)}{\Gamma_p(a+b+h)\Gamma_p(a)}, \ \mathrm{Re}(a+h) > \frac{1}{2}(p-1).$$

Similarly $E[\det(I_p - U)^h]$ can be derived. ∎

THEOREM 5.3.16. *Let* $U \sim B_p^I(a,b)$. *Then,*

(i) $E[C_\kappa(U)] = \dfrac{(a)_\kappa}{(a+b)_\kappa} C_\kappa(I_p),$

and

(ii) $E[C_\kappa(U^{-1})] = \dfrac{\Gamma_p(a+b)\Gamma_p(a,-\kappa)}{\Gamma_p(a)\Gamma_p(a+b,-\kappa)} C_\kappa(I_p), \ \mathrm{Re}(a) > \dfrac{1}{2}(p-1) + k_1.$

Proof: From the density of U, we have

$$E[C_\kappa(U)] = \{\beta_p(a,b)\}^{-1} \int_{0<U<I_p} C_\kappa(U) \det(U)^{a-\frac{1}{2}(p+1)} \det(I_p - U)^{b-\frac{1}{2}(p+1)}\, dU$$

$$= \frac{1}{\beta_p(a,b)} \frac{\Gamma_p(a,\kappa)\Gamma_p(b)}{\Gamma_p(a+b,\kappa)} C_\kappa(I_p)$$

$$= \frac{(a)_\kappa}{(a+b)_\kappa} C_\kappa(I_p),$$

where the integral has been evaluated by using (1.5.16).

The $E[C_\kappa(U^{-1})]$ is similarly derived by applying (1.5.17). ∎

THEOREM 5.3.17. *Let* $V \sim B_p^{II}(a,b)$. *Then,*

(i) $E[\det(V)^h] = \dfrac{\Gamma_p(a+h)\Gamma_p(b-h)}{\Gamma_p(a)\Gamma_p(b)}, \ -a + \dfrac{1}{2}(p-1) < \mathrm{Re}(h) < b - \dfrac{1}{2}(p-1),$

and

(ii) $E[\det((I_p + V)^{-1}V)^h] = \dfrac{\Gamma_p(a+h)\Gamma_p(a+b)}{\Gamma_p(a+b+h)\Gamma_p(a)}, \ \mathrm{Re}(a+h) > \dfrac{1}{2}(p-1).$

Proof: By definition,

$$E[\det(V)^h] = \{\beta_p(a,b)\}^{-1} \int_{V>0} \det(V)^{a+h-\frac{1}{2}(p+1)} \det(I_p + V)^{-(a+b)}\, dV$$

$$= \frac{\beta_p(a+h,b-h)}{\beta_p(a,b)}, \ \mathrm{Re}(a+h) > \frac{1}{2}(p-1), \ \mathrm{Re}(b-h) > \frac{1}{2}(p-1)$$

$$= \frac{\Gamma_p(a+h)\Gamma_p(b-h)}{\Gamma_p(a)\Gamma_p(b)}, \ -a + \frac{1}{2}(p-1) < \mathrm{Re}(h) < b - \frac{1}{2}(p-1).$$

Notice that $(I_p + V)^{-1}V \sim B_p^I(a,b)$ and hence, $E[\det((I_p+V)^{-1}V)^h]$ is obtained from Theorem 5.3.15. ∎

THEOREM 5.3.18. *Let* $V \sim B_p^{II}(a,b)$. *Then,*

(i) $E[C_\kappa(V)] = \dfrac{(-1)^k(a)_\kappa}{(-b+\frac{1}{2}(p+1))_\kappa}C_\kappa(I_p)$, $\mathrm{Re}(b) > \dfrac{1}{2}(p-1)+k_1,$

and

(ii) $E[C_\kappa(V^{-1})] = \dfrac{(-1)^k(b)_\kappa}{(-a+\frac{1}{2}(p+1))_\kappa}C_\kappa(I_p)$, $\mathrm{Re}(a) > \dfrac{1}{2}(p-1)+k_1.$

Proof: By definition,

$$E[C_\kappa(V)] = \{\beta_p(a,b)\}^{-1}\int_{V>0} C_\kappa(V)\det(V)^{a-\frac{1}{2}(p+1)}\det(I_p+V)^{-(a+b)}\,dV$$

$$= \frac{1}{\beta_p(a,b)}\frac{\Gamma_p(a,\kappa)\Gamma_p(b,-\kappa)}{\Gamma_p(a+b)}C_\kappa(I_p),\ \mathrm{Re}(b) > \frac{1}{2}(p-1)+k_1$$

$$= \frac{(-1)^k(a)_\kappa}{(-b+\frac{1}{2}(p+1))_\kappa}C_\kappa(I_p),\ \mathrm{Re}(b) > \frac{1}{2}(p-1)+k_1,$$

where the integral has been evaluated by using Lemma 1.5.4 and simplification has been done using (1.5.9).

By noting that $V^{-1} \sim B_p^{II}(b,a)$, $E[C_\kappa(V^{-1})]$ is obtained from $E[C_\kappa(V)]$. ∎

Konno (1988), using Haff's (1979) method, derived identities for expectations of certain functions of beta type I and type II matrices. When $U \sim GB_p^I(\frac{1}{2}n_1,\frac{1}{2}n_2;\Omega,0)$, he gave an identity for $E[g(U)\,\mathrm{tr}((\Omega-U)^{-1}T)]$, where $g(U)$ is a scalar function and $T\,(p\times p)$ is a matrix valued function of U and Ω. From this identity, the following results are obtained.

THEOREM 5.3.19. *Let* $U \sim GB_p^I\left(\frac{1}{2}n_1,\frac{1}{2}n_2;\Omega,0\right)$, *then*

(i) $E(u_{ij}) = \dfrac{n_1}{n}\omega_{ij}$

(ii) $E(u_{ij}u_{k\ell}) = \dfrac{n_1}{n(n-1)(n+2)}[\{n_1(n+1)-2\}\omega_{ij}\omega_{k\ell}+n_2(\omega_{j\ell}\omega_{ki}+\omega_{i\ell}\omega_{kj})],$

where $n = n_1 + n_1$, $U = (u_{ij})$ *and* $\Omega = (\omega_{ij})$.

Proof: See Konno (1988). ∎

From Theorem 5.3.19, we immediately get

$$\mathrm{cov}(u_{ij},u_{k\ell}) = \frac{n_1 n_2}{n(n-1)(n+2)}\left[-\frac{2}{n}\omega_{ij}\omega_{k\ell}+\omega_{j\ell}\omega_{ik}+\omega_{i\ell}\omega_{kj}\right], \tag{5.3.10}$$

and

$$E(UAU) = \frac{n_1}{n(n-1)(n+2)}[\{n_1(n+1)-2\}\Omega A\Omega+n_2\{(\Omega A\Omega)'+\mathrm{tr}(\Omega A)\Omega\}], \tag{5.3.11}$$

where $A\,(p\times p)$ is a fixed matrix.

When $V \sim GB_p^{II}(\frac{1}{2}n_1, \frac{1}{2}n_2; \Omega, 0)$, $h(V)$ is a scalar function of V and $T\,(p \times p)$ is a matrix valued function of V and Ω, Konno (1988) also derived an identity for $E[h(V)\operatorname{tr}((\Omega + V)^{-1}T)]$, from which the following results were obtained.

THEOREM 5.3.20. Let $V \sim GB_p^{II}(\frac{1}{2}n_1, \frac{1}{2}n_2; \Omega, 0)$, then

(i) $E(v_{ij}) = \dfrac{n_1}{n_2 - p - 1}\omega_{ij}$, $n_2 - p - 1 > 0$,

and

(ii) $E(v_{ij}v_{k\ell}) = \dfrac{n_1}{(n_2 - p)(n_2 - p - 1)(n_2 - p - 3)}[\{n_1(n_2 - p - 2) + 2\}\omega_{ij}\omega_{k\ell}$

$\qquad\qquad + (n - p - 1)(\omega_{j\ell}\omega_{ik} + \omega_{i\ell}\omega_{kj})]$, $n_2 - p - 3 > 0$.

Proof: See Konno (1988). ∎

From the above theorem, one can easily see that for $n_2 - p - 3 > 0$,

$$\operatorname{cov}(v_{ij}, v_{k\ell}) = \frac{n_1(n - p - 1)}{(n_2 - p)(n_2 - p - 1)(n_2 - p - 3)}$$
$$\left[\frac{2}{n_2 - p - 2}\omega_{ij}\omega_{k\ell} + \omega_{j\ell}\omega_{ik} + \omega_{i\ell}\omega_{kj}\right] \qquad (5.3.12)$$

and

$$E(VAV) = \frac{n_1}{(n_2 - p)(n_2 - p - 1)(n_2 - p - 3)}[\{n_1(n_2 - p - 2) + 2\}\Omega A\Omega$$
$$+ (n - p - 1)\{(\Omega A\Omega)' + \operatorname{tr}(\Omega A)\Omega\}], \qquad (5.3.13)$$

where $A\,(p \times p)$ is a fixed matrix.

Further by noting that the distributions of U^{-1} and $\Omega^{-1} + V^{-1}$ are identical, Konno (1988) derived

$$E(U^{-1}) = \frac{n - p - 1}{n_1 - p - 1}\Omega^{-1}, \quad n_1 - p - 1 > 0, \qquad (5.3.14)$$

$$E(u^{ij}u^{k\ell}) = \frac{n - p - 1}{(n_1 - p)(n_1 - p - 1)(n_1 - p - 3)}[\{(n - p)(n_1 - p - 3) + n_2\}\omega^{ij}\omega^{k\ell}$$
$$+ n_2(\omega^{j\ell}\omega^{ik} + \omega^{i\ell}\omega^{kj})], \quad n_1 - p - 3 > 0, \qquad (5.3.15)$$

$$\operatorname{cov}(u^{ij}, u^{k\ell}) = \frac{n_2(n - p - 1)}{(n_1 - p)(n_1 - p - 1)(n_1 - p - 3)}\left[\frac{2}{n_1 - p - 1}\omega^{ij}\omega^{k\ell}\right.$$
$$\left. + \omega^{j\ell}\omega^{ik} + \omega^{i\ell}\omega^{kj}\right], \quad n_1 - p - 3 > 0, \qquad (5.3.16)$$

and

$$E(U^{-1}AU^{-1}) = \frac{n - p - 1}{(n_1 - p)(n_1 - p - 1)(n_1 - p - 3)}[\{(n - p)(n_1 - p - 3) + n_2\}\Omega^{-1}A\Omega^{-1}$$
$$+ n_2\{(\Omega^{-1}A\Omega^{-1})' + \operatorname{tr}(\Omega^{-1}A)\Omega^{-1}\}], \quad n_1 - p - 3 > 0. \qquad (5.3.17)$$

where $A\,(p \times p)$ is a fixed matrix.

In the remaining part of this section we give various factorizations of beta type I and type II matrices. It is interesting to note that, like Wishart matrix which factorizes into normal matrices, the beta type I and type II matrices factorize into inverted t- and t- matrices.

THEOREM 5.3.21. *Let $U \sim B_p^I(a,b)$. If a is half an integer, then U can be factorized as $U = XX'$, where $X \sim IT_{p,2a}(2b - p + 1, 0, I_p, I_{2a})$.*

Proof: Let $2a = m$ and $L\,(p \times m)$ be a semiorthogonal ($LL' = I_p$) random matrix which is independent of U. Then, the joint density of L and U is given by

$$c^{-1}\left\{\beta_p\left(\tfrac{1}{2}m, b\right)\right\}^{-1} \det(U)^{\frac{1}{2}(m-p-1)} \det(I_p - U)^{b - \frac{1}{2}(p+1)} g_{m,p}(L), \qquad (5.3.18)$$

where $c = \frac{2^p \pi^{\frac{1}{2}mp}}{\Gamma_p(\frac{1}{2}m)}$ and $g_{m,p}(L)$ is defined in (1.3.26). Since $U > 0$, with probability one, we can write $U = TT'$ where $T\,(p \times p)$ is a lower triangular matrix with positive diagonal elements. Further, since $m \geq p$, we can write $TL = X$, where $X\,(p \times m)$ is a random matrix of rank p. Now transforming $U = TT'$, $TL = X$, with the Jacobian (from (1.3.14) and (1.3.25))

$$J(U, L \to X) = J(U \to TT')J(T, L \to X) = 2^p \prod_{i=1}^{p} t_{ii}^{p-i+1} \prod_{i=1}^{p} t_{ii}^{-m+i}\{g_{m,p}(L)\}^{-1},$$

the density of X is given by

$$\frac{\Gamma_p(\frac{1}{2}m + b)}{\pi^{\frac{1}{2}mp} \Gamma_p(b)} \det(I_p - XX')^{b - \frac{1}{2}(p-1)-1}, \ X \in \mathbb{R}^{p \times m}. \qquad (5.3.19)$$

From (5.3.19), it is clear that $X \sim IT_{p,2a}(2b - p + 1, 0, I_p, I_{2a})$. ∎

The result for beta type II matrix, corresponding to Theorem 5.3.21, is given below.

THEOREM 5.3.22. *Let $V \sim B_p^{II}(a,b)$. If a is half an integer, then V can be factorized as $V = YY'$, where $Y \sim T_{p,2a}(2b - p + 1, 0, I_p, I_{2a})$.*

Proof: Similar to the proof of Theorem 5.3.21. ∎

THEOREM 5.3.23. *Let $U \sim B_p^I(a,b)$ and $U = TT'$, where $T = (t_{ij})$ is an upper triangular matrix with $t_{ii} > 0$, $i = 1, \ldots, p$. Partition T as*

$$T = \begin{pmatrix} T_{11} & \boldsymbol{t} \\ \boldsymbol{0}' & t_{pp} \end{pmatrix} \begin{matrix} p-1 \\ 1 \end{matrix}. \qquad (5.3.20)$$

Then, t_{pp}, $\boldsymbol{y} = (1 - t_{pp}^2)^{-\frac{1}{2}}(I_{p-1} - T_{11}T_{11}')^{-\frac{1}{2}}\boldsymbol{t}$ and T_{11} are independently distributed, $t_{pp}^2 \sim B^I(a,b)$, $\boldsymbol{y} \sim It_{p-1}(2b - p + 1, 1, 0, I_{p-1})$ and the distribution of T_{11} is same as that of T with p and a replaced by $p-1$ and $a - \frac{1}{2}$ respectively.

Proof: Making the transformation $U = TT'$, with the Jacobian of transformation $J(U \to T) = 2^p \prod_{i=1}^p t_{ii}^i$ in the density of U, we get the p.d.f. of T as

$$2^p \{\beta_p(a,b)\}^{-1} \prod_{i=1}^p (t_{ii}^2)^{a - \frac{1}{2}(p-i+1)} \det(I_p - TT')^{b - \frac{1}{2}(p+1)}, \qquad (5.3.21)$$

where $-\infty < t_{ij} < \infty$, $i < j$, $i, j = 1, \ldots, p$, and $t_{ii} > 0$, $i = 1, \ldots, p$. From (5.3.20), we have

$$\det(I_p - TT') = \det \begin{pmatrix} I_{p-1} - T_{11}T_{11}' - tt' & -t_{pp}t \\ -t_{pp}t' & 1 - t_{pp}^2 \end{pmatrix}$$

$$= (1 - t_{pp}^2) \det(I_{p-1} - T_{11}T_{11}' - (1 + t_{pp}^2(1 - t_{pp}^2)^{-1})tt')$$

$$= (1 - t_{pp}^2) \det(I_{p-1} - T_{11}T_{11}')$$

$$\det(I_{p-1} - (1 - t_{pp}^2)^{-1}(I_{p-1} - T_{11}T_{11}')^{-1}tt')$$

$$= (1 - t_{pp}^2) \det(I_{p-1} - T_{11}T_{11}')$$

$$(1 - (1 - t_{pp}^2)^{-1}t'(I_{p-1} - T_{11}T_{11}')^{-1}t). \qquad (5.3.22)$$

Substituting (5.3.22) in (5.3.21) we get $f(T_{11}, t, t_{pp})$, the joint density of T_{11}, t and t_{pp} as

$$f(T_{11}, t, t_{pp}) = f_1(T_{11}) f_2(t_{pp}) f_3(t | T_{11}, t_{pp}),$$

where

$$f_1(T_{11}) = 2^{p-1} \left\{ \beta_{p-1} \left(a - \frac{1}{2}, b\right) \right\}^{-1} \prod_{i=1}^{p-1} (t_{ii}^2)^{a - \frac{1}{2} - \frac{1}{2}(p-i)} \det(I_{p-1} - T_{11}T_{11}')^{b - \frac{1}{2}p}, \quad (5.3.23)$$

$$f_2(t_{pp}) = 2\{\beta(a,b)\}^{-1}(t_{pp}^2)^{a - \frac{1}{2}}(1 - t_{pp}^2)^{b-1} \qquad (5.3.24)$$

and

$$f_3(t | T_{11}, t_{pp}) = \frac{\pi^{-\frac{1}{2}(p-1)} \Gamma(b)}{\Gamma[b - \frac{1}{2}(p-1)]} (1 - t_{pp}^2)^{-\frac{1}{2}(p-1)} \det(I_{p-1} - T_{11}T_{11}')^{-\frac{1}{2}}$$

$$(1 - (1 - t_{pp}^2)^{-1}t'(I_{p-1} - T_{11}T_{11}')^{-1}t)^{b - \frac{1}{2}(p+1)}. \qquad (5.3.25)$$

Now transforming $x_p = t_{pp}^2$ and $y = (1 - t_{pp}^2)^{-\frac{1}{2}}(I_{p-1} - T_{11}T_{11}')^{-\frac{1}{2}}t$, with the Jacobian $J(t, t_{pp} \to y, x_p) = (2x_p^{\frac{1}{2}})^{-1}(1 - x_p)^{\frac{1}{2}(p-1)} \det(I_{p-1} - T_{11}T_{11}')^{\frac{1}{2}}$, we get the desired result. ∎

Results similar to the above were derived by Kshirsagar (1961b, 1972) when T is a lower triangular matrix.

THEOREM 5.3.24. *Let $U \sim B_p^I(a,b)$ and $U = TT'$, where $T = (t_{ij})$ is an upper triangular matrix with $t_{ii} > 0$, $i = 1, \ldots, p$. Then, $t_{11}^2, \ldots, t_{pp}^2$ are independently distributed, $t_{ii}^2 \sim B^I(a - \frac{1}{2}(p-i), b)$, $i = 1, \ldots, p$.*

Proof: From Theorem 5.3.23, it is known that $t_{pp}^2 \sim B^I(a,b)$ and is independent of T_{11}, which has the same distribution as T with p and a replaced by $p-1$ and $a - \frac{1}{2}$ respectively. Further partitioning T_{11} yields $t_{p-1,p-1}^2 \sim B^I(a - \frac{1}{2}, b)$. Repeated application of this procedure completes the proof of the theorem. ∎

The next result, derived by Javier and Gupta (1985a), is a matrix variate generalization of a result given in Rao (1952).

THEOREM 5.3.25. *If* $X \sim B_p^I(a,b)$ *and* $Y \sim B_p^I(a+b,c)$ *are independent, then* $U = Y^{\frac{1}{2}} X (Y^{\frac{1}{2}})' \sim B_p^I(a, b+c)$.

Proof: The joint density of X and Y is

$$\{\beta_p(a,b)\beta_p(a+b,c)\}^{-1} \det(X)^{a-\frac{1}{2}(p+1)} \det(I_p - X)^{b-\frac{1}{2}(p+1)}$$

$$\det(Y)^{a+b-\frac{1}{2}(p+1)} \det(I_p - Y)^{c-\frac{1}{2}(p+1)}, \, 0 < X < I_p, \, 0 < Y < I_p. \quad (5.3.26)$$

Making the transformation $U = Y^{\frac{1}{2}} X (Y^{\frac{1}{2}})'$ with the Jacobian $J(X, Y \to U, Y) = \det(Y)^{-\frac{1}{2}(p+1)}$ in (5.3.26) we get the joint density of U and Y as

$$\frac{\Gamma_p(a+b+c)}{\Gamma_p(a)\Gamma_p(b)\Gamma_p(c)} \det(U)^{a-\frac{1}{2}(p+1)} \det(I_p - Y^{-\frac{1}{2}} U (Y^{-\frac{1}{2}})')^{b-\frac{1}{2}(p+1)} \det(Y)^{b-\frac{1}{2}(p+1)}$$

$$\det(I_p - Y)^{c-\frac{1}{2}(p+1)}, \, 0 < U < Y < I_p. \quad (5.3.27)$$

Now to obtain the marginal density of U, we need to evaluate

$$\int_{U < Y < I_p} \det(I_p - Y^{-\frac{1}{2}} U (Y^{-\frac{1}{2}})')^{b-\frac{1}{2}(p+1)} \det(Y)^{b-\frac{1}{2}(p+1)} \det(I_p - Y)^{c-\frac{1}{2}(p+1)} \, dY. \quad (5.3.28)$$

Substituting in (5.3.28), $W = (I_p - U)^{-\frac{1}{2}} (Y - U) ((I_p - U)^{-\frac{1}{2}})'$ with the Jacobian $J(Y \to W) = \det(I_p - U)^{\frac{1}{2}(p+1)}$, we get

$$\det(I_p - U)^{b+c-\frac{1}{2}(p+1)} \int_{0 < W < I_p} \det(W)^{b-\frac{1}{2}(p+1)} \det(I_p - W)^{c-\frac{1}{2}(p+1)} \, dW$$

$$= \det(I_p - U)^{b+c-\frac{1}{2}(p+1)} \frac{\Gamma_p(b)\Gamma_p(c)}{\Gamma_p(b+c)}. \quad (5.3.29)$$

Integration of Y in (5.3.27), using (5.3.28) and (5.3.29), completes the proof of the theorem. ∎

A shorter proof of the above theorem can be given by using the m.g.f. of $Y^{\frac{1}{2}} X (Y^{\frac{1}{2}})'$.

5.4. RELATED DISTRIBUTIONS

In this section, we study some distributions related to the matrix variate beta type I and type II distributions.

THEOREM 5.4.1. *Let $S_i \sim W_p(n_i, I_p)$, $i = 1, 2$ be independent.*

(i) If $S_2 = TT'$, where T is a lower triangular matrix with positive diagonal elements, then the distribution of $U = T'(S_1 + S_2)^{-1}T$ is given by

$$\left\{\beta_p\left(\frac{1}{2}n_2, \frac{1}{2}n_1\right)\right\}^{-1} \frac{\det(U)^{\frac{1}{2}n_2} \det(I_p - U)^{\frac{1}{2}(n_1-p-1)}}{\prod_{i=1}^{p} \det(U_{[i]})}, \quad 0 < U < I_p. \tag{5.4.1}$$

Further U and $S_1 + S_2$ are independently distributed.

(ii) If $S_2 = TT'$, where T is an upper triangular matrix with positive diagonal elements, then the distribution of $U = T'(S_1 + S_2)^{-1}T$ is given by

$$\left\{\beta_p\left(\frac{1}{2}n_2, \frac{1}{2}n_1\right)\right\}^{-1} \frac{\det(U)^{\frac{1}{2}n_2} \det(I_p - U)^{\frac{1}{2}(n_1-p-1)}}{\prod_{i=1}^{p} \det(U^{[i]})}, \quad 0 < U < I_p. \tag{5.4.2}$$

Further U and $S_1 + S_2$ are independently distributed.

Proof: (i) The joint p.d.f. of S_1 and S_2 is

$$\left\{2^{\frac{1}{2}(n_1+n_2)p}\Gamma_p\left(\frac{1}{2}n_1\right)\Gamma_p\left(\frac{1}{2}n_2\right)\right\}^{-1} \text{etr}\left\{-\frac{1}{2}(S_1 + S_2)\right\} \det(S_1)^{\frac{1}{2}(n_1-p-1)} \det(S_2)^{\frac{1}{2}(n_2-p-1)}.$$

Transforming $S = S_1 + S_2$, and $S_2 = TT'$ with the Jacobian of transformation $J(S_1, S_2 \to S, T) = 2^p \prod_{i=1}^{p} t_{ii}^{p-i+1}$, we get the joint p.d.f. of S and T as

$$\left\{2^{\frac{1}{2}(n_1+n_2-2)p}\Gamma_p\left(\frac{1}{2}n_1\right)\Gamma_p\left(\frac{1}{2}n_2\right)\right\}^{-1} \text{etr}\left(-\frac{1}{2}S\right) \det(S - TT')^{\frac{1}{2}(n_1-p-1)} \prod_{i=1}^{p} t_{ii}^{n_2-i},$$

where $S - TT' > 0$, $t_{ii} > 0$ and $-\infty < t_{ij} < \infty$, $i > j$. Further, transforming $U = T'S^{-1}T$ with the Jacobian $J(T \to U) = \left\{2^p \prod_{i=1}^{p} t_{ii}^i \det((S^{-1})_{[i]})\right\}^{-1}$, we get the joint p.d.f. of S and U,

$$\left\{2^{\frac{1}{2}(n_1+n_2)p}\Gamma_p\left(\frac{1}{2}n_1\right)\Gamma_p\left(\frac{1}{2}n_2\right)\right\}^{-1} \text{etr}\left(-\frac{1}{2}S\right) \det(S)^{\frac{1}{2}(n_1+n_2-p-1)}$$

$$\det(I_p - U)^{\frac{1}{2}(n_1-p-1)} \det(U)^{\frac{1}{2}n_2} \prod_{i=1}^{p} \det(U_{[i]})^{-1}, \tag{5.4.3}$$

since $\prod_{i=1}^{p} t_{ii}^{2i} = \prod_{i=1}^{p} \frac{\det(U_{[i]})}{\det((S^{-1})_{[i]})}$. From (5.4.3), it is easy to see that S, i.e., $S_1 + S_2$ and U are independent, $S \sim W_p(n_1 + n_2, I_p)$ and the distribution of U is given by (5.4.1).

(ii) Proof is similar to part (i). ∎

THEOREM 5.4.2. *Let $S_i \sim W_p(n_i, \Sigma)$, $i = 1, 2$ be independent, $S_2 = TT'$, where T is a triangular matrix with positive diagonal elements, and $V = T^{-1}S_1(T^{-1})'$. Then, V and $S_1 + S_2$ are independently distributed. Further*

(i) if T is a lower triangular matrix, then the p.d.f. of V is

$$\left\{\beta_p\left(\frac{1}{2}n_1, \frac{1}{2}n_2\right)\right\}^{-1} \frac{\det(V)^{\frac{1}{2}(n_1-p-1)} \det(I_p + V)^{-\frac{1}{2}(n_1+n_2-p-1)}}{\prod_{i=1}^{p} \det((I_p + V)_{[i]})}, \quad V > 0, \tag{5.4.4}$$

and

(ii) if T is an upper triangular matrix, then the p.d.f. of V is

$$\left\{\beta_p\left(\frac{1}{2}n_1, \frac{1}{2}n_2\right)\right\}^{-1} \frac{\det(V)^{\frac{1}{2}(n_1-p-1)} \det(I_p + V)^{-\frac{1}{2}(n_1+n_2-p-1)}}{\prod_{i=1}^{p} \det((I_p + V)_{[i]})}, \quad V > 0. \qquad (5.4.5)$$

Proof: (i) The joint p.d.f. of S_1 and S_2 is

$$\left\{2^{\frac{1}{2}(n_1+n_2)p}\Gamma_p\left(\frac{1}{2}n_1\right)\Gamma_p\left(\frac{1}{2}n_2\right)\det(\Sigma)^{\frac{1}{2}(n_1+n_2)}\right\}^{-1} \text{etr}\left\{-\frac{1}{2}\Sigma^{-1}(S_1 + S_2)\right\}$$

$$\det(S_1)^{\frac{1}{2}(n_1-p-1)}\det(S_2)^{\frac{1}{2}(n_2-p-1)}. \qquad (5.4.6)$$

Transforming $S_2 = TT'$ and $V = T^{-1}S_1(T^{-1})'$, with the Jacobian of transformation $J(S_1, S_2 \to V, T) = 2^p \prod_{i=1}^{p} t_{ii}^{2(p+1)-i}$, the joint density of V and T is given by

$$\left\{2^{\frac{1}{2}(n_1+n_2-2)p}\Gamma_p\left(\frac{1}{2}n_1\right)\Gamma_p\left(\frac{1}{2}n_2\right)\det(\Sigma)^{\frac{1}{2}(n_1+n_2)}\right\}^{-1} \text{etr}\left\{-\frac{1}{2}\Sigma^{-1}T(I_p + V)T'\right\}$$

$$\det(V)^{\frac{1}{2}(n_1-p-1)} \prod_{i=1}^{p} t_{ii}^{n_1+n_2-i}. \qquad (5.4.7)$$

Further transforming $S = T(I_p+V)T'$, with the Jacobian $J(T \to S) = \{2^p \prod_{i=1}^{p} t_{ii}^{p-i+1} \det((I_p + V)^{[i]})\}^{-1}$, the joint density of V and S, obtained from (5.4.7), is given by

$$\left\{2^{\frac{1}{2}(n_1+n_2)p}\Gamma_p\left(\frac{1}{2}n_1\right)\Gamma_p\left(\frac{1}{2}n_2\right)\det(\Sigma)^{\frac{1}{2}(n_1+n_2)}\right\}^{-1} \text{etr}\left(-\frac{1}{2}\Sigma^{-1}S\right)\det(S)^{\frac{1}{2}(n_1+n_2-p-1)}$$

$$\det(V)^{\frac{1}{2}(n_1-p-1)}\det(I_p + V)^{-\frac{1}{2}(n_1+n_2-p-1)} \prod_{i=1}^{p} \det((I_p + V)^{[i]})^{-1}. \qquad (5.4.8)$$

From (5.4.8) it is clear that V and $S = S_1 + S_2$ are independently distributed, $S \sim W_p(n_1 + n_2, \Sigma)$ and the p.d.f. of V is given by (5.4.4).

(ii) The proof is similar to part (i). ∎

Theorems 5.4.1 and 5.4.2 were proved by Olkin and Rubin (1964). In Theorem 5.4.2, we notice that $V = S_2^{-\frac{1}{2}}S_1(S_2^{-\frac{1}{2}})'$ and $S_1 + S_2$ are independent for all $\Sigma > 0$ when $S_2^{\frac{1}{2}}$ is a triangular square root. However this is not the case when $S_2^{\frac{1}{2}}$ is a symmetric square root as shown by Olkin and Rubin (1964) and given in the next theorem.

THEOREM 5.4.3. Let $S_i \sim W_p(n_i, \Sigma)$, $i = 1, 2$ be independent. Define $S = S_1 + S_2$ and $V = S_2^{-\frac{1}{2}}S_1 S_2^{-\frac{1}{2}}$, where $S_2^{\frac{1}{2}}S_2^{\frac{1}{2}} = S_2$. Then S and V are not independent and their joint p.d.f. is given by

$$\left\{2^{\frac{1}{2}(n_1+n_2)p}\Gamma_p\left(\frac{1}{2}n_1\right)\Gamma_p\left(\frac{1}{2}n_2\right)\det(\Sigma)^{\frac{1}{2}(n_1+n_2)}\right\}^{-1} \text{etr}\left(-\frac{1}{2}\Sigma^{-1}S\right)\det(S)^{\frac{1}{2}(n_1+n_2-p-1)}$$

$$\det(V)^{\frac{1}{2}(n_1-p-1)}\det(I_p + V)^{-\frac{1}{2}(n_1+n_2-p-1)} \prod_{i \le j} \left(\frac{\delta_i + \delta_j}{\lambda_i + \lambda_j}\right), \quad S > 0, V > 0,$$

where λ_i and δ_i $(i = 1, \ldots, p)$ are the eigenvalues of $\{(I_p + V)^{\frac{1}{2}} S (I_p + V)^{\frac{1}{2}}\}^{\frac{1}{2}}$ and $(I_p + V)^{-\frac{1}{2}} \{(I_p + V)^{\frac{1}{2}} X (I_p + V)^{\frac{1}{2}}\}^{\frac{1}{2}} (I_p + V)^{-\frac{1}{2}}$ respectively.

Proof: The joint p.d.f. of S_1 and S_2 is given by (5.4.6). Let $S_2 = X^2$, $V = X^{-1} S_1 X^{-1}$, and $\delta_1, \ldots, \delta_p$ be the eigenvalues of X. Then, from (1.3.5) and (1.3.20), the Jacobian of transformation is $J(S_1, S_2 \to V, X) = \prod_{i \leq j} (\delta_i + \delta_j) \det(X)^{p+1}$, and the joint p.d.f. of V and X is obtained as

$$\left\{ 2^{\frac{1}{2}(n_1 + n_2)p} \Gamma_p\left(\frac{1}{2} n_1\right) \Gamma_p\left(\frac{1}{2} n_2\right) \det(\Sigma)^{\frac{1}{2}(n_1 + n_2)} \right\}^{-1} \operatorname{etr}\left\{ -\frac{1}{2} \Sigma^{-1} X (I_p + V) X \right\}$$

$$\det(X^2)^{\frac{1}{2}(n_1 + n_2 - p - 1)} \prod_{i \leq j} (\delta_i + \delta_j) \det(V)^{\frac{1}{2}(n_1 - p - 1)}.$$

Now transforming $S = X(I_p + V)X$ with Jacobian $J(X \to S) = \prod_{i \leq j} (\lambda_i + \lambda_j)^{-1}$, where $\lambda_1, \ldots, \lambda_p$ are the eigenvalues of $(I_p + V)^{\frac{1}{2}} X (I_p + V)^{\frac{1}{2}}$, the joint p.d.f. of S and V is

$$\left\{ 2^{\frac{1}{2}(n_1 + n_2)p} \Gamma_p\left(\frac{1}{2} n_1\right) \Gamma_p\left(\frac{1}{2} n_2\right) \det(\Sigma)^{\frac{1}{2}(n_1 + n_2)} \right\}^{-1} \operatorname{etr}\left(-\frac{1}{2} \Sigma^{-1} S \right) \det(S)^{\frac{1}{2}(n_1 + n_2 - p - 1)}$$

$$\det(V)^{\frac{1}{2}(n_1 - p - 1)} \det(I_p + V)^{-\frac{1}{2}(n_1 + n_2 - p - 1)} \prod_{i \leq j} \left(\frac{\delta_i + \delta_j}{\lambda_i + \lambda_j} \right), \quad S > 0,\ V > 0, \quad (5.4.9)$$

Now the independence of V and S depends on the factorization of (5.4.9) into two functions, one of V alone and the other of S alone. However, this factorization depends on the factorization of $\prod_{i \leq j} \left(\frac{\delta_i + \delta_j}{\lambda_i + \lambda_j} \right)$. For $p = 2$, from (1.3.20) let

$$\prod_{i \leq j} \left(\frac{\delta_i + \delta_j}{\lambda_i + \lambda_j} \right) = \frac{h(\delta_1 + \delta_2)}{h(\lambda_1 + \lambda_2)} \tag{5.4.10}$$

where now

$$h(\delta_1 + \delta_2) = 2^2 a_1 a_2 = 4 \det(X) \operatorname{tr}(X), \tag{5.4.11}$$

$$h(\lambda_1 + \lambda_2) = 4 \det((I_p + V)X) \operatorname{tr}(X(I_p + V)), \tag{5.4.12}$$

and

$$X = (I_p + V)^{-\frac{1}{2}} \{(I_p + V)^{\frac{1}{2}} S (I_p + V)^{\frac{1}{2}}\}^{\frac{1}{2}} (I_p + V)^{-\frac{1}{2}}. \tag{5.4.13}$$

From (5.4.10)–(5.4.13), we get

$$\frac{h(\delta_1 + \delta_2)}{h(\lambda_1 + \lambda_2)} = \frac{\operatorname{tr}[(I_p + V)^{-1} \{(I_p + V)^{\frac{1}{2}} S (I_p + V)^{\frac{1}{2}}\}^{\frac{1}{2}}]}{\det(I_p + V) \operatorname{tr}[\{(I_p + V)^{\frac{1}{2}} S (I_p + V)^{\frac{1}{2}}\}^{\frac{1}{2}}]}. \tag{5.4.14}$$

Now, (5.4.14) does not factorize to give the independence of V and S. ∎

Next, we derive the p.d.f. of U when the Wishart matrices S_1 and S_2 have different covariance matrices.

THEOREM 5.4.4. *Let* $S_i \sim W_p(n_i, \Sigma_i)$, $i = 1, 2$ *be independent. If*

$$U = (S_1 + S_2)^{-\frac{1}{2}} S_1 ((S_1 + S_2)^{-\frac{1}{2}})',$$

where $(S_1 + S_2)^{\frac{1}{2}}((S_1 + S_2)^{\frac{1}{2}})'$ *is a reasonable nonsingular factorization of* $S_1 + S_2$, *then the p.d.f. of* U *is given by*

$$\left\{ 2^{\frac{1}{2}(n_1+n_2)p} \Gamma_p\left(\frac{1}{2}n_1\right)\Gamma_p\left(\frac{1}{2}n_2\right) \det(\Sigma_1)^{\frac{1}{2}n_1} \det(\Sigma_2)^{\frac{1}{2}n_2} \right\}^{-1}$$

$$\det(U)^{\frac{1}{2}(n_1-p-1)} \det(I_p - U)^{\frac{1}{2}(n_2-p-1)} \int_{S>0} \det(S)^{\frac{1}{2}(n_1+n_2-p-1)}$$

$$\text{etr}\left\{ -\frac{1}{2}\Sigma_2^{-1}S - \frac{1}{2}(\Sigma_1^{-1} - \Sigma_2^{-1})S^{\frac{1}{2}}U(S^{\frac{1}{2}})' \right\} dS, \quad 0 < U < I_p. \qquad (5.4.15)$$

Proof: The joint density of S_1 and S_2 is

$$\prod_{i=1}^{2} \left[\left\{ 2^{\frac{1}{2}n_i p} \Gamma_p\left(\frac{1}{2}n_i\right) \det(\Sigma_i)^{\frac{1}{2}n_i} \right\}^{-1} \text{etr}\left(-\frac{1}{2}\Sigma_i^{-1}S_i \right) \det(S_i)^{\frac{1}{2}(n_i-p-1)} \right].$$

Making the transformation $S_1 + S_2 = S$, $S_1 = S^{\frac{1}{2}}U(S^{\frac{1}{2}})'$ with Jacobian $J(S_1, S_2 \to U, S) = \det(S)^{\frac{1}{2}(p+1)}$, we get the joint p.d.f. of U and S as

$$\left\{ 2^{\frac{1}{2}(n_1+n_2)p} \Gamma_p\left(\frac{1}{2}n_1\right)\Gamma_p\left(\frac{1}{2}n_2\right) \det(\Sigma_1)^{\frac{1}{2}n_1} \det(\Sigma_2)^{\frac{1}{2}n_2} \right\}^{-1}$$

$$\det(U)^{\frac{1}{2}(n_1-p-1)} \det(I_p - U)^{\frac{1}{2}(n_2-p-1)} \det(S)^{\frac{1}{2}(n_1+n_2-p-1)}$$

$$\text{etr}\left\{ -\frac{1}{2}\Sigma_2^{-1}S^{\frac{1}{2}}(I_p - U)(S^{\frac{1}{2}})' - \frac{1}{2}\Sigma_1^{-1}S^{\frac{1}{2}}U(S^{\frac{1}{2}})' \right\}, \quad 0 < U < I_p, \; S > 0. \quad (5.4.16)$$

To find the marginal density of U, we integrate (5.4.16) with respect to S, obtaining (5.4.15). ∎

When $\Sigma_1 = \Sigma_2 = \Sigma$, the integral in the p.d.f. (5.4.15) can be easily evaluated as

$$\int_{S>0} \det(S)^{\frac{1}{2}(n_1+n_2-p-1)} \text{etr}\left(-\frac{1}{2}\Sigma^{-1}S \right) dS$$

$$= 2^{\frac{1}{2}(n_1+n_2)p} \Gamma_p\left[\frac{1}{2}(n_1 + n_2)\right] \det(\Sigma)^{\frac{1}{2}(n_1+n_2)},$$

and in this case $U \sim B_p^I(\frac{1}{2}n_1, \frac{1}{2}n_2)$.

THEOREM 5.4.5. *If U is distributed as (5.4.15), then*

$$E[\det(U)^h] = \frac{\Gamma_p(\frac{1}{2}n_1 + h)\Gamma_p[\frac{1}{2}(n_1 + n_2)]}{\Gamma_p(\frac{1}{2}n_1)\Gamma_p[\frac{1}{2}(n_1 + n_2) + h]} \det(\Sigma_1^{-1}\Sigma_2)^{\frac{1}{2}n_1}$$

$$_2F_1\left(\frac{1}{2}n_1 + h, \frac{1}{2}(n_1 + n_2); \frac{1}{2}(n_1 + n_2) + h; I_p - \Sigma_1^{-\frac{1}{2}}\Sigma_2\Sigma_1^{-\frac{1}{2}} \right),$$

$$\text{Re}(h) > -\frac{1}{2}(n_1 - p + 1),$$

and

$$E[\det(I_p - U)^h] = \frac{\Gamma_p(\frac{1}{2}n_2 + h)\Gamma_p[\frac{1}{2}(n_1 + n_2)]}{\Gamma_p(\frac{1}{2}n_2)\Gamma_p[\frac{1}{2}(n_1 + n_2) + h]} \det(\Sigma_1^{-1}\Sigma_2)^{\frac{1}{2}n_1}$$

$$_2F_1\Big(\frac{1}{2}n_1, \frac{1}{2}(n_1 + n_2); \frac{1}{2}(n_1 + n_2) + h; I_p - \Sigma_1^{-\frac{1}{2}}\Sigma_2\Sigma_1^{-\frac{1}{2}}\Big),$$

$$\mathrm{Re}(h) > -\frac{1}{2}(n_2 - p + 1).$$

Proof: From (5.4.15), we get

$$E[\det(U)^h] = \Big\{2^{\frac{1}{2}(n_1+n_2)p}\Gamma_p\Big(\frac{1}{2}n_1\Big)\Gamma_p\Big(\frac{1}{2}n_2\Big)\det(\Sigma_1)^{\frac{1}{2}n_1}\det(\Sigma_2)^{\frac{1}{2}n_2}\Big\}^{-1}$$

$$\int_{0<U<I_p}\det(U)^{\frac{1}{2}(n_1-p-1)+h}\det(I_p - U)^{\frac{1}{2}(n_2-p-1)}\int_{S>0}\det(S)^{\frac{1}{2}(n_1+n_2-p-1)}$$

$$\mathrm{etr}\Big\{-\frac{1}{2}\Sigma_2^{-1}S - \frac{1}{2}(\Sigma_1^{-1} - \Sigma_2^{-1})S^{\frac{1}{2}}U(S^{\frac{1}{2}})'\Big\}\, dU\, dS. \qquad (5.4.17)$$

Since $\mathrm{etr}\{\frac{1}{2}(\Sigma_2^{-1} - \Sigma_1^{-1})S^{\frac{1}{2}}U(S^{\frac{1}{2}})'\} = {}_0F_0(\frac{1}{2}(\Sigma_2^{-1} - \Sigma_1^{-1})S^{\frac{1}{2}}U(S^{\frac{1}{2}})')$, we use the integral (1.6.6) to obtain

$$\int_{0<U<I_p}\det(U)^{\frac{1}{2}(n_1-p-1)+h}\det(I_p - U)^{\frac{1}{2}(n_2-p-1)}{}_0F_0\Big(\frac{1}{2}(\Sigma_2^{-1} - \Sigma_1^{-1})S^{\frac{1}{2}}U(S^{\frac{1}{2}})'\Big)\, dU$$

$$= \frac{\Gamma_p(\frac{1}{2}n_1 + h)\Gamma_p(\frac{1}{2}n_2)}{\Gamma_p[\frac{1}{2}(n_1 + n_2) + h]}{}_1F_1\Big(\frac{1}{2}n_1 + h; \frac{1}{2}(n_1 + n_2) + h; \frac{1}{2}(\Sigma_2^{-1} - \Sigma_1^{-1})S\Big). \quad (5.4.18)$$

Now, substituting (5.4.18) in (5.4.17), we get

$$E[\det(U)^h] = \Big\{2^{\frac{1}{2}(n_1+n_2)p}\Gamma_p\Big(\frac{1}{2}n_1\Big)\Gamma_p\Big[\frac{1}{2}(n_1 + n_2) + h\Big]\det(\Sigma_1)^{\frac{1}{2}n_1}\det(\Sigma_2)^{\frac{1}{2}n_2}\Big\}^{-1}$$

$$\Gamma_p\Big(\frac{1}{2}n_1 + h\Big)\int_{S>0}\det(S)^{\frac{1}{2}(n_1+n_2-p-1)}\mathrm{etr}\Big(-\frac{1}{2}\Sigma_2^{-1}S\Big)$$

$$_1F_1\Big(\frac{1}{2}n_1 + h; \frac{1}{2}(n_1 + n_2) + h; \frac{1}{2}(\Sigma_2^{-1} - \Sigma_1^{-1})S\Big)\, dS. \qquad (5.4.19)$$

The integral in (5.4.19) can easily be evaluated by using (1.6.4) to give the desired result. The derivation of $E[\det(I_p - U)^h]$ is similar. ∎

Theorems 5.4.4 and 5.4.5 are special cases of the results derived by de Waal (1970) for Dirichlet matrices and are given in Theorems 6.4.1 and 6.4.3.

The statistic $\det(U) = \frac{\det(S_1)}{\det(S_1+S_2)}$ is used to test the null hypothesis $H : \Sigma_1 = \Sigma_2$. The distribution of U given in (5.4.15) is needed to study the power of this test.

5.5.   NONCENTRAL MATRIX VARIATE BETA DISTRIBUTION

In Section 5.2, we defined the matrix variate beta type I distribution and subsequently derived it using the representation $U = (S_1 + S_2)^{-\frac{1}{2}}S_1((S_1 + S_2)^{-\frac{1}{2}})'$, where $S_i \sim W_p(n_i, \Sigma)$, $i = 1, 2$ are independent. In case $S_2 \sim W_p(n_2, \Sigma, \Theta)$, the corresponding distribution of U is called the noncentral matrix variate beta type I(A), (Hart and Money, 1976) and is derived in the next theorem (de Waal, 1968).

THEOREM 5.5.1. *Let $S_1 \sim W_p(n_1, \Sigma)$, and $S_2 \sim W_p(n_2, \Sigma, \Theta)$, be independently distributed. Define $U = (S_1 + S_2)^{-\frac{1}{2}}S_1((S_1 + S_2)^{-\frac{1}{2}})'$, where $(S_1 + S_2)^{\frac{1}{2}}((S_1 + S_2)^{\frac{1}{2}})'$ is a reasonable nonsingular factorization of $S_1 + S_2$, then the p.d.f. of U is given by*

$$\left\{2^{\frac{1}{2}(n_1+n_2)p}\Gamma_p\left(\frac{1}{2}n_1\right)\Gamma_p\left(\frac{1}{2}n_2\right)\det(\Sigma)^{\frac{1}{2}(n_1+n_2)}\right\}^{-1}\text{etr}\left(-\frac{1}{2}\Theta\right)$$

$$\det(U)^{\frac{1}{2}(n_1-p-1)}\det(I_p - U)^{\frac{1}{2}(n_2-p-1)}\int_{S>0}\det(S)^{\frac{1}{2}(n_1+n_2-p-1)}$$

$$\text{etr}\left(-\frac{1}{2}\Sigma^{-1}S\right){}_0F_1\left(\frac{1}{2}n_2; \frac{1}{4}\Sigma^{-1}\Theta S^{\frac{1}{2}}(I_p - U)(S^{\frac{1}{2}})'\right)dS, \ 0 < U < I_p. \quad (5.5.1)$$

Proof: The joint density of S_1 and S_2 is

$$\left\{2^{\frac{1}{2}(n_1+n_2)p}\Gamma_p\left(\frac{1}{2}n_1\right)\Gamma_p\left(\frac{1}{2}n_2\right)\det(\Sigma)^{\frac{1}{2}(n_1+n_2)}\right\}^{-1}\text{etr}\left(-\frac{1}{2}\Theta\right)\text{etr}\left\{-\frac{1}{2}\Sigma^{-1}(S_1 + S_2)\right\}$$

$$\det(S_1)^{\frac{1}{2}(n_1-p-1)}\det(S_2)^{\frac{1}{2}(n_2-p-1)}{}_0F_1\left(\frac{1}{2}n_2; \frac{1}{4}\Sigma^{-1}\Theta S_2\right).$$

Using the transformation $S_1 + S_2 = S$, $S_1 = S^{\frac{1}{2}}U(S^{\frac{1}{2}})'$ with Jacobian $J(S_1, S_2 \to U, S) = \det(S)^{\frac{1}{2}(p+1)}$, we get the joint p.d.f. of U and S as

$$\left\{2^{\frac{1}{2}(n_1+n_2)p}\Gamma_p\left(\frac{1}{2}n_1\right)\Gamma_p\left(\frac{1}{2}n_2\right)\det(\Sigma)^{\frac{1}{2}(n_1+n_2)}\right\}^{-1}\text{etr}\left(-\frac{1}{2}\Theta\right)$$

$$\det(U)^{\frac{1}{2}(n_1-p-1)}\det(I_p - U)^{\frac{1}{2}(n_2-p-1)}\det(S)^{\frac{1}{2}(n_1+n_2-p-1)}$$

$$\text{etr}\left(-\frac{1}{2}\Sigma^{-1}S\right){}_0F_1\left(\frac{1}{2}n_2; \frac{1}{4}\Sigma^{-1}\Theta S^{\frac{1}{2}}(I_p - U)(S^{\frac{1}{2}})'\right). \quad (5.5.2)$$

To find the marginal density of U, we integrate (5.5.2) with respect to S. ∎

Substituting $\Theta = 0$ gives the central matrix variate beta type I density. When the rank of Θ is unity, the linear case, the distribution of $U = (u_{ij})$ is given by Kshirsagar (1961b) as

$$\left\{\beta_p\left(\frac{1}{2}n_1, \frac{1}{2}n_2\right)\right\}^{-1}\det(U)^{\frac{1}{2}(n_1-p-1)}\det(I_p - U)^{\frac{1}{2}(n_2-p-1)}\exp\left(-\frac{1}{2}\theta^2\right)$$

$$ {}_1F_1\left(\frac{1}{2}(n_1 + n_2); \frac{1}{2}n_2; \frac{1}{2}\theta^2(1 - u_{11})\right), \ 0 < U < I_p,$$

where θ^2 is the only nonzero eigenvalue of Θ. He also discussed the planar case, $i.e.$, when the rank of O is two.

If $W = (S_1 + S_2)^{-\frac{1}{2}} S_2 ((S_1 + S_2)^{-\frac{1}{2}})'$, $S_1 \sim W_p(n, \Sigma)$ and $S_2 \sim W_p(n_2, \Sigma, \Theta)$, then $W = I_p - U$ and from Theorem 5.5.1, its p.d.f. is

$$\left\{ 2^{\frac{1}{2}(n_1+n_2)p} \Gamma_p\left(\frac{1}{2}n_1\right) \Gamma_p\left(\frac{1}{2}n_2\right) \det(\Sigma)^{\frac{1}{2}(n_1+n_2)} \right\}^{-1} \operatorname{etr}\left(-\frac{1}{2}\Theta\right)$$

$$\det(W)^{\frac{1}{2}(n_2-p-1)} \det(I_p - W)^{\frac{1}{2}(n_1-p-1)} \int_{S>0} \det(S)^{\frac{1}{2}(n_1+n_2-p-1)}$$

$$\operatorname{etr}\left(-\frac{1}{2}\Sigma^{-1}S\right) {}_0F_1\left(\frac{1}{2}n_2; \frac{1}{4}\Sigma^{-1}\Theta S^{\frac{1}{2}}W(S^{\frac{1}{2}})'\right) dS, \ 0 < W < I_p.$$

This distribution is known as noncentral matrix variate beta type I(B), and for $p = 1$, the distribution of W is the usual noncentral beta type I distribution.

The next theorem gives the moments of $\det(U)$ and $\det(I_p - U)$, when U has noncentral matrix variate beta type I(A) distribution.

THEOREM 5.5.2. *If U is distributed as (5.5.1), then*

(i) $E[\det(U)^h] = \dfrac{\Gamma_p(\frac{1}{2}n_1 + h)\Gamma_p[\frac{1}{2}(n_1 + n_2)]}{\Gamma_p(\frac{1}{2}n_1)\Gamma_p[\frac{1}{2}(n_1 + n_2) + h]} \operatorname{etr}\left(-\frac{1}{2}\Theta\right)$

$$\qquad {}_1F_1\left(\frac{1}{2}(n_1 + n_2); \frac{1}{2}(n_1 + n_2) + h; \frac{1}{2}\Theta\right),$$

$$\operatorname{Re}(h) > -\frac{1}{2}(n_1 - p + 1),$$

and

(ii) $E[\det(I_p - U)^h] = \dfrac{\Gamma_p(\frac{1}{2}n_2 + h)\Gamma_p[\frac{1}{2}(n_1 + n_2)]}{\Gamma_p(\frac{1}{2}n_1)\Gamma_p[\frac{1}{2}(n_1 + n_2) + h]} \operatorname{etr}\left(-\frac{1}{2}\Theta\right)$

$$\qquad {}_2F_2\left(\frac{1}{2}(n_1 + n_2), \frac{1}{2}n_2 + h; \frac{1}{2}n_2, \frac{1}{2}(n_1 + n_2) + h; \frac{1}{2}\Theta\right),$$

$$\operatorname{Re}(h) > -\frac{1}{2}(n_2 - p + 1),$$

Proof: From (5.5.1), we have

$$E[\det(U)^h] = \left\{ 2^{\frac{1}{2}(n_1+n_2)p} \Gamma_p\left(\frac{1}{2}n_1\right) \Gamma_p\left(\frac{1}{2}n_2\right) \det(\Sigma)^{\frac{1}{2}(n_1+n_2)} \right\}^{-1} \operatorname{etr}\left(-\frac{1}{2}\Theta\right)$$

$$\int_{0<U<I_p} \det(U)^{\frac{1}{2}(n_1-p-1)+h} \det(I_p - U)^{\frac{1}{2}(n_2-p-1)} \int_{S>0} \det(S)^{\frac{1}{2}(n_1+n_2-p-1)}$$

$$\operatorname{etr}\left(-\frac{1}{2}\Sigma^{-1}S\right) {}_0F_1\left(\frac{1}{2}n_2; \frac{1}{4}\Sigma^{-1}\Theta S^{\frac{1}{2}}(I_p - U)(S^{\frac{1}{2}})'\right) dU \, dS. \qquad (5.5.3)$$

Now using the integral (1.6.6) we can write

$$\int_{0<U<I_p} \det(U)^{\frac{1}{2}(n_1-p-1)+h} \det(I_p - U)^{\frac{1}{2}(n_2-p-1)} {}_0F_1\left(\frac{1}{2}n_2; \frac{1}{4}\Sigma^{-1}\Theta S^{\frac{1}{2}}(I_p - U)(S^{\frac{1}{2}})'\right) dU$$

$$= \frac{\Gamma_p(\frac{1}{2}n_1 + h)\Gamma_p(\frac{1}{2}n_2)}{\Gamma_p[\frac{1}{2}(n_1 + n_2) + h]} {}_0F_1\left(\frac{1}{2}(n_1 + n_2) + h; \frac{1}{4}\Sigma^{-1}\Theta S\right). \qquad (5.5.4)$$

Substituting (5.5.4) in (5.5.3), we get

$$E[\det(U)^h] = \left\{2^{\frac{1}{2}(n_1+n_2)p}\Gamma_p\left(\frac{1}{2}n_1\right)\Gamma_p\left[\frac{1}{2}(n_1+n_2)+h\right]\det(\Sigma)^{\frac{1}{2}(n_1+n_2)}\right\}^{-1}\Gamma_p\left(\frac{1}{2}n_1+h\right)$$

$$\text{etr}\left(-\frac{1}{2}\Theta\right) \int_{S>0} \det(S)^{\frac{1}{2}(n_1+n_2-p-1)} \text{etr}\left(-\frac{1}{2}\Sigma^{-1}S\right)$$

$$ {}_0F_1\left(\frac{1}{2}(n_1+n_2)+h; \frac{1}{4}\Sigma^{-1}\Theta S\right) dS. \qquad (5.5.5)$$

The integral in (5.5.5) can be evaluated by using (1.6.4).

Similarly one can derive $E[\det(I_p - U)^h]$. ∎

The distribution of U given in (5.5.1) is useful in studying power of the test statistic $\frac{\det(S_1)}{\det(S_1+S_2)}$ for testing certain hypothesis in multivariate statistical analysis, e.g., see Roy (1966), de Waal (1968), Pillai and Gupta (1969), Gupta (1971a, 1971b, 1971c), Das Gupta (1972), Nagarsenker (1979), and Gupta and Javier (1986).

Asoo (1969), following the univariate density, defined the noncentral matrix variate beta type I density as follows.

DEFINITION 5.5.1. *A symmetric positive definite random matrix U $(p \times p)$ is said to have noncentral matrix variate beta type I(B) distribution with parameters a, b and Θ, if its p.d.f. is given by*

$$\frac{\text{etr}(-\Theta)}{\beta_p(a,b)} \det(U)^{a-\frac{1}{2}(p+1)} \det(I_p - U)^{b-\frac{1}{2}(p+1)} {}_1F_1(a+b; a; \Theta U), \ 0 < U < I_p.$$

In Theorem 5.2.5, it was shown that $V = S_2^{-\frac{1}{2}} S_1 S_2^{-\frac{1}{2}} \sim B_p^{II}(\frac{1}{2}n_1, \frac{1}{2}n_2)$ where $S_i \sim W_p(n_i, I_p)$, $i = 1, 2$ are independent and $S_2^{\frac{1}{2}}$ is a symmetric square root of S_2. Here, we derive the p.d.f. of V when $S_2 \sim W_p(n_2, I_p, \Theta)$. This distribution of V is called the noncentral matrix variate beta type II(A), see de Waal (1969).

THEOREM 5.5.3. *Let $S_1 \sim W_p(n_1, I_p)$ and $S_2 \sim W_p(n_2, I_p, \Theta)$ be independently distributed. Then the p.d.f. of $V = S_2^{-\frac{1}{2}} S_1 S_2^{-\frac{1}{2}}$, where $S_2^{\frac{1}{2}}$ is a symmetric square root of S_2, is given by*

$$\left\{\beta_p\left(\frac{1}{2}n_1, \frac{1}{2}n_2\right)\right\}^{-1} \text{etr}\left(-\frac{1}{2}\Theta\right) \det(V)^{\frac{1}{2}(n_1-p-1)} \det(I_p + V)^{-\frac{1}{2}(n_1+n_2)}$$

$$ {}_1F_1\left(\frac{1}{2}(n_1 + n_2); \frac{1}{2}n_2; \frac{1}{2}\Theta(I_p + V)^{-1}\right), \ V > 0. \qquad (5.5.6)$$

Proof: The joint p.d.f. of S_1 and S_2 is

$$\left\{ 2^{\frac{1}{2}(n_1+n_2)p} \Gamma_p\left(\frac{1}{2}n_1\right)\Gamma_p\left(\frac{1}{2}n_2\right) \right\}^{-1} \operatorname{etr}\left(-\frac{1}{2}\Theta\right) \operatorname{etr}\left\{-\frac{1}{2}(S_1+S_2)\right\}$$

$$\det(S_1)^{\frac{1}{2}(n_1-p-1)} \det(S_2)^{\frac{1}{2}(n_2-p-1)} {}_0F_1\left(\frac{1}{2}n_2;\frac{1}{4}\Theta S_2\right).$$

Transforming $V = S_2^{-\frac{1}{2}} S_1 S_2^{-\frac{1}{2}}$, with the Jacobian $J(S_1 \to V) = \det(S_2)^{\frac{1}{2}(p+1)}$, the joint p.d.f. of V and S_2 is given by

$$\left\{ 2^{\frac{1}{2}(n_1+n_2)p} \Gamma_p\left(\frac{1}{2}n_1\right)\Gamma_p\left(\frac{1}{2}n_2\right) \right\}^{-1} \operatorname{etr}\left(-\frac{1}{2}\Theta\right) \det(V)^{\frac{1}{2}(n_1-p-1)} \operatorname{etr}\left\{-\frac{1}{2}(I_p+V)S_2\right\}$$

$$\det(S_2)^{\frac{1}{2}(n_1+n_2-p-1)} {}_0F_1\left(\frac{1}{2}n_2;\frac{1}{4}\Theta S_2\right), \; S_2 > 0, \; V > 0. \tag{5.5.7}$$

From (1.6.4), we have

$$\int_{S_2>0} \operatorname{etr}\left\{-\frac{1}{2}(I_p+V)S_2\right\} \det(S_2)^{\frac{1}{2}(n_1+n_2-p-1)} {}_0F_1\left(\frac{1}{2}n_2;\frac{1}{4}\Theta S_2\right) dS_2$$

$$= 2^{\frac{1}{2}(n_1+n_2)p} \Gamma_p\left[\frac{1}{2}(n_1+n_2)\right] \det(I_p+V)^{-\frac{1}{2}(n_1+n_2)}$$

$$ {}_1F_1\left(\frac{1}{2}(n_1+n_2);\frac{1}{2}n_2;\frac{1}{2}\Theta(I_p+V)^{-1}\right). \tag{5.5.8}$$

Now, integrating (5.5.7), using (5.5.8) we get the marginal p.d.f. of V as

$$\frac{\Gamma_p\left[\frac{1}{2}(n_1+n_2)\right]}{\Gamma_p\left(\frac{1}{2}n_1\right)\Gamma_p\left(\frac{1}{2}n_2\right)} \operatorname{etr}\left(-\frac{1}{2}\Theta\right) \det(V)^{\frac{1}{2}(n_1-p-1)} \det(I_p+V)^{-\frac{1}{2}(n_1+n_2)}$$

$$ {}_1F_1\left(\frac{1}{2}(n_1+n_2);\frac{1}{2}n_2;\frac{1}{2}\Theta(I_p+V)^{-1}\right), \; V > 0.$$

which completes the proof of the theorem. ∎

If we transform $U = (I_p+V)^{-1}$ with Jacobian $J(V \to U) = \det(U)^{-(p+1)}$ in the above theorem, then the p.d.f. of U is

$$\left\{ \beta_p\left(\frac{1}{2}n_1,\frac{1}{2}n_2\right) \right\}^{-1} \operatorname{etr}\left(-\frac{1}{2}\Theta\right) \det(U)^{\frac{1}{2}(n_2-p-1)} \det(I_p-U)^{\frac{1}{2}(n_1-p-1)}$$

$$ {}_1F_1\left(\frac{1}{2}(n_1+n_2);\frac{1}{2}n_2;\frac{1}{2}\Theta U\right), \; 0 < U < I_p.$$

which is the noncentral matrix variate beta type I(B) distribution defined by Asoo (1969).

In the special case when $\Theta = \operatorname{diag}(\theta_{11},0,\ldots,0)$, the density (5.5.6) simplifies to

$$\left\{ \beta_p\left(\frac{1}{2}n_1,\frac{1}{2}n_2\right) \right\}^{-1} \operatorname{etr}\left(-\frac{1}{2}\theta_{11}\right) \det(V)^{\frac{1}{2}(n_1-p-1)} \det(I_p+V)^{-\frac{1}{2}(n_1+n_2)}$$

$$ {}_1F_1\left(\frac{1}{2}(n_1+n_2);\frac{1}{2}n_2;\frac{1}{2}\theta_{11}T^{11}\right), \; V > 0.$$

where $(I_p + V)^{-1} = (\tau^{ij})$, and $_1F_1$ is the hypergeometric function of scalar argument (Rainville, 1970).

If $F = S_2^{\frac{1}{2}} S_1^{-1} S_2^{\frac{1}{2}}$, $S_1 \sim W_p(n_1, I_p)$ and $S_2 \sim W_p(n_2, I_p, \Theta)$ are independent, then $F = V^{-1}$ and from Theorem 5.5.3, its p.d.f. is

$$\left\{\beta_p\left(\frac{1}{2}n_1, \frac{1}{2}n_2\right)\right\}^{-1} \operatorname{etr}\left(-\frac{1}{2}\Theta\right) \det(F)^{\frac{1}{2}(n_2-p-1)} \det(I_p + F)^{-\frac{1}{2}(n_1+n_2)}$$

$$_1F_1\left(\frac{1}{2}(n_1 + n_2); \frac{1}{2}n_2; \frac{1}{2}\Theta F(I_p + F)^{-1}\right), \quad F > 0.$$

This distribution is known as noncentral matrix variate beta type II(B) and for $p = 1$, the distribution of F is the usual noncentral beta type II distribution.

Following the univariate density, Asoo (1969) has defined the noncentral matrix variate beta type II distribution as follows:

DEFINITION 5.5.2. *A symmetric positive definite random matrix V $(p \times p)$ is said to have noncentral matrix variate beta type II(B) distribution with parameters a, b and Θ, if it p.d.f. is given by*

$$\{\beta_p(a, b)\}^{-1} \det(V)^{a-\frac{1}{2}(p+1)} \det(I_p + V)^{-(a+b)} \operatorname{etr}(-\Theta)$$

$$_1F_1(a + b; a; \Theta V(I_p + V)^{-1}), \quad V > 0.$$

In the above density by transforming $U = (I_p + V)^{-1}V$, we get the noncentral matrix variate beta type I(B) distribution given in Definition 5.5.1.

THEOREM 5.5.4. *If V is distributed as noncentral matrix variate beta type II(A) with p.d.f. (5.5.6) then*

(i) $E[\det(V)^h] = \dfrac{\Gamma_p(\frac{1}{2}n_1 + h)\Gamma_p(\frac{1}{2}n_2 - h)}{\Gamma_p(\frac{1}{2}n_1)\Gamma_p(\frac{1}{2}n_2)} \operatorname{etr}\left(-\frac{1}{2}\Theta\right) {}_1F_1\left(\frac{1}{2}n_2 - h; \frac{1}{2}n_2; \frac{1}{2}\Theta\right),$

$$-\frac{1}{2}(n_1 - p + 1) < \operatorname{Re}(h) < \frac{1}{2}(n_2 - p + 1),$$

and

(ii) $E[\det(I_p + V)^{-h}] = \dfrac{\Gamma_p(\frac{1}{2}n_2 + h)\Gamma_p[\frac{1}{2}(n_1 + n_2)]}{\Gamma_p(\frac{1}{2}n_2)\Gamma_p[\frac{1}{2}(n_1 + n_2) + h]} \operatorname{etr}\left(-\frac{1}{2}\Theta\right)$

$$_2F_2\left(\frac{1}{2}(n_1 + n_2), \frac{1}{2}n_2 + h; \frac{1}{2}n_2, \frac{1}{2}(n_1 + n_2) + h; \frac{1}{2}\Theta\right),$$

$$\operatorname{Re}(h) > -\frac{1}{2}(n_2 - p + 1).$$

Proof: By definition,

$$E[\det(V)^h] = \left\{\beta_p\left(\frac{1}{2}n_1, \frac{1}{2}n_2\right)\right\}^{-1} \operatorname{etr}\left(-\frac{1}{2}\Theta\right) \int_{V>0} \det(V)^{\frac{1}{2}(n_1-p-1)+h}$$

$$\det(I_p + V)^{-\frac{1}{2}(n_1+n_2)} {}_1F_1\left(\frac{1}{2}(n_1 + n_2); \frac{1}{2}n_2; \frac{1}{2}\Theta(I_p + V)^{-1}\right) dV. \quad (5.5.9)$$

Substituting $U = (I_p + V)^{-1}$ with Jacobian $J(V \to U) = \det(U)^{-(p+1)}$, in (5.5.9), and using (1.6.6), we get

$$E[\det(V)^h] = \left\{ \beta_p\left(\frac{1}{2}n_1, \frac{1}{2}n_2\right) \right\}^{-1} \text{etr}\left(-\frac{1}{2}\Theta\right) \int_{0 < U < I_p} \det(U)^{\frac{1}{2}(n_2 - 2h - p - 1)}$$

$$\det(I_p - U)^{\frac{1}{2}(n_1 + 2h - p - 1)} \, {}_1F_1\left(\frac{1}{2}(n_1 + n_2); \frac{1}{2}n_2; \frac{1}{2}\Theta U\right) dU$$

$$= \left\{ \beta_p\left(\frac{1}{2}n_1, \frac{1}{2}n_2\right) \right\}^{-1} \frac{\Gamma_p(\frac{1}{2}n_1 + h)\Gamma_p(\frac{1}{2}n_2 - h)}{\Gamma_p[\frac{1}{2}(n_1 + n_2)]} \text{etr}\left(-\frac{1}{2}\Theta\right)$$

$$\, {}_2F_2\left(\frac{1}{2}n_2 - h, \frac{1}{2}(n_1 + n_2); \frac{1}{2}(n_1 + n_2), \frac{1}{2}n_2; \frac{1}{2}\Theta\right)$$

$$= \frac{\Gamma_p(\frac{1}{2}n_1 + h)\Gamma_p(\frac{1}{2}n_2 - h)}{\Gamma_p(\frac{1}{2}n_1)\Gamma_p(\frac{1}{2}n_2)} \text{etr}\left(-\frac{1}{2}\Theta\right) {}_1F_1\left(\frac{1}{2}n_2 - h; \frac{1}{2}n_2; \frac{1}{2}\Theta\right).$$

Similarly one can prove second part. ∎

PROBLEMS

5.1. Let $S|\Sigma \sim W_p(n, \Sigma)$ where $S\,(p \times p) > 0$ and $\Sigma\,(p \times p) > 0$. Assume that a priori $S \sim IW_p(m, \Psi)$. Prove that the marginal distribution of S is a generalized matrix variate beta type II distribution.

5.2. Let $X \sim B_p^{II}(a, b)$ and $Y \sim B_p^{I}(a + b, c)$. Prove that $(I_p + X)^{-\frac{1}{2}} Y (I_p + X)^{-\frac{1}{2}} \sim B_p^{I}(b, a + c)$.

5.3. Let $S \sim W_p(n_1, I_p)$ and $X \sim N_{p,n_2}(0, I_p \otimes I_{n_2})$, $n_2 < p$ be independently distributed. Then show that,

(i) $F = X'(S + XX')^{-1}X \sim B_{n_2}^{I}\left(\frac{1}{2}p, \frac{1}{2}(n_1 + n_2 - p)\right)$, and

(ii) $X'S^{-1}X \sim B_{n_2}^{II}\left(\frac{1}{2}p, \frac{1}{2}(n_1 + n_2 - p)\right)$.

5.4. Let $S \sim W_p(n, \Sigma)$ and $A\,(p \times r)$ be given matrix of rank $r\,(< \frac{1}{2}p)$. Define $W = (A'S^{-1}A)^{-1}$, $\Delta = (A'\Sigma^{-1}A)^{-1}$, and $B = \Delta^{\frac{1}{2}}W^{-1}(A'S^{-1}\Sigma S^{-1}A)^{-1}W^{-1}\Delta^{\frac{1}{2}}$. Then,
(i) W and B are independent,
(ii) $W \sim W_r(n - p + r, \Delta)$, and

(iii) $B \sim B_r^{I}\left(\frac{1}{2}(n - p + 2r), \frac{1}{2}(p - r)\right)$.

(Khatri and Rao, 1987)

5.5. Let $S \sim W_p(n_1 + n_2, \Sigma)$ and $U \sim B_p^{I}(\frac{1}{2}n_1, \frac{1}{2}n_2)$ be independently distributed. Show that $S_1 = S^{\frac{1}{2}}U(S^{\frac{1}{2}})'$ and $S_2 = S^{\frac{1}{2}}(I_p - U)(S^{\frac{1}{2}})'$ are independent, $S_1 \sim W_p(n_1, \Sigma)$ and $S_2 \sim W_p(n_2, \Sigma)$.

5.6.   Let $T \sim T_{p,m}(n, M, \Sigma, \Omega)$. Then prove that $\Sigma^{-1}(T - M)\Omega^{-1}(T - M)' \sim B_p^{II}(\frac{1}{2}m, \frac{1}{2}(n + p - 1))$.

5.7.   Prove Theorem 5.3.4.

5.8.   Prove Theorem 5.3.7(ii).

5.9.   Prove Theorem 5.3.11.

5.10. Derive the characteristic function of Y, where $Y \sim GB_p^{II}(a, b; \Omega, \Psi)$.

5.11. Let $X \sim GB_p^I(a, b; \Omega, 0)$ and partition X and Ω as

$$X = \begin{pmatrix} X_{11} & X_{12} \\ X_{21} & X_{22} \end{pmatrix} \begin{matrix} p_1 \\ p_2 \end{matrix}, \quad \Omega = \begin{pmatrix} \Omega_{11} & \Omega_{12} \\ \Omega_{21} & \Omega_{22} \end{pmatrix} \begin{matrix} p_1 \\ p_2 \end{matrix}, \quad p_1 + p_2 = p.$$
$$\quad\;\; p_1 \quad p_2 \qquad\qquad\quad p_1 \quad p_2$$

Then, prove that (i) X_{11} and $X_{22 \cdot 1}$ are independent, $X_{11} \sim GB_{p_1}^I(a, b; \Omega_{11}, 0)$, $X_{22 \cdot 1} \sim GB_{p_2}^I(a - \frac{1}{2}p_1, b; \Omega_{22 \cdot 1}, 0)$, and (ii) $X_{21} | X_{11}, X_{22 \cdot 1} \sim IT_{p_2, p_1}(2b - p + 1, 0, \Omega_{21}\Omega_{11}^{-1}X_{11}, \Omega_{22 \cdot 1} - X_{22 \cdot 1}, (I_{p_1} - \Omega_{11}^{-1}X_{11})X_{11})$, where $X_{22 \cdot 1} = X_{22} - X_{21}X_{11}^{-1}X_{12}$ and $\Omega_{22 \cdot 1} = \Omega_{22} - \Omega_{21}\Omega_{11}^{-1}\Omega_{12}$.

5.12. Let $Y \sim GB_p^{II}(a, b; \Sigma, 0)$ and partition Y and Σ as

$$Y = \begin{pmatrix} Y_{11} & Y_{12} \\ Y_{21} & Y_{22} \end{pmatrix} \begin{matrix} p_1 \\ p_2 \end{matrix}, \quad \Sigma = \begin{pmatrix} \Sigma_{11} & \Sigma_{12} \\ \Sigma_{21} & \Sigma_{22} \end{pmatrix} \begin{matrix} p_1 \\ p_2 \end{matrix}, \quad p_1 + p_2 = p.$$
$$\quad\;\; p_1 \quad p_2 \qquad\qquad\quad p_1 \quad p_2$$

Then, prove that (i) Y_{11} and $Y_{22 \cdot 1}$ are independent, $Y_{11} \sim GB_{p_1}^{II}(a, b - \frac{1}{2}p_2; \Sigma_{11}, 0)$, $Y_{22 \cdot 1} \sim GB_{p_2}^{II}(a - \frac{1}{2}p_1, b; \Sigma_{22 \cdot 1}, 0)$, and (ii) $Y_{21} | Y_{11}, Y_{22 \cdot 1} \sim T_{p_2, p_1}(2a + 2b - p + 1, \Sigma_{21}\Sigma_{11}^{-1}Y_{11}, \Sigma_{22 \cdot 1} + Y_{22 \cdot 1}, (I_{p_1} + \Sigma_{11}^{-1}Y_{11})Y_{11})$, where $Y_{22 \cdot 1} = Y_{22} - Y_{21}Y_{11}^{-1}Y_{12}$ and $\Sigma_{22 \cdot 1} = \Sigma_{22} - \Sigma_{21}\Sigma_{11}^{-1}\Sigma_{12}$.

5.13. Prove Theorem 5.3.13.

5.14. Prove Theorem 5.3.14(i).

5.15. Prove Theorem 5.3.15(ii).

5.16. Prove Theorem 5.3.16(ii).

5.17. Let $X \sim GB_p^I(a, b; \Omega, 0)$, then prove that

$$E[\det(X)^h] = \det(\Omega)^h E[\det(U)^h]$$

$$E[\det(\Omega - X)^h] = \det(\Omega)^h E[\det(I_p - U)^h]$$

$$E(C_\kappa(X)) = \frac{C_\kappa(\Omega)}{C_\kappa(I_p)} E(C_\kappa(U))$$

and

$$E(C_\kappa(X^{-1})) = \frac{C_\kappa(\Omega^{-1})}{C_\kappa(I_p)} E(C_\kappa(U^{-1})),$$

where $U \sim B_p^I(a, b)$.

5.18. Let $Y \sim GB_p^{II}(a, b; \Sigma, 0)$, then prove that

$$E[\det(Y)^h] = \det(\Sigma)^h E[\det(V)^h]$$

$$E(C_\kappa(Y)) = \frac{C_\kappa(\Sigma)}{C_\kappa(I_p)} E(C_\kappa(V))$$

and

$$E(C_\kappa(Y^{-1})) = \frac{C_\kappa(\Sigma^{-1})}{C_\kappa(I_p)} E(C_\kappa(V^{-1})),$$

where $V \sim B_p^{II}(a, b)$.

5.19. Let $U \sim B_p^I(a, b)$ and $U = TT'$, where $T = (t_{ij})$ is a lower triangular matrix with positive diagonal elements. Show that $t_{11}^2, \ldots, t_{pp}^2$ are independently distributed, $t_{ii}^2 \sim B^I(a - \frac{1}{2}(i-1), b)$, $i = 1, \ldots, p$.

5.20. Let $U \sim GB_p^I(a, b)$ and $U^{[\alpha]} = (u_{ij})$, $1 \le i, j \le \alpha$, then find

$$E\Big(\prod_{\alpha=1}^{p} \det(U^{[\alpha]})^{-h} \det(U)^h \Big).$$

5.21. Let $U \sim B_p^I(a, b)$. Then prove that $\frac{\det(U^{[r]})}{\det(U^{[r-1]})}$, $r = 1, \ldots, p$ are independently distributed as $B^I(a - \frac{1}{2}(r-1), b)$, $r = 1, ..., p$.

5.22. Let $V \sim B_p^{II}(a, b)$ and $V = TT'$, where $T = (t_{ij})$ is a lower triangular matrix with positive diagonal elements. Show that $t_{11}^2, \ldots, t_{pp}^2$ are independently distributed, $t_{ii}^2 \sim B^{II}(a - \frac{1}{2}(i-1), b - \frac{1}{2}(p-i))$, $i = 1, \ldots, p$.

5.23. Let $U \sim B_p^I(\frac{1}{2}n_1, \frac{1}{2}n_2)$. Then show that

(i) $E((\operatorname{tr} U)U) = \dfrac{n_1\{n_1 p(n+1) + 2(n_2 - p)\}}{n(n-1)(n+2)} I_p,$

(ii) $E((\operatorname{tr} U^{-1})U^{-1}) = \dfrac{(n-p-1)\{p(n-p)(n_1-p-3) + n_2(p+2)\}}{(n_1-p)(n_1-p-1)(n_1-p-3)} I_p,$

$$n_1 - p - 3 > 0.$$

where $n_1 + n_2 = n$.

5.24. Let $G = T'(I_p + V)^{-1}T = (g_{ij})$, where $V = TT' \sim B_p^{II}(\frac{1}{2}n_1, \frac{1}{2}n_2)$ and T is a lower triangular matrix with positive diagonal elements. Then show that $E(g_{ii}) = \frac{n_1 + p - 2i + 1}{n_1 + n_2}$.

(Bilodeau and Srivastava, 1992)

5.25. Prove Theorem 5.4.1(ii).

5.26. Prove Theorem 5.4.2(ii).

5.27. Prove Theorem 5.4.5(ii).

5.28. Let $S_i \sim W_p(n_i, \Sigma_i)$, $i = 1, 2$ be independently distributed. Derive the distribution of $V = S_2^{-\frac{1}{2}} S_1 S_2^{-\frac{1}{2}}$.

5.29. In Problem 5.28 derive $E[\det(V)^h]$.

5.30. Prove Theorem 5.5.2(ii).

5.31. Prove Theorem 5.5.4(ii).

5.32. Let $S_1 \sim W_p(n_1, \Sigma_1)$ and $S_2 \sim W_p(n_2, \Sigma_2, \Theta)$ be independently distributed. Prove that the p.d.f. of $V = S_2^{-\frac{1}{2}} S_1 S_2^{-\frac{1}{2}}$ is given by

$$\left\{ 2^{\frac{1}{2}(n_1+n_2)p} \Gamma_p\left(\frac{1}{2}n_1\right) \Gamma_p\left(\frac{1}{2}n_2\right) \det(\Sigma_1)^{\frac{1}{2}n_1} \det(\Sigma_2)^{\frac{1}{2}n_2} \right\}^{-1} \text{etr}\left(-\frac{1}{2}\Theta\right)$$

$$\det(V)^{\frac{1}{2}(n_1-p-1)} \int_{S_2>0} \det(S_2)^{\frac{1}{2}(n_1+n_2-p-1)} \text{etr}\left\{ -\frac{1}{2}(\Sigma_2^{-1}S_2 + S_2^{\frac{1}{2}}\Sigma_1^{-1}S_2^{\frac{1}{2}}V) \right\}$$

$$_0F_1\left(\frac{1}{2}n_2; \frac{1}{4}\Theta\Sigma_2^{-1}S_2\right) dS_2, \ V > 0.$$

(de Waal, 1969)

5.33. Let $S|\Sigma \sim W_p(n, \Sigma)$ and a priori $\Sigma \sim IW_p(m, \Psi, \Theta)$. Then, prove that the marginal density of S is

$$\left\{ \beta_p\left(\frac{1}{2}(m-p-1), \frac{1}{2}n\right) \right\}^{-1} \text{etr}\left(-\frac{1}{2}\Theta\right) \det(\Psi)^{\frac{1}{2}(m-p-1)}$$

$$\det(S)^{\frac{1}{2}(n-p-1)} \det(S+\Psi)^{-\frac{1}{2}(m+n-p-1)}$$

$$_1F_1\left(\frac{1}{2}(m+n-p-1); \frac{1}{2}(m-p-1); \frac{1}{2}\Theta\Psi(S+\Psi)^{-1}\right), \ S > 0$$

which is the generalized noncentral matrix variate beta type II(B) density with parameters $\frac{1}{2}n$, $\frac{1}{2}(m-p-1)$, Ψ and $\frac{1}{2}\Theta$, where $\frac{1}{2}\Theta$ is the noncentrality parameter.

5.34. Let $S \sim W_p(n, I_p)$ and partition S as $S = (S_{ij})$, $i, j = 1, 2$, S_{11} $(q \times q)$. Then prove that S_{11}, S_{22} and $R_{12} = S_{11}^{-\frac{1}{2}} S_{12} S_{11}^{-\frac{1}{2}}$ are independently distributed. Further, show that the p.d.f. of R_{12} is

$$\frac{\pi^{-\frac{1}{2}(p-q)q} \Gamma_q\left(\frac{1}{2}n\right)}{\Gamma_q\left[\frac{1}{2}(n-p+q)\right]} \det(I_q - R_{12}R_{12}')^{\frac{1}{2}(n-p-1)}.$$

5.35. In Problem 5.34, derive the distribution of $R = R_{12}R_{12}'$ for $p \geq 2q$.

5.36. Let $S \sim W_p(n, \Sigma)$ and partition S and Σ as

$$S = \begin{pmatrix} S_{11} & S_{12} \\ S_{21} & S_{22} \end{pmatrix} \begin{matrix} q \\ p-q \end{matrix}, \quad \Sigma = \begin{pmatrix} \Sigma_{11} & \Sigma_{12} \\ \Sigma_{21} & \Sigma_{22} \end{pmatrix} \begin{matrix} q \\ p-q \end{matrix}$$
$$\quad\quad q \quad p-q \quad\quad\quad\quad q \quad p-q$$

where $p \geq 2q$. Further, let $\Sigma_{12} = 0$. Prove that (i) matrices $S_{11\cdot 2} = S_{11} - S_{12}S_{22}^{-1}S_{21}$, S_{22} and $S_{12}S_{22}^{-1}S_{21}$ are independently distributed, (ii) $S_{12}S_{22}^{-1}S_{21} \sim W_q(p - q, \Sigma_{11})$, and (iii) $\frac{\det(S)}{\det(S_{11})\det(S_{22})}$ is distributed as the product of independent beta variables.

5.37. Let $S \sim W_{p+q}(n, \Sigma)$ and partition

$$S = \begin{pmatrix} S_{11} & S_{12} \\ S_{21} & S_{22} \end{pmatrix} \begin{matrix} p \\ q \end{matrix}, \quad \Sigma = \begin{pmatrix} \Sigma_{11} & \Sigma_{12} \\ \Sigma_{21} & \Sigma_{22} \end{pmatrix} \begin{matrix} p \\ q \end{matrix}.$$
$$\qquad\quad p \qquad q \qquad\qquad\quad p \qquad q$$

Define $G = S_{12}S_{22}^{-1}S_{21}$. Then, show that the p.d.f. of G, for $p \leq q$, is given by

$$\left\{ 2^{\frac{1}{2}pq}\Gamma_p\left(\frac{1}{2}q\right)\det(\Sigma_{11\cdot 2})^{\frac{1}{2}q} \right\}^{-1} \det(I_p - P)^{\frac{1}{2}n}\,\mathrm{etr}\left(-\frac{1}{2}\Sigma_{11\cdot 2}^{-1}G\right)$$

$$\det(G)^{\frac{1}{2}(q-p-1)} {}_1F_1\left(\frac{1}{2}n; \frac{1}{2}q; \frac{1}{2}P^*G\right), \ G > 0,$$

where $P = \Sigma_{11}^{-\frac{1}{2}}\Sigma_{12}\Sigma_{22}^{-1}\Sigma_{21}\Sigma_{11}^{-\frac{1}{2}}$ and $P^* = \Delta_{12}\Delta_{22}^{-1}\Delta_{21}$, with $\Delta = \Sigma^{-1}$.

5.38. Let $S_1 \sim W_p(n_1, \Sigma)$ and $S_2 \sim W_p(n_2, \Sigma)$ be independently distributed. Prove that $\frac{\det(S_1)}{\det(S_1+S_2)}$ and $S_1 + S_2$ are independent and hence, deduce that if $S_i \sim W_p(n_i, \Sigma)$, $i = 1, 2, \ldots, ..$ are all independent, then

$$\frac{\det(S_1)}{\det(S_1 + S_2)}, \frac{\det(S_1 + S_2)}{\det(S_1 + S_2 + S_3)}, \ldots, ..$$

are all independent.

5.39. Let S and Σ be defined as in Problem 5.37. Define $R = S_{11}^{-\frac{1}{2}}S_{12}S_{22}^{-1}S_{21}S_{11}^{-\frac{1}{2}}$. Then, show that
(i) for $p = 1$, the p.d.f. of R is given by

$$\frac{\Gamma(\frac{1}{2}n)(1-\rho^2)^{\frac{1}{2}n}}{\Gamma(\frac{1}{2}q)\Gamma[\frac{1}{2}(n-q)]} R^{\frac{1}{2}q-1}(1-R)^{\frac{1}{2}(n-q)-1} {}_2F_1\left(\frac{1}{2}n, \frac{1}{2}n; \frac{1}{2}q; \rho^2 R\right),$$

where $P = \rho^2$,

(Mathai, 1981)

(ii) for arbitrary p, the p.d.f. of R is

$$\left\{ \Gamma_p\left(\frac{1}{2}q\right)\Gamma_p\left[\frac{1}{2}(n-q)\right] \right\}^{-1} \det(2\Sigma_{11\cdot 2})^{-\frac{1}{2}n}\det(I_p - P)^{\frac{1}{2}n}$$

$$\det(I_p - R)^{\frac{1}{2}(n-p-q-1)}\det(R)^{\frac{1}{2}(q-p-1)} \int_{S>0} \det(S)^{\frac{1}{2}(n-p-1)}$$

$$\mathrm{etr}\left(-\frac{1}{2}\Sigma_{11\cdot 2}^{-1}S\right) {}_1F_1\left(\frac{1}{2}n; \frac{1}{2}q; \frac{1}{2}P^*S^{\frac{1}{2}}RS^{\frac{1}{2}}\right) dS, \ 0 < R < I_p.$$

(Troskie, 1969)

5.40. Let R be distributed as in Problem 5.39(ii), then prove that

$$E[\det(I_p - R)^h] = \frac{\Gamma_p(\frac{1}{2}n)\Gamma_p[\frac{1}{2}(n-q)+h]}{\Gamma_p(\frac{1}{2}n+h)\Gamma_p[\frac{1}{2}(n-q)]}\det(I_p - P)^{\frac{1}{2}n}$$

$$ {}_2F_1\left(\frac{1}{2}n, \frac{1}{2}n; \frac{1}{2}n + h; P\right),\ \mathrm{Re}(h) > -\frac{1}{2}(n-p-q+1).$$

CHAPTER 6

MATRIX VARIATE DIRICHLET DISTRIBUTIONS

6.1. INTRODUCTION

The random variables $u_1, \ldots, u_r$ are said to have Dirichlet type I distribution with parameters $a_1, \ldots, a_{r+1}$, if their joint p.d.f. is given by

$$\frac{\Gamma(\sum_{i=1}^{r+1} a_i)}{\prod_{i=1}^{r+1} \Gamma(a_i)} \prod_{i=1}^{r} u_i^{a_i-1} \left(1 - \sum_{i=1}^{r} u_i\right)^{a_{r+1}-1}, 0 < u_i < 1, \sum_{i=1}^{r} u_i < 1, \qquad (6.1.1)$$

where $a_i > 0$, $i = 1, \ldots, r+1$. The random variables $v_1, \ldots, v_r$ are said to follow Dirichlet type II distribution with parameters $b_1, \ldots, b_{r+1}$, if their joint p.d.f. is given by

$$\frac{\Gamma(\sum_{i=1}^{r+1} b_i)}{\prod_{i=1}^{r+1} \Gamma(b_i)} \prod_{i=1}^{r} v_i^{b_i-1} \left(1 + \sum_{i=1}^{r} v_i\right)^{-\sum_{i=1}^{r+1} b_i}, v_i > 0, \qquad (6.1.2)$$

where $b_i > 0$, $i = 1, \ldots, r+1$. Letting $v_i = u_i \left(1 - \sum_{i=1}^{r} u_i\right)^{-1}$, $i = 1, \ldots, r$, we can obtain the p.d.f. (6.1.2) from (6.1.1) with parameters $a_1, \ldots, a_{r+1}$. For this reason, (6.1.2) is also known as inverted Dirichlet distribution (Tiao and Guttman, 1965).

In this chapter we study matrix variate generalizations of (6.1.1) and (6.1.2).

6.2. DENSITY FUNCTIONS

The matrix variate Dirichlet type I and II distributions are defined as follows.

DEFINITION 6.2.1. *The $p \times p$ symmetric positive definite random matrices $U_1, \ldots, U_r$ are said to have the matrix variate Dirichlet type I distribution with parameters $a_1, \ldots, a_{r+1}$, denoted by $(U_1, \ldots, U_r) \sim D_p^I(a_1, \ldots, a_r; a_{r+1})$, if their joint p.d.f. is given by*

$$\{\beta_p(a_1, \ldots, a_r; a_{r+1})\}^{-1} \prod_{i=1}^{r} \det(U_i)^{a_i - \frac{1}{2}(p+1)} \det\left(I_p - \sum_{i=1}^{r} U_i\right)^{a_{r+1} - \frac{1}{2}(p+1)},$$

$$0 < U_i < I_p, 0 < \sum_{i=1}^{r} U_i < I_p, \quad (6.2.1)$$

where $a_i > \frac{1}{2}(p-1)$, $i = 1, \ldots, r+1$, and

$$\beta_p(a_1, \ldots, a_r; a_{r+1}) = \frac{\prod_{i=1}^{r+1} \Gamma_p(a_i)}{\Gamma_p(\sum_{i=1}^{r+1} a_i)}. \tag{6.2.2}$$

DEFINITION 6.2.2. *The $p \times p$ symmetric positive definite random matrices $V_1, \ldots, V_r$ are said to have matrix variate Dirichlet type II distribution with parameters $b_1, \ldots, b_{r+1}$, denoted by $(V_1, \ldots, V_r) \sim D_p^{II}(b_1, \ldots, b_r; b_{r+1})$, if their joint p.d.f. is given by*

$$\{\beta_p(b_1, \ldots, b_r; b_{r+1})\}^{-1} \prod_{i=1}^{r} \det(V_i)^{b_i - \frac{1}{2}(p+1)} \det\left(I_p + \sum_{i=1}^{r} V_i\right)^{-\sum_{i=1}^{r+1} b_i}, \ V_i > 0, \tag{6.2.3}$$

where $b_i > \frac{1}{2}(p-1)$, $i = 1, \ldots, r+1$, and $\beta_p(b_1, \ldots, b_r; b_{r+1})$ is defined in (6.2.2).

For $r = 1$, (6.2.1) reduces to the p.d.f. of matrix variate beta type I and (6.2.3) to the p.d.f. of matrix variate beta type II given in Chapter 5. For $p = 1$, (6.2.1) and (6.2.3) reduce to (6.1.1) and (6.1.2) respectively. The matrix variate Dirichlet distributions are special cases of the matrix variate Liouville distributions discussed in Chapter 9.

In univariate distribution theory, if x_i, $i = 1, \ldots, r$ and y are independent random variables having chi-square distributions with n_i, $i = 1, \ldots, r$ and m degrees of freedom respectively, then the joint p.d.f. of $u_1, \ldots, u_r$, where

$$u_i = \frac{x_i}{\sum_{j=1}^{r} x_j + y}, \ i = 1, \ldots, r, \tag{6.2.4}$$

is Dirichlet type I with parameters $\frac{1}{2}n_1, \ldots, \frac{1}{2}n_r, \frac{1}{2}m$. Further, if

$$v_i = \frac{x_i}{y}, \ i = 1, \ldots, r, \tag{6.2.5}$$

then the joint density of $v_1, \ldots, v_r$ is Dirichlet type II with parameters $\frac{1}{2}n_1, \ldots, \frac{1}{2}n_r, \frac{1}{2}m$. In the matrix variate case Wishart distribution plays the role of chi-square distribution. Let $S_i \sim W_p(n_i, \Sigma)$, $i = 1, \ldots, r$ and $B \sim W_p(m, \Sigma)$ be independent random matrices. Then, natural generalizations of the ratios (6.2.4) and (6.2.5) are

$$U_i = \left(\sum_{j=1}^{r} S_j + B\right)^{-\frac{1}{2}} S_i \left(\sum_{j=1}^{r} S_j + B\right)^{-\frac{1}{2}}, \ i = 1, \ldots, r \tag{6.2.6}$$

and

$$V_i = B^{-\frac{1}{2}} S_i B^{-\frac{1}{2}}, \ i = 1, \ldots, r, \tag{6.2.7}$$

where $A^{\frac{1}{2}}$ denotes a square root of the matrix A.

The distributions of U_i (V_i) depend on the definition of the root in (6.2.6) ((6.2.7)). Also, certain independence properties depend on the choice of the root, *e.g.*, see Problems 3.11–3.14. Here we derive densities (6.2.1) and (6.2.3) by suitably choosing the root in (6.2.6) and (6.2.7). The densities of $U_i(V_i)$ for other choices of the roots which do not yield (6.2.1) and (6.2.3) are derived in Section 4.

THEOREM 6.2.1. *Let $S_i \sim W_p(n_i, \Sigma)$, $i = 1, \ldots, r$ and $B \sim W_p(m, \Sigma)$ be independently distributed. Define*

$$U_i = S^{-\frac{1}{2}} S_i (S^{-\frac{1}{2}})', \ i = 1, \ldots, r, \qquad (6.2.8)$$

where $S = \sum_{i=1}^r S_i + B$ and $S^{\frac{1}{2}}(S^{\frac{1}{2}})'$ is any reasonable factorization of S. Then $(U_1, \ldots, U_r) \sim D_p^I(\frac{1}{2}n_1, \ldots, \frac{1}{2}n_r; \frac{1}{2}m)$.

Proof: The joint density of $S_1, \ldots, S_r$ and B is given by

$$\prod_{i=1}^r \left[\left\{ 2^{\frac{1}{2}n_i p} \Gamma_p\left(\frac{1}{2}n_i\right) \det(\Sigma)^{\frac{1}{2}n_i} \right\}^{-1} \text{etr}\left(-\frac{1}{2}\Sigma^{-1} S_i\right) \det(S_i)^{\frac{1}{2}(n_i - p - 1)} \right]$$

$$\left\{ 2^{\frac{1}{2}mp} \Gamma_p\left(\frac{1}{2}m\right) \det(\Sigma)^{\frac{1}{2}m} \right\}^{-1} \text{etr}\left(-\frac{1}{2}\Sigma^{-1} B\right) \det(B)^{\frac{1}{2}(m - p - 1)}. \qquad (6.2.9)$$

Making the transformation $\sum_{i=1}^r S_i + B = S$, $S_i = S^{\frac{1}{2}} U_i (S^{\frac{1}{2}})'$, $i = 1, \ldots, r$ with Jacobian $J(S_1, \ldots, S_r, B \to U_1, \ldots, U_r, S) = \det(S)^{\frac{1}{2}r(p+1)}$ in (6.2.9), the joint density of $U_1, \ldots, U_r$ and S is

$$\frac{2^{-\frac{1}{2}(m+n)p} \det(\Sigma)^{-\frac{1}{2}(m+n)}}{\Gamma_p(\frac{1}{2}m) \prod_{i=1}^r \Gamma_p(\frac{1}{2}n_i)} \prod_{i=1}^r \det(U_i)^{\frac{1}{2}(n_i - p - 1)} \det\left(I_p - \sum_{i=1}^r U_i\right)^{\frac{1}{2}(m - p - 1)}$$

$$\det(S)^{\frac{1}{2}(m+n-p-1)} \text{etr}\left(-\frac{1}{2}\Sigma^{-1} S\right), \qquad (6.2.10)$$

where $n = \sum_{i=1}^r n_i$. From (6.2.10), it is easily seen that $(U_1, \ldots, U_r)$ and S are independent and the density of $(U_1, \ldots, U_r)$ is given by

$$\frac{\Gamma_p[\frac{1}{2}(m + n)]}{\Gamma_p(\frac{1}{2}m) \prod_{i=1}^r \Gamma_p(\frac{1}{2}n_i)} \prod_{i=1}^r \det(U_i)^{\frac{1}{2}(n_i - p - 1)} \det\left(I_p - \sum_{i=1}^r U_i\right)^{\frac{1}{2}(m - p - 1)}. \ \blacksquare$$

For r = 1, the above theorem gives the matrix variate beta type I distribution discussed in Chapter 5.

THEOREM 6.2.2. *Let $S_i \sim W_p(n_i, I_p)$, $i = 1, \ldots, r$, and $B \sim W_p(m, I_p)$ be independently distributed. Define*

$$V_i = B^{-\frac{1}{2}} S_i B^{-\frac{1}{2}}, \ i = 1, \ldots, r, \qquad (6.2.11)$$

where $B^{\frac{1}{2}} B^{\frac{1}{2}} = B$. Then $(V_1, \ldots, V_r) \sim D_p^{II}(\frac{1}{2}n_1, \ldots, \frac{1}{2}n_r; \frac{1}{2}m)$.

Proof: The joint density of $S_1, \ldots, S_r$ and B is given by (6.2.9) with $\Sigma = I_p$. Making the transformation $S_i = B^{\frac{1}{2}} V_i B^{\frac{1}{2}}$, $i = 1, \ldots, r$ with $J(S_1, \ldots, S_r \to V_1, \ldots, V_r) = \det(B)^{\frac{1}{2}r(p+1)}$, we obtain the joint density of $V_1, \ldots, V_r$ and B as

$$\left\{ 2^{\frac{1}{2}(m+n)p} \Gamma_p\left(\frac{1}{2}m\right) \prod_{i=1}^r \Gamma_p\left(\frac{1}{2}n_i\right) \right\}^{-1} \prod_{i=1}^r \det(V_i)^{\frac{1}{2}(n_i - p - 1)}$$

$$\text{etr}\left\{ -\frac{1}{2}\left(I_p + \sum_{i=1}^r V_i\right) B \right\} \det(B)^{\frac{1}{2}(m+n-p-1)}, \qquad (6.2.12)$$

where $n = \sum_{i=1}^{r} n_i$. Integrating out B using

$$\int_{B>0} \text{etr}\left\{-\frac{1}{2}\Big(I_p + \sum_{i=1}^{r} V_i\Big)B\right\} \det(B)^{\frac{1}{2}(m+n-p-1)} \, dB$$

$$= 2^{\frac{1}{2}p(m+n)}\Gamma_p\Big[\frac{1}{2}(m+n)\Big] \det\Big(I_p + \sum_{i=1}^{r} V_i\Big)^{-\frac{1}{2}(m+n)},$$

we get $(V_1, \ldots, V_r) \sim D_p^{II}(\frac{1}{2}n_1, \ldots, \frac{1}{2}n_r; \frac{1}{2}m)$. ∎

For $r = 1$, the above theorem gives the matrix variate beta type II distribution. The p.d.f.'s (6.2.1) and (6.2.3) are called the standard matrix variate Dirichlet type I and II distributions. Next we derive what are known as generalized matrix variate Dirichlet type I and II distributions.

THEOREM 6.2.3. *Let* $(U_1, \ldots, U_r) \sim D_p^I(a_1, \ldots, a_r; a_{r+1})$ *and* $\Psi_1, \ldots, \Psi_r, \Omega$ *be symmetric matrices such that* $\Omega > 0$ *and* $\Omega - \sum_{i=1}^{r} \Psi_i > 0$. *Define*

$$Z_i = \Big(\Omega - \sum_{i=1}^{r} \Psi_i\Big)^{\frac{1}{2}} U_i \Big(\Omega - \sum_{i=1}^{r} \Psi_i\Big)^{\frac{1}{2}} + \Psi_i, \ i = 1, \ldots, r. \qquad (6.2.13)$$

Then $(Z_1, \ldots, Z_r)$ *have the generalized matrix variate Dirichlet type I distribution with p.d.f.*

$$\frac{\prod_{i=1}^{r} \det(Z_i - \Psi_i)^{a_i - \frac{1}{2}(p+1)} \det(\Omega - \sum_{i=1}^{r} Z_i)^{a_{r+1} - \frac{1}{2}(p+1)}}{\beta_p(a_1, \ldots, a_r; a_{r+1}) \det(\Omega - \sum_{i=1}^{r} \Psi_i)^{\sum_{i=1}^{r+1} a_i - \frac{1}{2}(p+1)}},$$

$$\Psi_i < Z_i < \Omega, \ i = 1, \ldots, r, \ \sum_{i=1}^{r} Z_i < \Omega. \qquad (6.2.14)$$

Proof: Making the transformation

$$U_i = \Big(\Omega - \sum_{i=1}^{r} \Psi_i\Big)^{-\frac{1}{2}} (Z_i - \Psi_i)\Big(\Omega - \sum_{i=1}^{r} \Psi_i\Big)^{-\frac{1}{2}}, \ i = 1, \ldots, r,$$

with Jacobian $J(U_1, \ldots, U_r \to Z_1, \ldots, Z_r) = \det(\Omega - \sum_{i=1}^{r} \Psi_i)^{-\frac{1}{2}r(p+1)}$ in (6.2.1), we get (6.2.14). ∎

If $(Z_1, \ldots, Z_r)$ has p.d.f. (6.2.14), then we will write $(Z_1, \ldots, Z_r) \sim GD_p^I(a_1, \ldots, a_r; a_{r+1}; \Omega; \Psi_1, \ldots, \Psi_r)$. Note that $GD_p^I(a_1, \ldots, a_r; a_{r+1}; I_p; 0, \ldots, 0) \equiv D_p^I(a_1, \ldots, a_r; a_{r+1})$.

THEOREM 6.2.4. *Let* $(V_1, \ldots, V_r) \sim D_p^{II}(b_1, \ldots, b_r; b_{r+1})$ *and* $\Psi_1, \ldots, \Psi_r, \Omega$ *be symmetric matrices such that* $\Omega > 0$ *and* $\Omega + \sum_{i=1}^{r} \Psi_i > 0$. *Define*

$$Y_i = \Big(\Omega + \sum_{i=1}^{r} \Psi_i\Big)^{\frac{1}{2}} V_i \Big(\Omega + \sum_{i=1}^{r} \Psi_i\Big)^{\frac{1}{2}}, \ i = 1, \ldots, r.$$

Then, $(Y_1, \ldots, Y_r)$ *have the generalized matrix variate Dirichlet type II distribution with p.d.f.*

$$\frac{\det(\Omega + \sum_{i=1}^{r} \Psi_i)^{b_{r+1}}}{\beta_p(b_1, \ldots, b_r; b_{r+1})} \prod_{i=1}^{r} \det(Y_i - \Psi_i)^{b_i - \frac{1}{2}(p+1)} \det\left(\Omega + \sum_{i=1}^{r} Y_i\right)^{-\sum_{i=1}^{r+1} b_i},$$

$$Y_i > \Psi_i, \ i = 1, \ldots, r, \quad (6.2.15)$$

Proof: Making the transformation $V_i = (\Omega + \sum_{i=1}^{r} \Psi_i)^{-\frac{1}{2}}(Y_i - \Psi_i)(\Omega + \sum_{i=1}^{r} \Psi_i)^{-\frac{1}{2}}$, $i = 1, \ldots, r$, with the Jacobian $J(V_1, \ldots, V_r \to Y_1, \ldots, Y_r) = \det(\Omega + \sum_{i=1}^{r} \Psi_i)^{-\frac{1}{2}r(p+1)}$ in (6.2.3), we get (6.2.15). ∎

If $(Y_1, \ldots, Y_r)$ has p.d.f. (6.2.15), then we will write $(Y_1, \ldots, Y_r) \sim GD_p^{II}(b_1, \ldots, b_r; b_{r+1}; \Omega; \Psi_1, \ldots, \Psi_r)$. In this case $GD_p^{II}(b_1, \ldots, b_r; b_{r+1}; I_p; 0, \ldots, 0) \equiv D_p^{II}(b_1, \ldots, b_r; b_{r+1})$.

Next we define and derive the inverse Dirichlet distribution. The inverse Dirichlet distribution can be obtained from the matrix variate Dirichlet Type I distribution by means of an inverse transformation.

THEOREM 6.2.5. *Let* $(U_1, \ldots, U_r) \sim D_p^{I}(a_1, \ldots, a_r; a_{r+1})$. *Define* $X_i = U_i^{-1}$, $i = 1, \ldots, r$. *Then the joint p.d.f. of* $X_1, \ldots, X_r$ *is given by*

$$\{\beta_p(a_1, \ldots, a_r; a_{r+1})\}^{-1} \prod_{i=1}^{r} \det(X_i)^{-a_i - \frac{1}{2}(p+1)} \det\left(I_p - \sum_{i=1}^{r} X_i^{-1}\right)^{a_{r+1} - \frac{1}{2}(p+1)},$$

$$X_i > I_p, \ i = 1, \ldots, r, \ 0 < \sum_{i=1}^{r} X_i^{-1} < I_p, \quad (6.2.16)$$

where $a_i > \frac{1}{2}(p-1)$, $i = 1, \ldots, r+1$ *and* $\beta_p(a_1, \ldots, a_r; a_{r+1})$ *is defined in (6.2.2).*

Proof: The transformation $X_i = U_i^{-1}$, $i = 1, \ldots, r$, with Jacobian $J(U_1, \ldots, U_r \to X_1, \ldots, X_r) = \prod_{i=1}^{r} \det(X_i)^{-p-1}$, in the p.d.f. of $(U_1, \ldots, U_r)$ yields the joint p.d.f. of $X_1, \ldots, X_r$ as given above. ∎

The distribution of $(X_1, \ldots, X_r)$ given in the above theorem is called *inverse Dirichlet distribution*, Xu (1987). If $(X_1, \ldots, X_r)$ has p.d.f. (6.2.16), then we will write $(X_1, \ldots, X_r) \sim ID_p(a_1, \ldots, a_r; a_{r+1})$. For $r = 1$, this distribution reduces to the inverse beta distribution given in Theorem 5.3.5. Like the matrix variate Dirichlet distribution this distribution can also be derived using Wishart matrices.

THEOREM 6.2.6. *Let* $W_i \sim IW_p(n_i + p + 1, \Psi)$, $i = 1, \ldots, r$ *and* $V \sim IW_p(m + p + 1, \Psi)$ *be independently distributed. Define*

$$X_i = W^{\frac{1}{2}} W_i W^{\frac{1}{2}}, \ i = 1, \ldots, r \quad (6.2.17)$$

where $W = \sum_{i=1}^{r} W_i^{-1} + V^{-1}$ *and* $W^{\frac{1}{2}}(W^{\frac{1}{2}})'$ *is any reasonable factorization of* W. *Then* $(X_1, \ldots, X_r) \sim ID_p(\frac{1}{2}n_1, \ldots, \frac{1}{2}n_r; \frac{1}{2}m)$.

Proof: From Theorem 3.4.1, $S_i = W_i^{-1} \sim W_p(n_i, \Psi^{-1})$, $i = 1, \ldots, r$ and $B = V^{-1} \sim W_p(m, \Psi^{-1})$. Now from Theorem 6.2.1, the joint density of $X_i^{-1} = (W^{\frac{1}{2}} W_i W^{\frac{1}{2}})^{-1} = S^{-\frac{1}{2}} S_i S^{-\frac{1}{2}}$, $i = 1, \ldots, r$, where $S = \sum_{i=1}^{r} S_i + B$ is $D_p^{I}(\frac{1}{2}n_1, \ldots, \frac{1}{2}n_r; \frac{1}{2}m)$. Finally, using Theorem 6.2.5, the result follows. ∎

6.3.   PROPERTIES

In this section, we will study certain properties of matrix variate Dirichlet type I and II distributions. It may be noted that densities (6.2.1) and (6.2.3) are orthogonally invariant, that is, for any fixed orthogonal matrix Γ $(p \times p)$, the distribution of $(\Gamma U_1 \Gamma', \Gamma U_2 \Gamma', \ldots, \Gamma U_r \Gamma')$ is the same as the distribution of $(U_1, \ldots, U_r)$, and similarly the distribution of $(\Gamma V_1 \Gamma', \Gamma V_2 \Gamma', \ldots, \Gamma V_r \Gamma')$ is the same as that of $(V_1, \ldots, V_r)$.

THEOREM 6.3.1. (i) If $(U_1, \ldots, U_r) \sim D_p^I(a_1, \ldots, a_r; a_{r+1})$ and

$$V_i = \left(I_p - \sum_{i=1}^r U_i\right)^{-\frac{1}{2}} U_i \left(I_p - \sum_{i=1}^r U_i\right)^{-\frac{1}{2}}, \ i = 1, \ldots, r, \quad (6.3.1)$$

then $(V_1, \ldots, V_r) \sim D_p^{II}(a_1, \ldots, a_r; a_{r+1})$.
  (ii) If $(V_1, \ldots, V_r) \sim D_p^{II}(b_1, \ldots, b_r; b_{r+1})$ and

$$U_i = \left(I_p + \sum_{i=1}^r V_i\right)^{-\frac{1}{2}} V_i \left(I_p + \sum_{i=1}^r V_i\right)^{-\frac{1}{2}}, \ i = 1, \ldots, r, \quad (6.3.2)$$

then $(U_1, \ldots, U_r) \sim D_p^I(b_1, \ldots, b_r; b_{r+1})$.

Proof: (i) Let $Z = I_p - \sum_{i=1}^r U_i$ and $V_i = Z^{-\frac{1}{2}} U_i Z^{-\frac{1}{2}}, \ i = 1, \ldots, r-1$, then $V_r = Z^{-1} - (I_p + \sum_{i=1}^{r-1} V_i)$. The Jacobian of transformation (6.3.1) is given by

$$J(U_1, \ldots, U_r \to V_1, \ldots, V_r)$$

$$= J(U_1, \ldots, U_r \to V_1, \ldots, V_{r-1}, Z) J(V_1, \ldots, V_{r-1}, Z \to V_1, \ldots, V_r)$$

$$= \det(Z)^{\frac{1}{2}(r-1)(p+1)} \det(Z)^{p+1}$$

$$= \det\left(I_p - \sum_{i=1}^r U_i\right)^{\frac{1}{2}(r+1)(p+1)}$$

$$= \det\left(I_p + \sum_{i=1}^r V_i\right)^{-\frac{1}{2}(r+1)(p+1)}.$$

Now, making the transformation and substituting for the Jacobian in the joint density of $U_1, \ldots, U_r$ given in (6.2.1), we get the desired result.
  (ii) The proof of the second part follows similarly. ∎

THEOREM 6.3.2. If $(U_1, \ldots, U_r) \sim D_p^I(a_1, \ldots, a_r; a_{r+1})$, then $(U_1, \ldots, U_s) \sim D_p^I(a_1, \ldots, a_s; \sum_{i=s+1}^{r+1} a_i)$, $s \leq r$, and the density of $(U_{s+1}, \ldots, U_r)|(U_1, \ldots, U_s)$ is given by

$$\frac{\prod_{i=s+1}^r \det(U_i)^{a_i - \frac{1}{2}(p+1)} \det(I_p - \sum_{i=1}^s U_i - \sum_{i=s+1}^r U_i)^{a_{r+1} - \frac{1}{2}(p+1)}}{\beta_p(a_{s+1}, \ldots, a_r; a_{r+1}) \det(I_p - \sum_{i=1}^s U_i)^{\sum_{i=s+1}^{r+1} a_i - \frac{1}{2}(p+1)}},$$

$$0 < U_i < I_p - \sum_{i=1}^s U_i, \ i = s+1, \ldots, r, \quad \sum_{i=s+1}^r U_i < I_p - \sum_{i=1}^s U_i.$$

Proof: First we find the marginal density of $U_1, \ldots, U_{r-1}$ by integrating out U_r from the joint density of $U_1, \ldots, U_r$ as

$$\{\beta_p(a_1, \ldots, a_r; a_{r+1})\}^{-1} \int_{0 < U_r < I_p - \sum_{i=1}^{r-1} U_i} \prod_{i=1}^{r} \det(U_i)^{a_i - \frac{1}{2}(p+1)}$$

$$\det\left(I_p - \sum_{i=1}^{r} U_i\right)^{a_{r+1} - \frac{1}{2}(p+1)} dU_r. \tag{6.3.3}$$

Now, substituting $Z_r = (I_p - \sum_{i=1}^{r-1} U_i)^{-\frac{1}{2}} U_r (I_p - \sum_{i=1}^{r-1} U_i)^{-\frac{1}{2}}$ with Jacobian $J(U_r \rightarrow Z_r) = \det(I_p - \sum_{i=1}^{r-1} U_i)^{\frac{1}{2}(p+1)}$ in (6.3.3), we get

$$\{\beta_p(a_1, \ldots, a_r; a_{r+1})\}^{-1} \prod_{i=1}^{r-1} \det(U_i)^{a_i - \frac{1}{2}(p+1)} \det\left(I_p - \sum_{i=1}^{r-1} U_i\right)^{a_r + a_{r+1} - \frac{1}{2}(p+1)}$$

$$\int_{0 < Z_r < I_p} \det(Z_r)^{a_r - \frac{1}{2}(p+1)} \det(I_p - Z_r)^{a_{r+1} - \frac{1}{2}(p+1)} dZ_r.$$

But

$$\{\beta_p(a_1, \ldots, a_r; a_{r+1})\}^{-1} \int_{0 < Z_r < I_p} \det(Z_r)^{a_r - \frac{1}{2}(p+1)} \det(I_p - Z_r)^{a_{r+1} - \frac{1}{2}(p+1)} dZ_r$$

$$= \{\beta_p(a_1, \ldots, a_{r-1}; a_r + a_{r+1})\}^{-1}.$$

Hence, we get $(U_1, \ldots, U_{r-1}) \sim D_p^I(a_1, \ldots, a_{r-1}; a_r + a_{r+1})$. Repeating this procedure $r - s$ times gives the marginal density of $(U_1, \ldots, U_s)$.

Now, the second part of the theorem follows immediately. ∎

COROLLARY 6.3.2.1. *If* $(U_1, \ldots, U_r) \sim D_p^I(a_1, \ldots, a_r; a_{r+1})$, *then* $U_i \sim B_p^I(a_i, \sum_{j=1(\neq i)}^{r+1} a_j)$, $i = 1, \ldots, r$.

THEOREM 6.3.3. *If* $(V_1, \ldots, V_r) \sim D_p^{II}(b_1, \ldots, b_r; b_{r+1})$, *then* $(V_1, \ldots, V_s) \sim D_p^{II}(b_1, \ldots, b_s; b_{r+1})$, $s \leq r$, *and the density of* $(V_{s+1}, \ldots, V_r)|(V_1, \ldots, V_s)$ *is given by*

$$\frac{\det(I_p + \sum_{i=1}^{s} V_i)^{\sum_{i=1}^{s} b_i + b_{r+1}} \prod_{i=s+1}^{r} \det(V_i)^{b_i - \frac{1}{2}(p+1)}}{\beta_p(b_{s+1}, \ldots, b_r; \sum_{i=1}^{s} b_i + b_{r+1})}$$

$$\det\left(I_p + \sum_{i=1}^{s} V_i + \sum_{i=s+1}^{r} V_i\right)^{-\sum_{i=1}^{r+1} b_i}, \quad V_i > 0.$$

Proof: The proof is similar to the one given for Theorem 6.3.2. In this case, to obtain the marginal density of $V_1, \ldots, V_{r-1}$, we substitute $W_r = (I_p + \sum_{i=1}^{r-1} V_i)^{-\frac{1}{2}} V_r (I_p + \sum_{i=1}^{r-1} V_i)^{-\frac{1}{2}}$ with the Jacobian $J(U_r \rightarrow W_r) = \det(I_p + \sum_{i=1}^{r-1} V_i)^{\frac{1}{2}(p+1)}$. ∎

COROLLARY 6.3.3.1. *If* $(V_1, \ldots, V_r) \sim D_p^{II}(b_1, \ldots, b_r; b_{r+1})$, *then* $V_i \sim B_p^{II}(b_i, b_{r+1})$, $i = 1, \ldots, r$.

THEOREM 6.3.4. *Let* $(U_1, \ldots, U_r) \sim D_p^I(a_1, \ldots, a_r; a_{r+1})$ *and define*

$$W_i = \left(I_p - \sum_{i=1}^s U_i\right)^{-\frac{1}{2}} U_i \left(I_p - \sum_{i=1}^s U_i\right)^{-\frac{1}{2}}, \; i = s+1, \ldots, r.$$

Then (i) $(W_{s+1}, \ldots, W_r)$ *and* $(U_1, \ldots, U_s)$ *are independent, and*
(ii) $(W_{s+1}, \ldots, W_r) \sim D_p^I(a_{s+1}, \ldots, a_r; a_{r+1})$.

Proof: Transforming $W_i = \left(I_p - \sum_{i=1}^s U_i\right)^{-\frac{1}{2}} U_i \left(I_p - \sum_{i=1}^s U_i\right)^{-\frac{1}{2}}, \; i = s+1, \ldots, r$ with Jacobian $J(U_{s+1}, \ldots, U_r \to W_{s+1}, \ldots, W_r) = \det(I_p - \sum_{i=1}^s U_i)^{\frac{1}{2}(r-s)(p+1)}$, in the joint density of $(U_1, \ldots, U_r)$, we get the desired result. ∎

THEOREM 6.3.5. *Let* $(V_1, \ldots, V_r) \sim D_p^{II}(b_1, \ldots, b_r; b_{r+1})$ *and define*

$$Z_i = \left(I_p + \sum_{i=1}^s V_i\right)^{-\frac{1}{2}} V_i \left(I_p + \sum_{i=1}^s V_i\right)^{-\frac{1}{2}}, \; i = s+1, \ldots, r,$$

Then (i) $(Z_{s+1}, \ldots, Z_r)$ *and* $(V_1, \ldots, V_s)$ *are independent, and*
(ii) $(Z_{s+1}, \ldots, Z_r) \sim D_p^{II}(b_{s+1}, \ldots, b_r; \sum_{j=1}^s b_j + b_{r+1})$.

Proof: Similar to the proof of Theorem 6.3.4. ∎

In next two theorems, we derive the joint p.d.f.'s of partial sums of random matrices distributed as matrix variate Dirichlet type I or II.

THEOREM 6.3.6. *Let* $(U_1, \ldots, U_r) \sim D_p^I(a_1, \ldots, a_r; a_{r+1})$ *and define*

$$U_{(i)} = \sum_{j=r_{i-1}^*+1}^{r_i^*} U_j, \; a_{(i)} = \sum_{j=r_{i-1}^*+1}^{r_i^*} a_j, \; r_0^* = 0, \; r_i^* = \sum_{j=1}^i r_j, \; i = 1, \ldots, \ell.$$

Then $(U_{(1)}, \ldots, U_{(\ell)}) \sim D_p^I(a_{(1)}, \ldots, a_{(\ell)}; a_{r+1})$.

Proof: Make the transformation

$$U_{(i)} = \sum_{j=r_{i-1}^*+1}^{r_i^*} U_j, \; \text{and} \; W_j = U_{(i)}^{-\frac{1}{2}} U_j U_{(i)}^{-\frac{1}{2}}, \tag{6.3.4}$$

$j = r_{i-1}^* + 1, \ldots, r_i^* - 1, \; i = 1, \ldots, \ell$. The Jacobian of this transformation is given by

$$J(U_1, \ldots, U_r \to W_1, \ldots, W_{r_1-1}, U_{(1)}, \ldots, W_{r_{\ell-1}^*+1}, \ldots, W_{r-1}, U_{(\ell)})$$

$$= \prod_{i=1}^{\ell} J(U_{r_{i-1}^*+1}, \ldots, U_{r_i^*} \to W_{r_{i-1}^*+1}, \ldots, W_{r_i^*-1}, U_{(i)})$$

$$= \prod_{i=1}^{\ell} \det(U_{(i)})^{\frac{1}{2}(r_i-1)(p+1)}. \tag{6.3.5}$$

Now, substituting from (6.3.4) and (6.3.5) in the joint density of $(U_1, \ldots, U_r)$ given by (6.2.1), we get the joint density of $W_{r^*_{i-1}+1}, \ldots, W_{r^*_i-1}, U_{(i)}$, $i = 1, \ldots, \ell$ as

$$\{\beta_p(a_1, \ldots, a_r; a_{r+1})\}^{-1} \prod_{i=1}^{\ell} \det(U_{(i)})^{a_{(i)} - \frac{1}{2}(p+1)} \det \left(I_p - \sum_{i=1}^{\ell} U_{(i)}\right)^{a_{r+1} - \frac{1}{2}(p+1)}$$

$$\prod_{i=1}^{\ell} \left\{ \prod_{j=r^*_{i-1}+1}^{r^*_i-1} \det(W_j)^{a_j - \frac{1}{2}(p+1)} \det \left(I_p - \sum_{j=r^*_{i-1}+1}^{r^*_i-1} W_j\right)^{a_{r^*_i} - \frac{1}{2}(p+1)} \right\}, \quad (6.3.6)$$

where $0 < U_{(i)} < I_p$, $\sum_{i=1}^{\ell} U_{(i)} < I_p$, $0 < W_j < I_p$, $j = r^*_{i-1} + 1, \ldots, r^*_i - 1$, $\sum_{j=r^*_{i-1}+1}^{r^*_i-1} W_j < I_p$, $i = 1, \ldots, \ell$. From (6.3.6), it is easy to see that $(U_{(1)}, \ldots, U_{(\ell)})$ and $(W_{r^*_{i-1}+1}, \ldots, W_{r^*_i-1})$, $i = 1, \ldots, \ell$, are independently distributed. Further, $(U_{(1)}, \ldots, U_{(\ell)}) \sim D^I_p(a_{(1)}, \ldots, a_{(\ell)}; a_{r+1})$ and $(W_{r^*_{i-1}+1}, \ldots, W_{r^*_i-1}) \sim D^I_p(a_{r^*_{i-1}+1}, \ldots, a_{r^*_i-1}; a_{r^*_i})$, $i = 1, \ldots, \ell$. $\blacksquare$

When $\ell = 1$, $\sum_{i=1}^{r} U_i \sim B^I_p(\sum_{i=1}^{r} a_i, a_{r+1})$.

THEOREM 6.3.7. *Let* $(V_1, \ldots, V_r) \sim D^{II}_p(b_1, \ldots, b_r; b_{r+1})$ *and define*

$$V_{(i)} = \sum_{j=r^*_{i-1}+1}^{r^*_i} V_j, \quad b_{(i)} = \sum_{j=r^*_{i-1}+1}^{r^*_i} b_j, \quad r^*_0 = 0, \quad r^*_i = \sum_{j=1}^{i} r_j, \quad i = 1, \ldots, \ell.$$

Then $(V_{(1)}, \ldots, V_{(\ell)}) \sim D^{II}_p(b_{(1)}, \ldots, b_{(\ell)}; b_{r+1})$.

Proof: Make the transformation

$$V_{(i)} = \sum_{j=r^*_{i-1}+1}^{r^*_i} V_j, \quad \text{and } Z_j = V_{(i)}^{-\frac{1}{2}} V_j V_{(i)}^{-\frac{1}{2}}, \quad (6.3.7)$$

$j = r^*_{i-1} + 1, \ldots, r^*_i - 1$, $i = 1, \ldots, \ell$. The Jacobian of this transformation is given by

$$J(V_1, \ldots, V_r \rightarrow Z_1, \ldots, Z_{r_1-1}, V_{(1)}, \ldots, Z_{r^*_{\ell-1}+1}, \ldots, Z_{r-1}, V_{(\ell)})$$

$$= \prod_{i=1}^{\ell} J(V_{r^*_{i-1}+1}, \ldots, V_{r^*_i} \rightarrow Z_{r^*_{i-1}+1}, \ldots, Z_{r^*_i-1}, V_{(i)})$$

$$= \prod_{i=1}^{\ell} \det(V_{(i)})^{\frac{1}{2}(r_i-1)(p+1)}. \quad (6.3.8)$$

Now, substituting from (6.3.7) and (6.3.8) in the joint density of $(V_1, \ldots, V_r)$ given by (6.2.3), it can easily be shown that $(V_{(1)}, \ldots, V_{(\ell)})$ and $(Z_{r^*_{i-1}+1}, \ldots, Z_{r^*_i-1})$, $i = 1, \ldots, \ell$, are independently distributed. Further, $(V_{(1)}, \ldots, V_{(\ell)}) \sim D^{II}_p(b_{(1)}, \ldots, b_{(\ell)}; b_{r+1})$ and $(Z_{r^*_{i-1}+1}, \ldots, Z_{r^*_i-1}) \sim D^I_p(b_{r^*_{i-1}+1}, \ldots, b_{r^*_i-1}; b_{r^*_i})$, $i = 1, \ldots, \ell$. $\blacksquare$

When $\ell = 1$, the distribution of $\sum_{i=1}^{r} V_i$ is beta type II with parameters $\sum_{i=1}^{r} b_i$ and b_{r+1}.

Next, we give generalizations of the results for marginal and conditional distributions of beta random matrix.

THEOREM 6.3.8. *Let* $(U_1, \ldots, U_r) \sim D_p^I(a_1, \ldots, a_r; a_{r+1})$ *and define*

$$U_i = \begin{pmatrix} U_{11(i)} & U_{12(i)} \\ U_{21(i)} & U_{22(i)} \end{pmatrix} \begin{matrix} p_1 \\ p_2 \end{matrix}, \quad p_1 + p_2 = p,$$
$$\qquad\quad p_1 \qquad p_2$$

$$U_{22 \cdot 1(i)} = U_{22(i)} - U_{21(i)} U_{11(i)}^{-1} U_{12(i)},$$

$$A_{i0} = \Big(I_{p_1} - \sum_{j=i}^{r} U_{11(j)}\Big)^{-1},$$

$$A_i = U_{11(i)}^{-1} \Big(I_{p_1} - \sum_{j=i+1}^{r} U_{11(j)}\Big)\Big(I_{p_1} - \sum_{j=i}^{r} U_{11(j)}\Big)^{-1}$$

and

$$Z_{21(i)} = U_{21(i)} + \Big(\sum_{j=i+1}^{r} U_{21(j)}\Big) A_{i0} A_i^{-1}$$

for $i = 1, 2, \ldots, r$. *Then,*
   (i) $(U_{11(1)}, \ldots, U_{11(r)}) \sim D_{p_1}^I(a_1, \ldots, a_r; a_{r+1})$,
   (ii) $(U_{22 \cdot 1(1)}, \ldots, U_{22 \cdot 1(r)}) \sim D_{p_2}^I(a_1 - \frac{1}{2}p_1, \ldots, a_r - \frac{1}{2}p_1; a_{r+1} + \frac{1}{2}p_1(r-1))$,
and
   (iii) $(Z_{21(1)}, \ldots, Z_{21(r)}) | (U_{11(i)}, U_{22 \cdot 1(i)}, i = 1, \ldots, r) \sim IT_{p_2, rp_1}(2a_{r+1} - p + 1, 0, I_{p_2} - \sum_{j=1}^{r} U_{22 \cdot 1(j)}, A^{-1})$,
where $A = \mathrm{diag}(A_1, \ldots, A_r)$.

Proof: See Tan (1968, 1969c). ∎

THEOREM 6.3.9. *Let* $(V_1, \ldots, V_r) \sim D_p^{II}(b_1, \ldots, b_r; b_{r+1})$ *and define*

$$V_i = \begin{pmatrix} V_{11(i)} & V_{12(i)} \\ V_{21(i)} & V_{22(i)} \end{pmatrix} \begin{matrix} p_1 \\ p_2 \end{matrix}, \quad p_1 + p_2 = p,$$
$$\qquad\quad p_1 \qquad p_2$$

$$V_{22 \cdot 1(i)} = V_{22(i)} - V_{21(i)} V_{11(i)}^{-1} V_{12(i)},$$

$$B_{i0} = \Big(I_{p_1} + \sum_{j=i}^{r} V_{11(j)}\Big)^{-1},$$

$$B_i = V_{11(i)}^{-1} \Big(I_{p_1} + \sum_{j=i+1}^{r} V_{11(j)}\Big)\Big(I_{p_1} + \sum_{j=i}^{r} V_{11(j)}\Big)^{-1}$$

and

$$W_{21(i)} = V_{21(i)} - \Big(\sum_{j=i+1}^{r} V_{21(j)}\Big) B_{i0} B_i^{-1}$$

for $i = 1, 2, \ldots, r$. *Then,*
   (i) $(V_{11(1)}, \ldots, V_{11(r)}) \sim D_{p_1}^{II}(b_1, \ldots, b_r; b_{r+1} - \frac{1}{2}p_2)$,

(ii) $(V_{22\cdot1(1)}, \ldots, V_{22\cdot1(r)}) \sim D_{p_2}^{II}(b_1 - \frac{1}{2}p_1, \ldots, b_r - \frac{1}{2}p_1; b_{r+1})$,

and

(iii) $(W_{21(1)}, \ldots, W_{21(r)})|(V_{11(i)}, V_{22\cdot1(i)}, i = 1, \ldots, r) \sim T_{p_2, rp_1}(2\sum_{i=1}^{r+1} b_i - p_2 - rp_1 + 1, 0, I_{p_2} + \sum_{j=1}^{r} V_{22\cdot1(j)}, B^{-1})$,

where $B = \mathrm{diag}(B_1, \ldots, B_r)$.

Proof: See Tan (1968, 1969c). ■

Next we derive factorizations of the matrix variate Dirichlet density.

THEOREM 6.3.10. *Let* $(U_1, \ldots, U_r) \sim D_p^I(a_1, \ldots, a_r; a_{r+1})$ *and define*

$$X_1 = U_1$$
$$X_2 = (I_p - U_1)^{-\frac{1}{2}} U_2 (I_p - U_1)^{-\frac{1}{2}}$$
$$\vdots$$
$$X_r = (I_p - U_1 - \cdots - U_{r-1})^{-\frac{1}{2}} U_r (I_p - U_1 - \cdots - U_{r-1})^{-\frac{1}{2}}. \qquad (6.3.9)$$

Then $X_1, \ldots, X_r$ *are independently distributed,* $X_i \sim B_p^I(a_i, \sum_{j=i+1}^{r+1} a_j)$, $i = 1, \ldots, r$.

Proof: The density of $(U_1, \ldots, U_r)$ is given by (6.2.1). From the above transformation it is easy to see that

$$\det(I_p - U_1) = \det(I_p - X_1)$$
$$\det(I_p - U_1 - U_2) = \det(I_p - X_1 - (I_p - X_1)^{\frac{1}{2}} X_2 (I_p - X_1)^{\frac{1}{2}})$$
$$= \det(I_p - X_1)\det(I_p - X_2)$$
$$\vdots$$
$$\det(I_p - U_1 - \cdots - U_r) = \det(I_p - X_1) \cdots \det(I_p - X_r)$$
$$\det(U_1) = \det(X_1)$$
$$\det(U_2) = \det(X_2)\det(I_p - X_1)$$
$$\vdots$$
$$\det(U_r) = \det(X_r)\det(I_p - X_1) \cdots \det(I_p - X_{r-1})$$

and

$$J(U_1, \ldots, U_r \to X_1, \ldots, X_r) = \prod_{i=1}^{r-1} \det\left(I_p - \sum_{j=1}^{i} U_j\right)^{\frac{1}{2}(p+1)}$$
$$= \prod_{i=1}^{r-1} \det(I_p - X_i)^{\frac{1}{2}(r-i)(p+1)}.$$

Substituting appropriately in the density of $(U_1, \ldots, U_r)$, one obtains

$$\{\beta(a_1,\ldots,a_r;a_{r+1})\}^{-1} \det(X_1)^{a_1-\frac{1}{2}(p+1)} \prod_{i=2}^{r} \left\{ \det(X_i) \prod_{j=1}^{i-1} \det(I_p - X_j) \right\}^{a_i-\frac{1}{2}(p+1)}$$

$$\prod_{i=1}^{r} \det(I_p - X_i)^{a_{r+1}-\frac{1}{2}(p+1)} \prod_{i=1}^{r-1} \det(I_p - X_i)^{\frac{1}{2}(r-i)(p+1)},$$

$$0 < X_i < I_p, \ i = 1,\ldots,r. \qquad (6.3.10)$$

Combining factors containing X_i together and using the result

$$\beta_p(a_1,\ldots,a_r;a_{r+1}) = \prod_{i=1}^{r} \beta_p\left(a_i, \sum_{j=i+1}^{r+1} a_j\right) \qquad (6.3.11)$$

we obtain the desired result. ∎

THEOREM 6.3.11. Let $(U_1,\ldots,U_r) \sim D_p^I(a_1,\ldots,a_r;a_{r+1})$ and define

$$X_r = U_r$$
$$X_{r-1} = (I_p - U_r)^{-\frac{1}{2}} U_{r-1} (I_p - U_r)^{-\frac{1}{2}}$$
$$\vdots$$
$$X_1 = (I_p - U_r - \cdots - U_2)^{-\frac{1}{2}} U_1 (I_p - U_r - \cdots - U_2)^{-\frac{1}{2}}. \qquad (6.3.12)$$

Then $X_1,\ldots,X_r$ are independently distributed, $X_i \sim B_p^I(a_i, \sum_{j=1}^{i-1} a_j + a_{r+1})$, $i = 1,\ldots,r$.

Proof: Similar to the proof of Theorem 6.3.10. ∎

THEOREM 6.3.12. Let $(V_1,\ldots,V_r) \sim D_p^{II}(b_1,\ldots,b_r;b_{r+1})$ and define

$$Y_r = V_r$$
$$Y_{r-1} = (I_p + V_r)^{-\frac{1}{2}} V_{r-1} (I_p + V_r)^{-\frac{1}{2}}$$
$$\vdots$$
$$Y_1 = (I_p + V_r + \cdots + V_2)^{-\frac{1}{2}} V_1 (I_p + V_r + \cdots + V_2)^{-\frac{1}{2}}. \qquad (6.3.13)$$

Then $Y_1,\ldots,Y_r$ are independently distributed, $Y_i \sim B_p^{II}(b_i, \sum_{j=i+1}^{r+1} b_j)$, $i = 1,\ldots,r$.

Proof: Observe that

$$\det(I_p + V_r) = \det(I_p + Y_r)$$
$$\det(I_p + V_r + V_{r-1}) = \det(I_p + Y_r)\det(I_p + Y_{r-1})$$
$$\vdots$$
$$\det(I_p + V_r + \cdots + V_1) = \det(I_p + Y_r)\cdots\det(I_p + Y_1)$$

$$\det(V_r) = \det(Y_r)$$

$$\det(V_{r-1}) = \det(Y_{r-1})\det(I_p + Y_r)$$

$$\vdots$$

$$\det(V_1) = \det(Y_1)\det(I_p + Y_r)\cdots\det(I_p + Y_2).$$

Substituting these together with the Jacobian of transformation

$$J(V_1,\ldots,V_r \to Y_1,\ldots,Y_r) = \prod_{j=2}^{r} \det(I_p + Y_j)^{\frac{1}{2}(j-1)(p+1)}$$

in the density of $(V_1,\ldots,V_r)$ and simplifying, one obtains

$$\{\beta_p(b_1,\ldots,b_r;b_{r+1})\}^{-1}\prod_{i=1}^{r}\det(Y_i)^{b_i-\frac{1}{2}(p+1)}\prod_{i=1}^{r}\det(I_p + Y_i)^{-\sum_{j=i}^{r+1} b_j}, \quad Y_i > 0. \quad (6.3.14)$$

Now using (6.3.11) the desired result follows. ∎

THEOREM 6.3.13. Let $(V_1,\ldots,V_r) \sim D_p^{II}(b_1,\ldots,b_r;b_{r+1})$ and define

$$Y_1 = V_1$$

$$Y_2 = (I_p + V_1)^{-\frac{1}{2}}V_2(I_p + V_2)^{-\frac{1}{2}}$$

$$\vdots$$

$$Y_r = (I_p + V_1 + \cdots + V_{r-1})^{-\frac{1}{2}}V_r(I_p + V_1 + \cdots + V_{r-1})^{-\frac{1}{2}}. \quad (6.3.15)$$

Then $Y_1,\ldots,Y_r$ are independently distributed, $Y_i \sim B_p^{II}(b_i, \sum_{j=1}^{i-1} b_j + b_{r+1})$, $i = 1,\ldots,r$.

Proof: Similar to the proof of Theorem 6.3.12. ∎

The above results have been derived using matrix transformations. Tan (1969c) has derived Theorem 6.3.11 and Theorem 6.3.12 using certain results on marginal and conditional distributions.

Likewise, using suitable inverse transformations, one can derive the matrix variate Dirichlet distribution from the independent beta matrices as given in the following theorem.

THEOREM 6.3.14. Let $X_1,\ldots,X_r$ be independent $p \times p$ random matrices, $X_i \sim B_p^I(\alpha_i,\beta_i)$, $i = 1,\ldots,r$. Define

$$U_1 = X_1$$

$$U_2 = (I_p - X_1)^{\frac{1}{2}}X_2(I_p - X_1)^{\frac{1}{2}}$$

$$\vdots$$

$$U_r = (I_p - X_1)^{\frac{1}{2}}\cdots(I_p - X_{r-1})^{\frac{1}{2}}X_r(I_p - X_{r-1})^{\frac{1}{2}}\cdots(I_p - X_1)^{\frac{1}{2}}.$$

Then $(U_1,\ldots,U_r) \sim D_p^I(\alpha_1,\ldots,\alpha_r;\beta_r)$ iff $\beta_i = \alpha_{i+1} + \beta_{i+1}$, $i = 1,\ldots,r-1$.

THEOREM 6.3.15. *Let $X_1, \ldots, X_r$ be independent $p \times p$ random matrices, and $X_i \sim B_p^I(\alpha_i, \beta_i)$, $i = 1, \ldots, r$. Define*

$$U_r = X_r$$

$$U_{r-1} = (I_p - X_r)^{\frac{1}{2}} X_{r-1} (I_p - X_r)^{\frac{1}{2}}$$

$$\vdots$$

$$U_1 = (I_p - X_r)^{\frac{1}{2}} \cdots (I_p - X_2)^{\frac{1}{2}} X_1 (I_p - X_2)^{\frac{1}{2}} \cdots (I_p - X_r)^{\frac{1}{2}}.$$

Then $(U_1, \ldots, U_r) \sim D_p^I(\alpha_1, \ldots, \alpha_r; \beta_1)$ iff $\beta_{i+1} = \alpha_i + \beta_i$, $i = 1, \ldots, r-1$.

THEOREM 6.3.16. *Let $Y_1, \ldots, Y_r$ be independent $p \times p$ random matrices, $Y_i \sim B_p^{II}(\alpha_i, \beta_i)$, $i = 1, \ldots, r$. Define*

$$V_r = Y_r$$

$$V_{r-1} = (I_p + Y_r)^{\frac{1}{2}} Y_{r-1} (I_p + Y_r)^{\frac{1}{2}}$$

$$\vdots$$

$$V_1 = (I_p + Y_r)^{\frac{1}{2}} \cdots (I_p + Y_2)^{\frac{1}{2}} Y_1 (I_p + Y_2)^{\frac{1}{2}} \cdots (I_p + Y_r)^{\frac{1}{2}}.$$

Then $(V_1, \ldots, V_r) \sim D_p^{II}(\alpha_1, \ldots, \alpha_r; \beta_1)$ iff $\beta_i = \alpha_{i+1} + \beta_{i+1}$, $i = 1, \ldots, r-1$.

THEOREM 6.3.17. *Let $Y_1, \ldots, Y_r$ be independent $p \times p$ random matrices, and $Y_i \sim B_p^{II}(\alpha_i, \beta_i)$, $i = 1, \ldots, r$. Define*

$$V_1 = Y_1$$

$$V_2 = (I_p + Y_1)^{\frac{1}{2}} Y_2 (I_p + Y_1)^{\frac{1}{2}}$$

$$\vdots$$

$$V_r = (I_p + Y_1)^{\frac{1}{2}} \cdots (I_p + Y_{r-1})^{\frac{1}{2}} Y_r (I_p + Y_{r-1})^{\frac{1}{2}} \cdots (I_p + Y_1)^{\frac{1}{2}}.$$

Then $(V_1, \ldots, V_r) \sim D_p^{II}(\alpha_1, \ldots, \alpha_r; \beta_1)$ iff $\beta_{i+1} = \alpha_i + \beta_i$, $i = 1, \ldots, r-1$.

From the transformations given in Theorem 6.3.14, one can see that

$$I_p - \sum_{i=1}^r U_i = (I_p - X_1)^{\frac{1}{2}} \cdots (I_p - X_{r-1})^{\frac{1}{2}} (I_p - X_r)(I_p - X_{r-1})^{\frac{1}{2}} \cdots (I_p - X_1)^{\frac{1}{2}}$$

where $I_p - X_1, \ldots, I_p - X_r$ are independent, $I_p - X_i \sim B_p^I(\beta_i, \alpha_i)$, $i = 1, \ldots, r$ and $I_p - \sum_{i=1}^r U_i \sim B_p^I(\beta_r, \sum_{i=1}^r \alpha_i)$ iff $\beta_i = \alpha_{i+1} + \beta_{i+1}$, $i = 1, \ldots, r-1$.

Similarly, from Theorem 6.3.15, one obtains

$$I_p - \sum_{i=1}^r U_i = (I_p - X_r)^{\frac{1}{2}} \cdots (I_p - X_2)^{\frac{1}{2}} (I_p - X_1)(I_p - X_2)^{\frac{1}{2}} \cdots (I_p - X_r)^{\frac{1}{2}}$$

where $I_p - X_1, \ldots, I_p - X_r$ are independent, $I_p - X_i \sim B_p^I(\beta_i, \alpha_i)$, $i = 1, \ldots, r$ and $I_p - \sum_{i=1}^{r} U_i \sim B_p^I(\beta_1, \sum_{i=1}^{r} \alpha_i)$ iff $\beta_{i+1} = \alpha_i + \beta_i$, $i = 1, \ldots, r - 1$.

Thus we obtain the following result generalizing a result given by Javier and Gupta (1985a)(see Theorem 5.3.25), and Rao (1952).

THEOREM 6.3.18. *Let $W_1, \ldots, W_r$ be independent $p \times p$ random matrices, $W_i \sim B_p^I(c_i, d_i)$, $i = 1, \ldots, r$. Then*

$$W_1^{\frac{1}{2}} \ldots W_{r-1}^{\frac{1}{2}} W_r W_{r-1}^{\frac{1}{2}} \ldots W_1^{\frac{1}{2}} \sim B_p^I\left(c_r, \sum_{i=1}^{r} d_i\right)$$

iff $c_i = d_{i+1} + c_{i+1}$, $i = 1 \ldots, r - 1$ and

$$W_r^{\frac{1}{2}} \ldots W_2^{\frac{1}{2}} W_1 W_2^{\frac{1}{2}} \ldots W_r^{\frac{1}{2}} \sim B_p^I\left(c_1, \sum_{i=1}^{r} d_i\right)$$

iff $c_{i+1} = d_i + c_i$, $i = 1 \ldots, r - 1$.

Tiao and Guttman (1965) derived certain asymptotic distribution for the univariate Dirichlet type I distribution. Here we give the matrix variate generalization of their result due to Javier and Gupta (1985a).

THEOREM 6.3.19. *Let $(U_1, \ldots, U_r) \sim D_p^I(a_1, \ldots, a_r; a_{r+1})$ and $W = (W_1, \ldots, W_r)$ be defined by $W_i = a_{r+1} U_i$, $i = 1, \ldots, r$. Then W is asymptotically distributed as a product of independent matrix variate gamma densities; more specifically*

$$\lim_{a_{r+1} \to \infty} f(W) = \prod_{i=1}^{r} \frac{\det(W_i)^{a_i - \frac{1}{2}(p+1)} \operatorname{etr}(-W_i)}{\Gamma_p(a_i)},$$

where $f(W)$ denotes the density of the matrix W.

Proof: In the joint density of $(U_1, \ldots, U_r)$ given by (6.2.1) transform $W_i = a_{r+1} U_i$, $i = 1, \ldots, r$ with the Jacobian $J(U_1, \ldots, U_r \to W_1, \ldots, W_r) = a_{r+1}^{-\frac{1}{2} r p(p+1)}$. The density of $W = (W_1, \ldots, W_r)$ is given by

$$f(W) = \frac{\Gamma_p(\sum_{i=1}^{r+1} a_i)}{\Gamma_p(a_{r+1})} a_{r+1}^{-p \sum_{i=1}^{r} a_i} \left\{ \prod_{i=1}^{r} \frac{\det(W_i)^{a_i - \frac{1}{2}(p+1)}}{\Gamma_p(a_i)} \right\} \det\left(I_p - \frac{1}{a_{r+1}} \sum_{i=1}^{r} W_i\right)^{a_{r+1} - \frac{1}{2}(p+1)}.$$

The result follows, since

$$\lim_{a_{r+1} \to \infty} \frac{\Gamma_p(\sum_{i=1}^{r+1} a_i)}{\Gamma_p(a_{r+1})} a_{r+1}^{-p \sum_{i=1}^{r} a_i} = 1$$

and

$$\lim_{a_{r+1} \to \infty} \det\left(I_p - \frac{1}{a_{r+1}} \sum_{i=1}^{r} W_i\right)^{a_{r+1} - \frac{1}{2}(p+1)} = \operatorname{etr}\left(-\sum_{i=1}^{r} W_i\right). \blacksquare$$

An analogous result for Dirichlet type II distribution is easily shown to be the following.

THEOREM 6.3.20. *Let* $(V_1, \ldots, V_r) \sim D_p^{II}(b_1, \ldots, b_r; b_{r+1})$ *and* $W = (W_1, \ldots, W_r)$ *be defined by* $W_i = b_{r+1}V_i$, $i = 1, \ldots, r$. *Then,* W *is asymptotically distributed as a product of independent matrix variate gamma densities; more specifically*

$$\lim_{b_{r+1} \to \infty} g(W) = \prod_{i=1}^{r} \frac{\det(W_i)^{b_i - \frac{1}{2}(p+1)} \operatorname{etr}(-W_i)}{\Gamma_p(b_i)},$$

where $g(W)$ *denotes the density of matrix* W.

6.4.   RELATED DISTRIBUTIONS

In this section, we study distributions that are closely related to the matrix variate Dirichlet type I and type II distributions by generalizing the ratios (6.2.4) and (6.2.5).

THEOREM 6.4.1. *Let* $S_i \sim W_p(n_i, \Sigma_1)$, $i = 1, \ldots, r$ *and* $B \sim W_p(m, \Sigma_2)$ *be independently distributed. Define*

$$U_i = S^{-\frac{1}{2}} S_i (S^{-\frac{1}{2}})', \ i = 1, \ldots, r,$$

where $S^{\frac{1}{2}}(S^{\frac{1}{2}})' = \sum_{i=1}^{r} S_i + B$ *is a reasonable factorization of* S. *Then, the density of* $U_1, \ldots, U_r$ *is given by*

$$\frac{2^{-\frac{1}{2}(m+n)p} \det(\Sigma_1)^{-\frac{1}{2}n} \det(\Sigma_2)^{-\frac{1}{2}m}}{\Gamma_p(\frac{1}{2}m) \prod_{i=1}^{r} \Gamma_p(\frac{1}{2}n_i)} \prod_{i=1}^{r} \det(U_i)^{\frac{1}{2}(n_i-p-1)} \det\left(I_p - \sum_{i=1}^{r} U_i\right)^{\frac{1}{2}(m-p-1)}$$

$$\int_{S>0} \det(S)^{\frac{1}{2}(m+n-p-1)} \operatorname{etr}\left\{-\frac{1}{2}\Sigma_2^{-1}S + \frac{1}{2}(\Sigma_2^{-1} - \Sigma_1^{-1})S^{\frac{1}{2}}\left(\sum_{i=1}^{r} U_i\right)(S^{\frac{1}{2}})'\right\} dS,$$

$$0 < U_i < I_p, \ 0 < \sum_{i=1}^{r} U_i < I_p, \tag{6.4.1}$$

where $n = \sum_{j=1}^{r} n_j$.

Proof: The joint density of $S_1, \ldots, S_r$ and B is given by

$$\prod_{i=1}^{r} \left[\left\{2^{\frac{1}{2}n_i p} \Gamma_p\left(\frac{1}{2}n_i\right) \det(\Sigma_1)^{\frac{1}{2}n_i}\right\}^{-1} \det(S_i)^{\frac{1}{2}(n_i-p-1)} \operatorname{etr}\left(-\frac{1}{2}\Sigma_1^{-1}S_i\right)\right]$$

$$\left\{2^{\frac{1}{2}mp} \Gamma_p\left(\frac{1}{2}m\right) \det(\Sigma_2)^{\frac{1}{2}m}\right\}^{-1} \det(B)^{\frac{1}{2}(m-p-1)} \operatorname{etr}\left(-\frac{1}{2}\Sigma_2^{-1}B\right).$$

Making the transformation $\sum_{i=1}^{r} S_i + B = S$, $S_i = S^{\frac{1}{2}}U_i(S^{\frac{1}{2}})'$, $i = 1, \ldots, r$ with Jacobian $J(S_1, \ldots, S_r, B \to U_1, \ldots, U_r, S) = \det(S)^{\frac{1}{2}r(p+1)}$, we get the joint density of $U_1, \ldots, U_r$ and S as

$$\frac{2^{-\frac{1}{2}(m+n)p}\det(\Sigma_1)^{-\frac{1}{2}n}\det(\Sigma_2)^{-\frac{1}{2}m}}{\Gamma_p(\frac{1}{2}m)\prod_{i=1}^r\Gamma_p(\frac{1}{2}n_i)}\prod_{i=1}^r\det(U_i)^{\frac{1}{2}(n_i-p-1)}\det\left(I_p-\sum_{i=1}^rU_i\right)^{\frac{1}{2}(m-p-1)}$$

$$\det(S)^{\frac{1}{2}(m+n-p-1)}\,\mathrm{etr}\left\{-\frac{1}{2}\Sigma_2^{-1}S+\frac{1}{2}(\Sigma_2^{-1}-\Sigma_1^{-1})S^{\frac{1}{2}}\left(\sum_{i=1}^rU_i\right)(S^{\frac{1}{2}})'\right\}\quad(6.4.2)$$

Now, integrating (6.4.2) with respect to S we get the desired result (6.4.1). ∎

In Theorem 6.4.1, if $\Sigma_1=\Sigma_2$, then $(U_1,\dots,U_r)\sim D_p^I(\frac{1}{2}n_1,\dots,\frac{1}{2}n_r;\frac{1}{2}m)$. For $r=1$, the distribution of U_1 is given in Theorem 5.4.4.

THEOREM 6.4.2. *If the joint density of symmetric positive definite random matrices $U_1,\dots,U_r$ is (6.4.1), then the density of $Z=\sum_{i=1}^rU_i$ is given by*

$$\frac{2^{-\frac{1}{2}(m+n)p}\det(\Sigma_1)^{-\frac{1}{2}n}\det(\Sigma_2)^{-\frac{1}{2}m}}{\Gamma_p(\frac{1}{2}m)\Gamma_p(\frac{1}{2}n)}\det(Z)^{\frac{1}{2}(n-p-1)}\det(I_p-Z)^{\frac{1}{2}(m-p-1)}$$

$$\int_{S>0}\det(S)^{\frac{1}{2}(m+n-p-1)}\,\mathrm{etr}\left\{-\frac{1}{2}\Sigma_2^{-1}S+\frac{1}{2}(\Sigma_2^{-1}-\Sigma_1^{-1})S^{\frac{1}{2}}Z(S^{\frac{1}{2}})'\right\}dS,$$

$$0<Z<I_p.\quad(6.4.3)$$

Proof: Substituting $\sum_{i=1}^rU_i=Z$, $W_i=Z^{-\frac{1}{2}}U_iZ^{-\frac{1}{2}}$, $i=1,\dots,r-1$, where $Z^{\frac{1}{2}}Z^{\frac{1}{2}}=Z$ in (6.4.1) with Jacobian of transformation $J(U_1,\dots,U_{r-1},U_r\to W_1,\dots,W_{r-1},Z)=\det(Z)^{\frac{1}{2}(r-1)(p+1)}$, we get the joint density of $W_1,\dots,W_{r-1}$ and Z as

$$\left\{\beta\left(\frac{1}{2}n_1,\dots,\frac{1}{2}n_{r-1};\frac{1}{2}n_r\right)\right\}^{-1}\prod_{i=1}^{r-1}\det(W_i)^{\frac{1}{2}(n_i-p-1)}\det\left(I_p-\sum_{i=1}^{r-1}W_i\right)^{\frac{1}{2}(n_r-p-1)}$$

$$\left\{2^{\frac{1}{2}(m+n)p}\Gamma_p\left(\frac{1}{2}m\right)\Gamma_p\left(\frac{1}{2}n\right)\det(\Sigma_1)^{\frac{1}{2}n}\det(\Sigma_2)^{\frac{1}{2}m}\right\}^{-1}\det(Z)^{\frac{1}{2}(n-p-1)}\det(I_p-Z)^{\frac{1}{2}(m-p-1)}$$

$$\int_{S>0}\det(S)^{\frac{1}{2}(m+n-p-1)}\,\mathrm{etr}\left\{-\frac{1}{2}\Sigma_2^{-1}S+\frac{1}{2}(\Sigma_2^{-1}-\Sigma_1^{-1})S^{\frac{1}{2}}Z(S^{\frac{1}{2}})'\right\}dS,\quad(6.4.4)$$

where $0<W_i<I_p$, $i=1,\dots,r-1$, $\sum_{i=1}^{r-1}W_i<I_p$ and $0<Z<I_p$. Now, from (6.4.4), it is clear that $(W_1,\dots,W_{r-1})$ and Z are independently distributed, $(W_1,\dots,W_{r-1})\sim D_p^I(\frac{1}{2}n_1,\dots,\frac{1}{2}n_{r-1};\frac{1}{2}n_r)$ and the density of Z is given by (6.4.3). ∎

Next we give moments of $\prod_{i=1}^r\det(U_i)^{n_i}$ and $\det(I_p-\sum_{i=1}^rU_i)$.

THEOREM 6.4.3. *If the joint density of symmetric positive definite random matrices $U_1,\dots,U_r$ is (6.4.1), then*

(i) $E\left[\prod_{i=1}^r\det(U_i)^{n_i}\right]^h=\dfrac{\Gamma_p[\frac{1}{2}(m+n)]\prod_{i=1}^r\Gamma_p(\frac{1}{2}n_i+hn_i)}{\prod_{i=1}^r\Gamma_p(\frac{1}{2}n_i)\Gamma_p[\frac{1}{2}(m+n)+hn]}\det(\Sigma_1^{-1}\Sigma_2)^{\frac{1}{2}n}$

$$_2F_1\left(\frac{1}{2}n+nh,\frac{1}{2}(m+n);\frac{1}{2}(m+n)+hn;I_p-\Sigma_1^{-\frac{1}{2}}\Sigma_2\Sigma_1^{-\frac{1}{2}}\right),$$

$$\mathrm{Re}(n_ih)>-\frac{1}{2}(n_i-p+1),\,i=1,\dots,r,$$

and

$$(ii)\ E\Big[\det\Big(I_p - \sum_{i=1}^{r} U_i\Big)^h\Big] = \frac{\Gamma_p(\frac{1}{2}m + h)\Gamma_p[\frac{1}{2}(m + n)]}{\Gamma_p(\frac{1}{2}m)\Gamma_p[\frac{1}{2}(m + n) + h]} \det(\Sigma_1^{-1}\Sigma_2)^{\frac{1}{2}n}$$

$$_2F_1\Big(\tfrac{1}{2}n, \tfrac{1}{2}(m + n); \tfrac{1}{2}(m + n) + h; I_p - \Sigma_1^{-\frac{1}{2}}\Sigma_2\Sigma_1^{-\frac{1}{2}}\Big),$$

$$\mathrm{Re}(h) > -\frac{1}{2}(m - p + 1).$$

Proof: (i) From (6.4.1), we have

$$E\Big[\prod_{i=1}^{r} \det(U_i)^{n_i}\Big]^h = \frac{2^{-\frac{1}{2}(m+n)p}\det(\Sigma_1)^{-\frac{1}{2}n}\det(\Sigma_2)^{-\frac{1}{2}m}}{\Gamma_p(\frac{1}{2}m)\prod_{i=1}^{r}\Gamma_p(\frac{1}{2}n_i)}$$

$$\int\cdots\int_{\substack{0<U_i<I_p \\ 0<\sum_{i=1}^{r} U_i<I_p}} \prod_{i=1}^{r}\det(U_i)^{\frac{1}{2}(2hn_i+n_i-p-1)}\det\Big(I_p - \sum_{i=1}^{r} U_i\Big)^{\frac{1}{2}(m-p-1)}$$

$$\int_{S>0}\mathrm{etr}\Big\{-\frac{1}{2}\Sigma_2^{-1}S + \frac{1}{2}(\Sigma_2^{-1} - \Sigma_1^{-1})S^{\frac{1}{2}}\Big(\sum_{i=1}^{r} U_i\Big)(S^{\frac{1}{2}})'\Big\}$$

$$\det(S)^{\frac{1}{2}(m+n-p-1)}\,dS\,dU_1\cdots dU_r. \qquad (6.4.5)$$

Writing $\mathrm{etr}\{\frac{1}{2}(\Sigma_2^{-1} - \Sigma_1^{-1})S^{\frac{1}{2}}(\sum_{i=1}^{r} U_i)(S^{\frac{1}{2}})'\} = {}_0F_0(\frac{1}{2}(\Sigma_2^{-1} - \Sigma_1^{-1})S^{\frac{1}{2}}(\sum_{i=1}^{r} U_i)(S^{\frac{1}{2}})')$, and using the integral given in Problem 1.10 we get

$$\int\cdots\int_{\substack{0<U_i<I_p \\ 0<\sum_{i=1}^{r} U_i<I_p}} \prod_{i=1}^{r}\det(U_i)^{\frac{1}{2}(2hn_i+n_i-p-1)}\det\Big(I_p - \sum_{i=1}^{r} U_i\Big)^{\frac{1}{2}(m-p-1)}$$

$$_0F_0\Big(\frac{1}{2}(\Sigma_2^{-1} - \Sigma_1^{-1})S^{\frac{1}{2}}(\sum_{i=1}^{r} U_i)(S^{\frac{1}{2}})'\Big)\,dU_1\cdots dU_r$$

$$= \frac{\prod_{i=1}^{r}\Gamma_p[\frac{1}{2}n_i(1+2h)]\,\Gamma_p(\frac{1}{2}m)}{\Gamma_p[\frac{1}{2}(m+n) + hn]}$$

$$_1F_1\Big(\frac{1}{2}n(1+2h); \frac{1}{2}(m+n) + hn; \frac{1}{2}(\Sigma_2^{-1} - \Sigma_1^{-1})S\Big),$$

$$\mathrm{Re}(n_i h) > -\frac{1}{2}(n_i - p + 1). \qquad (6.4.6)$$

Now, substituting (6.4.6) in (6.4.5) we have

$$E\Big[\prod_{i=1}^{r} \det(U_i)^{n_i}\Big]^h = \frac{2^{-\frac{1}{2}(m+n)p}\det(\Sigma_1)^{-\frac{1}{2}n}\det(\Sigma_2)^{-\frac{1}{2}m}\prod_{i=1}^{r}\Gamma_p(\frac{1}{2}n_i + hn_i)}{\Gamma_p[\frac{1}{2}(m+n) + hn]\prod_{i=1}^{r}\Gamma_p(\frac{1}{2}n_i)}$$

$$\int_{S>0}\mathrm{etr}\Big(-\frac{1}{2}\Sigma_2^{-1}S\Big)\det(S)^{\frac{1}{2}(m+n-p-1)}$$

$$_1F_1\Big(\frac{1}{2}n(1+2h); \frac{1}{2}(m+n) + hn; \frac{1}{2}(\Sigma_2^{-1} - \Sigma_1^{-1})S\Big)\,dS.$$

Using the integral (1.6.4), we get the desired result.

(ii) The derivation of $E[\det(I_p - \sum_{i=1}^r U_i)^h]$ is similar. ■

The above results were derived by de Waal (1970) when $S^{\frac{1}{2}}$ is a lower triangular matrix. The next theorem gives the results derived by Olkin and Rubin (1964).

THEOREM 6.4.4. *Let* $S_i \sim W_p(n_i, \Sigma)$, $i = 1, \ldots, r$ *and* $B \sim W_p(m, \Sigma)$ *be independent.*

(i) If $B = TT'$ *where* T *is a lower triangular matrix with positive diagonal matrix, then the joint density of* $V_j = T^{-1}S_j(T^{-1})'$, $j = 1, \ldots, r$ *is given by*

$$\left\{\beta_p\left(\frac{1}{2}n_1, \ldots, \frac{1}{2}n_r; \frac{1}{2}m\right)\right\}^{-1} \frac{\prod_{j=1}^r \det(V_j)^{\frac{1}{2}(n_j-p-1)} \det(I_p + \sum_{j=1}^r V_j)^{-\frac{1}{2}(m+n-p-1)}}{\prod_{i=1}^p \det((I_p + \sum_{j=1}^r V_j)^{[i]})}.$$
$$(6.4.7)$$

(ii) If $B = TT'$ *where* T *is an upper triangular matrix with positive diagonal matrix, then the joint density of* $V_j = T^{-1}S_j(T^{-1})'$, $j = 1, \ldots, r$ *is given by*

$$\left\{\beta_p\left(\frac{1}{2}n_1, \ldots, \frac{1}{2}n_r; \frac{1}{2}m\right)\right\}^{-1} \frac{\prod_{j=1}^r \det(V_j)^{\frac{1}{2}(n_j-p-1)} \det(I_p + \sum_{j=1}^r V_j)^{-\frac{1}{2}(m+n-p-1)}}{\prod_{i=1}^p \det((I_p + \sum_{j=1}^r V_j)_{[i]})}.$$
$$(6.4.8)$$

Proof: In the joint density of $S_1, \ldots, S_r$ and B given by (6.2.9), making the transformation $B = TT'$ ($T = (t_{ij})$, $t_{ii} > 0$, is lower triangular) and $V_j = T^{-1}S_j(T^{-1})'$, $j = 1, \ldots, r$, with Jacobian $J(S_1, \ldots, S_r, B \to V_1, \ldots, V_r, T) = 2^p \det(TT')^{\frac{1}{2}r(p+1)}$ $\prod_{j=1}^p t_{jj}^{p+1-j}$, we get the joint density of $V_1, \ldots, V_r$ and T as

$$\left\{2^{\frac{1}{2}(m+n)p}\Gamma_p\left(\frac{1}{2}m\right)\prod_{i=1}^r \Gamma_p\left(\frac{1}{2}n_i\right)\right\}^{-1} \det(\Sigma)^{-\frac{1}{2}(m+n)} \prod_{i=1}^r \det(V_i)^{\frac{1}{2}(n_i-p-1)}$$

$$\text{etr}\left\{-\frac{1}{2}\Sigma^{-1}T\left(I_p + \sum_{i=1}^r V_i\right)T'\right\} \det(TT')^{\frac{1}{2}(m+n-p-1)}2^p \prod_{j=1}^p t_{jj}^{p+1-j}. \quad (6.4.9)$$

Now, in order to obtain the joint density of $V_1, \ldots, V_r$ we need to integrate (6.4.9) with respect to T. For this, consider the integral

$$\int_T \text{etr}\left\{-\frac{1}{2}\Sigma^{-1}T\left(I_p + \sum_{i=1}^r V_i\right)T'\right\} \det(TT')^{\frac{1}{2}(m+n-p-1)}2^p \prod_{j=1}^p t_{jj}^{p+1-j}\, dT. \quad (6.4.10)$$

Substituting $W = T(I_p + \sum_{i=1}^r V_i)T'$ with the Jacobian $J(T \to W) = \{2^p \prod_{j=1}^p t_{jj}^{p+1-j}$ $\prod_{i=1}^p \det((I_p + \sum_{j=1}^r V_j)^{[i]})\}^{-1}$ the above integral becomes

$$\det\left(I_p + \sum_{j=1}^r V_j\right)^{-\frac{1}{2}(m+n-p-1)}\left\{\prod_{i=1}^p \det\left(\left(I_p + \sum_{j=1}^r V_j\right)^{[i]}\right)\right\}^{-1}$$

$$\int_{W>0} \det(W)^{\frac{1}{2}(m+n-p-1)} \text{etr}\left(-\frac{1}{2}\Sigma^{-1}W\right) dW$$

$$= \det \Big(I_p + \sum_{j=1}^{r} V_j\Big)^{-\frac{1}{2}(m+n-p-1)} \Big\{ \prod_{i=1}^{p} \det \Big(\big(I_p + \sum_{j=1}^{r} V_j\big)^{[i]}\Big)\Big\}^{-1}$$

$$\Gamma_p\Big[\frac{1}{2}(m+n)\Big] \det \Big(\frac{1}{2}\Sigma^{-1}\Big)^{-\frac{1}{2}(m+n)}. \tag{6.4.11}$$

Now, from (6.4.9) and (6.4.11) we get (6.4.7). The proof for the case when $B = TT'$ (T upper triangular) is similar. ∎

In the above theorem, without loss of generality, Σ can be taken as I_p since the densities (6.4.7) and (6.4.8) do not depend on it. Now, it may be noted that we have three different joint densities of V_j, $j = 1, \ldots, r$, (6.2.11), (6.4.7), and (6.4.8), depending whether the root of matrix B is symmetric, lower triangular or upper triangular respectively.

6.5. NONCENTRAL MATRIX VARIATE DIRICHLET DISTRIBUTIONS

Here, we derive the distribution of $(U_1, \ldots, U_r)$ and $(V_1, \ldots, V_r)$, defined by (6.2.6) and (6.2.7) respectively when the matrix B has noncentral Wishart distribution.

THEOREM 6.5.1. *Let $S_i \sim W_p(n_i, \Sigma)$, $i = 1, \ldots, r$ and $B \sim W_p(m, \Sigma, \Theta)$ be independently distributed. Define*

$$U_i = S^{-\frac{1}{2}} S_i (S^{-\frac{1}{2}})', \ i = 1, \ldots, r,$$

where $S = \sum_{i=1}^{r} S_i + B$ and $S^{\frac{1}{2}}(S^{\frac{1}{2}})'$ is any reasonable factorization of S. Then, the joint density of $U_1, \ldots, U_r$ is given by

$$\frac{2^{-\frac{1}{2}(m+n)p} \det(\Sigma)^{-\frac{1}{2}(m+n)}}{\Gamma_p(\frac{1}{2}m) \prod_{i=1}^{r} \Gamma_p(\frac{1}{2}n_i)} \operatorname{etr}\Big(-\frac{1}{2}\Theta\Big) \prod_{i=1}^{r} \det(U_i)^{\frac{1}{2}(n_i-p-1)} \det\Big(I_p - \sum_{i=1}^{r} U_i\Big)^{\frac{1}{2}(m-p-1)}$$

$$\int_{S>0} \operatorname{etr}\Big(-\frac{1}{2}\Sigma^{-1}S\Big) \det(S)^{\frac{1}{2}(m+n-p-1)} {}_0F_1\Big(\frac{1}{2}m; \frac{1}{4}\Theta\Sigma^{-1}S^{\frac{1}{2}}\Big(I_p - \sum_{i=1}^{r} U_i\Big)(S^{\frac{1}{2}})'\Big) dS,$$

$$0 < U_i < I_p, \ i = 1, \ldots, r, \ 0 < \sum_{i=1}^{r} U_i < I_p. \tag{6.5.1}$$

Proof: The joint density of $S_1, \ldots, S_r$ and B is given by

$$\prod_{i=1}^{r} \Big[\Big\{2^{\frac{1}{2}n_i p} \Gamma_p\Big(\frac{1}{2}n_i\Big) \det(\Sigma)^{\frac{1}{2}n_i}\Big\}^{-1} \operatorname{etr}\Big(-\frac{1}{2}\Sigma^{-1}S_i\Big) \det(S_i)^{\frac{1}{2}(n_i-p-1)}\Big]$$

$$\Big\{2^{\frac{1}{2}mp} \Gamma_p\Big(\frac{1}{2}m\Big) \det(\Sigma)^{\frac{1}{2}m}\Big\}^{-1} \operatorname{etr}\Big(-\frac{1}{2}\Theta\Big) \det(B)^{\frac{1}{2}(m-p-1)}$$

$$\operatorname{etr}\Big(-\frac{1}{2}\Sigma^{-1}B\Big) {}_0F_1\Big(\frac{1}{2}m; \frac{1}{4}\Theta\Sigma^{-1}B\Big). \tag{6.5.2}$$

Making the transformation $\sum_{i-1}^{r} S_i + B = S$, $S_i = S^{\frac{1}{2}} U_i (S^{\frac{1}{2}})'$, $i = 1, \ldots, r$ with Jacobian $J(S_1, \ldots, S_r, B \to U_1, \ldots, U_r, S) = \det(S)^{\frac{1}{2}r(p+1)}$ in (6.5.2), we get the joint density of $U_1, \ldots, U_r$ and S as

$$\frac{2^{-\frac{1}{2}(m+n)p} \det(\Sigma)^{-\frac{1}{2}(m+n)}}{\Gamma_p(\frac{1}{2}m) \prod_{i=1}^{r} \Gamma_p(\frac{1}{2}n_i)} \, \text{etr}\left(-\frac{1}{2}\Theta\right) \prod_{i=1}^{r} \det(U_i)^{\frac{1}{2}(n_i-p-1)} \det\left(I_p - \sum_{i=1}^{r} U_i\right)^{\frac{1}{2}(m-p-1)}$$

$$\text{etr}\left(-\frac{1}{2}\Sigma^{-1}S\right) \det(S)^{\frac{1}{2}(m+n-p-1)} \, {}_0F_1\left(\frac{1}{2}m; \frac{1}{4}\Theta\Sigma^{-1}S^{\frac{1}{2}}\left(I_p - \sum_{i=1}^{r} U_i\right)(S^{\frac{1}{2}})'\right),$$

$$S > 0, \; 0 < U_i < I_p, \; i = 1, \ldots, r, \; 0 < \sum_{i=1}^{r} U_i < I_p. \qquad (6.5.3)$$

Integrating (6.5.3) with respect to S we get (6.5.1). ∎

The above result was derived by de Waal (1972b) for a triangular root of S. Substituting $\Theta = 0$ in (6.5.1), we get the results of Theorem 6.2.1. When $\Sigma = I_p$ and $\Theta = \text{diag}(\theta, 0, \ldots, 0)$, the joint distribution of $U_i = (u_{jk(i)})$, $i = 1, \ldots, r$, is given by, Troskie (1967), as

$$\left\{\beta_p\left(\frac{1}{2}n_1, \ldots, \frac{1}{2}n_r; \frac{1}{2}m\right)\right\}^{-1} \exp\left(-\frac{1}{2}\theta\right) \prod_{i=1}^{r} \det(U_i)^{\frac{1}{2}(n_i-p-1)} \det\left(I_p - \sum_{i=1}^{r} U_i\right)^{\frac{1}{2}(m-p-1)}$$

$$ {}_1F_1\left(\frac{1}{2}(m+n); \frac{1}{2}m; \frac{1}{2}\theta\left(1 - \sum_{i=1}^{r} u_{11(i)}\right)\right). \qquad (6.5.4)$$

For $r = 1$, Theorem 6.5.1 reduces to Theorem 5.5.1 and the result (6.5.1) simplifies to (5.5.1).

THEOREM 6.5.2. *Let $S_i \sim W_p(n_i, \Sigma)$, $i = 1, \ldots, r$ and $B \sim W_p(m, \Sigma, \Theta)$ be independently distributed. Define*

$$V_i = B^{-\frac{1}{2}} S_i B^{-\frac{1}{2}}, \; i = 1, \ldots, r,$$

where $B^{\frac{1}{2}} B^{\frac{1}{2}} = B$. Then, the joint density of $V_1, \ldots, V_r$ is given by

$$\frac{2^{-\frac{1}{2}(m+n)p} \det(\Sigma)^{-\frac{1}{2}(m+n)}}{\Gamma_p(\frac{1}{2}m) \prod_{i=1}^{r} \Gamma_p(\frac{1}{2}n_i)} \, \text{etr}\left(-\frac{1}{2}\Theta\right) \prod_{i=1}^{r} \det(V_i)^{\frac{1}{2}(n_i-p-1)}$$

$$\int_{B>0} \text{etr}\left\{-\frac{1}{2}\Sigma^{-1}B^{\frac{1}{2}}\left(I_p + \sum_{i=1}^{r} V_i\right)B^{\frac{1}{2}}\right\} \det(B)^{\frac{1}{2}(m+n-p-1)}$$

$$ {}_0F_1\left(\frac{1}{2}m; \frac{1}{4}\Theta\Sigma^{-1}B\right) dB, \qquad (6.5.5)$$

where $V_i > 0$, $i = 1, \ldots, r$.

Proof: The joint density of $S_1, \ldots, S_r$ and B is given by (6.5.2). Making the transformation $S_i = B^{\frac{1}{2}} V_i B^{\frac{1}{2}}$, $i = 1, \ldots, r$ with $J(S_1, \ldots, S_r, \to V_1, \ldots, V_r) = \det(B)^{\frac{1}{2}r(p+1)}$, we get the joint density of $V_1, \ldots, V_r$ and B as

$$\frac{2^{-\frac{1}{2}(m+n)p} \det(\Sigma)^{-\frac{1}{2}(m+n)}}{\Gamma_p(\frac{1}{2}m) \prod_{i=1}^{r} \Gamma_p(\frac{1}{2}n_i)} \operatorname{etr}\left(-\frac{1}{2}\Theta\right) \prod_{i=1}^{r} \det(V_i)^{\frac{1}{2}(n_i-p-1)}$$

$$\operatorname{etr}\left\{-\frac{1}{2}\Sigma^{-1}B^{\frac{1}{2}}\left(I_p + \sum_{i=1}^{r} V_i\right)B^{\frac{1}{2}}\right\} \det(B)^{\frac{1}{2}(m+n-p-1)} {}_0F_1\left(\frac{1}{2}m; \frac{1}{4}\Theta\Sigma^{-1}B\right). \quad (6.5.6)$$

Integrating (6.5.6) with respect to B, we get the desired result. ∎

The above result was derived by Troskie (1972). For $\Sigma = I_p$, the integral in (6.5.5), using (1.6.4), becomes

$$\int_{B>0} \operatorname{etr}\left\{-\frac{1}{2}B\left(I_p + \sum_{i=1}^{r} V_i\right)\right\} \det(B)^{\frac{1}{2}(m+n-p-1)} {}_0F_1\left(\frac{1}{2}m; \frac{1}{4}\Theta B\right) dB$$

$$= 2^{\frac{1}{2}(m+n)p} \Gamma_p\left[\frac{1}{2}(m+n)\right] \det\left(I_p + \sum_{i=1}^{r} V_i\right)^{-\frac{1}{2}(m+n)}$$

$$_1F_1\left(\frac{1}{2}(m+n); \frac{1}{2}m; \frac{1}{2}\Theta\left(I_p + \sum_{i=1}^{r} V_i\right)^{-1}\right),$$

and the density (6.5.5) simplifies to

$$\left\{\beta_p\left(\frac{1}{2}n_1, \ldots, \frac{1}{2}n_r; \frac{1}{2}m\right)\right\}^{-1} \operatorname{etr}\left(-\frac{1}{2}\Theta\right) \prod_{i=1}^{r} \det(V_i)^{\frac{1}{2}(n_i-p-1)} \det\left(I_p + \sum_{i=1}^{r} V_i\right)^{-\frac{1}{2}(m+n)}$$

$$_1F_1\left(\frac{1}{2}(m+n); \frac{1}{2}m; \frac{1}{2}\Theta\left(I_p + \sum_{i=1}^{r} V_i\right)^{-1}\right), \quad V_i > 0, \ i = 1, \ldots, r. \quad (6.5.7)$$

For $\Theta = 0$, the density (6.5.7) reduces to the Dirichlet type II density. When $\Theta = \operatorname{diag}(\theta, 0, \ldots, 0)$, the density (6.5.7) simplifies to

$$\left\{\beta_p\left(\frac{1}{2}n_1, \ldots, \frac{1}{2}n_r; \frac{1}{2}m\right)\right\}^{-1} \exp\left(-\frac{1}{2}\theta\right) \prod_{i=1}^{r} \det(V_i)^{\frac{1}{2}(n_i-p-1)} \det\left(I_p + \sum_{i=1}^{r} V_i\right)^{-\frac{1}{2}(m+n)}$$

$$_1F_1\left(\frac{1}{2}(m+n); \frac{1}{2}m; \frac{1}{2}\theta\lambda^{11}\right), \quad V_i > 0, \ i = 1, \ldots, r.$$

where $(I_p + \sum_{i=1}^{r} V_i)^{-1} = (\lambda^{jk})$.

THEOREM 6.5.3. *If the joint density of symmetric positive definite random matrices $U_1, \ldots, U_r$ is (6.5.1), then*

$$(i) \ E\left[\prod_{i=1}^{r} \det(U_i)^{n_i}\right]^h = \frac{\Gamma_p[\frac{1}{2}(m+n)] \prod_{i=1}^{r} \Gamma_p(\frac{1}{2}n_i + hn_i)}{\Gamma_p[\frac{1}{2}(m+n) + hn] \prod_{i=1}^{r} \Gamma_p(\frac{1}{2}n_i)} \operatorname{etr}\left(-\frac{1}{2}\Theta\right)$$

$$_1F_1\left(\frac{1}{2}(m+n); \frac{1}{2}(m+n) + hn; \frac{1}{2}\Theta\right),$$

$$\operatorname{Re}(n_i h) > -\frac{1}{2}(n_i - p + 1),$$

and

$$(ii)\ E\Big[\det\Big(I_p - \sum_{i=1}^{r} U_i\Big)^h\Big] = \frac{\Gamma_p(\frac{1}{2}m + h)\Gamma_p[\frac{1}{2}(m+n)]}{\Gamma_p(\frac{1}{2}m)\Gamma_p[\frac{1}{2}(m+n)+h]} \operatorname{etr}\Big(-\frac{1}{2}\Theta\Big)$$

$$_2F_2\Big(\frac{1}{2}m+h, \frac{1}{2}(m+n); \frac{1}{2}m, \frac{1}{2}(m+n)+h; \frac{1}{2}\Theta\Big),$$

$$\operatorname{Re}(h) > -\frac{1}{2}(m-p+1).$$

Proof: (i) From (6.5.1), we have

$$E\Big[\prod_{i=1}^{r} \det(U_i)^{n_i}\Big]^h = \frac{2^{-\frac{1}{2}(m+n)p} \det(\Sigma)^{-\frac{1}{2}(m+n)}}{\Gamma_p(\frac{1}{2}m)\prod_{i=1}^{r}\Gamma_p(\frac{1}{2}n_i)} \operatorname{etr}\Big(-\frac{1}{2}\Theta\Big)$$

$$\int\cdots\int_{\substack{0<U_i<I_p\\0<\sum_{i=1}^{r}U_i<I_p}} \prod_{i=1}^{r} \det(U_i)^{\frac{1}{2}(2hn_i+n_i-p-1)} \det\Big(I_p - \sum_{i=1}^{r} U_i\Big)^{\frac{1}{2}(m-p-1)}$$

$$\int_{S>0} \operatorname{etr}\Big(-\frac{1}{2}\Sigma^{-1}S\Big) \det(S)^{\frac{1}{2}(m+n-p-1)}$$

$$_0F_1\Big(\frac{1}{2}m; \frac{1}{4}\Theta\Sigma^{-1}S^{\frac{1}{2}}\Big(I_p - \sum_{i=1}^{r} U_i\Big)(S^{\frac{1}{2}})'\Big)\, dS\, dU_1 \cdots dU_r. \quad (6.5.8)$$

Now, using the integral given in Problem 1.10 we can write

$$\int\cdots\int_{\substack{0<U_i<I_p\\0<\sum_{i=1}^{r}U_i<I_p}} \prod_{i=1}^{r} det(U_i)^{\frac{1}{2}(2hn_i+n_i-p-1)} \det\Big(I_p - \sum_{i=1}^{r} U_i\Big)^{\frac{1}{2}(m-p-1)}$$

$$_0F_1\Big(\frac{1}{2}m; \frac{1}{4}\Theta\Sigma^{-1}S^{\frac{1}{2}}\Big(I_p - \sum_{i=1}^{r} U_i\Big)(S^{\frac{1}{2}})'\Big)\, dU_1 \cdots dU_r$$

$$= \frac{\prod_{i=1}^{r}\Gamma_p[\frac{1}{2}n_i(1+2h)]\Gamma_p(\frac{1}{2}m)}{\Gamma_p[\frac{1}{2}(m+n)+hn]}\, _0F_1\Big(\frac{1}{2}(m+n)+hn; \frac{1}{4}\Theta\Sigma^{-1}S\Big),$$

$$\operatorname{Re}(n_ih) > -\frac{1}{2}(n_i-p+1). \quad (6.5.9)$$

Substituting (6.5.9) in (6.5.8), we have

$$E\Big[\prod_{i=1}^{r} \det(U_i)^{n_i}\Big]^h = \frac{2^{-\frac{1}{2}(m+n)p} \det(\Sigma)^{-\frac{1}{2}(m+n)}}{\prod_{i=1}^{r}\Gamma_p(\frac{1}{2}n_i)} \frac{\prod_{i=1}^{r}\Gamma_p[\frac{1}{2}n_i(1+2h)]}{\Gamma_p[\frac{1}{2}(m+n)+hn]} \operatorname{etr}\Big(-\frac{1}{2}\Theta\Big)$$

$$\int_{S>0} \operatorname{etr}\Big(-\frac{1}{2}\Sigma^{-1}S\Big) \det(S)^{\frac{1}{2}(m+n-p-1)}$$

$$_0F_1\Big(\frac{1}{2}(m+n)+hn; \frac{1}{4}\Theta\Sigma^{-1}S\Big)\, dS. \quad (6.5.10)$$

Now, using integral (1.6.4), we obtain the desired result.

(ii) The derivation of $E[\det(I_p - \sum_{i=1}^r U_i)^h]$ is similar. ∎

de Waal (1972b) and Gupta and Nagar (1987) have derived asymptotic expansions of suitable functions of $-2\ln\prod_{i=1}^r \det(U_i)^{n_i}$ and $-2\ln\det(I_p - \sum_{i=1}^r U_i)$.

PROBLEMS

6.1. Let the random matrix T $(p \times m)$ be partitioned as $T = (T_1, \ldots, T_k)$, $T_i(p \times m_i)$, $m_i \geq p$, $i = 1, \ldots, k$ and $m_1 + \cdots + m_k = m$. Define $B_i = T_i T_i'$, $i = 1, \ldots, k$. Prove that
(i) $(B_1, \ldots, B_k) \sim D_p^{II}(\frac{1}{2}m_1, \ldots, \frac{1}{2}m_k; \frac{1}{2}(n+p-1))$ if $T \sim T_{p,m}(n, 0, I_p, I_m)$ and
(ii) $(B_1, \ldots, B_k) \sim D_p^I(\frac{1}{2}m_1, \ldots, \frac{1}{2}m_k; \frac{1}{2}(n+p-1))$ if $T \sim IT_{p,m}(n, 0, I_p, I_m)$.

6.2. Let $(U_1, \ldots, U_r) \sim D_p^I(a_1, \ldots, a_r; b)$ and $X \sim B_p^I(\sum_{i=1}^r a_i + b, c)$ be independent. Prove that $(X^{\frac{1}{2}} U_1 X^{\frac{1}{2}}, \ldots, X^{\frac{1}{2}} U_r X^{\frac{1}{2}}) \sim D_p^I(a_1, \ldots, a_r; b + c)$.

6.3. Let $(U_1, \ldots, U_r) \sim D_p^I(a_1, \ldots, a_r; b)$. Prove that for any nonzero $\boldsymbol{a} \in \mathbb{R}^p$,

$$\left(\frac{\boldsymbol{a}'U_1\boldsymbol{a}}{\boldsymbol{a}'\boldsymbol{a}}, \ldots, \frac{\boldsymbol{a}'U_r\boldsymbol{a}}{\boldsymbol{a}'\boldsymbol{a}}\right) \sim D_1^I(a_1, \ldots, a_r; b).$$

6.4. Let $(U_1, \ldots, U_r) \sim D_p^I(a_1, \ldots, a_r; b)$ and $X \sim B_p^{II}(c, \sum_{i=1}^r a_i + b)$ be independent. Prove that $((I_p + X)^{-\frac{1}{2}} U_1 (I_p + X)^{-\frac{1}{2}}, \ldots, (I_p + X)^{-\frac{1}{2}} U_r (I_p + X)^{-\frac{1}{2}}) \sim D_p^I(a_1, \ldots, a_r; b + c)$.

6.5. Let $(U_1, \ldots, U_r) \sim D_p^I(a_1, \ldots, a_r; a_{r+1})$. Prove that for any nonzero $\boldsymbol{a} \in \mathbb{R}^p$,

$$\left(\frac{\boldsymbol{a}'\boldsymbol{a}}{\boldsymbol{a}'U_1^{-1}\boldsymbol{a}}, \ldots, \frac{\boldsymbol{a}'\boldsymbol{a}}{\boldsymbol{a}'U_r^{-1}\boldsymbol{a}}\right) \sim D_1^I\Big(a_1 - \frac{1}{2}(p-1), \ldots, a_r - \frac{1}{2}(p-1);$$

$$a_{r+1} + \frac{1}{2}(p-1)(r-1)\Big).$$

6.6. If $(U_1, \ldots, U_r) \sim D_p^I(a_1, \ldots, a_r; a_{r+1})$, then show that

$$\Big(U_1, \ldots, U_{i-1}, I_p - \sum_{j=1}^r U_j, U_{i+1}, \ldots, U_r\Big) \sim D_p^I(a_1, \ldots, a_{i-1}, a_{r+1},$$

$$a_{i+1}, \ldots, a_r; a_i).$$

6.7. Let $(U_1, \ldots, U_r) \sim D_p^I(a_1, \ldots, a_r; a_{r+1})$ and $S \sim W_p(2a, \Sigma)$, $a = \sum_{j=1}^{r+1} a_j$, be independent. Define $W_i = S^{\frac{1}{2}} U_i S^{\frac{1}{2}}$, $i = 1, \ldots, r-1$, and $W_r = S^{\frac{1}{2}}(I_p - \sum_{j=1}^r U_j) S^{\frac{1}{2}}$. Then show that $W_1, \ldots, W_r$ are independent, $W_i \sim W_p(2a_i, \Sigma)$, $i = 1, \ldots, r-1$ and $W_r \sim W_p(2a_{r+1}, \Sigma)$.

6.8. Let $(Z_1, \ldots, Z_r) \sim GD_p^I(a_1, \ldots, a_r; a_{r+1}; \Omega; \Psi_1, \ldots, \Psi_r)$.
(i) Show that, for $s \leq r$, $(Z_1, \ldots, Z_s) \sim GD_p^I(a_1, \ldots, a_s; \sum_{j=s+1}^{r+1} a_j; \Omega - \sum_{j=s+1}^r \Psi_j; \Psi_1, \ldots, \Psi_s)$.
(ii) Prove that $Z_i \sim GB_p^I(a_i, \sum_{j=1(\neq i)}^{r+1} a_j; \Omega - \sum_{j=1(\neq i)}^r \Psi_j; \Psi_i)$, $i = 1, \ldots, r$.
(iii) Derive the conditional distribution of $(Z_{s+1}, \ldots, Z_r)|(Z_1, \ldots, Z_s)$.

6.9. Let $(Y_1, \ldots, Y_r) \sim GD_p^{II}(b_1, \ldots, b_r; b_{r+1}; \Omega; \Psi_1, \ldots, \Psi_r)$. Derive the marginal distribution of $(Y_1, \ldots, Y_s)$, $s \leq r$, and the conditional distribution of $(Y_{s+1}, \ldots, Y_r) | (Y_1, \ldots, Y_s)$.

6.10. Prove Theorem 6.3.1(ii).

6.11. Prove Theorem 6.3.20.

6.12. Prove that the inverse matrix variate Dirichlet distribution is orthogonally invariant, that is, if $(X_1, \ldots, X_r) \sim ID_p(a_1, \ldots, a_r; a_{r+1})$ then for any fixed orthogonal matrix Γ $(p \times p)$, the distribution of $(\Gamma X_1 \Gamma', \Gamma X_2 \Gamma', \ldots, \Gamma X_r \Gamma')$ is same as that of $(X_1, \ldots, X_r)$.

6.13. Let $(X_1, \ldots, X_r) \sim ID_p(a_1, \ldots, a_r; a_{r+1})$. Prove that for any nonzero $\boldsymbol{a} \in \mathbb{R}^p$,

$$\left(\frac{\boldsymbol{a}'\boldsymbol{a}}{\boldsymbol{a}'X_1^{-1}\boldsymbol{a}}, \ldots, \frac{\boldsymbol{a}'\boldsymbol{a}}{\boldsymbol{a}'X_r^{-1}\boldsymbol{a}} \right) \sim ID_1(a_1, \ldots, a_r; a_{r+1}).$$

6.14. Let $(X_1, \ldots, X_r) \sim ID_p(a_1, \ldots, a_r; b)$ and $Y \sim IB_p(\sum_{i=1}^r a_i + b, c)$ be independent. Prove that $(Y^{\frac{1}{2}}X_1 Y^{\frac{1}{2}}, \ldots, Y^{\frac{1}{2}}X_r Y^{\frac{1}{2}}) \sim ID_p(a_1, \ldots, a_r; b + c)$.

6.15. Let $(X_1, \ldots, X_r) \sim ID_p(a_1, \ldots, a_r; a_{r+1})$. Show that, for $s \leq r$, $(X_1, \ldots, X_s) \sim ID_p(a_1, \ldots, a_s; \sum_{i=s+1}^{r+1} a_i)$, and the density of $(X_{s+1}, \ldots, X_r) | (X_1, \ldots, X_s)$ is given by

$$\frac{\prod_{i=s+1}^r \det(X_i)^{-a_i - \frac{1}{2}(p+1)} \det(I_p - \sum_{i=1}^s X_i^{-1} - \sum_{i=s+1}^r X_i^{-1})^{a_{r+1} - \frac{1}{2}(p+1)}}{\beta_p(a_{s+1}, \ldots, a_r; a_{r+1}) \det(I_p - \sum_{i=1}^s X_i^{-1})^{\sum_{i=s+1}^{r+1} a_i - \frac{1}{2}(p+1)}},$$

$$0 < X_i^{-1} < I_p,\ i = s+1, \ldots, r,\ \sum_{i=s+1}^r X_i^{-1} < I_p - \sum_{i=1}^s X_i^{-1}.$$

Hence or otherwise show that $X_i \sim IB_p(a_i, \sum_{j(\neq i)=1}^{r+1} a_j)$.

6.16. If $(X_1, \ldots, X_r) \sim ID_p(a_1, \ldots, a_r; a_{r+1})$, then show that

$$\left(X_1, \ldots, X_{i-1}, \left(I_p - \sum_{j=1}^r X_j^{-1}\right)^{-1}, X_{i+1}, \ldots, X_r \right)$$

$$\sim ID_p(a_1, \ldots, a_{i-1}, a_{r+1}, a_{i+1}, \ldots, a_r; a_i).$$

6.17. Let $(X_1, \ldots, X_r) \sim ID_p(a_1, \ldots, a_r; a_{r+1})$. Define $Y_i = (I_p - \sum_{j=1}^s X_j^{-1})^{\frac{1}{2}} X_i (I_p - \sum_{j=1}^s X_j^{-1})^{\frac{1}{2}}$, $i = s+1, \ldots, r$. Then show that
(i) $(Y_{s+1}, \ldots, Y_r) \sim ID_p(a_{s+1}, \ldots, a_r; a_{r+1})$, and
(ii) $(Y_{s+1}, \ldots, Y_r)$ and $(X_1, \ldots, X_s)$ are independent.

6.18. Let $(X_1, \ldots, X_r) \sim ID_p(a_1, \ldots, a_r; a_{r+1})$ and $V \sim IW_p(2a + p + 1, \Psi)$, $a = \sum_{j=1}^{r+1} a_j$, be independent. Define $Y_i = V^{\frac{1}{2}} X_i V^{\frac{1}{2}}$, $i = 1, \ldots, r-1$, and $Y_r = V^{\frac{1}{2}}(I_p - \sum_{j=1}^r X_j^{-1})^{-1} V^{\frac{1}{2}}$. Then show that $Y_1, \ldots, Y_r$ are independent, $Y_i \sim IW_p(2a_i + p + 1, \Psi)$, $i = 1, \ldots, r-1$ and $Y_r \sim W_p(2a_{r+1} + p + 1, \Psi)$.

6.19. Let $(X_1, \ldots, X_r) \sim ID_p(a_1, \ldots, a_r; a_{r+1})$ and the random matrix X_i be partitioned as $X_i = \begin{pmatrix} X_{11(i)} & X_{21(i)} \\ X_{12(i)} & X_{22(i)} \end{pmatrix}$, where $X_{11(i)}$ is a matrix of order $q \times q$. Then show that

(i) $(X_{22(1)}, \ldots, X_{22(r)}) \sim ID_{p-q}(a_1 - \frac{1}{2}q, \ldots, a_r - \frac{1}{2}q; a_{r+1} + \frac{1}{2}q(r-1))$, and

(ii) $(X_{11\cdot2(1)}, \ldots, X_{11\cdot2(r)}) \sim ID_p(a_1, \ldots, a_r; a_{r+1})$ where $X_{11\cdot2(i)} = X_{11(i)} - X_{12(i)} X_{22(i)}^{-1} X_{21(i)}$.

6.20. Let $(X_1, \ldots, X_r) \sim ID_p(a_1, \ldots, a_r; a_{r+1})$ and $A\,(q \times p)$ be a constant matrix such that $AA' = I_q$. Then show that

$$(AX_1 A', \ldots, AX_r A') \sim ID_q\Big(a_1 - \frac{1}{2}(p-q), \ldots, a_r - \frac{1}{2}(p-q);$$

$$a_{r+1} + \frac{1}{2}(p-q)(r-1)\Big).$$

6.21. Prove Theorem 6.4.3(ii).

6.22. Prove Theorem 6.4.4(ii).

6.23. Let $(U_1, \ldots, U_r)$ be distributed as in Theorem 6.5.1. Derive the distribution of $Z = \sum_{i=1}^r U_i$.

6.24. Let $(V_1, \ldots, V_r)$ be distributed as in Theorem 6.5.2. Derive the distribution of $Z = \sum_{i=1}^r V_i$.

6.25. Prove Theorem 6.5.3(ii).

6.26. Let $\{x_{ij}, j = 1, \ldots, N_i\}$ be a random sample from a p-variate normal population with mean vector μ_i and covariance matrix Σ_i, $i = 1, \ldots, k$. Then show that for testing $H : \Sigma_1 = \cdots = \Sigma_k; \mu_1 = \cdots = \mu_k$, the likelihood ratio criterion is a function of Dirichlet type I matrices.

6.27. Let $(V_1, \ldots, V_r) \sim D_p^{II}(b_1, \ldots, b_r; b_{r+1})$ and $W_i = b_{r+1} V_i$, $i = 1, \ldots, r$. Then show that the p.d.f. of $W = (W_1, \ldots, W_r)$ can be expanded as

$$g(W) = \prod_{i=1}^r \frac{\det(W_i)^{b_i - \frac{1}{2}(p+1)} \operatorname{etr}(-W_i)}{\Gamma_p(b_i)} \left[1 + \frac{a_1}{2b_{r+1}} + \frac{3a_1^2 + 4a_2}{24b_{r+1}^2} + O(b_{r+1}^{-3}) \right],$$

$$W_i > 0, \ i = 1, \ldots, r,$$

where $a_1 = \operatorname{tr}[(-\sum_{i=1}^r W_i)^2] + 2m \operatorname{tr}(-\sum_{i=1}^r W_i) + m^2 p - \frac{1}{2}mp(p+1)$, $a_2 = 2\operatorname{tr}[(-\sum_{i=1}^r W_i)^3] + 3m \operatorname{tr}[(-\sum_{i=1}^r W_i)^2] - m^3 p + \frac{3}{4}m^2 p(p+1) - \frac{1}{8}mp(2p^2 + 3p - 1)$ and $m = \sum_{j=1}^r b_j$.

(Gupta and Song, 1990)

CHAPTER 7

DISTRIBUTION OF QUADRATIC FORMS

7.1. INTRODUCTION

Let x $(n \times 1)$ be a random vector and A $(n \times n)$ be a symmetric matrix. The quadratic form in x associated with A is defined as

$$s = x'Ax. \qquad (7.1.1)$$

The distribution of s, assuming $x \sim N_p(\mu, \Sigma)$, has been studied extensively by many authors, e.g., Kotz, Johnson and Boyd (1967a, 1967b), Johnson and Kotz (1970, 1972), Khatri (1980), Konishi, Niki and Gupta (1988), and Mathai and Provost (1992).

When X $(p \times n)$ is a random matrix, the matrix quadratic form in X associated with A is defined by

$$S = XAX'. \qquad (7.1.2)$$

In this chapter we study the distribution of S assuming X has matrix variate normal distribution. The distribution of S has been studied by Khatri (1959b, 1962, 1963, 1966, 1971, 1975, 1977, 1980), Hogg (1963), Hayakawa (1966, 1972), Shah (1970), Crowther (1975), and Gupta and Varga (1991, 1992, 1993, 1994d).

7.2. DENSITY FUNCTION

In this section we derive the density of S when $E(X) = 0$. First we give the derivation given by Khatri (1966).

THEOREM 7.2.1. *Let* $X \sim N_{p,n}(0, \Sigma \otimes \Psi)$, $n \geq p$, $\Sigma > 0$, *and* $\Psi > 0$. *Then the density function of* $S = XAX'$, *where* A $(n \times n) > 0$, *is given by*

$$\left\{ 2^{\frac{1}{2}np} \Gamma_p\left(\frac{1}{2}n\right) \right\}^{-1} \det(A\Psi)^{-\frac{1}{2}p} \det(\Sigma)^{-\frac{1}{2}n} \det(S)^{\frac{1}{2}(n-p-1)} \operatorname{etr}\left(-\frac{1}{2}q^{-1}\Sigma^{-1}S\right)$$

$$_0F_0^{(n)}\left(B, \frac{1}{2}q^{-1}\Sigma^{-1}S\right), \ S > 0, \qquad (7.2.1)$$

where $B = I_n - qA^{-\frac{1}{2}}\Psi^{-1}A^{-\frac{1}{2}}$, $q > 0$ is an arbitrary constant and $A^{\frac{1}{2}}A^{\frac{1}{2}} = A$.

Proof: The density of X is

$$(2\pi)^{-\frac{1}{2}np}\det(\Sigma)^{-\frac{1}{2}n}\det(\Psi)^{-\frac{1}{2}p}\operatorname{etr}\left(-\frac{1}{2}\Sigma^{-1}X\Psi^{-1}X'\right), \ X \in \mathbb{R}^{p\times n}.$$

Transforming $Y = XA^{\frac{1}{2}}$, with the Jacobian $J(X \to Y) = \det(A)^{-\frac{1}{2}p}$, we obtain the density of Y as

$$(2\pi)^{-\frac{1}{2}np}\det(\Sigma)^{-\frac{1}{2}n}\det(A\Psi)^{-\frac{1}{2}p}\operatorname{etr}\left(-\frac{1}{2}\Sigma^{-1}YA^{-\frac{1}{2}}\Psi^{-1}A^{-\frac{1}{2}}Y'\right)$$

$$= (2\pi)^{-\frac{1}{2}np}\det(\Sigma)^{-\frac{1}{2}n}\det(A\Psi)^{-\frac{1}{2}p}\operatorname{etr}\left(-\frac{1}{2}q^{-1}\Sigma^{-1}YY' + \frac{1}{2}q^{-1}\Sigma^{-1}YBY'\right), Y \in \mathbb{R}^{p\times n}.$$

Now using the Definition 1.6.1, we can write

$$\operatorname{etr}\left(\frac{1}{2}q^{-1}\Sigma^{-1}YBY'\right) = {}_0F_0^{(n)}\left(\frac{1}{2}q^{-1}Y'\Sigma^{-1}YB\right),$$

and integrating out the density of Y over the surface $YY' = S$, we get the density of S as

$$(2\pi)^{-\frac{1}{2}np}\det(\Sigma)^{-\frac{1}{2}n}\det(A\Psi)^{-\frac{1}{2}p}g(B), \tag{7.2.2}$$

where

$$g(B) = \int_{YY'=S}\operatorname{etr}\left(-\frac{1}{2}q^{-1}\Sigma^{-1}YY'\right){}_0F_0^{(n)}\left(\frac{1}{2}q^{-1}Y'\Sigma^{-1}YB\right)dY. \tag{7.2.3}$$

Since B is a symmetric matrix, the integral (7.2.3) is invariant under the transformation $B \to HBH'$, $H \in O(n)$, and integration with respect to H over orthogonal group $O(n)$. Hence using (1.6.3), we get

$$g(B) = \int_{YY'=S}\operatorname{etr}\left(-\frac{1}{2}q^{-1}\Sigma^{-1}YY'\right)\int_{H\in O(n)}{}_0F_0^{(n)}\left(\frac{1}{2}q^{-1}Y'\Sigma^{-1}YHBH'\right)[dH]\,dY$$

$$= \int_{YY'=S}\operatorname{etr}\left(-\frac{1}{2}q^{-1}\Sigma^{-1}YY'\right){}_0F_0^{(n)}\left(B,\frac{1}{2}q^{-1}\Sigma^{-1}YY'\right)dY$$

$$= \frac{\pi^{\frac{1}{2}np}}{\Gamma_p(\frac{1}{2}n)}\det(S)^{\frac{1}{2}(n-p-1)}\operatorname{etr}\left(-\frac{1}{2}q^{-1}\Sigma^{-1}S\right){}_0F_0^{(n)}\left(B,\frac{1}{2}q^{-1}\Sigma^{-1}S\right). \tag{7.2.4}$$

The last step is obtained by using (1.4.24). Now substituting for $g(B)$ from (7.2.4) in (7.2.2) we get the desired result. ∎

We will write $S \sim Q_{p,n}(A, \Sigma, \Psi)$ if the density of S is (7.2.1). It may be noted here that for $A\Psi = I_n$ the density (7.2.1) reduces to the Wishart density $W_p(n, \Sigma)$.

The density of S, in an equivalent form, can also be written as

$$\left\{2^{\frac{1}{2}np}\Gamma_p\left(\frac{1}{2}n\right)\right\}^{-1}\det(A\Psi)^{-\frac{1}{2}p}\det(\Sigma)^{-\frac{1}{2}n}\det(S)^{\frac{1}{2}(n-p-1)}$$

$$_0F_0^{(n)}\left(\Psi^{-1}A^{-1}, -\frac{1}{2}\Sigma^{-1}S\right), \ S > 0. \tag{7.2.5}$$

By substituting from

$$_0F_0(R, S) = \sum_{k=0}^{\infty} \sum_{\kappa} \frac{C_\kappa(R)C_\kappa(S)}{C_\kappa(I)k!} \tag{7.2.6}$$

in (7.2.5) we obtain the expansion in terms of zonal polynomials, which, however, is only slowly convergent. The following expansion in terms of Laguerre polynomials may be preferable for computational purposes

$$\left\{ (2q)^{\frac{1}{2}np} \Gamma_p\left(\tfrac{1}{2}n\right) \right\}^{-1} \det(\Sigma)^{-\frac{1}{2}n} \det(S)^{\frac{1}{2}(n-p-1)} \operatorname{etr}\left(-\tfrac{1}{2}q^{-1}\Sigma^{-1}S \right)$$

$$\sum_{k=0}^{\infty} \sum_{\kappa} \frac{C_\kappa(I_n - q^{-1}A^{-\frac{1}{2}}\Psi^{-1}A^{-\frac{1}{2}})}{C_\kappa(I_n)\,k!} L_\kappa^{\frac{1}{2}(n-p-1)}\left(\tfrac{1}{2}q\Sigma^{-1}S \right), \; S > 0. \tag{7.2.7}$$

The forms of the density (7.2.5) and (7.2.7), for $\Psi = I_n$, were given by Hayakawa (1966) and Shah (1970) respectively.

Khatri (1966) also derived the following form for the density of S, which is useful in obtaining expected values of $C_\kappa(S)$.

THEOREM 7.2.2. *Let* $X \sim N_{p,n}(0, \Sigma \otimes \Psi)$, $n \geq p$, *and* $\Sigma > 0$, $\Psi > 0$. *Then the density of* $S = XAX'$, *where* $A\,(n \times n) > 0$, *is given by*

$$\left\{ 2^{\frac{1}{2}np} \Gamma_p\left(\tfrac{1}{2}n\right) \right\}^{-1} \det(\Sigma)^{-\frac{1}{2}n} \det(Q)^{\frac{1}{2}p} \det(S)^{\frac{1}{2}(n-p-1)}$$

$$\int_{H \in O(n)} \operatorname{etr}\left(-\tfrac{1}{2}\Sigma^{-\frac{1}{2}}H_1 Q H_1' \Sigma^{-\frac{1}{2}} S \right) [dH], \tag{7.2.8}$$

where $H' = (\, H_1' \quad H_2' \,)$ *is an* $n \times n$ *orthogonal matrix with* $H_1\,(p \times n)$ *and* $H_2\,((n-p) \times n)$ *and* $Q^{-1} = A^{\frac{1}{2}}\Psi A^{\frac{1}{2}}$.

Proof: By using (7.2.6), and (1.5.11), we have

$$_0F_0^{(n)}\left(B, \tfrac{1}{2}q^{-1}\Sigma^{-1}S \right) = \sum_{k=0}^{\infty} \sum_{\kappa} \frac{C_\kappa(B)C_\kappa(\tfrac{1}{2}q^{-1}\Sigma^{-1}S)}{C_\kappa(I_n)k!}$$

$$= \sum_{k=0}^{\infty} \sum_{\kappa} \frac{1}{k!} \int_{H \in O(n)} C_\kappa\left(BH'\begin{pmatrix} \tfrac{1}{2}q^{-1}\Sigma^{-1}S & 0 \\ 0 & 0 \end{pmatrix} H \right) [dH]$$

$$= \sum_{k=0}^{\infty} \sum_{\kappa} \frac{1}{k!} \int_{H \in O(n)} C_\kappa\left(BH_1'\left(\tfrac{1}{2}q^{-1}\Sigma^{-1}S\right)H_1 \right) [dH]$$

$$= \int_{H \in O(n)} \operatorname{etr}\left(BH_1'\left(\tfrac{1}{2}q^{-1}\Sigma^{-1}S\right)H_1 \right) [dH], \tag{7.2.9}$$

where $[dH]$ is the unit invariant Haar measure defined on $O(n)$.

Now using (7.2.9) it is easy to see that

$$\operatorname{etr}\left(-\tfrac{1}{2}q^{-1}\Sigma^{-1}S \right) {}_0F_0^{(n)}\left(B, \tfrac{1}{2}q^{-1}\Sigma^{-1}S \right) = \int_{H \in O(n)} \operatorname{etr}\left(-\tfrac{1}{2}\Sigma^{-1}SH_1QH_1' \right) [dH]. \tag{7.2.10}$$

Finally, by substituting (7.2.10) and $Q^{-1} = A^{\frac{1}{2}}\Psi A^{\frac{1}{2}}$ in (7.2.1) we get (7.2.8). ∎

From the p.d.f. (7.2.5) we get the c.d.f. of S as

$$P(S < \Omega) = \left\{2^{\frac{1}{2}np}\Gamma_p\left(\frac{1}{2}n\right)\right\}^{-1}\det(A\Psi)^{-\frac{1}{2}p}\det(\Sigma)^{-\frac{1}{2}n}$$

$$\int_{0<S<\Omega}\det(S)^{\frac{1}{2}(n-p-1)}{}_0F_0^{(n)}\left(\Psi^{-1}A^{-1}, -\frac{1}{2}\Sigma^{-1}S\right)dS. \quad (7.2.11)$$

Expanding ${}_0F_0^{(n)}(\Psi^{-1}A^{-1}, -\frac{1}{2}\Sigma^{-1}S)$, using (7.2.6), we can write

$$\int_{0<S<\Omega}\det(S)^{\frac{1}{2}(n-p-1)}{}_0F_0^{(n)}\left(\Psi^{-1}A^{-1}, -\frac{1}{2}\Sigma^{-1}S\right)dS$$

$$= \sum_{k=0}^{\infty}\sum_{\kappa}\frac{C_\kappa(\Psi^{-1}A^{-1})}{C_\kappa(I_n)k!}\int_{0<S<\Omega}\det(S)^{\frac{1}{2}(n-p-1)}C_\kappa\left(-\frac{1}{2}\Sigma^{-1}S\right)dS$$

$$= \det(\Omega)^{\frac{1}{2}n}\sum_{k=0}^{\infty}\sum_{\kappa}\frac{C_\kappa(\Psi^{-1}A^{-1})}{C_\kappa(I_n)k!}\int_{0<W<I_p}\det(W)^{\frac{1}{2}(n-p-1)}C_\kappa\left(-\frac{1}{2}\Omega^{\frac{1}{2}}\Sigma^{-1}\Omega^{\frac{1}{2}}W\right)dW$$

$$= \det(\Omega)^{\frac{1}{2}n}\sum_{k=0}^{\infty}\sum_{\kappa}\frac{C_\kappa(\Psi^{-1}A^{-1})}{C_\kappa(I_n)k!}\frac{\Gamma_p(\frac{1}{2}n,\kappa)\Gamma_p[\frac{1}{2}(p+1)]}{\Gamma_p(\frac{1}{2}(n+p+1),\kappa)}C_\kappa\left(-\frac{1}{2}\Omega^{\frac{1}{2}}\Sigma^{-1}\Omega^{\frac{1}{2}}\right)$$

$$= \det(\Omega)^{\frac{1}{2}n}\frac{\Gamma_p(\frac{1}{2}n)\Gamma_p[\frac{1}{2}(p+1)]}{\Gamma_p[\frac{1}{2}(n+p+1)]}{}_1F_1^{(n)}\left(\frac{1}{2}n; \frac{1}{2}(n+p+1); \Psi^{-1}A^{-1}, -\frac{1}{2}\Sigma^{-1}\Omega\right). \quad (7.2.12)$$

The last two expressions have been obtained by using (1.5.16), and (1.6.2), respectively. Substituting from (7.2.12) in (7.2.11), we get the c.d.f. of S as

$$P(S < \Omega) = \frac{\Gamma_p[\frac{1}{2}(p+1)]}{\Gamma_p[\frac{1}{2}(n+p+1)]}\det(A\Psi)^{-\frac{1}{2}p}\det(2\Sigma)^{-\frac{1}{2}n}\det(\Omega)^{\frac{1}{2}n}$$

$${}_1F_1^{(n)}\left(\frac{1}{2}n; \frac{1}{2}(n+p+1); \Psi^{-1}A^{-1}, -\frac{1}{2}\Sigma^{-1}\Omega\right). \quad (7.2.13)$$

The corresponding results for the p.d.f.'s (7.2.1) and (7.2.7) can also be derived. However, they are quite involved.

7.3.   PROPERTIES

In this section we first derive the m.g.f. of S and then study some properties of its distribution.

THEOREM 7.3.1. *If $S \sim Q_{p,n}(A, \Sigma, \Psi)$, then the moment generating function of S is*

$$M_S(Z) = \det(q^{-1}A\Psi)^{-\frac{1}{2}p}\det(\Lambda)^{-\frac{1}{2}n}{}_1F_0^{(n)}\left(\frac{1}{2}n; B, \Lambda^{-1}\right), \quad (7.3.1)$$

$$= \det(q^{-1}A\Psi)^{-\frac{1}{2}p}\det(\Lambda)^{-\frac{1}{2}n}\prod_{j=1}^{n}\det(I_p - b_j\Lambda^{-1})^{-\frac{1}{2}}, \quad (7.3.2)$$

where b_j, $j = 1, \ldots, n$ are the roots of $B = I_n - qA^{-\frac{1}{2}}\Psi^{-1}A^{-\frac{1}{2}}$, and $\Lambda = I_p - 2q\Sigma Z$.

Proof: From (7.2.1), we have

$$M_S(Z) = \left\{2^{\frac{1}{2}np}\Gamma_p\left(\frac{1}{2}n\right)\right\}^{-1}\det(A\Psi)^{-\frac{1}{2}p}\det(\Sigma)^{-\frac{1}{2}n}\int_{S>0}\det(S)^{\frac{1}{2}(n-p-1)}\operatorname{etr}(ZS)$$

$$\operatorname{etr}\left(-\frac{1}{2}q^{-1}\Sigma^{-1}S\right){}_0F_0^{(n)}\left(B,\frac{1}{2}q^{-1}\Sigma^{-1}S\right)dS$$

$$= \left\{2^{\frac{1}{2}np}\Gamma_p\left(\frac{1}{2}n\right)\right\}^{-1}\det(A\Psi)^{-\frac{1}{2}p}\det(\Sigma)^{-\frac{1}{2}n}\int_{S>0}\det(S)^{\frac{1}{2}(n-p-1)}$$

$$\operatorname{etr}\left(-\frac{1}{2}q^{-1}S\Sigma^{-1}\Lambda\right){}_0F_0^{(n)}\left(B,\frac{1}{2}q^{-1}\Sigma^{-1}S\right)dS.$$

Now, using (1.6.5), we get (7.3.1). To derive (7.3.2), consider

$$E[\operatorname{etr}(ZXAX')] = (2\pi)^{-\frac{1}{2}np}\det(\Sigma)^{-\frac{1}{2}n}\det(\Psi)^{-\frac{1}{2}p}$$

$$\int_{X\in\mathbb{R}^{p\times n}}\operatorname{etr}\left(ZXAX' - \frac{1}{2}\Sigma^{-1}X\Psi^{-1}X'\right)dX. \qquad (7.3.3)$$

Now making the transformation $Y = q^{-\frac{1}{2}}\Sigma^{-\frac{1}{2}}XA^{\frac{1}{2}}$, with Jacobian $J(X \to Y) = q^{\frac{1}{2}np}\det(\Sigma)^{\frac{1}{2}n}\det(A)^{-\frac{1}{2}p}$, we obtain

$$M_S(Z) = (2\pi)^{-\frac{1}{2}np}\det(q^{-1}A\Psi)^{-\frac{1}{2}p}\int_{Y\in\mathbb{R}^{p\times n}}\operatorname{etr}\left(\frac{1}{2}YBY' - \frac{1}{2}\Lambda YY'\right)dY. \qquad (7.3.4)$$

Since YY' is invariant under post multiplication of Y by an orthogonal matrix, we can take B in (7.3.4) to be a diagonal matrix with b_j's as diagonal elements. Hence (7.3.4) can be written as

$$M_S(Z) = (2\pi)^{-\frac{1}{2}np}\det(q^{-1}A\Psi)^{-\frac{1}{2}p}\prod_{i=1}^{n}\int_{\boldsymbol{y}_i\in\mathbb{R}^p}\exp\left\{-\frac{1}{2}\boldsymbol{y}_i'(\Lambda - b_iI_p)\boldsymbol{y}_i\right\}d\boldsymbol{y}_i, \qquad (7.3.5)$$

where $\boldsymbol{y}_i$ $(p \times 1)$ is the ith column of Y. Now (7.3.2) follows immediately by evaluating the integral. ∎

An alternate expression for $M_S(Z)$, given in (7.3.2), is

$$M_S(Z) = \prod_{j=1}^{n}\det(I_p - 2\ell_j\Sigma^{\frac{1}{2}}Z\Sigma^{\frac{1}{2}})^{-\frac{1}{2}}, \qquad (7.3.6)$$

where ℓ_j, $j = 1, \ldots, n$, are the characteristic roots of $A\Psi$.

If in (7.3.3), we transform $Y = \Sigma^{-\frac{1}{2}}X\Psi^{-\frac{1}{2}}$, with the Jacobian $J(X \to Y) = \det(\Sigma)^{\frac{1}{2}n}\det(\Psi)^{\frac{1}{2}p}$, and use the expansion

$$\operatorname{etr}(Y'\Sigma^{\frac{1}{2}}Z\Sigma^{\frac{1}{2}}Y\Psi^{\frac{1}{2}}A\Psi^{\frac{1}{2}}) = \sum_{k=0}^{\infty}\sum_{\kappa}\frac{1}{k!}C_\kappa(Y'\Sigma^{\frac{1}{2}}Z\Sigma^{\frac{1}{2}}Y\Psi^{\frac{1}{2}}A\Psi^{\frac{1}{2}}),$$

where $\kappa = (k_1, \ldots, k_n)$, $k_1 \geq \cdots \geq k_n \geq 0$, $k_1 + \cdots + k_n = k$, we get

$$M_S(Z) = (2\pi)^{-\frac{1}{2}np} \sum_{k=0}^{\infty} \sum_{\kappa} \frac{1}{k!} \int_{Y \in \mathbb{R}^{p \times n}} \operatorname{etr}\left(-\frac{1}{2}YY'\right)$$

$$C_\kappa(Y'\Sigma^{\frac{1}{2}}Z\Sigma^{\frac{1}{2}}Y\Psi^{\frac{1}{2}}A\Psi^{\frac{1}{2}})\, dY. \tag{7.3.7}$$

Since (7.3.7) is homogeneous and symmetric function in $\Psi^{\frac{1}{2}}A\Psi^{\frac{1}{2}}$, we have as in the proof of Theorem 7.2.1,

$$M_S(Z) = (2\pi)^{-\frac{1}{2}np} \sum_{k=0}^{\infty} \sum_{\kappa} \frac{C_\kappa(A\Psi)}{C_\kappa(I_n)k!} \int_{Y \in \mathbb{R}^{p \times n}} \operatorname{etr}\left(-\frac{1}{2}YY'\right) C_\kappa(\Sigma^{\frac{1}{2}}Z\Sigma^{\frac{1}{2}}YY')\, dY$$

$$= (2\pi)^{-\frac{1}{2}np} \sum_{k=0}^{\infty} \sum_{\kappa} \frac{C_\kappa(A\Psi)}{C_\kappa(I_n)k!} \int_{S>0} \int_{YY'=S} \operatorname{etr}\left(-\frac{1}{2}YY'\right) C_\kappa(\Sigma^{\frac{1}{2}}Z\Sigma^{\frac{1}{2}}YY')\, dY\, dS$$

$$= \sum_{k=0}^{\infty} \sum_{\kappa} \frac{(\frac{1}{2}n)_\kappa}{k!} \frac{C_\kappa(A\Psi)C_\kappa(2\Sigma Z)}{C_\kappa(I_n)}, \tag{7.3.8}$$

where the last two expressions have been obtained by using Theorem 1.4.10 and Lemma 1.5.2.

Another expression for the m.g.f. of S can be given by using the density (7.2.8) as follows.

$$M_S(Z) = \left\{2^{\frac{1}{2}np} \Gamma_p\left(\frac{1}{2}n\right)\right\}^{-1} \det(\Sigma)^{-\frac{1}{2}n} \det(Q)^{\frac{1}{2}p} \int_{H \in O(n)} \int_{S>0} \det(S)^{\frac{1}{2}(n-p-1)}$$

$$\operatorname{etr}\left(ZS - \frac{1}{2}\Sigma^{-\frac{1}{2}}H_1QH_1'\Sigma^{-\frac{1}{2}}S\right) dS\, [dH]$$

$$= \det(Q)^{\frac{1}{2}p} \int_{H \in O(n)} \det(H_1QH_1')^{-\frac{1}{2}n} \det(I_p - 2\Sigma^{\frac{1}{2}}Z\Sigma^{\frac{1}{2}}(H_1QH_1')^{-1})^{-\frac{1}{2}n}\, [dH]. \tag{7.3.9}$$

Using the expansion, in terms of zonal polynomials,

$$\det(I_p - 2\Sigma^{\frac{1}{2}}Z\Sigma^{\frac{1}{2}}(H_1QH_1')^{-1})^{-\frac{1}{2}n} = \sum_{k=0}^{\infty} \sum_{\kappa} \frac{2^k}{k!}\left(\frac{1}{2}n\right)_\kappa C_\kappa(\Sigma^{\frac{1}{2}}Z\Sigma^{\frac{1}{2}}(H_1QH_1')^{-1}),$$

in (7.3.9) we get

$$M_S(Z) = \det(Q)^{\frac{1}{2}p} \sum_{k=0}^{\infty} \sum_{\kappa} \frac{2^k}{k!}\left(\frac{1}{2}n\right)_\kappa g(\Sigma^{\frac{1}{2}}Z\Sigma^{\frac{1}{2}}), \tag{7.3.10}$$

where

$$g(\Sigma^{\frac{1}{2}}Z\Sigma^{\frac{1}{2}}) = \int_{H \in O(n)} \det(H_1QH_1')^{-\frac{1}{2}n} C_\kappa(\Sigma^{\frac{1}{2}}Z\Sigma^{\frac{1}{2}}(H_1QH_1')^{-1})\, [dH]. \tag{7.3.11}$$

Note that $g(\Sigma^{\frac{1}{2}}Z\Sigma^{\frac{1}{2}})$ is a homogeneous and symmetric function in $\Sigma^{\frac{1}{2}}Z\Sigma^{\frac{1}{2}}$. Proceeding as in the proof of Theorem 7.2.1, we get

$$g(\Sigma^{\frac{1}{2}}Z\Sigma^{\frac{1}{2}}) = \frac{C_\kappa(Z\Sigma)}{C_\kappa(I_n)} \int_{H \in O(n)} \det(H_1QH_1')^{-\frac{1}{2}n} C_\kappa((H_1QH_1')^{-1})\, [dH]. \tag{7.3.12}$$

Since (7.3.8) and (7.3.10) are both m.g.f.'s of S, by comparing the coefficients of $C_\kappa(ZS)$, we get

$$\det(Q)^{\frac{1}{2}p} \int_{H \in O(n)} \det(H_1 Q H_1')^{-\frac{1}{2}n} C_\kappa((H_1 Q H_1')^{-1}) \, [dH] = \frac{C_\kappa(I_p)}{C_\kappa(I_n)} C_\kappa(Q^{-1}). \quad (7.3.13)$$

Substituting (7.3.13) in (7.3.12), we get

$$g(\Sigma^{\frac{1}{2}} Z \Sigma^{\frac{1}{2}}) = \frac{C_\kappa(Z\Sigma) C_\kappa(Q^{-1})}{C_\kappa(I_n)} \det(Q)^{-\frac{1}{2}p}. \quad (7.3.14)$$

THEOREM 7.3.2. *Let $S \sim Q_{p,n}(A, \Sigma, \Psi)$, and $B \,(p \times p)$ be any constant nonsingular matrix. Then, $BSB' \sim Q_{p,n}(A, B\Sigma B', \Psi)$.*

Proof: The result follows by transforming $W = BSB'$, with the Jacobian $J(S \to W) = \det(B)^{-p-1}$ in the density (7.2.1). ∎

COROLLARY 7.3.2.1. *Let $S \sim Q_{p,n}(A, \Sigma, \Psi)$, and $\Sigma = (C'C)^{-1}$. Then, $CSC' \sim Q_{p,n}(A, I_p, \Psi)$.*

THEOREM 7.3.3. *Let $S \sim Q_{p,n}(A, I_p, \Psi)$ and $H \,(p \times p)$ be an orthogonal matrix, whose elements are either constants or random variables distributed independently of S. Then, the distribution of S is invariant under the transformation $S \to HSH'$, and is independent of H in the latter case.*

Proof: First, let H be a constant matrix. Then, from Theorem 7.3.2, $HSH' \sim Q_{p,n}(A, I_p, \Psi)$ since $HH' = I_p$. If, however, H is a random orthogonal matrix, then the conditional distribution of $HSH'|H \sim Q_{p,n}(A, I_p, \Psi)$. Since this distribution does not depend on H, $HSH' \sim Q_{p,n}(A, I_p, \Psi)$. ∎

THEOREM 7.3.4. *Let $S = (S_{ij})$, and $\Sigma = (\Sigma_{ij})$ where $S_{ij} \,(p_i \times p_j)$ and $\Sigma_{ij} \,(p_i \times p_j)$, $i, j = 1, \ldots, k$, $p_1 + \cdots + p_k = p$. If $S \sim Q_{p,n}(A, \Sigma, \Psi)$, then $S_{ii} \sim Q_{p_i,n}(A, \Sigma_{ii}, \Psi)$, $i = 1, \ldots, k$. Moreover, if $\Sigma_{ij} = 0$, $i \neq j$, then they are independent.*

Proof: By using the definition $S = XAX'$, where $X \sim N_{p,n}(0, \Sigma \otimes \Psi)$, we can write $S_{ij} = X_i A X_j'$, with $X = (X_1', \ldots, X_k')$, $X_i \,(p_i \times n)$. Then $X_i \sim N_{p_i,n}(0, \Sigma_{ii} \otimes \Psi)$, and consequently $S_{ii} \sim Q_{p_i,n}(A, \Sigma_{ii}, \Psi)$. Further, if $\Sigma_{ij} = 0$, $i \neq j$, then X_i's are independent. Therefore S_{ii}'s are independent. ∎

Next we give results on expected values of functions of quadratic forms. For proofs and other details the reader is referred to Section 7.7.

THEOREM 7.3.5. *Let $X \sim N_{p,n}(0, \Sigma \otimes \Psi)$, and define $S_A = XAX'$ and $S_B = XBX'$, where the constant matrices $A \,(n \times n)$ and $B \,(m \times m)$ need not be symmetric. Then*

(i) $E(S_A) = \operatorname{tr}(A\Psi)\Sigma$,

(ii) $E(S_A C S_B) = \operatorname{tr}(\Psi B' \Psi A') \operatorname{tr}(C\Sigma)\Sigma + \operatorname{tr}(A\Psi) \operatorname{tr}(B\Psi)\Sigma C \Sigma$
   $+ \operatorname{tr}(A\Psi B'\Psi)\Sigma C'\Sigma$,

(iii) $E(\operatorname{tr}(S_B C) S_A) = \operatorname{tr}(A'\Psi B\Psi)\Sigma C'\Sigma + \operatorname{tr}(A\Psi) \operatorname{tr}(B\Psi) \operatorname{tr}(C\Sigma)\Sigma$
   $+ \operatorname{tr}(A'\Psi B'\Psi)\Sigma C\Sigma$,

and

(iv) $\operatorname{cov}(\operatorname{vec}(S_A), \operatorname{vec}(S_B)) = \operatorname{tr}(A\Psi B'\Psi)\Sigma \otimes \Sigma + \operatorname{tr}(A'\Psi B'\Psi)(\Sigma \otimes \Sigma)K_{pp}.$

By substituting $A = B$ in (iv), we get the covariance matrix of $\operatorname{vec}(S_A)$ as

$$\operatorname{cov}(\operatorname{vec}(S_A)) = \operatorname{tr}(A\Psi A'\Psi)\Sigma \otimes \Sigma + \operatorname{tr}(A'\Psi A'\Psi)(\Sigma \otimes \Sigma)K_{pp}.$$

THEOREM 7.3.6. *Let* $S \sim Q_{p,n}(A, \Sigma, \Psi)$. *Then*

$$E(C_\kappa(ZS)) = 2^k \left(\frac{1}{2}n\right)_\kappa \frac{C_\kappa(A\Psi)C_\kappa(Z\Sigma)}{C_\kappa(I_n)}. \qquad (7.3.15)$$

Proof: From the p.d.f. (7.2.8), we obtain

$$E(C_\kappa(ZS)) = \left\{2^{\frac{1}{2}np}\Gamma_p\left(\frac{1}{2}n\right)\right\}^{-1} \det(\Sigma)^{-\frac{1}{2}n} \det(Q)^{\frac{1}{2}p} \int_{H \in O(n)} \int_{S>0} C_\kappa(ZS)$$

$$\det(S)^{\frac{1}{2}(n-p-1)} \operatorname{etr}\left(-\frac{1}{2}\Sigma^{-\frac{1}{2}}H_1QH_1'\Sigma^{-\frac{1}{2}}S\right) dS \, [dH].$$

Next, use of Lemma 1.5.2 yields

$$E(C_\kappa(ZS)) = 2^k \left(\frac{1}{2}n\right)_\kappa \det(Q)^{\frac{1}{2}p} \int_{H \in O(n)} \det(H_1QH_1')^{-\frac{1}{2}n}$$

$$C_\kappa(\Sigma^{\frac{1}{2}}Z\Sigma^{\frac{1}{2}}(H_1QH_1')^{-1}) \, [dH].$$

Finally, substituting from (7.3.14) in the above expression gives the desired result. ∎

When $A\Psi = I_n$, $S \sim W_p(n, \Sigma)$, and (7.3.15) simplifies to

$$E(C_\kappa(ZS)) = 2^k \left(\frac{1}{2}n\right)_\kappa C_\kappa(Z\Sigma). \qquad (7.3.16)$$

The result (7.3.16) was derived by Constantine (1963). Similarly the expectation of a zonal polynomial in S^{-1} is given by

$$E(C_\kappa(ZS^{-1})) = \left\{2^{\frac{1}{2}np}\Gamma_p\left(\frac{1}{2}n\right)\right\}^{-1} \det(\Sigma)^{-\frac{1}{2}n} \det(Q)^{\frac{1}{2}p} \int_{H \in O(n)} \int_{S>0} C_\kappa(ZS^{-1})$$

$$\det(S)^{\frac{1}{2}(n-p-1)} \operatorname{etr}\left(-\frac{1}{2}\Sigma^{-\frac{1}{2}}H_1QH_1'\Sigma^{-\frac{1}{2}}S\right) dS \, [dH]$$

$$= 2^{-k} \frac{\Gamma_p(\frac{1}{2}n, -\kappa)}{\Gamma_p(\frac{1}{2}n)} \det(Q)^{\frac{1}{2}p} \int_{H \in O(n)} \det(H_1QH_1')^{-\frac{1}{2}n}$$

$$C_\kappa(\Sigma^{-\frac{1}{2}}Z\Sigma^{-\frac{1}{2}}H_1QH_1') \, [dH], \quad \frac{1}{2}(n-p+1) > k_1. \quad (7.3.17)$$

The last expression is obtained by using Lemma 1.5.2. The integral in (7.3.17) is a homogeneous and symmetric function in $\Sigma^{-\frac{1}{2}}Z\Sigma^{-\frac{1}{2}}$. Therefore,

$$E(C_\kappa(ZS^{-1})) = 2^{-k} \frac{\Gamma_p(\frac{1}{2}n, -\kappa)}{\Gamma_p(\frac{1}{2}n)} \frac{C_\kappa(Z\Sigma^{-1})}{C_\kappa(I_p)} \det(Q)^{\frac{1}{2}p} \int_{H \in O(n)} \det(H_1QH_1')^{-\frac{1}{2}n}$$

$$C_\kappa(H_1QH_1') \, [dH], \quad \frac{1}{2}(n-p+1) > k_1. \quad (7.3.18)$$

The above integral is not available in the literature. However, for $A^{\frac{1}{2}}\Psi A^{\frac{1}{2}} = Q^{-1} = I_n$, i.e., when $S \sim W_p(n, \Sigma)$, its value is $C_\kappa(I_p)$. Hence,

$$E(C_\kappa(ZS^{-1})) = 2^{-k}\frac{\Gamma_p(\frac{1}{2}n, -\kappa)}{\Gamma_p(\frac{1}{2}n)}C_\kappa(Z\Sigma^{-1}), \quad \frac{1}{2}(n-p+1) > k_1. \qquad (7.3.19)$$

The above expression can be further simplified by using (1.5.9).

7.4. FUNCTIONS OF QUADRATIC FORMS

In this section, we derive some distributions of functions of positive definite matrix quadratic forms. These distributions are useful in deriving the distribution of characteristic roots, which are fundamental in the study of multivariate tests.

THEOREM 7.4.1. Let X $(p \times n)$ and Y $(p \times m)$ be independent, $X \sim N_{p,n}(0, \Sigma \otimes \Psi)$ and $Y \sim N_{p,m}(0, \Sigma \otimes I_m)$. Then the p.d.f. of $F = Y'(XAX')^{-1}Y$, for $m \le p \le n$, is given by

$$\frac{\Gamma_p[\frac{1}{2}(m+n)]}{\Gamma_p(\frac{1}{2}n)\Gamma_m(\frac{1}{2}p)}q^{\frac{1}{2}(m+n)p}\det(A\Psi)^{-\frac{1}{2}p}\det(F)^{\frac{1}{2}(p-m-1)}\det(I_m+qF)^{-\frac{1}{2}(m+n)}$$

$$_1F_0^{(n)}\left(\frac{1}{2}(m+n); B, R^*\right), \quad F > 0, \qquad (7.4.1)$$

where $R^* = \begin{pmatrix} (I_m + qF)^{-1} & 0 \\ 0 & I_{p-m} \end{pmatrix}$.

Proof: Since F is invariant under the transformation $X \to \Sigma^{-\frac{1}{2}}X$, and $Y \to \Sigma^{-\frac{1}{2}}Y$, we can assume $\Sigma = I_p$. Hence the joint p.d.f. of $S = XAX'$ and Y is

$$\left\{2^{\frac{1}{2}(m+n)p}\pi^{\frac{1}{2}mp}\Gamma_p\left(\frac{1}{2}n\right)\right\}^{-1}\det(A\Psi)^{-\frac{1}{2}p}\det(S)^{\frac{1}{2}(n-p-1)}\operatorname{etr}\left\{-\frac{1}{2}(YY'+q^{-1}S)\right\}$$

$$_0F_0^{(n)}\left(B, \frac{1}{2}q^{-1}S\right), \quad S > 0, \ Y \in \mathbb{R}^{p \times m}. \qquad (7.4.2)$$

Now making the transformation $Z = S^{-\frac{1}{2}}Y$, with Jacobian $J(Y \to Z) = \det(S)^{\frac{1}{2}m}$, we get the joint p.d.f. of Z and S as

$$\left\{2^{\frac{1}{2}(m+n)p}\pi^{\frac{1}{2}mp}\Gamma_p\left(\frac{1}{2}n\right)\right\}^{-1}\det(A\Psi)^{-\frac{1}{2}p}\det(S)^{\frac{1}{2}(m+n-p-1)}\operatorname{etr}\left\{-\frac{1}{2}q^{-1}(I_p+qZZ')S\right\}$$

$$_0F_0^{(n)}\left(B, \frac{1}{2}q^{-1}S\right), \quad S > 0, \ Z \in \mathbb{R}^{p \times m}.$$

Integrating this joint p.d.f. with respect to S, using (1.6.5), we get the p.d.f. of Z as

$$\frac{\Gamma_p[\frac{1}{2}(m+n)]}{\pi^{\frac{1}{2}mp}\Gamma_p(\frac{1}{2}n)}q^{\frac{1}{2}(m+n)p}\det(A\Psi)^{-\frac{1}{2}p}\det(I_p+qZZ')^{-\frac{1}{2}(m+n)}$$

$$_1F_0^{(n)}\left(\frac{1}{2}(m+n); B, (I_p+qZZ')^{-1}\right), \quad Z \in \mathbb{R}^{p \times m}.$$

Note that $C_\kappa (I_p + qZZ')^{-1} = C_\kappa \begin{pmatrix} (I_m + qZ'Z)^{-1} & 0 \\ 0 & I_{p-m} \end{pmatrix}$, and $\det(I_p + qZZ') = \det(I_m + qZ'Z)$. Therefore, the density of Z can be written as

$$\frac{\Gamma_p[\frac{1}{2}(m+n)]}{\pi^{\frac{1}{2}mp}\Gamma_p(\frac{1}{2}n)} q^{\frac{1}{2}(m+n)p} \det(A\Psi)^{-\frac{1}{2}p} \det(I_m + qZ'Z)^{-\frac{1}{2}(m+n)}$$

$$_1F_0^{(n)}\left(\frac{1}{2}(m+n); B, \begin{pmatrix} (I_m + qZ'Z)^{-1} & 0 \\ 0 & I_{p-m} \end{pmatrix}\right), Z \in \mathbb{R}^{p \times m}.$$

Now, by using Theorem 1.4.10, we get the density of $F = Z'Z$, for $m \le p$. ∎

COROLLARY 7.4.1.1. *If $A\Psi = I_n$, then the random matrix S has $W_p(n, \Sigma)$ and $F \sim B_m^{II}(\frac{1}{2}p, \frac{1}{2}(m+n-p))$.*

THEOREM 7.4.2. *Let X $(p \times n)$ and Y $(p \times m)$ be independent, $X \sim N_{p,n}(0, \Sigma \otimes \Psi)$ and $Y \sim N_{p,m}(0, \Sigma_1 \otimes I_m)$. Then*
(i) the p.d.f. of $F_1 = X'(YY')^{-1}X$, for $n \le p \le m$, is given by

$$\frac{\Gamma_p[\frac{1}{2}(m+n)]}{\Gamma_p(\frac{1}{2}m)\Gamma_n(\frac{1}{2}p)} \det(\Omega)^{\frac{1}{2}n} \det(\Psi)^{-\frac{1}{2}p} \det(F_1)^{\frac{1}{2}(p-n-1)} \det(I_m + (q\Psi)^{-1}F_1)^{-\frac{1}{2}(m+n)}$$

$$_1F_0^{(p)}\left(\frac{1}{2}(m+n); \Omega^*, F_1(q\Psi + F_1)^{-1}\right), F_1 > 0,$$

where $q > 0$, $\Omega = \Sigma^{-\frac{1}{2}}\Sigma_1\Sigma^{-\frac{1}{2}}$ and $\Omega^* = I_p - q\Omega$, and
(ii) the p.d.f. of $F_2 = (YY')^{-\frac{1}{2}}XX'(YY')^{-\frac{1}{2}}$, for $n \ge p$, $m \ge p$, is given by

$$\frac{\Gamma_p[\frac{1}{2}(m+n)]}{\Gamma_p(\frac{1}{2}m)\Gamma_p(\frac{1}{2}n)} \det(\Omega)^{\frac{1}{2}n} \det(\Psi)^{-\frac{1}{2}p} \det(F_2)^{\frac{1}{2}(n-p-1)} \det(I_p + q^{-1}\Omega F_2)^{-\frac{1}{2}(m+n)}$$

$$_1F_0^{(n)}\left(\frac{1}{2}(m+n); B, F_2(q\Omega^{-1} + F_2)^{-1}\right), F_2 > 0,$$

where $B = I_n - q\Psi^{-1}$.

Proof: (i) Since F_1 is invariant under the transformation $X \to \Sigma^{-\frac{1}{2}}X$, and $Y \to \Sigma^{-\frac{1}{2}}Y$, we can take $X \sim N_{p,n}(0, I_p \otimes \Psi)$ and $Y \sim N_{p,m}(0, \Omega \otimes I_m)$, where $\Omega = \Sigma^{-\frac{1}{2}}\Sigma_1\Sigma^{-\frac{1}{2}}$. Further, for $m \ge p$, $YY' = S \sim W_p(m, \Omega)$. Now transforming $Z = S^{-\frac{1}{2}}X$, with the Jacobian $J(X \to Z) = \det(S)^{\frac{1}{2}n}$, in the joint density of S and X we obtain the joint density of Z and S as

$$\left\{ 2^{\frac{1}{2}(m+n)p} \pi^{\frac{1}{2}np} \Gamma_p\left(\frac{1}{2}m\right) \right\}^{-1} \det(\Omega)^{-\frac{1}{2}m} \det(\Psi)^{-\frac{1}{2}p} \det(S)^{\frac{1}{2}(m+n-p-1)}$$

$$\text{etr}\left\{ -\frac{1}{2}S(\Omega^{-1} + Z\Psi^{-1}Z') \right\}, S > 0, Z \in \mathbb{R}^{p \times n}.$$

Integrating S in the above density, we get the marginal density of Z as

$$\frac{\Gamma_p[\frac{1}{2}(m+n)]}{\pi^{\frac{1}{2}np}\Gamma_p(\frac{1}{2}m)} \det(\Omega)^{\frac{1}{2}n} \det(\Psi)^{-\frac{1}{2}p} \det(I_p + \Omega Z\Psi^{-1}Z')^{-\frac{1}{2}(m+n)}, Z \in \mathbb{R}^{p \times n}. \quad (7.4.3)$$

Since,

$$\det(I_p + \Omega Z \Psi^{-1} Z') = \det(I_n + \Psi^{-1} Z' \Omega Z)$$
$$= \det(\Psi)^{-1} \det(\Psi + q^{-1} Z' Z - q^{-1} Z' \Omega^* Z)$$
$$= q^{-n} \det(\Psi)^{-1} \det(q\Psi + Z'Z) \det(I_n - Z'\Omega^* Z (q\Psi + Z'Z)^{-1})$$
$$= q^{-n} \det(\Psi)^{-1} \det(q\Psi + Z'Z) \det(I_p - Z(q\Psi + Z'Z)^{-1} Z'\Omega^*),$$

we get

$$\det(I_p + \Omega Z \Psi^{-1} Z')^{-\frac{1}{2}(m+n)} = \{q^{-n} \det(\Psi)^{-1} \det(q\Psi + Z'Z)\}^{-\frac{1}{2}(m+n)}$$
$${}_1F_0^{(p)}\left(\frac{1}{2}(m+n); Z(q\Psi + Z'Z)^{-1} Z'\Omega^*\right). \quad (7.4.4)$$

Now substituting (7.4.4) in (7.4.3), and integrating over $Z'Z = F_1$, the density of F_1 is

$$\frac{\Gamma_p[\frac{1}{2}(m+n)]}{\pi^{\frac{1}{2}np}\Gamma_p(\frac{1}{2}m)} \det(\Omega)^{\frac{1}{2}n} \det(\Psi)^{-\frac{1}{2}p} \int_{Z'Z=F_1} \det(I_n + q^{-1}\Psi^{-1} Z'Z)^{-\frac{1}{2}(m+n)}$$
$${}_1F_0^{(p)}\left(\frac{1}{2}(m+n); Z(q\Psi + Z'Z)^{-1} Z'\Omega^*\right) dZ.$$

The integral in above density is a homogeneous and symmetric function in Ω^*. Transforming $\Omega^* \to H\Omega^* H'$, $H \in O(p)$, and integrating with respect to H over $O(p)$, by using Theorem 1.6.1, we get

$$\frac{\Gamma_p[\frac{1}{2}(m+n)]}{\pi^{\frac{1}{2}np}\Gamma_p(\frac{1}{2}m)} \det(\Omega)^{\frac{1}{2}n} \det(\Psi)^{-\frac{1}{2}p} \int_{Z'Z=F_1} \det(I_n + q^{-1}\Psi^{-1} Z'Z)^{-\frac{1}{2}(m+n)}$$
$${}_1F_0^{(p)}\left(\frac{1}{2}(m+n); (q\Psi + Z'Z)^{-1} Z'Z, \Omega^*\right) dZ.$$

Finally, the result follows from Theorem 1.4.10.

(ii) Proof is similar to (i). ∎

COROLLARY 7.4.2.1. *If in the above theorem $\Sigma = \Sigma_1$, then the p.d.f. of F_1, for $n \le p \le m$, is given by*

$$\frac{\Gamma_p[\frac{1}{2}(m+n)]}{\Gamma_p(\frac{1}{2}m)\Gamma_n(\frac{1}{2}p)} \det(\Psi)^{-\frac{1}{2}p} \det(F_1)^{\frac{1}{2}(p-n-1)} \det(I_m + \Psi^{-1} F_1)^{-\frac{1}{2}(m+n)}, \quad F_1 > 0.$$

COROLLARY 7.4.2.2. *If in the above theorem (ii), $\Psi = I_n$, then the p.d.f. of F_2, for $n \ge p$, $m \ge p$, is given by*

$$\frac{\Gamma_p[\frac{1}{2}(m+n)]}{\Gamma_p(\frac{1}{2}m)\Gamma_p(\frac{1}{2}n)} \det(\Omega)^{\frac{1}{2}n} \det(F_2)^{\frac{1}{2}(n-p-1)} \det(I_p + \Omega F_2)^{-\frac{1}{2}(m+n)}, \quad F_2 > 0.$$

It may be noted that the density of $F_3 = (XX')^{-\frac{1}{2}}YY'(XX')^{-\frac{1}{2}}$, $n \geq p$, $m \geq p$, where X and Y are as in Theorem 7.4.1, can be obtained from the density of F_2, by the relation of the transformation $F_2 \to F_3^{-1}$. Hence, the densities of F_3 and $F_4 = (YY')^{\frac{1}{2}}(XX')^{-1}(YY')^{\frac{1}{2}}$ are identical.

THEOREM 7.4.3. *Let $S\,(p \times p)$ and $Y\,(p \times m)$ be independent, $S \sim Q_{p,n}(A, I_p, \Psi)$ and $Y \sim N_{p,m}(0, qI_p \otimes I_m)$. Then*

(i) the p.d.f. of $F_5 = (S + YY')^{-\frac{1}{2}}YY'(S + YY')^{-\frac{1}{2}}$, for $m \geq p$, is given by

$$\frac{q^{\frac{1}{2}np}\Gamma_p[\frac{1}{2}(m+n)]}{\Gamma_p(\frac{1}{2}m)\Gamma_p(\frac{1}{2}n)} \det(A\Psi)^{-\frac{1}{2}p}\det(F_5)^{\frac{1}{2}(m-p-1)}\det(I_p - F_5)^{\frac{1}{2}(n-p-1)}$$

$$_1F_0^{(n)}\Big(\frac{1}{2}(m+n); B, I_p - F_5\Big),\ 0 < F_5 < I_p,$$

where $q > 0$, $B = I_n - qA^{-\frac{1}{2}}\Psi^{-1}A^{-\frac{1}{2}}$, and

(ii) the p.d.f. of $F_6 = Y'(S + YY')^{-1}Y$, for $p \geq m$, is given by

$$\frac{q^{\frac{1}{2}np}\Gamma_p[\frac{1}{2}(m+n)]}{\Gamma_m(\frac{1}{2}p)\Gamma_p(\frac{1}{2}n)} \det(A\Psi)^{-\frac{1}{2}p}\det(F_6)^{\frac{1}{2}(p-m-1)}\det(I_p - F_6)^{\frac{1}{2}(n-p-1)}$$

$$_1F_0^{(n)}\Big(\frac{1}{2}(m+n); B, I_m - F_6\Big),\ 0 < F_6 < I_p,$$

Proof: The joint density of S and Y is

$$\Big\{2^{\frac{1}{2}(m+n)p}(q\pi)^{\frac{1}{2}mp}\Gamma_p\Big(\frac{1}{2}n\Big)\Big\}^{-1}\det(A\Psi)^{-\frac{1}{2}p}\,\mathrm{etr}\Big\{-\frac{1}{2}q^{-1}(S + YY')\Big\}$$

$$\det(S)^{\frac{1}{2}(n-p-1)}\,_0F_0^{(n)}\Big(B, \frac{1}{2}q^{-1}S\Big),\ S > 0, Y \in \mathbb{R}^{p \times m}. \qquad (7.4.5)$$

Now making the transformation $G = S + YY'$ and $Z = G^{-\frac{1}{2}}Y$, with the Jacobian $J(Y, S \to Z, G) = \det(G)^{\frac{1}{2}m}$, in (7.4.5), we get the joint density of G and Z as

$$\Big\{2^{\frac{1}{2}(m+n)p}(q\pi)^{\frac{1}{2}mp}\Gamma_p\Big(\frac{1}{2}n\Big)\Big\}^{-1}\det(A\Psi)^{-\frac{1}{2}p}\,\mathrm{etr}\Big(-\frac{1}{2}q^{-1}G\Big)\det(G)^{\frac{1}{2}(m+n-p-1)}$$

$$\det(I_p - ZZ')^{\frac{1}{2}(n-p-1)}\,_0F_0^{(n)}\Big(B, \frac{1}{2}q^{-1}G(I_p - ZZ')\Big),\ G > 0, Z \in \mathbb{R}^{p \times m}.$$

Integrating this joint density with respect to G, by using Theorem 1.6.2, we get the marginal density of Z as

$$\frac{q^{\frac{1}{2}np}\Gamma_p[\frac{1}{2}(m+n)]}{\pi^{\frac{1}{2}mp}\Gamma_p(\frac{1}{2}n)}\det(A\Psi)^{-\frac{1}{2}p}\det(I_p - ZZ')^{\frac{1}{2}(n-p-1)}$$

$$_1F_0^{(n)}\Big(\frac{1}{2}(m+n); B, I_p - ZZ'\Big), Z \in \mathbb{R}^{p \times m}.$$

Finally, by using Theorem 1.4.10, we get the density of $F_5 = ZZ'$ if $m \geq p$, and the density of $F_6 = Z'Z$ if $m \leq p$. ∎

The above theorem is a generalization of Theorems 5.2.3 and 5.2.4. If we let $A\Psi = I_n$ here, we get $F_5 \sim B_p^I(\frac{1}{2}m, \frac{1}{2}n)$ and $F_6 \sim B_m^I(\frac{1}{2}p, \frac{1}{2}(m+n-p))$.

Next, we study the density of $\text{tr}(S)$.

THEOREM 7.4.4. *Let $S \sim Q_{p,n}(A, \Sigma, \Psi)$. Then, the p.d.f. of $u = \text{tr}(S)$, for $n \geq p$, is*

$$\left\{2^{\frac{1}{2}np}\Gamma\left(\frac{1}{2}np\right)\right\}^{-1} \det(A\Psi)^{-\frac{1}{2}p}\det(\Sigma)^{-\frac{1}{2}n}u^{\frac{1}{2}(np-2)}\exp\left(-\frac{1}{2}q^{-1}u\right)$$

$$_0F_0^{(np)}\left(B_1, \frac{1}{2}q^{-1}u\right), \ u > 0, \tag{7.4.6}$$

where $B_1 = I_{np} - q(\Sigma^{-1} \otimes A^{-\frac{1}{2}}\Psi^{-1}A^{-\frac{1}{2}})$.

Proof: Writing $S = XAX'$, with $X \sim N_{p,n}(0, \Sigma \otimes \Psi)$, and by using Theorem 1.2.22, we can write

$$u = (\text{vec}(X'))'(I_p \otimes A)\text{vec}(X'), \tag{7.4.7}$$

where $\text{vec}(X') \sim N_{np}(0, \Sigma \otimes \Psi)$. Since (7.4.7) is a quadratic form in $\text{vec}(X')$, the density of u, from Theorem 7.2.1, is given by

$$\left\{2^{\frac{1}{2}np}\Gamma\left(\frac{1}{2}np\right)\right\}^{-1}\det((I_p \otimes A)(\Sigma \otimes \Psi))^{-\frac{1}{2}}u^{\frac{1}{2}(np-2)}\exp\left(-\frac{1}{2}q^{-1}u\right)$$

$$_0F_0^{(np)}\left(I_{np} - q(I_p \otimes A^{-\frac{1}{2}})(\Sigma \otimes \Psi)^{-1}(I_p \otimes A^{-\frac{1}{2}}), \frac{1}{2}q^{-1}u\right), \ u > 0,$$

Substituting $\det((I_p \otimes A)(\Sigma \otimes \Psi)) = \det(\Sigma)^n \det(A\Psi)^p$ and $(I_p \otimes A^{-\frac{1}{2}})(\Sigma \otimes \Psi)^{-1}(I_p \otimes A^{-\frac{1}{2}}) = \Sigma^{-1} \otimes A^{-\frac{1}{2}}\Psi^{-1}A^{-\frac{1}{2}}$ in the above expression we get the final result. ∎

The c.d.f. of u, by using p.d.f. (7.4.6), is

$$P(u \leq w) = \left\{2^{\frac{1}{2}np}\Gamma\left(\frac{1}{2}np\right)\right\}^{-1}\det(A\Psi)^{-\frac{1}{2}p}\det(\Sigma)^{-\frac{1}{2}n}\int_{u \leq w}u^{\frac{1}{2}(np-2)}$$

$$\exp\left(-\frac{1}{2}q^{-1}u\right)_0F_0^{(np)}\left(B_1, \frac{1}{2}q^{-1}u\right)du$$

$$= \left\{2^{\frac{1}{2}np}\Gamma\left(\frac{1}{2}np\right)\right\}^{-1}\det(A\Psi)^{-\frac{1}{2}p}\det(\Sigma)^{-\frac{1}{2}n}\sum_{k=0}^{\infty}\sum_{\kappa}\frac{(\frac{1}{2}q^{-1})^k}{k!}\frac{C_\kappa(B_1)}{C_\kappa(I_{np})}$$

$$\int_{u \leq w}\exp\left(-\frac{1}{2}q^{-1}u\right)u^{\frac{1}{2}(np+2k-2)}du$$

$$= \left\{\Gamma\left(\frac{1}{2}np\right)\right\}^{-1}q^{\frac{1}{2}np}\det(A\Psi)^{-\frac{1}{2}p}\det(\Sigma)^{-\frac{1}{2}n}$$

$$\sum_{k=0}^{\infty}\sum_{\kappa}\frac{C_\kappa(B_1)}{C_\kappa(I_{np})\,k!}\gamma\left(k + \frac{1}{2}np, (2q)^{-1}w\right),$$

where $\gamma(a, x) = \int_0^x \exp(-t)t^{a-1}\,dt$ is the incomplete gamma function.

7.5. SERIES REPRESENTATION OF THE DENSITY

Let $X = (\boldsymbol{x}_1, \ldots, \boldsymbol{x}_n) \sim N_{p,n}(M, \Sigma \otimes \Psi)$. The quadratic form XAX', $A = (a_{ij}) > 0$, can be written as

$$\sum_{i=1}^{n} \sum_{j=1}^{n} a_{ij} \boldsymbol{x}_i \boldsymbol{x}_j'.$$

For $M = 0$, the density of XAX', in various forms, using different methods, has been derived by Khatri (1966), Hayakawa (1966), and Shah (1970). These are given in Section 7.1. Using suitable transformations, we can easily show that the density of XAX' is the same as that of

$$S = \sum_{i=1}^{n} \boldsymbol{y}_i \boldsymbol{y}_i' = YY', \tag{7.5.1}$$

where $Y = (\boldsymbol{y}_1, \ldots, \boldsymbol{y}_n) \sim N_{p,n}(\Delta, \Sigma \otimes D)$, $\Delta = (\boldsymbol{\delta}_1, \ldots, \boldsymbol{\delta}_n)$, $D = \operatorname{diag}(\alpha_1, \ldots, \alpha_n)$ and $\alpha_1 \geq \cdots \geq \alpha_n > 0$ are the characteristic roots of $A\Psi$. If $\alpha_1 - \cdots = \alpha_n = \alpha_0$, then $S \sim W_p(n, \alpha_0 \Sigma, \alpha_0^{-1} \Sigma^{-1} \Delta \Delta')$.

In this section we study series representations of the density of the quadratic form for $\Delta = 0$ (*i.e.*, central case), given by Khatri (1971) who generalized the results of Kotz, Johnson and Boyd (1967a). The series representation for $\Delta \neq 0$ will be discussed in Section 7.6. Write the density $f(S)$ of S as

$$f(S) = \sum_{k=0}^{\infty} \sum_{\kappa} a_\kappa h_\kappa(S). \tag{7.5.2}$$

Then Khatri (1971) has studied two types of representations:

(i) Power-series and Wishart type representations:

$$f_1(S) = \sum_{k=0}^{\infty} \sum_{\kappa} a_\kappa^{(1)} h_\kappa^{(1)}(S), \tag{7.5.3}$$

where

$$h_\kappa^{(1)}(S) = w_p\left(\frac{1}{2}n, \gamma\Sigma^{-1}; S\right) \left\{ \left(\frac{1}{2}n\right)_\kappa \right\}^{-1} C_\kappa(\Sigma^{-1}S), \tag{7.5.4}$$

with

$$w_p\left(\frac{1}{2}n, \gamma\Sigma^{-1}; S\right) = \left\{ \Gamma_p\left(\frac{1}{2}n\right) \right\}^{-1} \det(S)^{\frac{1}{2}(n-p-1)} \operatorname{etr}(-\gamma\Sigma^{-1}S), \quad n \geq p.$$

For $\gamma = 0$, (7.5.3) reduces to the power-series representation of Hayakawa (1966), for $\gamma > 0$, it is the Wishart type representation (or a mixture of Wishart densities), and for $\gamma = q^{-1}$, $q > 0$, it is the representation (7.2.1) given by Khatri (1966).

(ii) Laguerre type representation:

$$f_2(S) = \sum_{k=0}^{\infty} \sum_{\kappa} a_\kappa^{(2)} h_\kappa^{(2)}(S), \tag{7.5.5}$$

where

$$h_\kappa^{(2)}(S) = w_p\left(\frac{1}{2}n, \gamma\Sigma^{-1}; S\right)\left\{\left(\frac{1}{2}n\right)_\kappa\right\}^{-1} L_\kappa^{\frac{1}{2}(n-p-1)}(\alpha\Sigma^{-1}S), \ \alpha \neq 0. \quad (7.5.6)$$

For $\alpha = \gamma = \frac{1}{2}q^{-1}$, $q > 0$, Shah (1970) obtained this representation and is given in Section 7.2.

To derive representations (7.5.3), and (7.5.5), and to study their convergence properties, we shall need the following results.

LEMMA 7.5.1. *Let $S\,(p \times p)$ be a positive definite matrix and $\Theta\,(q \times q)$, $q > p$, be a real symmetric matrix. Let $\theta_1, \ldots, \theta_q$ be the characteristic roots of Θ, such that $|\omega\theta_i| < 1$, $i = 1, \ldots, q$, where ω is any real or complex number. Then, for $H \in O(q)$, $H' = (\,H_1'\ \ H_2'\,)$, $H_1\,(p \times q)$,*

$$\sum_{k=0}^\infty \sum_\kappa L_\kappa^{\beta-\frac{1}{2}(p+1)}(S)\frac{C_\kappa(\omega\Theta)}{k!\,C_\kappa(I_q)} = \int_{H\in O(q)} \det(I_p - \omega H_1\Theta H_1')^{-\beta}$$

$$\operatorname{etr}\{-\omega S H_1\Theta H_1'(I_p - \omega H_1\Theta H_1')^{-1}\}\,[dH], \quad (7.5.7)$$

and

$$\left|\sum_\kappa L_\kappa^{\beta-\frac{1}{2}(p+1)}(S)\frac{C_\kappa(\Theta)}{k!\,C_\kappa(I_q)}\right| \leq \rho^{-k}\int_{H\in O(q)} \det(I_p - \rho H_1\Theta_0 H_1')^{-\beta}$$

$$\operatorname{etr}\{\rho S H_1\Theta_0 H_1'(I_p + \rho H_1\Theta_0 H_1')^{-1}\}\,[dH]$$

$$< \rho^{-k}(1-\rho\epsilon)^{-p\beta}\exp\{\rho\epsilon(1+\rho\epsilon)^{-1}\operatorname{tr}(S)\}, \quad (7.5.8)$$

where $\Theta_0 = \operatorname{diag}(|\theta_1|, \ldots, |\theta_q|)$, $\epsilon = \max_i |\theta_i|$, and ρ is any number such that $0 < \rho\epsilon < 1$.

Proof: From Lemma 1.5.1, for $H \in O(q)$, we have

$$\frac{C_\kappa(I_p)C_\kappa(\Theta)}{C_\kappa(I_q)} = \int_{H\in O(q)} C_\kappa(H_1\Theta H_1')\,[dH]. \quad (7.5.9)$$

Substituting from (7.5.9) in the left hand side of (7.5.7), we get

$$\sum_{k=0}^\infty \sum_\kappa L_\kappa^{\beta-\frac{1}{2}(p+1)}(S)\frac{C_\kappa(\omega\Theta)}{k!\,C_\kappa(I_q)}$$

$$= \int_{H\in O(q)} \sum_{k=0}^\infty \sum_\kappa L_\kappa^{\beta-\frac{1}{2}(p+1)}(S)\frac{C_\kappa(\omega H_1\Theta H_1')}{k!\,C_\kappa(I_p)}\,[dH]. \quad (7.5.10)$$

Now by using (1.7.7) in the integrand of (7.5.10), we get

$$\int_{H\in O(q)}\int_{H_3\in O(p)} \det(I_p - \omega H_3 H_1\Theta H_1' H_3')^{-\beta}$$

$$\operatorname{etr}\{-\omega S H_3 H_1\Theta H_1'(I_p - \omega H_1\Theta H_1')^{-1}H_3'\}\,[dH]\,[dH_3], \quad (7.5.11)$$

where $H_3 \in O(p)$. Next transforming $\begin{pmatrix} H_3 & 0 \\ 0 & I_{q-p} \end{pmatrix} H = H_4$, i.e., $\begin{pmatrix} H_3 H_1 \\ H_2 \end{pmatrix} = H_4 = \begin{pmatrix} H_{41} \\ H_{42} \end{pmatrix}$, and $[dH] = [dH_4]$, (7.5.11) becomes

$$\int_{H_4 \in O(q)} \int_{H_3 \in O(p)} \det(I_p - \omega H_{41} \Theta H_{41}')^{-\beta}$$

$$\mathrm{etr}\{-\omega S H_{41} \Theta H_{41}' (I_p - \omega H_{41} \Theta H_{41}')^{-1}\} [dH_4] [dH_3].$$

Integrating with respect to H_3 and replacing H_{41} by H_1, (7.5.7) follows.

To prove (7.5.8), note that

$$\left| \sum_\kappa L_\kappa^{\beta-\frac{1}{2}(p+1)}(S) \frac{C_\kappa(\omega\Theta)}{k! \, C_\kappa(I_q)} \right| \leq \sum_\kappa L_\kappa^{\beta-\frac{1}{2}(p+1)}(S) \frac{|C_\kappa(\omega\Theta)|}{k! \, C_\kappa(I_p)}$$

$$\leq \sum_{k=0}^\infty \sum_\kappa L_\kappa^{\beta-\frac{1}{2}(p+1)}(S) \frac{C_\kappa(\rho\Theta_0)}{k! \, C_\kappa(I_p)}. \qquad (7.5.12)$$

By using (7.5.7) in (7.5.12), the inequalities are easily obtained. ∎

Using the representation (7.5.1) and results on normal distributions, we get the Laplace transform of the p.d.f. of S.

LEMMA 7.5.2. *Let $Z\,(p \times p)$ be a complex symmetric matrix such that $\mathrm{Re}(Z) > 0$. Then, the Laplace transform of the density of S, for $\Delta \neq 0$, is*

$$E[\mathrm{etr}(-ZS)] = \prod_{j=1}^n \det(I_p + 2\alpha_j \Sigma Z)^{-\frac{1}{2}} \exp\left\{ -\sum_{j=1}^n \delta_j' Z(I_p + 2\alpha_j \Sigma Z)^{-1} \delta_j \right\}.$$

When $\Delta = 0$, this Laplace transform reduces to

$$E[\mathrm{etr}(-ZS)] = \prod_{j=1}^n \det(I_p + 2\alpha_j \Sigma Z)^{-\frac{1}{2}},$$

which is the Laplace transform of the density of S in the central case.

Next lemma follows from an application of Lebesgue dominated convergence theorem.

LEMMA 7.5.3. *Let $\{h_\kappa\}$ be a sequence of complex valued measurable functions on the space of positive definite matrices such that*

$$\sum_{k=0}^\infty \left| \sum_\kappa a_\kappa h_\kappa(S) \right| < a\, \mathrm{etr}(BS), \quad \text{for almost all } S > 0,$$

where $\{a_\kappa\}$ is a sequence of complex numbers, a is a real number, and B is a symmetric matrix. Define

$$f(S) = \sum_{k=0}^\infty \sum_\kappa a_\kappa h_\kappa(S),$$

(well defined a.e. for $S > 0$). Then, the Laplace transforms $\hat{h}_\kappa(Z)$ and $\hat{f}(Z)$ of $h_\kappa(S)$ and $f(S)$, respectively exist for $\mathrm{Re}(Z) > B$, and

$$\hat{f}(Z) = \sum_{k=0}^{\infty} \sum_{\kappa} a_\kappa \hat{h}_\kappa(Z) \ for \ \mathrm{Re}(Z) > B. \tag{7.5.13}$$

The definition and existence of Laplace transform are given in Chapter 1. In order to obtain explicit expressions for $a_\kappa^{(1)}$ of (7.5.3), and $a_\kappa^{(2)}$ of (7.5.5), we use the following method.

Let us write

$$\hat{h}_\kappa(Z) = \xi(Z) C_\kappa(G(Z)), \tag{7.5.14}$$

where $\xi(Z) \neq 0$, is analytic for $\mathrm{Re}(Z) > B$, and $G(Z)$ is a one to one function. Further let $\Theta = G(Z)$. Then $G^{-1}[G(Z)] = G^{-1}(\Theta) = Z$. Define

$$M(\Theta) = \frac{L_0(G^{-1}(\Theta))}{\xi(G^{-1}(\Theta))}, \tag{7.5.15}$$

where $L_0(Z) = E\{\mathrm{etr}(-ZS)\}$, *i.e.*, $\hat{f}(Z) = L_0(Z)$. Hence, from (7.5.15), we get

$$M(\Theta) = \frac{\hat{f}(G^{-1}(\Theta))}{\xi(G^{-1}(\Theta))}, \tag{7.5.16}$$

$$= \sum_{k=0}^{\infty} \sum_{\kappa} a_\kappa \frac{\hat{h}_\kappa(G^{-1}(\Theta))}{\xi(G^{-1}(\Theta))},$$

$$= \sum_{k=0}^{\infty} \sum_{\kappa} a_\kappa C_\kappa(\Theta), \tag{7.5.17}$$

where the last two steps have been obtained by using (7.5.13) and (7.5.14). Now equating the coefficients of $C_\kappa(\Theta)$ in (7.5.16) and (7.5.17), we get the explicit form for a_κ.

THEOREM 7.5.1. *For the power series and Wishart type representation (7.5.3),*

$$a_\kappa^{(1)} = a_0^{(1)} \frac{(\frac{1}{2}n)_\kappa}{k!} \frac{C_\kappa(A_1)}{C_\kappa(I_n)}, \tag{7.5.18}$$

where $a_0^{(1)} = \det(D)^{-\frac{1}{2}p} \det(2\Sigma)^{-\frac{1}{2}n}$, $D = \mathrm{diag}(\alpha_1, \ldots, \alpha_n)$, $A_1 = \mathrm{diag}(\beta_1, \ldots, \beta_n)$, $\beta_j = \frac{2\alpha_j\gamma - 1}{2\alpha_j}$, $j = 1, \ldots, n$.

Proof: From (7.5.4), using Lemma 1.5.2, we get

$$\hat{h}_\kappa^{(1)}(Z) = \left\{ \left(\frac{1}{2}n\right)_\kappa \Gamma_p\left(\frac{1}{2}n\right) \right\}^{-1} \int_{S>0} \mathrm{etr}\{-(Z + \gamma\Sigma^{-1})S\} \det(S)^{\frac{1}{2}(n-p-1)} C_\kappa(\Sigma^{-1}S) \, dS$$

$$= \det(\Sigma)^{\frac{1}{2}n} \det(\gamma I_p + \Sigma Z)^{-\frac{1}{2}n} C_\kappa((\gamma I_p + \Sigma Z)^{-1}), \ \mathrm{Re}(Z + \gamma\Sigma^{-1}) > 0.$$

Now let $\xi(Z) = \det(\Sigma)^{\frac{1}{2}n} \det(\gamma I_p + \Sigma Z)^{-\frac{1}{2}n}$, and $G(Z) = \Theta = (\gamma I_p + \Sigma Z)^{-1}$. Then $Z = \Sigma^{-1}(\Theta^{-1} - \gamma I_p) = G^{-1}(\Theta)$. Using Lemma 7.5.2 and (7.5.16), we have

$$
\begin{aligned}
M(\Theta) &= \frac{\prod_{i=1}^n \det(I_p + 2\alpha_i \Sigma G^{-1}(\Theta))^{-\frac{1}{2}}}{\det(\Sigma)^{\frac{1}{2}n} \det(\gamma I_p + \Sigma G^{-1}(\Theta))^{-\frac{1}{2}n}} \\
&= \det(\Sigma)^{-\frac{1}{2}n} \det(2D)^{-\frac{1}{2}p} \prod_{i=1}^n \det(I_p - \beta_i \Theta)^{-\frac{1}{2}} \\
&= \det(\Sigma)^{-\frac{1}{2}n} \det(2D)^{-\frac{1}{2}p} \sum_{k=0}^\infty \sum_\kappa \frac{(\frac{1}{2}n)_\kappa}{k!} \frac{C_\kappa(A_1) C_\kappa(\Theta)}{C_\kappa(I_n)}, \quad (7.5.19)
\end{aligned}
$$

where the last equality is written by comparing (7.3.6) and (7.3.8). The series expansion in (7.5.19) is valid if and only if

$$
\max_i |\operatorname{ch}_i \Theta| < \frac{1}{\epsilon}, \text{ or } \min_i |\operatorname{ch}_i(\Sigma Z) + \gamma| > \epsilon, \quad (7.5.20)
$$

where $\epsilon = \max_j |\beta_j| = \max_j |\gamma - (2\alpha_j)^{-1}|$.

Now comparing coefficients of $C_\kappa(\Theta)$ in (7.5.17) and (7.5.19), we have

$$
a_\kappa^{(1)} = a_0^{(1)} \frac{(\frac{1}{2}n)_\kappa}{k!} \frac{C_\kappa(A_1)}{C_\kappa(I_n)}. \quad \blacksquare
$$

Using $a_\kappa^{(1)}$ from (7.5.18), we get the power series and Wishart type representation (7.5.3) of the density of S as

$$
f_1(S) = a_0^{(1)} \sum_{k=0}^\infty \sum_\kappa \frac{C_\kappa(A_1) C_\kappa(\Sigma^{-1}S)}{k! C_\kappa(I_n)} w_p\left(\frac{1}{2}n, \gamma \Sigma^{-1}; S\right), \quad (7.5.21)
$$

where γ is any real positive number. For this series to be a density, it should satisfy conditions of Lemma 7.5.3. Here we have

$$
\begin{aligned}
\sum_{k=0}^\infty \left| \sum_\kappa a_\kappa^{(1)} h_\kappa^{(1)}(S) \right| &= a_0^{(1)} \sum_{k=0}^\infty \left| \sum_\kappa \frac{C_\kappa(A_1) C_\kappa(\Sigma^{-1}S)}{k! C_\kappa(I_n)} w_p\left(\frac{1}{2}n, \gamma \Sigma^{-1}; S\right) \right| \\
&\leq a_0^{(1)} \sum_{k=0}^\infty \sum_\kappa \frac{|C_\kappa(A_1)| C_\kappa(\Sigma^{-1}S)}{k! C_\kappa(I_n)} w_p\left(\frac{1}{2}n, \gamma \Sigma^{-1}; S\right).
\end{aligned}
$$

Now, using (1.5.4) and $\epsilon = \max_j |\beta_j| = \max_j |\gamma - (2\alpha_j)^{-1}|$, we have $|C_\kappa(A_1)| \leq C_\kappa(A_{10}) \leq \epsilon^k C_\kappa(I_n)$, where $A_{10} = \operatorname{diag}(|\beta_1|, \ldots, |\beta_n|)$. Hence

$$
\sum_{k=0}^\infty \left| \sum_\kappa a_\kappa^{(1)} h_\kappa^{(1)}(S) \right| \leq a_0^{(1)} w_p\left(\frac{1}{2}n, (\gamma - \epsilon)\Sigma^{-1}; S\right).
$$

For $B = -(\gamma - \epsilon)\Sigma^{-1}$, $\operatorname{Re}(Z) > B$ satisfies the condition (7.5.20) and therefore from the uniqueness property of Laplace transform we get the density (7.5.21). The series

(7.5.21) is uniformly convergent if $\gamma > \epsilon$. For choosing γ, and for rapid convergence of the series (7.5.21), we give upper bound for

$$e_N^{(1)}(S) = \left| \sum_{k=N+1}^{\infty} \sum_{\kappa} a_\kappa^{(1)} h_\kappa^{(1)}(S) \right|. \tag{7.5.22}$$

From (7.5.21), we have

$$k!\, a_\kappa^{(1)} h_\kappa^{(1)}(S) = a_0^{(1)} \left\{ \Gamma_p\!\left(\frac{1}{2}n\right) \right\}^{-1} \frac{C_\kappa(A_1) C_\kappa(\Sigma^{-1}S)}{k!\, C_\kappa(I_n)}$$

$$\det(S)^{\frac{1}{2}(n-p-1)} \operatorname{etr}(-\gamma \Sigma^{-1} S). \tag{7.5.23}$$

Next, using Theorem 1.4.10, we can write

$$C_\kappa(\Sigma^{-1}S) \det(S)^{\frac{1}{2}(n-p-1)} \operatorname{etr}(-\gamma \Sigma^{-1} S)$$

$$= \pi^{-\frac{1}{2}np} \Gamma_p\!\left(\frac{1}{2}n\right) \int_{YY'=S} C_\kappa(\Sigma^{-1} YY') \operatorname{etr}(-\gamma \Sigma^{-1} YY')\, dY. \tag{7.5.24}$$

Substituting from (7.5.24), and then using the results

$$\frac{C_\kappa(A_1) C_\kappa(\Sigma^{-1} YY')}{k!\, C_\kappa(I_n)} = \int_{H \in O(n)} C_\kappa(\Sigma^{-1} Y H A_1 H' Y')\, [dH],$$

we get

$$k!\, a_\kappa^{(1)} h_\kappa^{(1)}(S) = \pi^{-\frac{1}{2}np} a_0^{(1)} \int_{YY'=S} \int_{H \in O(n)} \operatorname{etr}(-\gamma \Sigma^{-1} YY')$$

$$C_\kappa(\Sigma^{-1} Y H A_1 H' Y')\, [dH]\, dY.$$

Hence,

$$e_N^{(1)}(S) = \pi^{-\frac{1}{2}np} a_0^{(1)} \left| \int_{YY'=S} \int_{H \in O(n)} \operatorname{etr}(-\gamma \Sigma^{-1} YY') \right.$$

$$\left. \sum_{k=N+1}^{\infty} \sum_{\kappa} \frac{C_\kappa(\Sigma^{-1} Y H A_1 H' Y')}{k!}\, [dH]\, dY \right|$$

$$= \pi^{-\frac{1}{2}np} a_0^{(1)} \left| \int_{YY'=S} \int_{H \in O(n)} \operatorname{etr}(-\gamma \Sigma^{-1} YY') \right.$$

$$\left. \sum_{k=N+1}^{\infty} \frac{\{\operatorname{tr}(\Sigma^{-1} Y H A_1 H' Y')\}^k}{k!}\, [dH]\, dY \right|. \tag{7.5.25}$$

(i) Let A_1 be negative semidefinite, i.e., $\gamma \leq \frac{1}{2\alpha_1}$, where $\alpha_1 > \cdots > \alpha_n > 0$. Since,

$$\left| \sum_{k=N+1}^{\infty} \frac{(-x)^k}{k!} \right| = \left| \exp(-x) - \sum_{k=0}^{N} \frac{(-x)^k}{k!} \right| \leq \frac{x^{N+1}}{(N+1)!},$$

we can write

$$\left| \sum_{k=N+1}^{\infty} \frac{\{-\operatorname{tr}(\Sigma^{-1}YH(-A_1)H'Y')\}^k}{k!} \right| \le \frac{\{\operatorname{tr}(\Sigma^{-1}YH(-A_1)H'Y')\}^{N+1}}{(N+1)!}$$

$$= \sum_{\lambda} \frac{C_\lambda(\Sigma^{-1}YH(-A_1)H'Y')}{(N+1)!}, \quad (7.5.26)$$

where C_λ is a zonal polynomial, $\lambda = (\ell_1, \ldots, \ell_n)$, $\ell_1 \ge \cdots \ge \ell_n \ge 0$, $\ell_1 + \cdots + \ell_n = N+1$. Hence

$$e_N^{(1)}(S) \le \pi^{-\frac{1}{2}np} a_0^{(1)} \int_{YY'=S} \int_{H \in O(n)} \operatorname{etr}(-\gamma \Sigma^{-1} YY')$$

$$\sum_{\lambda} \frac{C_\lambda(\Sigma^{-1}YH(-A_1)H'Y')}{(N+1)!} [dH] \, dY.$$

Now using

$$\int_{H \in O(n)} C_\lambda(\Sigma^{-1}YH(-A_1)H'Y') [dH] = \frac{C_\lambda(-A_1)C_\lambda(\Sigma^{-1}YY')}{C_\lambda(I_n)},$$

$$C_\lambda(-A_1) \le C_\lambda(A_{10}) \le C_\lambda(I_n),$$

and integrating over the surface $YY' = S$, we get

$$e_N^{(1)}(S) \le a_0^{(1)} w_p\left(\frac{1}{2}n, \gamma \Sigma^{-1}; S\right) \epsilon^{N+1} \sum_{\lambda} \frac{C_\lambda(\Sigma^{-1}S)}{(N+1)!}$$

$$= a_0^{(1)} w_p\left(\frac{1}{2}n, \gamma \Sigma^{-1}; S\right) \epsilon^{N+1} \frac{\{\operatorname{tr}(\Sigma^{-1}S)\}^{N+1}}{(N+1)!}, \quad (7.5.27)$$

where $\epsilon = \frac{1}{2\alpha_n} - \gamma$. For the uniform convergence of the series (7.5.21), we need $\frac{1}{4\alpha_n} < \gamma < \frac{1}{2\alpha_1}$. Note that the power series expansion does not converge rapidly and uniformly, because in this case $\gamma = 0$. The best choice of γ is $\gamma = \frac{1}{2\alpha_1}$, when A_1 is negative semidefinite, and $\epsilon = \frac{1}{2}(\frac{1}{\alpha_n} - \frac{1}{\alpha_1})$.

(ii) If $\gamma > \frac{1}{2\alpha_1}$, the matrix A_1 will have negative as well as positive elements. Let $A_{10} = \operatorname{diag}(|\beta_1|, \ldots, |\beta_n|)$. Then

$$\left| \sum_{k=N+1}^{\infty} \frac{\{\operatorname{tr}(\Sigma^{-1}YHA_1H'Y')\}^k}{k!} \right| \le \sum_{k=N+1}^{\infty} \frac{\{\operatorname{tr}(\Sigma^{-1}YHA_{10}H'Y')\}^k}{k!}$$

$$\le \frac{\{\operatorname{tr}(\Sigma^{-1}YHA_{10}H'Y')\}^{N+1}}{(N+1)!} \operatorname{etr}(\epsilon \Sigma^{-1}YY')$$

$$= \sum_{\lambda} \frac{C_\lambda(\Sigma^{-1}YHA_{10}H'Y')}{(N+1)!} \operatorname{etr}(\epsilon \Sigma^{-1}YY'). \quad (7.5.28)$$

Now substituting from (7.5.28) in (7.5.25), we get

$$
e_N^{(1)}(S) \leq \pi^{-\frac{1}{2}np} a_0^{(1)} \sum_\lambda \int_{YY'=S} \int_{H \in O(n)} \operatorname{etr}(-(\gamma - \epsilon)\Sigma^{-1}YY')
$$

$$
\frac{C_\lambda(\Sigma^{-1}YHA_{10}H'Y')}{(N+1)!} [dH] \, dY
$$

$$
= \pi^{-\frac{1}{2}np} a_0^{(1)} \sum_\lambda \frac{C_\lambda(A_{10})}{C_\lambda(I_n)(N+1)!} \int_{YY'=S} \operatorname{etr}(-(\gamma - \epsilon)\Sigma^{-1}YY')
$$

$$
C_\lambda(\Sigma^{-1}YY') \, dY
$$

$$
= a_0^{(1)} \left\{ \Gamma_p\left(\frac{1}{2}n\right) \right\}^{-1} \sum_\lambda \frac{C_\lambda(A_{10})}{C_\lambda(I_n)(N+1)!} \det(S)^{\frac{1}{2}(n-p-1)}
$$

$$
\operatorname{etr}\{-(\gamma - \epsilon)\Sigma^{-1}S\} C_\lambda(\Sigma^{-1}S)
$$

$$
\leq a_0^{(1)} w_p\left(\frac{1}{2}n, (\gamma - \epsilon)\Sigma^{-1}; S\right) \epsilon^{N+1} \frac{\{\operatorname{tr}(\Sigma^{-1}S)\}^{N+1}}{(N+1)!}, \qquad (7.5.29)
$$

where $\epsilon = \max_j |\beta_j| = \max_j |\gamma - (2\alpha_j)^{-1}|$. For uniform convergence of (7.5.21), $\gamma > \epsilon$ gives $\gamma > \frac{1}{4\alpha_n}$. The best choice for ϵ is $\epsilon = \inf_\gamma \max_j |\gamma - (2\alpha_j)^{-1}|$, which gives $\gamma = \frac{1}{4}(\frac{1}{\alpha_1} + \frac{1}{\alpha_n})$ and hence $\epsilon = \frac{1}{4}(\frac{1}{\alpha_n} - \frac{1}{\alpha_1})$ and $\gamma - \epsilon = \frac{1}{2\alpha_n}$. Therefore the choice $\gamma = \frac{1}{4}(\frac{1}{\alpha_1} + \frac{1}{\alpha_n})$ is a better choice than the best choice in (i).

THEOREM 7.5.2. *For the Laguerre type representation (7.5.5),*

$$
a_\kappa^{(2)} = a_0^{(2)} \frac{(\frac{1}{2}n)_\kappa}{k!} \frac{C_\kappa(A_2)}{C_\kappa(I_n)}, \qquad (7.5.30)
$$

where $a_0^{(2)} = \det(D)^{-\frac{1}{2}p} \det(2\Sigma)^{-\frac{1}{2}n} \det(I_n - A_2)^{\frac{1}{2}p}$, $D = \operatorname{diag}(\alpha_1, \ldots, \alpha_n)$, $A_2 = \operatorname{diag}(\phi_1, \ldots, \phi_n)$, *and* $\phi_j = \frac{1 - 2\alpha_j\gamma}{1 - 2\alpha_j\gamma + 2\alpha_j\alpha}$, $j = 1, \ldots, n$.

Proof: From (7.5.6), and Theorem 1.7.1, we have

$$
\hat{h}_\kappa^{(2)}(Z) = \det(\Sigma)^{\frac{1}{2}n} \det(\gamma I_p + \Sigma Z)^{-\frac{1}{2}n} C_\kappa(I_p - \alpha(\gamma I_p + \Sigma Z)^{-1}), \ \operatorname{Re}(\gamma I_p + \Sigma Z) > 0.
$$

Now let $\xi(Z) = \det(\Sigma)^{\frac{1}{2}n} \det(\gamma I_p + \Sigma Z)^{-\frac{1}{2}n}$, and $G(Z) = \Theta = I_p - \alpha(\gamma I_p + \Sigma Z)^{-1}$. Then $Z = \Sigma^{-1}[\alpha(I_p - \Theta)^{-1} - \gamma I_p] = G^{-1}(\Theta)$. Using Lemma 7.5.2 and (7.5.16), we have

$$
M(\Theta) = \frac{\prod_{i=1}^n \det(I_p + 2\alpha_i\{\alpha(I_p - \Theta)^{-1} - \gamma I_p\})^{-\frac{1}{2}}}{\det(\Sigma)^{\frac{1}{2}n} \det(\alpha(I_p - \Theta)^{-1})^{-\frac{1}{2}n}}
$$

$$
= a_0^{(2)} \prod_{i=1}^n \det(I_p - \phi_j\Theta)^{-\frac{1}{2}}
$$

$$
= a_0^{(2)} \sum_{k=0}^\infty \sum_\kappa \frac{(\frac{1}{2}n)_\kappa}{k!} \frac{C_\kappa(A_2)C_\kappa(\Theta)}{C_\kappa(I_n)}. \qquad (7.5.31)
$$

The series expansion in (7.5.31) is valid if and only if

$$\max_i |\operatorname{ch}_i \Theta| < \frac{1}{\epsilon_1}, \text{ or } \max_i \left|1 - \frac{\alpha}{\operatorname{ch}_i(\Sigma Z) + \gamma}\right| < \frac{1}{\epsilon_1}, \tag{7.5.32}$$

where $\epsilon_1 = \max_j |\phi_j|$.

Now comparing the coefficients of $C_\kappa(\Theta)$ in (7.5.17) and (7.5.31), we have

$$a_\kappa^{(2)} = a_0^{(2)} \frac{(\frac{1}{2}n)_\kappa}{k!} \frac{C_\kappa(A_2)}{C_\kappa(I_n)}. \quad \blacksquare$$

Using $a_\kappa^{(2)}$ from (7.5.30), we get the Laguerre type representation (7.5.5) of the density of S as

$$f_2(S) = a_0^{(2)} w_p\left(\frac{1}{2}n, \gamma\Sigma^{-1}; S\right) \sum_{k=0}^{\infty} \sum_\kappa \frac{L_\kappa^{\frac{1}{2}(n-p-1)}(\alpha\Sigma^{-1}S)C_\kappa(A_2)}{k! C_\kappa(I_n)}, \tag{7.5.33}$$

where γ is any real positive number. It may be noted here that the series (7.5.33) is convergent if and only if $\epsilon_1 = \max_j |\phi_j| < 1$. Therefore, we choose γ and α such that $\epsilon_1 < 1$. Then using Lemma 7.5.1,

$$\sum_{k=0}^{\infty} \left|\sum_\kappa a_\kappa^{(2)} h_\kappa^{(2)}(S)\right| \le (1-\rho)^{-\frac{1}{2}np}\left(1 - \frac{\epsilon_1}{\rho}\right)^{-1} a_0^{(2)} w_p\left(\frac{1}{2}n, \left(\gamma - \frac{\rho\alpha}{1+\rho}\right)\Sigma^{-1}; S\right),$$

where ρ is a real number, $\epsilon_1 < \rho < 1$. Since $\operatorname{Re}(Z) > -(\gamma - \frac{\rho\alpha}{1+\rho})\Sigma^{-1}$, satisfies the condition (7.5.32) and therefore from the uniqueness property of Laplace transform and Lemma 7.5.3, $f_2(S)$ defines the density of S. The series (7.5.33) is uniformly convergent if $\gamma > \frac{\rho\alpha}{1+\rho}$. $\blacksquare$

The results given in this section were derived by Khatri (1971). For the Laguerre series expansion, he has also given upper bound for

$$e_N^{(2)}(S) = \left| \sum_{k=N+1}^{\infty} \sum_\kappa a_\kappa^{(2)} h_\kappa^{(2)}(S)\right|.$$

7.6. NONCENTRAL DENSITY FUNCTION

Let $X \sim N_{p,n}(M, \Sigma \otimes \Psi)$. In Section 7.2, the density of $S = XAX'$, $A(n \times n) > 0$, $n \ge p$, for $M = 0$ has been derived. In this section, the density of S for $M \ne 0$, called the noncentral density, is derived.

THEOREM 7.6.1. Let $X \sim N_{p,n}(M, \Sigma \otimes I_n)$, $n \ge p$, and $\Sigma > 0$. Then the density of $S = XAX'$, $A(n \times n) > 0$, is given by

$$\left\{2^{\frac{1}{2}np}\Gamma_p\left(\frac{1}{2}n\right)\right\}^{-1} \det(\Sigma)^{-\frac{1}{2}n} \det(A)^{-\frac{1}{2}p} \operatorname{etr}\left(-\frac{1}{2}\Sigma^{-1}MM' - \frac{1}{2}q\Sigma^{-1}S\right) \det(S)^{\frac{1}{2}(n-p-1)}$$

$$\sum_{k=0}^{\infty} \sum_\kappa \frac{1}{(\frac{1}{2}n)_\kappa k!} P_\kappa\left(\frac{1}{\sqrt{2}}\Sigma^{-\frac{1}{2}}M(I_n - qA)^{-\frac{1}{2}}, A^{-1} - qI_n, \frac{1}{2}\Sigma^{-\frac{1}{2}}S\Sigma^{-\frac{1}{2}}\right), S > 0,$$

where $q > 0$, $I_n - qA$ is positive definite and $P_\kappa(\cdot, \cdot, \cdot)$ is the generalized Hayakawa polynomial defined in Section 1.8.

Proof: The density of $S = XAX'$, $X \sim N_{p,n}(M, \Sigma \otimes I_n)$, can be obtained from the density of $S = XAX'$, $X \sim N_{p,n}(\Sigma^{-\frac{1}{2}}M, I_p \otimes I_n)$, by transforming $S \to \Sigma^{\frac{1}{2}}S\Sigma^{\frac{1}{2}}$. Therefore we derive the density of $S = XAX'$, $X \sim N_{p,n}(\mu, I_p \otimes I_n)$, $\mu = \Sigma^{-\frac{1}{2}}M$, which is given by

$$f(S) = (2\pi)^{-\frac{1}{2}np} \int_{XAX'=S} \text{etr}\left\{-\frac{1}{2}(X-\mu)(X-\mu)'\right\} dX$$

$$= (2\pi)^{-\frac{1}{2}np} \int_{XAX'=S} \text{etr}\left[-\frac{1}{2}\{qXAX' + X(I_n - qA)X'\right.$$

$$\left. -2\mu X' + \mu\mu'\}\right] dX. \tag{7.6.1}$$

Note that the integral

$$\pi^{-\frac{1}{2}np} \int_U \text{etr}\left[-\left\{U - \frac{\iota}{\sqrt{2}}\left(X(I_n - qA)^{\frac{1}{2}} - \mu(I_n - qA)^{-\frac{1}{2}}\right)\right\}\right.$$

$$\left.\left\{U - \frac{\iota}{\sqrt{2}}\left(X(I_n - qA)^{\frac{1}{2}} - \mu(I_n - qA)^{-\frac{1}{2}}\right)\right\}'\right] dU \tag{7.6.2}$$

is unity since $U \sim N_{p,n}\left(\frac{\iota}{\sqrt{2}}\left(X(I_n - qA)^{\frac{1}{2}} - \mu(I_n - qA)^{-\frac{1}{2}}\right), \frac{1}{2}I_p \otimes I_n\right)$.

Next multiplying (7.6.2) and (7.6.1), and changing the order of integration we get

$$f(S) = (2\pi^2)^{-\frac{1}{2}np} \text{etr}\left\{-\frac{1}{2}\mu\mu' + \frac{1}{2}\mu(I_n - qA)^{-1}\mu'\right\}$$

$$\int_U \text{etr}\{-UU' - \sqrt{2}\,\iota U(I_n - qA)^{-\frac{1}{2}}\mu')\}$$

$$\int_{XAX'=S} \text{etr}\left[-\frac{1}{2}qXAX' + \sqrt{2}\,\iota U(I_n - qA)^{\frac{1}{2}}X'\right] dX\, dU. \tag{7.6.3}$$

Now substituting $Y = XA^{\frac{1}{2}}$, with the Jacobian $J(X \to Y) = \det(A)^{-\frac{1}{2}p}$, and using Theorem 1.6.6, we get

$$\int_{XAX'=S} \text{etr}\left[-\frac{1}{2}qXAX' + \sqrt{2}\,\iota U(I_n - qA)^{\frac{1}{2}}X'\right] dX$$

$$= \frac{\pi^{\frac{1}{2}np}}{\Gamma_p\left(\frac{1}{2}n\right)} \det(A)^{-\frac{1}{2}p} \text{etr}\left(-\frac{1}{2}qS\right) \det(S)^{\frac{1}{2}(n-p-1)}$$

$$_0F_1\left(\frac{1}{2}n, -\frac{1}{2}U(A^{-1} - qI_n)U'S\right)$$

$$= \frac{\pi^{\frac{1}{2}np}}{\Gamma_p\left(\frac{1}{2}n\right)} \det(A)^{-\frac{1}{2}p} \text{etr}\left(-\frac{1}{2}qS\right) \det(S)^{\frac{1}{2}(n-p-1)}$$

$$\sum_{k=0}^{\infty}\sum_{\kappa} \frac{1}{\left(\frac{1}{2}n\right)_\kappa k!} C_\kappa\left(-\frac{1}{2}U(A^{-1} - qI_n)U'S\right). \tag{7.6.4}$$

In (7.6.3), substitute from (7.6.4), and use (1.8.2) to get

$$\left\{2^{\frac{1}{2}np}\Gamma_p\left(\tfrac{1}{2}n\right)\right\}^{-1}\det(A)^{-\frac{1}{2}p}\,\mathrm{etr}\left(-\frac{1}{2}\mu\mu'-\frac{1}{2}qS\right)\det(S)^{\frac{1}{2}(n-p-1)}$$

$$\sum_{k=0}^{\infty}\sum_{\kappa}\frac{1}{(\frac{1}{2}n)_\kappa\,k!}P_\kappa\left(\frac{1}{\sqrt{2}}\mu(I_n-qA)^{-\frac{1}{2}},A^{-1}-qI_n,\frac{1}{2}S\right),\ S>0.$$

Finally transforming $S\to\Sigma^{\frac{1}{2}}S\Sigma^{\frac{1}{2}}$ we get the desired result. ∎

THEOREM 7.6.2. *Let $X\sim N_{p,n}(M,\Sigma\otimes\Psi)$, $n\ge p$, $\Sigma>0$ and $\Psi>0$. Then the density function of $S=XAX'$, $A\,(n\times n)>0$, is given by*

$$\left\{2^{\frac{1}{2}np}\Gamma_p\left(\tfrac{1}{2}n\right)\right\}^{-1}\det(\Sigma)^{-\frac{1}{2}n}\det(B)^{-\frac{1}{2}p}\,\mathrm{etr}\left(-\frac{1}{2}\Sigma^{-1}M\Psi^{-1}M'\right)$$

$$\mathrm{etr}\left(-\frac{1}{2}q\Sigma^{-1}S\right)\det(S)^{\frac{1}{2}(n-p-1)}\sum_{k=0}^{\infty}\sum_{\kappa}\frac{1}{(\frac{1}{2}n)_\kappa\,k!}$$

$$P_\kappa\left(\frac{1}{\sqrt{2}}\Sigma^{-\frac{1}{2}}M\Psi^{-\frac{1}{2}}(I_n-qB)^{-\frac{1}{2}},B^{-1}-qI_n,\frac{1}{2}\Sigma^{-\frac{1}{2}}S\Sigma^{-\frac{1}{2}}\right),\ S>0,$$

where $q>0$, $B=\Psi^{\frac{1}{2}}A\Psi^{\frac{1}{2}}$, I_n-qB is positive definite and $P_\kappa(\cdot,\cdot,\cdot)$ is the generalized Hayakawa polynomial defined in Section 1.8.

Proof: Note that $XAX'=YBY'$, where $Y\sim N_{p,n}(M\Psi^{-\frac{1}{2}},\Sigma\otimes I_n)$ and $B=\Psi^{\frac{1}{2}}A\Psi^{\frac{1}{2}}$. The result now follows from Theorem 7.6.1. ∎

COROLLARY 7.6.2.1. *For $M=0$, the above density of S reduces to (7.2.1).*

Proof: When $M=0$, from (1.8.3), we have

$$P_\kappa\left(0,B^{-1}-qI_n,\frac{1}{2}\Sigma^{-\frac{1}{2}}S\Sigma^{-\frac{1}{2}}\right)=\left(\frac{1}{2}n\right)_\kappa\frac{C_\kappa(I_n-q^{-1}B^{-1})C_\kappa(\frac{1}{2}q\Sigma^{-\frac{1}{2}}S\Sigma^{-\frac{1}{2}})}{C_\kappa(I_n)}.\quad(7.6.5)$$

Using (7.6.5) in Theorem 7.6.2 for M = 0, and simplifying we get the desired result. ∎

COROLLARY 7.6.2.2. *The density of $S=XX'$ is given by*

$$\left\{2^{\frac{1}{2}np}\Gamma_p\left(\tfrac{1}{2}n\right)\right\}^{-1}\det(\Sigma)^{-\frac{1}{2}n}\det(\Psi)^{-\frac{1}{2}p}\,\mathrm{etr}\left(-\frac{1}{2}\Sigma^{-1}M\Psi^{-1}M'\right)$$

$$\mathrm{etr}\left(-\frac{1}{2}q\Sigma^{-1}S\right)\det(S)^{\frac{1}{2}(n-p-1)}\sum_{k=0}^{\infty}\sum_{\kappa}\frac{1}{(\frac{1}{2}n)_\kappa\,k!}$$

$$P_\kappa\left(\frac{1}{\sqrt{2}}\Sigma^{-\frac{1}{2}}M\Psi^{-\frac{1}{2}}(I_n-q\Psi)^{-\frac{1}{2}},\Psi^{-1}-qI_n,\frac{1}{2}\Sigma^{-\frac{1}{2}}S\Sigma^{-\frac{1}{2}}\right),\ S>0.$$

COROLLARY 7.6.2.3. *For $\Psi^{\frac{1}{2}}A\Psi^{\frac{1}{2}}=I_n$, $S\sim W_p(n,\Sigma,\Sigma^{-1}M\Psi^{-1}M')$.*

COROLLARY 7.6.2.4. *For* $X\,(1 \times n) = \boldsymbol{x}'$, $M\,(1 \times n) = \boldsymbol{m}' = (m_1, \ldots, m_n)$, $A = I_n$, $\Sigma\,(1 \times 1) = 1$, $\Psi = \operatorname{diag}(\psi_1 I_{n_1}, \ldots, \psi_m I_{n_m})$, $\sum_{i=1}^{m} n_i = n$, $s = \boldsymbol{x}'\boldsymbol{x} \sim \sum_{i=1}^{m} \psi_i \chi_{n_i}'^2(\omega_i)$ *where* $\chi_{n_i}'^2$ *is a noncentral chi-square distribution with n_i degrees of freedom and noncentrality parameter* $\omega_i = \sum_{j=n_1+\cdots+n_{i-1}+1}^{n_1+\cdots+n_i} \frac{m_j^2}{\psi_j}$. *The density of s, from Theorem 7.6.2, is*

$$\left\{ 2^{\frac{1}{2}n} \Gamma\!\left(\tfrac{1}{2}n\right) \right\}^{-1} \left(\prod_{i=1}^{m} \psi_i^{n_i} \right)^{-\frac{1}{2}} \exp\left(-\frac{1}{2} \sum_{i=1}^{m} \omega_i - \frac{1}{2}qs \right) s^{\frac{1}{2}(n-2)}$$

$$\sum_{k=0}^{\infty} \frac{1}{(\frac{1}{2}n)_k\, k!} P_k\!\left(\frac{1}{\sqrt{2}} \boldsymbol{m}' \Psi^{-\frac{1}{2}} (I_n - q\Psi)^{-\frac{1}{2}}, \Psi^{-1} - qI_n, \frac{1}{2}s \right), \quad s > 0.$$

For calculating $P_\kappa(\boldsymbol{t}', A, B)$, Crowther (1975) has given a method of utilizing the cumulants of certain quadratic form involving A.

Khatri (1977), using Laplace transforms, has generalized the results of Shah (1971) to the noncentral case. The density of S, derived by Khatri (1977), is

$$\left\{ 2^{\frac{1}{2}np} \Gamma_p\!\left(\tfrac{1}{2}n\right) \right\}^{-1} \det(q^{-1}\Sigma)^{-\frac{1}{2}n} \operatorname{etr}\left(-\frac{1}{2}q\Sigma^{-1}S \right) \det(S)^{\frac{1}{2}(n-p-1)}$$

$$\sum_{k=0}^{\infty} \sum_{\kappa} \frac{1}{(\frac{1}{2}n)_\kappa\, k!} L_\kappa\!\left(\frac{1}{2}q\Sigma^{-\frac{1}{2}}S\Sigma^{-\frac{1}{2}}, I_n - qA, \frac{1}{\sqrt{2}}\Sigma^{-\frac{1}{2}}M((qA)^{-1} - I_n)^{-\frac{1}{2}} \right), \quad S > 0.$$

where $q > 0$ is a constant which governs the convergence of the series, and $L_\kappa^\tau(S, A, T)$ is the generalized Laguerre polynomial defined by Khatri (1977). When $A = I_n$, *i.e.*, $S \sim W_p(n, \Sigma, \Sigma^{-1}MM')$, he also obtained the following representation of noncentral Wishart density in terms of the generalized Laguerre polynomials,

$$\left\{ 2^{\frac{1}{2}np} \Gamma_p\!\left(\tfrac{1}{2}n\right) \right\}^{-1} \det(\Sigma)^{-\frac{1}{2}n} \operatorname{etr}\left(-\frac{1}{2}\Sigma^{-1}S \right) \det(S)^{\frac{1}{2}(n-p-1)}$$

$$\sum_{k=0}^{\infty} \sum_{\kappa} \frac{1}{(\frac{1}{2}n)_\kappa\, k!} L_\kappa^{\frac{1}{2}(n-p-1)}\!\left(\frac{1}{2}\Sigma^{-\frac{1}{2}}S\Sigma^{-\frac{1}{2}}, -\frac{1}{2}\Sigma^{-\frac{1}{2}}MM'\Sigma^{-\frac{1}{2}} \right), \quad S > 0.$$

Next, following the method similar to Section 7.5, we derive a series representation for the density of S in the noncentral case. Since S has $\frac{1}{2}p(p+1)$ distinct random variables, we can write the series form of $f(S)$ as

$$f(S) = \sum_{k=0}^{\infty} \sum_{K} a_K f_K(S), \tag{7.6.6}$$

where $K = (k_{11}, k_{12}, \ldots, k_{1p}, k_{22}, \ldots, k_{2p}, \ldots, k_{pp})$, $k_{ij} \geq 0$, $\sum_{i=1}^{p} \sum_{j=i}^{p} k_{ij} = k$, $\sum_K$ is the multinomial sum, a_K is a constant, and $f_K(S)$ is a suitable function of S. This series uses $\frac{1}{2}p(p+1)$ partitions whereas the series (7.5.2) uses p partitions of k, and structurally these two representations are different.

For the convergence of the series (7.6.6), we require

$$|f(S)| \leq b \operatorname{etr}(BS), \text{ for all } S > 0,$$

where b is a real constant and $-B$ is a positive semidefinite matrix. Since the density of S can be obtained from the density of $\Sigma^{-\frac{1}{2}} S \Sigma^{-\frac{1}{2}}$, (Khatri, 1975), we therefore without loss of generality take $\Sigma = I_p$ in the following derivation. From Lemma 7.5.2, the Laplace transform, $L_0(Z)$, of S is

$$L_0(Z) = \prod_{j=1}^{n} \det(I_p + 2\alpha_j Z)^{-\frac{1}{2}} \exp\left\{-\sum_{j=1}^{n} \delta'_j Z(I_p + 2\alpha_j Z)^{-1}\delta_j\right\}. \qquad (7.6.7)$$

Let $\hat{f}(Z)$ be the Laplace transform of (7.6.6). Then (7.6.6) is the p.d.f. of S if and only if

$$\hat{f}(Z) = L_0(Z), \qquad (7.6.8)$$

for almost all Z such that $\mathrm{Re}(Z) > 0$. Further let $\Theta = (\theta_{ij}) = (\gamma I_p + Z)^{-1}$, i.e., $Z = \Theta^{-1} - \gamma I_p$. Then from (7.6.7) we have

$$L_0(\Theta^{-1} - \gamma I_p) = a_0 \det(\Theta)^{\frac{1}{2}n} \prod_{j=1}^{n} \det(I_p - \beta_j\Theta)^{-\frac{1}{2}} \exp\left\{\sum_{j=1}^{n} \nu'_j\Theta(I_p - \beta_j\Theta)^{-1}\nu_j\right\}, \qquad (7.6.9)$$

where $a_0 = \det(2D)^{-\frac{1}{2}p} \exp\{-\frac{1}{2}\sum_{j=1}^{n}\frac{\delta'_j\delta_j}{\alpha_j}\}$, $\nu_j = \frac{\delta_j}{2\alpha_j}$, $D = \mathrm{diag}(\alpha_1,\ldots,\alpha_n)$, and $b_j = \frac{2\alpha_j\gamma-1}{2\alpha_j}$, $j = 1,\ldots,n$. Now (7.6.9) can be written as

$$L_0(\Theta^{-1} - \gamma I_p) = \det(\Theta)^{\frac{1}{2}n} \sum_{k=0}^{\infty} \sum_K a_K N_K(\Theta), \qquad (7.6.10)$$

where $N_K(\Theta) = \prod_{i=1}^{p}\prod_{j=i}^{p}\theta_{ij}^{k_{ij}}$, $\theta_{ij} = \theta_{ji}$, $i,j = 1,\ldots,p$. From (7.6.8) and (7.6.10), we get

$$\hat{f}(Z) = \det(\gamma I_p + Z)^{-\frac{1}{2}n} \sum_{k=0}^{\infty} \sum_K a_K N_K((\gamma I_p + Z)^{-1}). \qquad (7.6.11)$$

Comparing (7.6.11) with (7.6.6), it follows that the Laplace transform of $f_K(S)$ is

$$\hat{f}_K(Z) = \det(\gamma I_p + Z)^{-\frac{1}{2}n} N_K((\gamma I_p + Z)^{-1}). \qquad (7.6.12)$$

Thus, we need to find the function $f_K(S)$ whose Laplace transform is (7.6.12). To do so let us consider

$$g_k(S,Q) = \sum_\lambda \left\{\left(\tfrac{1}{2}n\right)_\lambda\right\}^{-1} C_\lambda(QS) w_p\left(\tfrac{1}{2}n, \gamma I_p; S\right), \qquad (7.6.13)$$

where

$$w_p\left(\tfrac{1}{2}n, \gamma I_p; S\right) = \left\{\Gamma_p\left(\tfrac{1}{2}n\right)\right\}^{-1} \det(S)^{\frac{1}{2}(n-p-1)} \mathrm{etr}(-\gamma S), \quad n \geq p,$$

$\lambda = (\ell_1, \ell_2, \ldots, \ell_p)$, $\ell_1 \geq \ell_2 \geq \cdots \geq \ell_p \geq 0$, and $\ell_1 + \ell_2 + \cdots + \ell_p = k$. The Laplace transform of (7.6.13) is

$$\hat{g}_k(S,Q) = \sum_\lambda \left\{\Gamma_p\left(\tfrac{1}{2}n\right)\left(\tfrac{1}{2}n\right)_\lambda\right\}^{-1} \int_{S>0} \mathrm{etr}(-ZS) C_\lambda(QS) w_p\left(\tfrac{1}{2}n, \gamma I_p; S\right) dS$$

$$= \sum_{\lambda} \left\{ \Gamma_p\left(\tfrac{1}{2}n\right)\left(\tfrac{1}{2}n\right)_{\lambda}\right\}^{-1} \int_{S>0} \mathrm{etr}\{-(\gamma I_p + Z)S\}C_{\lambda}(QS)\det(S)^{\frac{1}{2}(n-p-1)}\,dS$$

$$= \det(\gamma I_p + Z)^{-\frac{1}{2}n}\sum_{\lambda} C_{\lambda}(Q(\gamma I_p + Z)^{-1})$$

$$= \det(\gamma I_p + Z)^{-\frac{1}{2}n}\{\mathrm{tr}(Q(\gamma I_p + Z)^{-1})\}^k$$

$$= \sum_{K} \hat{f}_K(Z)N_K(Q)c_K, \tag{7.6.14}$$

where $c_K = 2^{k-\sum_{i=1}^p k_{ii}}\, k!\left\{\prod_{i=1}^p \prod_{j=i}^p \frac{1}{k_{ij}!}\right\}$. Then from the uniqueness of the Laplace transform, $\hat{f}_K(Z)$ is the coefficient of $N_K(Q)c_K$ in the expansion of $\hat{g}_k(S,Q)$, and $f_K(S)$ is the coefficient of $N_K(Q)c_K$ in the expansion of $g_k(S,Q)$. Thus

$$f(S) = \sum_{k=0}^{\infty} \sum_{K} a_K f_K(S), \tag{7.6.15}$$

where $f_K(S)$ can be obtained as described above. For convergence of the series (7.6.15), Khatri (1975) has obtained bounds for $|\sum_K a_K f_K(S)|$, for $k > 1$. He has also tabulated a_K and $f_K(S)$ for $k = 1, 2$.

7.7. EXPECTED VALUES

The matrix quadratic forms studied in this chapter are defined in terms of $X \sim N_{p,n}(M, \Sigma \otimes \Psi)$. The matrix variate normal distribution has been studied in Chapter 2. There we have also given several expected values of functions of XAX'. For the sake of completeness, we state those below (for proof the reader is referred to Chapter 2). Throughout this section the matrices of quadratic forms need not be symmetric.

THEOREM 7.7.1. *Let* $S_A = XAX'$ *and* $S_B = XBX'$, *where* $X \sim N_{p,n}(M, \Sigma \otimes \Psi)$, *and* $A\,(n \times n)$ *and* $B\,(n \times n)$ *be constant matrices. Then*

(i) $E(S_A) = \mathrm{tr}(A\Psi)\Sigma + MAM'$,

(ii) $E(S_A C S_B) = \mathrm{tr}(\Psi B'\Psi A')\,\mathrm{tr}(C\Sigma)\Sigma + \mathrm{tr}(A\Psi)\,\mathrm{tr}(B\Psi)\Sigma C\Sigma$
$\quad + \mathrm{tr}(A\Psi B'\Psi)\Sigma C'\Sigma + \mathrm{tr}(B\Psi)MAM'C\Sigma + MA\Psi B'M'C'\Sigma$
$\quad + \mathrm{tr}(AM'CMB\Psi)\Sigma + \mathrm{tr}(C\Sigma)MA\Psi BM' + \Sigma C'MA'\Psi BM'$
$\quad + \mathrm{tr}(A\Psi)\Sigma CMBM' + MAM'CMBM'$,

and

(iii) $E(\mathrm{tr}(S_B C)S_A) = \mathrm{tr}(A'\Psi B\Psi)\Sigma C'\Sigma + \mathrm{tr}(A\Psi)\,\mathrm{tr}(B\Psi)\,\mathrm{tr}(C\Sigma)\Sigma$
$\quad + \mathrm{tr}(A'\Psi B'\Psi)\Sigma C\Sigma + \mathrm{tr}(B\Psi)\,\mathrm{tr}(C\Sigma)MAM'$
$\quad + MA\Psi BM'C\Sigma + \Sigma C'MB'\Psi AM'$
$\quad + MA\Psi B'M'C'\Sigma + \Sigma CM'B\Psi AM'$
$\quad + \mathrm{tr}(A\Psi)\,\mathrm{tr}(M'CMB)\Sigma + \mathrm{tr}(BM'CM)MAM'$

where $C\,(p \times p)$ *is a constant matrix.*

Next we derive the covariance matrix of $\mathrm{vec}(XAX')$ and $\mathrm{vec}(XBX')$, a result due to Neudecker and Wansbeek (1987).

THEOREM 7.7.2. *Let $S_A = XAX'$ and $S_B = XBX'$, where $X \sim N_{p,n}(M, \Sigma \otimes \Psi)$. Then*

$$\mathrm{cov}(\mathrm{vec}(S_A), \mathrm{vec}(S_B)) = \mathrm{tr}(A\Psi B'\Psi)\Sigma \otimes \Sigma + MA'\Psi BM' \otimes \Sigma$$
$$+ \Sigma \otimes MA\Psi B'M' + \{\mathrm{tr}(A'\Psi B'\Psi)\Sigma \otimes \Sigma$$
$$+ MA'\Psi B'M' \otimes \Sigma + \Sigma \otimes MA\Psi BM'\}K_{pp},$$

where K_{pp} is the commutation matrix defined in Section 1.2.

Proof: From Theorem 7.7.1, we have

$$E(S_A C S_B) - E(S_A)CE(S_B) = \mathrm{tr}(\Psi B'\Psi A')\,\mathrm{tr}(C\Sigma)\Sigma + \mathrm{tr}(A\Psi B'\Psi)\Sigma C'\Sigma$$
$$+ MA\Psi B'M'C'\Sigma + \mathrm{tr}(AM'CMB\Psi)\Sigma$$
$$+ \mathrm{tr}(C\Sigma)MA\Psi BM' + \Sigma C'MA'\Psi BM'$$

$$= \sum_{i=1}^{3} P_i C'Q_i + \sum_{i=4}^{6} \mathrm{tr}(C'P_i)Q_i, \qquad (7.7.1)$$

where $P_1 = \mathrm{tr}(A\Psi B'\Psi)\Sigma$, $P_3 = MA\Psi B'M'$, $P_5 = MA'\Psi B'M'$, $P_2 = P_4 = P_6 = \Sigma = Q_1 = Q_3 = Q_5$, $Q_2 = MA'\Psi BM'$, $Q_4 = \mathrm{tr}(A'\Psi B'\Psi)\Sigma$ and $Q_6 = MA\Psi BM'$.

Now for $D\,(p \times p)$, we can write

$$\mathrm{tr}\{K_{pp}(C' \otimes D)\,\mathrm{cov}(\mathrm{vec}(S_A), \mathrm{vec}(S_B))\}$$
$$= \mathrm{tr}\left[K_{pp}(C' \otimes D)\{E(\mathrm{vec}(S_A)(\mathrm{vec}(S_B))') - E(\mathrm{vec}(S_A))E(\mathrm{vec}(S_B))'\}\right]$$
$$= \mathrm{tr}\left[(C' \otimes D)\{E(\mathrm{vec}(S_A)(\mathrm{vec}(S_B'))') - E(\mathrm{vec}(S_A))E(\mathrm{vec}(S_B'))'\}\right]$$
$$= \mathrm{tr}\left[E\{\mathrm{vec}(DS_A C)(\mathrm{vec}(S_B'))'\} - E\{\mathrm{vec}(DS_A C)\}E\{(\mathrm{vec}(S_B'))'\}\right]$$
$$= \mathrm{tr}\{E(S_A C S_B D)\} - \mathrm{tr}\{E(S_A)CE(S_B)D\}. \qquad (7.7.2)$$

The expression (7.7.2) has been obtained by using the properties of Kronecker product and the commutation matrix given in Section 1.2. Now, using (7.7.1) in (7.7.2), we get

$$\mathrm{tr}\{K_{pp}(C' \otimes D)\,\mathrm{cov}(\mathrm{vec}(S_A), \mathrm{vec}(S_B))\}$$
$$= \sum_{i=1}^{3} \mathrm{tr}(P_i C'Q_i D) + \sum_{i=4}^{6} \mathrm{tr}(C'P_i)\,\mathrm{tr}(DQ_i),$$
$$= \sum_{i=1}^{3} \mathrm{tr}\{K_{pp}(C' \otimes D)(Q_i \otimes P_i)\}$$
$$+ \sum_{i=4}^{6} \mathrm{tr}\{K_{pp}(C' \otimes D)K_{pp}(Q_i \otimes P_i)\}, \qquad (7.7.3)$$

since

$$\mathrm{tr}(C'Q_i DP_i) = \mathrm{tr}\{K_{pp}(C'Q_i \otimes DP_i)\} = \mathrm{tr}\{K_{pp}(C' \otimes D)(Q_i \otimes P_i)\}$$

and

$$\text{tr}(P_i C') \text{tr}(Q_i D) = \text{tr}(P_i C' \otimes Q_i D)$$

$$= \text{tr}\{(C' \otimes D)(P_i \otimes Q_i)\}$$

$$= \text{tr}\{(C' \otimes D)K_{pp}(Q_i \otimes P_i)K_{pp}\}.$$

The result (7.7.3) holds for any C and D. Hence

$$\text{cov}(\text{vec}(S_A), \text{vec}(S_B)) = \sum_{i=1}^{3}(Q_i \otimes P_i) + \sum_{i=4}^{6} K_{pp}(Q_i \otimes P_i), \qquad (7.7.4)$$

By substituting for P_i, Q_i, $i = 1, \dots, 6$, in (7.7.4) we get the desired result. ∎

Letting $A = B$ in the above theorem, we get the following result.

COROLLARY 7.7.2.1. *The covariance matrix of* $\text{vec}(S_A)$ *is given by*

$$\text{cov}(\text{vec}(S_A)) = \text{tr}(A\Psi A'\Psi)\Sigma \otimes \Sigma + MA'\Psi AM' \otimes \Sigma$$

$$+ \Sigma \otimes MA\Psi A'M' + \{\text{tr}(A'\Psi A'\Psi)\Sigma \otimes \Sigma$$

$$+ MA'\Psi A'M' \otimes \Sigma + \Sigma \otimes MA\Psi AM'\}K_{pp}, \qquad (7.7.5)$$

When $\Psi = I_n$, (7.7.5) reduces to the result given by Neudecker (1985). For $A = I_n$, $\Psi = I_n$, (7.7.5) gives the covariance matrix of noncentral Wishart matrix, as in Magnus and Neudecker (1979). By substituting $M = 0$ in Theorems 7.7.1 and 7.7.2 we get the results given in Theorem 7.3.5.

von Rosen (1988b) derived $E[(XAX') \otimes (XBX')]$ when $X \sim N_{p,n}(M, \Sigma \otimes \Psi)$. Tracy and Sultana (1993) derived $E[(XAX') \otimes (XBX') \otimes (XCX')]$ when $X \sim N_{p,n}(0, \Sigma \otimes \Psi)$. Kang and Kim (1996) gave general result for $E[\otimes_{i=1}^{N}(XA_i X')]$, for $X \sim N_{p,n}(0, \Sigma \otimes \Psi)$.

7.8. WISHARTNESS AND INDEPENDENCE OF QUADRATIC FORMS OF THE TYPE XAX'

So far we have studied the distribution of XAX', where $X \sim N_{p,n}(M, \Sigma \otimes \Psi)$ for $M = 0$ and $M \neq 0$, *i.e.*, the central and noncentral cases. Under certain conditions these quadratic forms follow Wishart or noncentral Wishart distribution, as noted in Chapter 3 and also in this chapter. In the present section we give conditions for Wishartness of quadratic forms of the type XAX'. Conditions for independence of two or more such quadratic forms are also given. In the next section we have derived similar conditions for the quadratic forms of the type $XAX' + \frac{1}{2}(LX' + XL') + C$. Most of the results derived in the present section can be obtained as special cases of the results in the next section. However, for the sake of completeness and readability, the results for two types of quadratic forms are presented sequentially.

Let us now consider the quadratic forms of the type XAX'. First we derive the m.g.f. of XAX' for $\Psi = I_n$.

THEOREM 7.8.1. *Let $X \sim N_{p,n}(M, \Sigma \otimes I_n)$. Then the m.g.f. of $S = XAX'$, where $A\,(n \times n)$ is a symmetric matrix of rank t, is given by*

$$M_S(Z) = \prod_{j=1}^{t} \det(I_p - 2\lambda_j \Sigma Z)^{-\frac{1}{2}} \exp\left\{ \sum_{j=1}^{t} \lambda_j q'_j M' Z (I_p - 2\lambda_j \Sigma Z)^{-1} M q_j \right\}, \quad (7.8.1)$$

where $A = Q \begin{pmatrix} D_\lambda & 0 \\ 0 & 0 \end{pmatrix} Q'$, $Q\,(n \times n)$ is an orthogonal matrix, $Q = (q_1, \ldots, q_n)$, $D_\lambda = \mathrm{diag}(\lambda_1, \ldots, \lambda_t)$ and $\lambda_1, \ldots, \lambda_t$ are the nonzero characteristic roots of A.

Proof: The m.g.f. of $S = XAX'$ is given by

$$M_S(Z) = (2\pi)^{-\frac{1}{2}np} \det(\Sigma)^{-\frac{1}{2}n}$$

$$\int_{X \in \mathbb{R}^{p \times n}} \mathrm{etr}\left\{ ZXAX' - \frac{1}{2}\Sigma^{-1}(X - M)(X - M)' \right\} dX. \quad (7.8.2)$$

Since A is symmetric matrix of rank $t\,(\leq n)$, we can write $A = Q \begin{pmatrix} D_\lambda & 0 \\ 0 & 0 \end{pmatrix} Q'$, where $D_\lambda = \mathrm{diag}(\lambda_1, \ldots, \lambda_t)$, λ_j, $j = 1, \ldots, t$ are the nonzero characteristic roots of A, and $Q\,(n \times n)$ is an orthogonal matrix. Using the transformation $Y = \Sigma^{-\frac{1}{2}} X Q$ in (7.8.2) with the Jacobian $J(X \to Y) = \det(\Sigma)^{\frac{1}{2}n}$, we have

$$M_S(Z) = (2\pi)^{-\frac{1}{2}np} \int_{Y \in \mathbb{R}^{p \times n}} \mathrm{etr}\left\{ \Sigma^{\frac{1}{2}} Z \Sigma^{\frac{1}{2}} Y Q' A Q Y' \right.$$

$$\left. - \frac{1}{2}(YQ' - \Sigma^{-\frac{1}{2}}M)(YQ' - \Sigma^{-\frac{1}{2}}M)' \right\} dY$$

$$= (2\pi)^{-\frac{1}{2}np} \int_{Y \in \mathbb{R}^{p \times n}} \mathrm{etr}\left\{ \Sigma^{\frac{1}{2}} Z \Sigma^{\frac{1}{2}} Y_1 D_\lambda Y_1' - \frac{1}{2}(Y - N)(Y - N)' \right\} dY$$

$$= (2\pi)^{-\frac{1}{2}tp} \int_{Y_1 \in \mathbb{R}^{p \times t}} \mathrm{etr}\left\{ \Sigma^{\frac{1}{2}} Z \Sigma^{\frac{1}{2}} Y_1 D_\lambda Y_1' - \frac{1}{2}(Y_1 - N_1)(Y_1 - N_1)' \right\} dY_1$$

$$= (2\pi)^{-\frac{1}{2}tp} \mathrm{etr}\left(-\frac{1}{2}N_1 N_1' \right)$$

$$\int_{Y_1 \in \mathbb{R}^{p \times t}} \mathrm{etr}\left\{ -\frac{1}{2}\left(Y_1 Y_1' - 2\Sigma^{\frac{1}{2}} Z \Sigma^{\frac{1}{2}} Y_1 D_\lambda Y_1' \right) + Y_1 N_1' \right\} dY_1, \quad (7.8.3)$$

where $N = \Sigma^{-\frac{1}{2}} M Q$, $Y = (Y_1 \quad Y_2)$, $Y_1\,(p \times t)$, and $N = (N_1 \quad N_2)$, $N_1\,(p \times t)$.
Writing $Y_1 = (y_{11}, \ldots, y_{1t})$ and $N_1 = (n_{11}, \ldots, n_{1t})$, we get

$$\mathrm{tr}(Y_1 N_1') = \sum_{j=1}^{t} n'_{1j} y_{1j}, \quad (7.8.4)$$

$$\mathrm{tr}(N_1 N_1') = \sum_{j=1}^{t} n'_{1j} n_{1j}, \quad (7.8.5)$$

and

$$\operatorname{tr}\left(Y_1 Y_1' - 2\Sigma^{\frac{1}{2}} Z \Sigma^{\frac{1}{2}} Y_1 D_\lambda Y_1'\right) = \operatorname{tr}\left\{\sum_{j=1}^{t}(I_p - 2\lambda_j \Sigma^{\frac{1}{2}} Z \Sigma^{\frac{1}{2}}) \boldsymbol{y}_{1j}\boldsymbol{y}_{1j}'\right\}. \tag{7.8.6}$$

By substituting from (7.8.4), (7.8.5) and (7.8.6) in (7.8.3) we get

$$M_S(Z) = (2\pi)^{-\frac{1}{2}tp}\exp\left(-\frac{1}{2}\sum_{j=1}^{t}\boldsymbol{n}_{1j}'\boldsymbol{n}_{1j}\right)$$

$$\prod_{j=1}^{t}\int_{\boldsymbol{y}_{1j}\in\mathbb{R}^p}\exp\left\{-\frac{1}{2}\boldsymbol{y}_{1j}'(I_p - 2\lambda_j\Sigma^{\frac{1}{2}}Z\Sigma^{\frac{1}{2}})\boldsymbol{y}_{1j} + \boldsymbol{n}_{1j}'\boldsymbol{y}_{1j}\right\}d\boldsymbol{y}_{1j}$$

$$= \exp\left(-\frac{1}{2}\sum_{j=1}^{t}\boldsymbol{n}_{1j}'\boldsymbol{n}_{1j}\right)\prod_{j=1}^{t}\det(I_p - 2\lambda_j\Sigma^{\frac{1}{2}}Z\Sigma^{\frac{1}{2}})^{-\frac{1}{2}}$$

$$\prod_{j=1}^{t}\left\{(2\pi)^{-\frac{1}{2}p}\det(I_p - 2\lambda_j\Sigma^{\frac{1}{2}}Z\Sigma^{\frac{1}{2}})^{\frac{1}{2}}\right.$$

$$\left.\int_{\boldsymbol{y}_{1j}\in\mathbb{R}^p}\exp\left\{-\frac{1}{2}\boldsymbol{y}_{1j}'(I_p - 2\lambda_j\Sigma^{\frac{1}{2}}Z\Sigma^{\frac{1}{2}})\boldsymbol{y}_{1j} + \boldsymbol{n}_{1j}'\boldsymbol{y}_{1j}\right\}d\boldsymbol{y}_{1j}\right\}$$

$$= \prod_{j=1}^{t}\det(I_p - 2\lambda_j\Sigma^{\frac{1}{2}}Z\Sigma^{\frac{1}{2}})^{-\frac{1}{2}}$$

$$\exp\left\{-\frac{1}{2}\sum_{j=1}^{t}\boldsymbol{n}_{1j}'\boldsymbol{n}_{1j} + \frac{1}{2}\sum_{j=1}^{t}\boldsymbol{n}_{1j}'(I_p - 2\lambda_j\Sigma^{\frac{1}{2}}Z\Sigma^{\frac{1}{2}})^{-1}\boldsymbol{n}_{1j}\right\}$$

$$= \prod_{j=1}^{t}\det(I_p - 2\lambda_j\Sigma^{\frac{1}{2}}Z\Sigma^{\frac{1}{2}})^{-\frac{1}{2}}$$

$$\exp\left\{\sum_{j=1}^{t}\lambda_j\boldsymbol{n}_{1j}'\Sigma^{\frac{1}{2}}Z\Sigma^{\frac{1}{2}}(I_p - 2\lambda_j\Sigma^{\frac{1}{2}}Z\Sigma^{\frac{1}{2}})^{-1}\boldsymbol{n}_{1j}\right\}. \tag{7.8.7}$$

The final result is obtained from (7.8.7) by noting that $N_1 = \Sigma^{-\frac{1}{2}}M(\boldsymbol{q}_1,\ldots,\boldsymbol{q}_t)$. ∎

COROLLARY 7.8.1.1. *If* $\lambda_i = 1$, $i = 1,\ldots,t$, $t \geq p$, *then* $S \sim W_p(t,\Sigma,\Sigma^{-1}MAM')$.

Proof: Substituting $\lambda_i = 1$, $i = 1,\ldots,t$ in (7.8.1) and noting that $A = Q\begin{pmatrix} I_t & 0 \\ 0 & 0 \end{pmatrix}Q'$ $= \sum_{j=1}^{t}\boldsymbol{q}_j\boldsymbol{q}_j'$, we have the m.g.f. of S as

$$M_S(Z) = \det(I_p - 2\Sigma Z)^{-\frac{1}{2}t}\operatorname{etr}\{Z(I_p - 2\Sigma Z)^{-1}MAM'\},$$

which is the m.g.f. of a noncentral Wishart matrix. ∎

From the above theorem we obtain the cumulant generating function of S.

THEOREM 7.8.2. *Let $X \sim N_{p,n}(M, \Sigma \otimes I_n)$. Then the c.g.f. of $S = XAX'$, where $A (n \times n)$ is a symmetric matrix of rank t, is given by*

$$\ln M_S(Z) = \sum_{s=1}^{\infty} \frac{2^{s-1}}{s} \operatorname{tr}(A^s) \operatorname{tr}((\Sigma Z)^s) + \sum_{s=0}^{\infty} 2^s \operatorname{tr}\{MA^{s+1}M'Z(\Sigma Z)^s\}. \qquad (7.8.8)$$

Proof: From Theorem 7.8.1, we get

$$\ln M_S(Z) = -\frac{1}{2} \sum_{j=1}^{t} \ln\{\det(I_p - 2\lambda_j \Sigma Z)\} + \sum_{j=1}^{t} \lambda_j \boldsymbol{q}_j' M' Z(I_p - 2\lambda_j \Sigma Z)^{-1} M \boldsymbol{q}_j. \quad (7.8.9)$$

Note that

$$\ln\{\det(I_p - 2\lambda_j \Sigma Z)\} = -\sum_{s=1}^{\infty} \frac{2^s}{s} \lambda_j^s \operatorname{tr}((\Sigma Z)^s), \qquad (7.8.10)$$

and

$$\sum_{j=1}^{t} \lambda_j \boldsymbol{q}_j' M' Z(I_p - 2\lambda_j \Sigma Z)^{-1} M \boldsymbol{q}_j = \sum_{j=1}^{t} \lambda_j \boldsymbol{q}_j' M' Z \Big(\sum_{s=0}^{\infty} (2\lambda_j \Sigma Z)^s \Big) M \boldsymbol{q}_j$$

$$= \sum_{s=0}^{\infty} 2^s \operatorname{tr} \Big\{ M'Z(SZ)^s M \sum_{j=1}^{t} \lambda_j^{s+1} \boldsymbol{q}_j \boldsymbol{q}_j' \Big\}, \quad (7.8.11)$$

where the series expansion in (7.8.10) and (7.8.11) are valid for $\max_i |\operatorname{ch}_i(\lambda_j \Sigma Z)| < \frac{1}{2}$, which can be met since Z is arbitrary. From Theorem 7.8.1, we have

$$A^s = \sum_{j=1}^{t} \lambda_j^s \boldsymbol{q}_j \boldsymbol{q}_j', \qquad (7.8.12)$$

and

$$\operatorname{tr}(A^s) = \sum_{j=1}^{t} \lambda_j^s. \qquad (7.8.13)$$

Finally substituting (7.8.10) and (7.8.11) in (7.8.9) and simplifying the resulting expression using (7.8.12) and (7.8.13), we get the desired result. ∎

COROLLARY 7.8.2.1. *If $\lambda_i = 1$, $i = 1, \ldots, t$, $t \geq p$, then $S \sim W_p(t, \Sigma, \Sigma^{-1}MAM')$ with the c.g.f.*

$$\ln M_S(Z) = t \sum_{s=1}^{\infty} \frac{2^{s-1}}{s} \operatorname{tr}((\Sigma Z)^s) + \sum_{s=0}^{\infty} 2^s \operatorname{tr}\{MAM'Z(\Sigma Z)^s\}. \qquad (7.8.14)$$

Proof: From the Corollary 7.8.1.1, $S \sim W_p(t, \Sigma, \Sigma^{-1}MAM')$. Now the result follows by substituting $\lambda_i = 1$, $i = 1, \ldots, t$, i.e., $A^2 = A$, and $\operatorname{tr}(A) = t$, in (7.8.8). ∎

Next we derive conditions for Wishartness of a matrix quadratic form XAX'.

THEOREM 7.8.3. *Let $S = XAX'$, where $X \sim N_{p,n}(M, \Sigma \otimes I_n)$. The necessary and sufficient condition for S to be distributed as $W_p(t, \Sigma, \Sigma^{-1}MAM')$ is that A is idempotent of rank $t \geq p$.*

Proof: The m.g.f. of XAX' is given in Theorem 7.8.1. Let A be idempotent of rank $t \geq p$. Then $\lambda_i = 1$, $i = 1, \ldots, t$ and thus from Corollary 7.8.1.1, $S \sim W_p(t, \Sigma, \Sigma^{-1} MAM')$ with the m.g.f.

$$M_S(Z) = \det(I_p - 2\Sigma Z)^{-\frac{1}{2}t} \operatorname{etr}\{Z(I_p - 2\Sigma Z)^{-1} MAM'\}. \tag{7.8.15}$$

Conversely, if XAX' is distributed as noncentral Wishart with parameters t, Σ and $\Sigma^{-1} MAM'$, then its m.g.f. given in (7.8.1) must be identical with (7.8.15). Hence, equating the logarithm of these two expressions, from (7.8.8) and (7.8.15), we get

$$\sum_{s=1}^{\infty} \frac{2^{s-1}}{s} \operatorname{tr}(A^s) \operatorname{tr}((\Sigma Z)^s) + \sum_{s=0}^{\infty} 2^s \operatorname{tr}\{MA^{s+1} M' Z (\Sigma Z)^s\}$$

$$= t \sum_{s=1}^{\infty} \frac{2^{s-1}}{s} \operatorname{tr}((\Sigma Z)^s) + \sum_{s=0}^{\infty} 2^s \operatorname{tr}\{MAM' Z (\Sigma Z)^s\}. \tag{7.8.16}$$

Since this holds for any linear function of Z, by equating the coefficient of $\operatorname{tr}((\Sigma Z)^s)$ we get $\operatorname{tr}(A^s) = t$, $s = 1, 2, \ldots, \ldots$, i.e. $\sum_{j=1}^{t} \lambda_j^s = t$, $s = 1, \ldots, t$. Consequently $\sum_{j=1}^{t} \lambda_j^{2s}(\lambda_j - 1)^2 = 0$, and hence $\lambda_1 = \lambda_2 = \cdots = \lambda_t = 1$ i.e. A is idempotent of rank t. It may be noted that for $A = A^s$, the identity (7.8.16) is satisfied. ∎

COROLLARY 7.8.3.1. *The necessary and sufficient condition for $S = XAX'$, where $X \sim N_{p,n}(0, \Sigma \otimes I_n)$, to be distributed as $W_p(t, \Sigma)$ is that A is idempotent of rank $t \geq p$.*

An alternate proof of Theorem 7.8.3 can be given by using the condition for chi-squaredness (noncentral) of the diagonal elements of XAX'.

The conditions for the Wishartness of a matrix quadratic form XAX', when the columns of X are correlated, are given next.

THEOREM 7.8.4. *Let $S = XAX'$, where $X \sim N_{p,n}(M, \Sigma \otimes \Psi)$. The necessary and sufficient condition for S to be distributed as $W_p(t, \Sigma, \Sigma^{-1} MAM')$ is that $A\Psi A = A$ and $\operatorname{rank}(A) = t \geq p$.*

Proof: Note that XAX' has same distribution as $Y(\Psi^{\frac{1}{2}} A \Psi^{\frac{1}{2}}) Y'$, where $Y \sim N_{p,n}(M\Psi^{-\frac{1}{2}}, \Sigma \otimes I_n)$. The condition for $Y(\Psi^{\frac{1}{2}} A \Psi^{\frac{1}{2}}) Y'$ to be distributed as $W_p(t, \Sigma, \Sigma^{-1} MAM')$ is that $\Psi^{\frac{1}{2}} A \Psi^{\frac{1}{2}}$ is idempotent of rank $t \geq p$, i.e., $\Psi^{\frac{1}{2}} A \Psi^{\frac{1}{2}} \Psi^{\frac{1}{2}} A \Psi^{\frac{1}{2}} = \Psi^{\frac{1}{2}} A \Psi^{\frac{1}{2}}$ or equivalently $A\Psi A = A$, and $\operatorname{rank}(\Psi^{\frac{1}{2}} A \Psi^{\frac{1}{2}}) = t (\geq p)$. ∎

COROLLARY 7.8.4.1. *The necessary and sufficient condition for $S = XAX'$, where $X \sim N_{p,n}(0, \Sigma \otimes \Psi)$, to be distributed as $W_p(t, \Sigma)$ is that $A\Psi A = A$ and $\operatorname{rank}(A) = t \geq p$.*

In the remainder of this section we prove some theorems about the stochastic independence of quadratic forms of the type XAX', where $X \sim N_{p,n}(0, \Sigma \otimes I_n)$. We first state a lemma (see Khatri, 1959b) which will be used in the proof of independence.

LEMMA 7.8.1. *Let $A\,(m \times n)$ be a matrix of rank $r \leq n\,(n \leq m)$. Then there exists an orthogonal matrix $Q\,(n \times n)$ such that*

$$A = \begin{pmatrix} T_1 & 0 \\ T_2 & 0 \end{pmatrix} Q,$$

where $T_1\,(r \times r)$ is a lower triangular matrix with positive diagonal elements, and $T_2\,((m-r) \times r)$ is a linear function of T_1.

THEOREM 7.8.5. *Let $S_A = XAX'$ and $S_B = XBX'$, where $X \sim N_{p,n}(M, \Sigma \otimes I_n)$, and $A\,(n \times n)$ and $B\,(m \times m)$ be the constant symmetric matrices. The necessary and sufficient condition for S_A and S_B to be stochastically independent is that $AB = 0$.*

Proof: The joint m.g.f. of S_A and S_B is

$$M_{S_A, S_B}(Z_1, Z_2) = (2\pi)^{-\frac{1}{2}np} \det(\Sigma)^{-\frac{1}{2}n} \int_{X \in \mathbb{R}^{p \times n}} \text{etr}\left\{ Z_1 XAX' + Z_2 XBX' \right.$$
$$\left. - \frac{1}{2}\Sigma^{-1}(X-M)(X-M)' \right\} dX. \quad (7.8.17)$$

If $AB = 0$, then from Theorem 1.2.17, there exists an orthogonal matrix Q such that

$$Q'AQ = \begin{pmatrix} D_\alpha & 0 \\ 0 & 0 \end{pmatrix} \begin{matrix} r \\ n-r \end{matrix},$$
$$\phantom{Q'AQ = \begin{pmatrix}} r \quad\; n-r$$

and

$$Q'BQ = \begin{pmatrix} 0 & 0 & 0 \\ 0 & D_\beta & 0 \\ 0 & 0 & 0 \end{pmatrix} \begin{matrix} r \\ s \\ n-r-s \end{matrix},$$
$$\phantom{Q'BQ = \begin{pmatrix}0} r \quad\; s \quad n-r-s$$

where $D_\alpha\,(r \times r)$ and $D_\beta\,(s \times s)$ are diagonal matrices of the nonzero characteristic roots of A and B, respectively. Using the transformation $Y = \Sigma^{-\frac{1}{2}} XQ$ in (7.8.17) with the Jacobian $J(X \to Y) = \det(\Sigma)^{\frac{1}{2}n}$, we have

$$M_{S_A, S_B}(Z_1, Z_2) = (2\pi)^{-\frac{1}{2}np} \int_{Y \in \mathbb{R}^{p \times n}} \text{etr}\left\{ \Sigma^{\frac{1}{2}} Z_1 \Sigma^{\frac{1}{2}} YQ'AQY' + \Sigma^{\frac{1}{2}} Z_2 \Sigma^{\frac{1}{2}} YQ'BQY' \right.$$
$$\left. - \frac{1}{2}\Sigma^{-1}(YQ' - \Sigma^{-\frac{1}{2}}M)(YQ' - \Sigma^{-\frac{1}{2}}M)' \right\} dY$$

$$= (2\pi)^{-\frac{1}{2}np} \int_{Y \in \mathbb{R}^{p \times n}} \text{etr}\left\{ \Sigma^{\frac{1}{2}} Z_1 \Sigma^{\frac{1}{2}} Y_1 D_\alpha Y_1' + \Sigma^{\frac{1}{2}} Z_2 \Sigma^{\frac{1}{2}} Y_2 D_\beta Y_2' \right.$$
$$\left. - \frac{1}{2}(Y-N)(Y-N)' \right\} dY$$

$$= (2\pi)^{-\frac{1}{2}(r+s)p} \int_{Y_1 \in \mathbb{R}^{p \times r}} \int_{Y_2 \in \mathbb{R}^{p \times s}} \text{etr}\left\{ \Sigma^{\frac{1}{2}} Z_1 \Sigma^{\frac{1}{2}} Y_1 D_\alpha Y_1' \right.$$

$$+ \Sigma^{\frac{1}{2}} Z_2 \Sigma^{\frac{1}{2}} Y_2 D_\beta Y_2' - \frac{1}{2}(Y_1 - N_1)(Y_1 - N_1)'$$

$$- \frac{1}{2}(Y_2 - N_2)(Y_2 - N_2)' \Big\} dY_1 Y_2$$

$$= M_{S_A}(Z_1) M_{S_B}(Z_2),$$

where $N = \Sigma^{-\frac{1}{2}} M Q$, $Y (p \times n) = (Y_1 \quad Y_2 \quad Y_3)$, $Y_1 (p \times r)$, $Y_2 (p \times s)$, and $N (p \times n) = (N_1 \quad N_2 \quad N_3)$, $N_1 (p \times r)$, $N_2 (p \times s)$. The last step follows from (7.8.3). Hence S_A and S_B are independent. This proves the sufficiency.

Conversely if S_A and S_B are independent, then

$$M_{S_A, S_B}(Z_1, Z_2) = M_{S_A}(Z_1) M_{S_B}(Z_2)$$

must hold for $Z_1 = Z$ and $Z_2 = \rho Z$, where $\rho \neq 0$. In this case

$$M_{S_A, S_B}(Z_1, Z_2) = M_{S_{A+\rho B}}(Z)$$

$$= M_{S_A}(Z) M_{S_B}(\rho Z). \qquad (7.8.18)$$

Let $Q_1 (n \times n)$, $Q_2 (n \times n)$ and $Q (n \times n)$ be orthogonal matrices such that $Q_1' A Q_1 = \text{diag}(\alpha_1, \ldots, \alpha_r, 0, \ldots, 0)$, $Q_2' B Q_2 = \text{diag}(\beta_1, \ldots, \beta_q, 0, \ldots, 0)$ and $Q'(A + \rho B)Q = \text{diag}(\lambda_1, \ldots, \lambda_t, 0, \ldots, 0)$, $r = \text{rank}(A)$, $q = \text{rank}(B)$ and $t = \text{rank}(A + \rho B)$. Then the m.g.f. and the c.g.f. of S_A, S_B and $S_{A+\rho B}$ can be obtained from (7.8.1) and (7.8.8) respectively by making appropriate substitutions. Thus we get

$$\ln M_{S_A}(Z) = \sum_{s=1}^{\infty} \frac{2^{s-1}}{s} \text{tr}(A^s) \text{tr}(\Sigma Z)^s + \sum_{s=0}^{\infty} 2^s \text{tr}\{M A^{s+1} M' Z(\Sigma Z)^s\},$$

$$\ln M_{S_B}(\rho Z) = \sum_{s=1}^{\infty} \frac{2^{s-1}}{s} \text{tr}(\rho B)^s \text{tr}(\Sigma Z)^s + \sum_{s=0}^{\infty} 2^s \text{tr}\{M (\rho B)^{s+1} M' Z(\Sigma Z)^s\},$$

and

$$\ln M_{S_{A+\rho B}}(Z) = \sum_{s=1}^{\infty} \frac{2^{s-1}}{s} \text{tr}(A + \rho B)^s \text{tr}(\Sigma Z)^s + \sum_{s=0}^{\infty} 2^s \text{tr}\{M (A + \rho B)^{s+1} M' Z(\Sigma Z)^s\}.$$

Now taking logarithm of (7.8.18) and substituting $\ln M_{S_A}(Z)$, $\ln M_{S_B}(\rho Z)$ and $\ln M_{S_{A+\rho B}}(Z)$ from the above equations, after simplifying, we have

$$\sum_{s=1}^{\infty} \frac{2^{s-1}}{s}\{\text{tr}(A + \rho B)^s - \text{tr}(A^s) - \text{tr}(\rho B)^s\} \text{tr}(\Sigma Z)^s$$

$$= \sum_{s=0}^{\infty} 2^s \text{tr}[M\{A^{s+1} + (\rho B)^{s+1} - (A + \rho B)^{s+1}\} M' Z(\Sigma Z)^s], \qquad (7.8.19)$$

which must be true for any linear function of Z and any value of ρ. Equating coefficients of $\text{tr}(\Sigma Z)^s$, we have

$$\text{tr}(A + \rho B)^s = \text{tr}(A^s) + \text{tr}(\rho B)^s, \ s = 1, 2, \ldots, \ldots \ .$$

For $s = 4$, equating the coefficients of ρ^2, we get

$$2\operatorname{tr}(CC') + \operatorname{tr}(C^2) = 0, \tag{7.8.20}$$

where $C = AB = (c_{ij})$, say. From (7.8.20), it is easy to see that

$$\operatorname{tr}(C + C')^2 = \operatorname{tr}(C')^2 = \operatorname{tr}(C^2). \tag{7.8.21}$$

Since $\operatorname{tr}(CC') = \sum_{i,j} c_{ij}^2$, from (7.8.20) we get

$$2\sum_{i,j} c_{ij}^2 + \operatorname{tr}(C + C')^2 = 0.$$

The second term in the above equation is the trace of the square of a symmetric matrix and hence is nonnegative. The first term is the sum of squared quantities. Thus $c_{ij} = 0$, i.e., $AB = 0$, which proves the necessity. ∎

COROLLARY 7.8.5.1. *Let* $S = XAX'$ *and* $V = XL$, *where* $L\,(n \times t)$ *is a constant matrix, and* $X \sim N_{p,n}(M, \Sigma \otimes I_n)$. *The necessary and sufficient condition for* S *and* V *to be stochastically independent is that* $AL = 0$.

Proof: Let $AL = 0$, then $ALL' = 0$. Now if S and VV' are independent, then S and V are also independent. From Theorem 7.8.5, S and VV' are independent if and only if $ALL' = 0$. Thus the sufficiency is proved. Conversely, if S and V are independent then, from Theorem 7.8.5, $ALL' = 0$. From Lemma 7.8.1, we can write

$$L = \begin{pmatrix} T_1 & 0 \\ T_2 & 0 \end{pmatrix} Q, \tag{7.8.22}$$

where $\operatorname{rank}(L) = r$, $T_1\,(r \times r)$ is a lower triangular matrix with positive diagonal elements, $T_2\,((n - r) \times r)$ is a linear function of T_1, and $Q\,(n \times n)$ is an orthogonal matrix. Using (7.8.22) in $ALL' = 0$, we get

$$A \begin{pmatrix} T_1 \\ T_2 \end{pmatrix} (T_1' \quad T_2') = 0.$$

Therefore

$$A \begin{pmatrix} T_1 \\ T_2 \end{pmatrix} T_1' = 0,$$

$$A \begin{pmatrix} T_1 & 0 \\ T_2 & 0 \end{pmatrix} = 0,$$

$$A \begin{pmatrix} T_1 & 0 \\ T_2 & 0 \end{pmatrix} Q = 0,$$

i.e., $AL = 0$. Hence $ALL' = 0$ if and only if $AL = 0$. This completes the proof of necessity. ∎

COROLLARY 7.8.5.2. *Let $S_i = XA_iX'$, $i = 1, \ldots, k$, where $X \sim N_{p,n}(M, \Sigma \otimes I_n)$. Then the quadratic forms $S_1, \ldots, S_k$ are stochastically independent if and only if $A_iA_j = 0$, $i \neq j$.*

In Chapter 2, we have proved the independence of $\bar{x} = \frac{1}{N}\sum_{i=1}^{N} x_i$ and $S = \sum_{i=1}^{N}(x_i - \bar{x})(x_i - \bar{x})'$, together with Wishartness of S, where $x_i \sim N_p(\mu, \Sigma)$, $i = 1, \ldots, N$ $(N > p)$. This result can now easily be obtained from the above theorem. Let $X = (x_1, \ldots, x_N)$, then $X \sim N_{p,N}(\mu e', \Sigma \otimes I_N)$. Note that $\bar{x} = \frac{1}{N}Xe$, and $S = X(I_N - \frac{1}{N}ee')X'$. From Corollary 7.8.5.1, it follows that $\bar{x}$ and S are independent since $(I_N - \frac{1}{N}ee')e = 0$. Also the matrix $(I_N - \frac{1}{N}ee')$ is idempotent of rank $N - 1$ and therefore from Corollary 7.8.3.1, $S \sim W_p(N - 1, \Sigma)$.

THEOREM 7.8.6. *Let $S_A = XAX'$ and $S_B = XBX'$, where $X \sim N_{p,n}(M, \Sigma \otimes \Psi)$, and $A\,(n \times n)$ and $B\,(n \times n)$ be the constant symmetric matrices. The necessary and sufficient condition for S_A and S_B to be stochastically independent is that $A\Psi B = 0$.*

Proof: Note that the quadratic forms XAX' and XBX' are stochastically independent if and only if the quadratic forms $Y(\Psi^{\frac{1}{2}}A\Psi^{\frac{1}{2}})Y'$ and $Y(\Psi^{\frac{1}{2}}B\Psi^{\frac{1}{2}})Y'$, where $Y \sim N_{p,n}(M\Psi^{-\frac{1}{2}}, \Sigma \otimes I_n)$, are stochastically independent. Hence from Theorem 7.8.5, we get the condition $\Psi^{\frac{1}{2}}A\Psi^{\frac{1}{2}}\Psi^{\frac{1}{2}}B\Psi^{\frac{1}{2}} = 0$, i.e., $A\Psi B = 0$. ∎

COROLLARY 7.8.6.1. *Let $S = XAX'$ and $V = XL$, where $L\,(n \times t)$ is a constant matrix, and $X \sim N_{p,n}(M, \Sigma \otimes \Psi)$. The necessary and sufficient condition for S and V to be stochastically independent is that $A\Psi L = 0$.*

Proof: Note that $S = XAX'$ and $V = XL$ are stochastically independent if and only if the quadratic form $Y(\Psi^{\frac{1}{2}}A\Psi^{\frac{1}{2}})Y'$ and $Y(\Psi^{\frac{1}{2}}L)$, where $Y \sim N_{p,n}(M\Psi^{-\frac{1}{2}}, \Sigma \otimes I_n)$, are stochastically independent. Therefore from Corollary 7.8.5.1, we get $\Psi^{\frac{1}{2}}A\Psi^{\frac{1}{2}}\Psi^{\frac{1}{2}}L = 0$, i.e., $A\Psi L = 0$. ∎

COROLLARY 7.8.6.2. *Let $S_i = XA_iX'$, $i = 1, \ldots, k$, where $X \sim N_{p,n}(M, \Sigma \otimes \Psi)$, and $A_i\,(n \times n)$ are constant symmetric matrices. Then the quadratic forms $S_1, \ldots, S_k$ are stochastically independent if and only if $A_i\Psi A_j = 0$, $i \neq j$.*

The above results are taken from Khatri (1959b). As mentioned in the beginning of this section, these results on quadratic forms have been derived by comparing moment generating functions. An alternate proof of Corollary 7.8.6.1 has been given by Hogg (1963).

It may be noted that the proof of Theorems 7.8.3. and 7.8.5 can also be given by first obtaining conditions on the diagonal elements of the quadratic forms, together with the additional conditions obtained from the moment generating function.

Next we give general results of Cochran theorem.

THEOREM 7.8.7. *Let $XAX' = \sum_{i=1}^{k} XA_iX'$, where $X \sim N_{p,n}(M, \Sigma \otimes I_n)$, $\text{rank}(A) = r\,(\geq p)$, and $\text{rank}(A_i) = r_i\,(\geq p)$, $i = 1, \ldots, k$. Consider the following four conditions:*

(i) $XA_iX' \sim W_p(r_i, \Sigma, \Sigma^{-1}MA_iM')$,
(ii) XA_iX' and XA_jX', $i \neq j$, are stochastically independent,
(iii) $XAX' \sim W_p(r, \Sigma, \Sigma^{-1}MAM')$, and
(iv) $r = \sum_{i=1}^k r_i$.

Then, (a) any two of the conditions (i), (ii), and (iii) imply the remaining, and (b) conditions (iii) and (iv) imply (i) and (ii).

Proof: Let any two of the conditions (i), (ii), and (iii) be satisfied. Then from Theorems 7.8.3 and 7.8.5, any two of the conditions (i), (ii), and (iii) of Theorem 1.2.20 will hold and consequently the remaining conditions of the theorem will also hold. Hence all four conditions of the Theorem 1.2.20 will hold. Therefore all the conditions of Theorem 7.8.7 hold which proves the part (a). Further let the conditions (iii) and (iv) hold. Then from Theorem 7.8.3, the conditions (iii) and (iv) of Theorem 1.2.20 hold. Consequently conditions (i) and (ii) of Theorem 1.2.20 also hold, and hence (i) and (ii) of Theorem 7.8.7 follow, which completes the proof of part (b). ∎

COROLLARY 7.8.7.1. Let $XAX' = \sum_{i=1}^k XA_iX'$, where $A^2 = A$, and $X \sim N_{p,n}(0, \Sigma \otimes I_n)$, $\text{rank}(A) = r \ (\geq p)$, and $\text{rank}(A_i) = r_i \ (\geq p)$, $i = 1, \ldots, k$. Then $XA_iX' \sim W_p(r_i, \Sigma)$, $i = 1, \ldots, k$ and are independent if and only if $r = \sum_{i=1}^k r_i$.

It is noticeable that the conditions for Wishartness and independence of quadratic forms of the type XAX', $X \sim N_{p,n}(M, \Sigma \otimes \Psi)$, do not depend on Σ, and hence they are valid even when Σ is singular, *i.e.*, when $X \sim N_{p,n}(0, \Sigma \otimes \Psi | p_1, n)$, $p_1 \leq p$. However when $X \sim N_{p,n}(0, \Sigma \otimes \Psi | p, n_1)$, $n_1 \leq n$, the conditions given in Theorems 7.8.4 and 7.8.6 are no more valid. For this case, in the following theorem, conditions for Wishartness of the quadratic form of the type XAX' are given without proof.

THEOREM 7.8.8. Let $S = XAX'$, where $X \sim N_{p,n}(M, \Sigma \otimes \Psi | p, n_1)$, $n_1 \leq n$. The necessary and sufficient conditions for S to be distributed as $W_p(t, \Sigma, \Sigma^{-1}MAM')$ are (i) $\Psi A \Psi A \Psi = \Psi A \Psi$ (ii) $MA\Psi = MA\Psi A\Psi$, and (iii) $MAM' = MA\Psi AM'$ where $\text{rank}(A\Psi) = t \geq p$.

7.9.   WISHARTNESS AND INDEPENDENCE OF QUADRATIC FORMS OF THE TYPE $XAX' + \frac{1}{2}(LX' + XL') + C$

In Section 7.8 we studied conditions for Wishartness and independence of quadratic forms of the type XAX', where $X \sim N_{p,n}(M, \Sigma \otimes I_n)$. In this section we study conditions for Wishartness and independence of quadratic forms which are generalizations of the quadratic forms of the type XAX'. The results derived here are more general and include results derived in Section 7.8 as special cases. The method of proofs of preceeding section can be used here for deriving conditions for Wishartness and independence, but we will follow a slightly different approach. The generalized quadratic form in X $(p \times n)$ is defined by

$$S = XAX' + \frac{1}{2}(LX' + XL') + C, \tag{7.9.1}$$

where $A\,(n \times n) = A'$, $L\,(p \times n)$ and $C\,(p \times p) = C'$ are constant matrices. We first derive the m.g.f. of S when $X \sim N_{p,n}(M, \Sigma \otimes \Psi)$. We begin with a lemma which expresses the m.g.f. of S in terms of m.g.f. of YBY', where the distribution of $Y\,(p \times n)$ is matrix variate normal and $B\,(n \times n)$ is a symmetric matrix.

LEMMA 7.9.1. Let $X \sim N_{p,n}(M, \Sigma \otimes \Psi)$. Then the m.g.f. of $S = XAX' + \frac{1}{2}(LX' + XL') + C$ can be expressed as

$$M_S(Z) = \text{etr}\left\{Z(C + ML') + \frac{1}{2}L\Psi L'Z\Sigma Z\right\}M_{YBY'}(Z), \tag{7.9.2}$$

where $Y \sim N_{p,n}(M\Psi^{-\frac{1}{2}} + \Sigma Z L\Psi^{\frac{1}{2}}, \Sigma \otimes I_n)$ and $B = \Psi^{\frac{1}{2}}A\Psi^{\frac{1}{2}}$.

Proof: The m.g.f. of S is

$$M_S(Z) = E[\text{etr}(ZS)]$$

$$= E\left[\text{etr}\left\{Z\left(XAX' + \frac{1}{2}(LX' + XL') + C\right)\right\}\right]$$

$$= E\left[\text{etr}\left\{Z\left(YBY' + \frac{1}{2}(NY' + YN') + C\right)\right\}\right], \; Y \sim N_{p,n}(M\Psi^{-\frac{1}{2}}, \Sigma \otimes I_n)$$

$$= (2\pi)^{-\frac{1}{2}np}\det(\Sigma)^{-\frac{1}{2}n}\int_{Y \in \mathbb{R}^{p \times n}}\text{etr}\left\{Z\left(YBY' + \frac{1}{2}(NY' + YN') + C\right)\right.$$

$$\left. - \frac{1}{2}\Sigma^{-1}(Y - \mu)(Y - \mu)'\right\}dY,$$

where $B = \Psi^{\frac{1}{2}}A\Psi^{\frac{1}{2}}$, $N = L\Psi^{\frac{1}{2}}$ and $\mu = M\Psi^{-\frac{1}{2}}$. Next writing the exponent inside the integral as

$$\text{tr}\left\{Z\left(YBY' + \frac{1}{2}(NY' + YN') + C\right) - \frac{1}{2}\Sigma^{-1}(Y - \mu)(Y - \mu)'\right\}$$

$$= \text{tr}\left\{Z(C + \mu N') + \frac{1}{2}NN'Z\Sigma Z\right\} + \text{tr}(ZYBY')$$

$$- \frac{1}{2}\text{tr}\{\Sigma^{-1}(Y - \mu - \Sigma ZN)(Y - \mu - \Sigma ZN)'\},$$

we get

$$M_S(Z) = (2\pi)^{-\frac{1}{2}np}\det(\Sigma)^{-\frac{1}{2}n}\text{etr}\left\{Z(C + \mu N') + \frac{1}{2}NN'Z\Sigma Z\right\}$$

$$\int_{Y \in \mathbb{R}^{p \times n}}\text{etr}\left\{ZYBY' - \frac{1}{2}\Sigma^{-1}(Y - \mu - \Sigma ZN)(Y - \mu - \Sigma ZN)'\right\}dY$$

$$= \text{etr}\left\{Z(C + \mu N') + \frac{1}{2}NN'Z\Sigma Z\right\}E[\text{etr}(ZYBY')],$$

where $Y \sim N_{p,n}(\mu + \Sigma ZN, \Sigma \otimes I_n)$. The proof is completed by substituting $\mu = M\Psi^{-\frac{1}{2}}$ and $N = L\Psi^{\frac{1}{2}}$. ∎

Now the m.g.f. of S is evaluated in the following theorem.

THEOREM 7.9.1. Let $X \sim N_{p,n}(M, \Sigma \otimes \Psi)$. Then the m.g.f. of $S = XAX' + \frac{1}{2}(LX' + XL') + C$, where $A (n \times n)$ is a symmetric matrix of rank t, is given by

$$M_S(Z) = \prod_{j=1}^{t} \det(I_p - 2\lambda_j \Sigma Z)^{-\frac{1}{2}} \operatorname{etr}\left\{ Z(C + ML') + \frac{1}{2} L\Psi L' Z\Sigma Z \right\}$$

$$\exp\left\{ \sum_{j=1}^{t} \lambda_j q_j' \Psi^{-\frac{1}{2}} (M + \Sigma ZL\Psi)' Z (I_p - 2\lambda_j \Sigma Z)^{-1} \right.$$

$$\left. (M + \Sigma ZL\Psi)\Psi^{-\frac{1}{2}} q_j \right\}, \tag{7.9.3}$$

where $\Psi^{\frac{1}{2}} A \Psi^{\frac{1}{2}} = Q \begin{pmatrix} D_\lambda & 0 \\ 0 & 0 \end{pmatrix} Q'$, $Q (n \times n)$ is an orthogonal matrix, $Q = (q_1, \ldots, q_n)$, $D_\lambda = \operatorname{diag}(\lambda_1, \ldots, \lambda_t)$ and $\lambda_1, \ldots, \lambda_t$ are the nonzero characteristic roots of $\Psi^{\frac{1}{2}} A \Psi^{\frac{1}{2}}$.

Proof: From Lemma 7.9.1, we have

$$M_S(Z) = \operatorname{etr}\left\{ Z(C + ML') + \frac{1}{2} L\Psi L' Z\Sigma Z \right\} M_{YBY'}(Z), \tag{7.9.4}$$

where $Y \sim N_{p,n}(M\Psi^{-\frac{1}{2}} + \Sigma ZL\Psi^{\frac{1}{2}}, \Sigma \otimes I_n)$ and $B = \Psi^{\frac{1}{2}} A \Psi^{\frac{1}{2}}$. Now appropriately substituting from Theorem 7.8.1, for $M_{YBY'}(Z)$ we get the desired result. ∎

COROLLARY 7.9.1.1. If $A\Psi A = A$, then the m.g.f. of S is given by

$$M_S(Z) = \det(I_p - 2\Sigma Z)^{-\frac{1}{2}t} \operatorname{etr}\left\{ Z(I_p - 2\Sigma Z)^{-1}(M + \Sigma ZL\Psi)A(M + \Sigma ZL\Psi)' \right.$$

$$\left. + Z(C + ML') + \frac{1}{2} L\Psi L' Z\Sigma Z \right\}.$$

Proof: If $A\Psi A = A$ holds, then $\Psi^{\frac{1}{2}} A \Psi^{\frac{1}{2}} = A$ is an idempotent matrix of rank t, i.e., $\lambda_i = 1$, $i = 1, \ldots, t$ and $\Psi^{\frac{1}{2}} A \Psi^{\frac{1}{2}} = Q \begin{pmatrix} I_t & 0 \\ 0 & 0 \end{pmatrix} Q' = \sum_{j=1}^{t} q_j q_j'$. Substituting these in (7.9.3) we get the result. ∎

The c.g.f. of S is derived in the next theorem.

THEOREM 7.9.2. Let $X \sim N_{p,n}(M, \Sigma \otimes \Psi)$. Then the c.g.f. of $XAX' + \frac{1}{2}(LX' + XL') + C$, where $A (n \times n)$ is a symmetric matrix of rank t, is given by

$$\ln M_S(Z) = \operatorname{tr}\left\{ Z(C + ML') + \frac{1}{2} L\Psi L' Z\Sigma Z \right\} + \sum_{s=1}^{\infty} \frac{2^{s-1}}{s} \operatorname{tr}(\Psi A)^s \operatorname{tr}(\Sigma Z)^s$$

$$+ \sum_{s=0}^{\infty} 2^s \operatorname{tr}\{Z(\Sigma Z)^s (M + \Sigma ZL\Psi)A(\Psi A)^s (M + \Sigma ZL\Psi)'\}.$$

Proof: From (7.9.2), we obtain

$$\ln M_S(Z) = \operatorname{tr}\left\{ Z(C + ML') + \frac{1}{2}L\Psi L'Z\Sigma Z \right\} + \ln M_{YBY'}(Z), \qquad (7.9.5)$$

where $Y \sim N_{p,n}(M\Psi^{-\frac{1}{2}} + \Sigma ZL\Psi^{\frac{1}{2}}, \Sigma \otimes I_n)$ and $B = \Psi^{\frac{1}{2}}A\Psi^{\frac{1}{2}}$ is of rank t with $\lambda_1, \ldots, \lambda_t$ being the nonzero characteristic roots. Using Theorem 7.8.2, the c.g.f. of $M_{YBY'}(Z)$ is obtained as

$$\ln M_{YBY'}(Z) = \sum_{s=1}^{\infty} \frac{2^{s-1}}{s} \operatorname{tr}(\Psi A)^s \operatorname{tr}(\Sigma Z)^s + \sum_{s=0}^{\infty} 2^s \operatorname{tr}\{Z(\Sigma Z)^s$$

$$(M + \Sigma ZL\Psi)A(\Psi A)^s(M + \Sigma ZL\Psi)'\}. \qquad (7.9.6)$$

Now using (7.9.6) in (7.9.5) we get the desired result. ∎

Alternately, a proof can be constructed parallel to the proof of the Theorem 7.8.2. The following lemmas, which are used in the sequel, give conditions for chi-squaredness and independence of second degree polynomials in n-variate normal vector.

LEMMA 7.9.2. *Let* $P(x) = x'Ax + \ell'x + c$, *where* $x \sim N_n(\mu, I_n)$, $A (n \times n)$ *is a symmetric matrix of rank* t, $\ell (n \times 1)$ *is a constant vector, and* c *is a scalar. Then the necessary and sufficient conditions for* $P(x)$ *to be distributed as noncentral chi-square with* t *degrees of freedom and noncentrality parameter* $\mu'\mu$ *are that (i)* $A^2 = A$, *(ii)* $A\ell = \ell$, *and (iii)* $c = \frac{1}{4}\ell'A\ell$.

LEMMA 7.9.3. *Let* $P_A(x) = x'Ax + \ell'x + c$ *and* $P_B(x) = x'Bx + n'x + d$, *where* $x \sim N_n(\mu, I_n)$, $A (n \times n)$ *and* $B (n \times n)$ *are a symmetric matrices,* $\ell (n \times 1)$ *and* $n (n \times 1)$ *are constant vectors, and* c *and* d *are scalars. Then the necessary and sufficient conditions for* $P_A(x)$ *and* $P_B(x)$ *to be distributed independently are (i)* $AB = 0$, *(ii)* $\ell'B = 0$, *(ii)* $n'A = 0$, *and (iv)* $\ell'n = 0$.

Lemma 7.9.2 was establish by Khatri (1962) and the proof of Lemma 7.9.3 was given by Laha (1956).

LEMMA 7.9.4. *Let* $A (n \times n)$ *and* $B (n \times n)$ *be symmetric matrices and* $L (p \times n)$ *and* $N (p \times n)$ *be matrices such that* $t = \operatorname{rank}(A \quad L')$, $u = \operatorname{rank}(B \quad N')$, $AB = 0$, $LB = 0$, $NA = 0$ *and* $LN' = 0$. *Then there exists a semiorthogonal matrix* $Q (n \times (t + u))$, $t + u \le n$, *such that* $L = (T \quad 0)Q'$, $M = (0 \quad U)Q'$, $A = Q\begin{pmatrix} E & 0 \\ 0 & 0 \end{pmatrix}Q'$, *and* $B = Q\begin{pmatrix} 0 & 0 \\ 0 & F \end{pmatrix}Q'$, *where* $E (t \times t)$, $F (u \times u)$ *are symmetric matrices,* A *and* U *are* $p \times t$ *and* $p \times u$ *respectively.*

Proof: See Khatri (1962). ∎

THEOREM 7.9.3. *Let* $S = XAX' + \frac{1}{2}(LX' + XL') + C$, *where* $X \sim N_{p,n}(M, \Sigma \otimes I_n)$. *The necessary and sufficient conditions for* S *to be distributed as* $W_p(t, \Sigma, \Sigma^{-1}(M + \frac{1}{2}L)A(M + \frac{1}{2}L)')$ *are that (i)* $A^2 = A$, $\operatorname{rank}(A) = t \ge p$, *(ii)* $LA = L$, *and (iii)* $C = \frac{1}{4}LAL'$.

Proof: Let us assume that the conditions (i)–(iii) hold. Then

$$S = XAX' + \frac{1}{2}(LAX' + XAL') + \frac{1}{4}LAL'$$

$$= \left(X + \frac{1}{2}L\right)A\left(X + \frac{1}{2}L\right)'$$

$$= YAY'$$

where $Y \sim N_{p,n}(M + \frac{1}{2}L, \Sigma \otimes I_n)$. Now from Theorem 7.8.3, $YAY' \sim W_p(t, \Sigma, \Sigma^{-1}(M + \frac{1}{2}L)A(M + \frac{1}{2}L)')$.

Conversely, let $S = (s_{ij}) \sim W_p(t, \Sigma, \Sigma^{-1}(M + \frac{1}{2}L)A(M + \frac{1}{2}L)')$,

$$s_{ii} = \boldsymbol{x}_i^{*'}A\boldsymbol{x}_i^* + \boldsymbol{\ell}_i^{*'}\boldsymbol{x}_i^* + c_{ii},$$

where $\boldsymbol{x}_i^{*'}$ and $\boldsymbol{\ell}_i^{*'}$ denote the ith row of the matrices X and L respectively, and c_{ii} is the $(i, i)^{\text{th}}$ diagonal element of the matrix C. Then, the diagonal elements s_{ii} are distributed as noncentral chi-square if and only if (Lemma 7.9.2)

$$A^2 = A, \; \boldsymbol{\ell}_i^{*'}A = \boldsymbol{\ell}_i^{*'}, \; \text{and} \; c_{ii} = \frac{1}{4}\boldsymbol{\ell}_i^{*'}A\boldsymbol{\ell}_i^*, \; i = 1, \ldots, p,$$

or equivalently, $A^2 = A$, $LA = L$, and the diagonal elements of $\frac{1}{4}LAL' - C$ are all zero. Therefore, under these conditions $S = (X + \frac{1}{2}L)A(X + \frac{1}{2}L)' + C - \frac{1}{4}LAL'$. Now, given that S is a noncentral Wishart matrix, and $(X + \frac{1}{2}L)A(X + \frac{1}{2}L)'$ is also a noncentral Wishart matrix, we must have $C - \frac{1}{4}LAL' = 0$. This completes the proof of necessity. ∎

THEOREM 7.9.4. Let $S = XAX' + \frac{1}{2}(LX' + XL') + C$, where $X \sim N_{p,n}(M, \Sigma \otimes \Psi)$. The necessary and sufficient conditions for S to be distributed as $W_p(t, \Sigma, \Sigma^{-1}(M + \frac{1}{2}L)A(M + \frac{1}{2}L)')$ are that (i) $A\Psi A = A$, $\text{rank}(A) = t \geq p$, (ii) $L\Psi A = L$, and (iii) $C = \frac{1}{4}L\Psi L'$.

Proof: Note that $S = XAX' + \frac{1}{2}(LX' + XL') + C$ has same distribution as $S^* = Y(\Psi^{\frac{1}{2}}A\Psi^{\frac{1}{2}})Y' + \frac{1}{2}(L\Psi^{\frac{1}{2}}Y' + Y\Psi^{\frac{1}{2}}L') + C$, where $Y \sim N_{p,n}(M\Psi^{-\frac{1}{2}}, \Sigma \otimes I_n)$. From Theorem 7.9.3, the necessary and sufficient conditions for S^* (and hence for S) to be distributed as $W_p(t, \Sigma, \Sigma^{-1}(M + \frac{1}{2}L)A(M + \frac{1}{2}L)')$ are that $\Psi^{\frac{1}{2}}A\Psi^{\frac{1}{2}}\Psi^{\frac{1}{2}}A\Psi^{\frac{1}{2}} = \Psi^{\frac{1}{2}}A\Psi^{\frac{1}{2}}$, $L\Psi^{\frac{1}{2}}\Psi^{\frac{1}{2}}A\Psi^{\frac{1}{2}} = L\Psi^{\frac{1}{2}}$ and $\frac{1}{4}L\Psi^{\frac{1}{2}}(\Psi^{\frac{1}{2}}A\Psi^{\frac{1}{2}})\Psi^{\frac{1}{2}}L' = C$. Now conditions (i), (ii) and (iii) are obtained from above conditions upon simplification. ∎

THEOREM 7.9.5. Let $S_1 = XAX' + \frac{1}{2}(LX' + XL') + C$ and $S_2 = XBX' + \frac{1}{2}(NX' + XN') + D$, where $X \sim N_{p,n}(0, \Sigma \otimes I_n)$, $A(n \times n) = A'$, $B(n \times n) = B'$, $C(p \times p) = C'$, $D(p \times p) = D'$, $L(p \times n)$, and $N(p \times n)$ are constant matrices. Then the necessary and sufficient conditions for S_1 and S_2 to be stochastically independent are (i) $AB = 0$, (ii) $LB = 0$, (iii) $NA = 0$ and (iv) $LN' = 0$.

Proof: The joint m.g.f. of S_1 and S_2 is

$$M_{S_1,S_2}(Z_1, Z_2) = (2\pi)^{-\frac{1}{2}np} \det(\Sigma)^{-\frac{1}{2}n} \int_{X \in \mathbb{R}^{p \times n}} \text{etr} \left\{ Z_1 \left(XAX' + \frac{1}{2}(LX' + XL') + C \right) \right.$$

$$+ Z_2 \left(XBX' + \frac{1}{2}(NX' + XN') + D \right)$$

$$\left. - \frac{1}{2}\Sigma^{-1}(X - M)(X - M)' \right\} dX. \tag{7.9.7}$$

If the conditions (i)–(iv) hold, then using Lemma 7.9.4, there exists a semi-orthogonal matrix $Q\,(n \times q) = (Q_1 \quad Q_2)$ such that $L = (T \quad 0)Q'$, $N = (0 \quad U)Q'$, $A = Q \begin{pmatrix} E & 0 \\ 0 & 0 \end{pmatrix} Q'$, and $B = Q \begin{pmatrix} 0 & 0 \\ 0 & F \end{pmatrix} Q'$, where $q = t + u\ (\leq n)$, $t = \text{rank}\,(A \quad L')$, $u = \text{rank}\,(B \quad N')$, $E = E'$, $F = F'$, T and U are matrices of order $t \times t$, $u \times u$, $p \times t$ and $p \times u$ respectively. Let $Q_3\,(n \times (n - q))$ be a semiorthogonal matrix such that $Q_0\,(n \times n) = (Q \quad Q_3)$ is orthogonal. Next using the transformation $Y = \Sigma^{-\frac{1}{2}}XQ_0$ in (7.9.7), with Jacobian $J(X \to Y) = \det(\Sigma)^{\frac{1}{2}n}$, we get

$$M_{S_1,S_2}(Z_1, Z_2) = (2\pi)^{-\frac{1}{2}np} \int_{Y \in \mathbb{R}^{p \times n}} \text{etr} \left\{ h(Z_1, Z_2, Y) \right.$$

$$\left. - \frac{1}{2}(YQ_0' - \Sigma^{-\frac{1}{2}}M)(YQ_0' - \Sigma^{-\frac{1}{2}}M)' \right\} dY \tag{7.9.8}$$

where

$$h(Z_1, Z_2, Y) = Z_1 \left[\Sigma^{\frac{1}{2}}YQ_0'AQ_0Y'\Sigma^{\frac{1}{2}} + \frac{1}{2}(T \quad 0)Q'Q_0Y'\Sigma^{\frac{1}{2}} \right.$$

$$+ \frac{1}{2}\Sigma^{\frac{1}{2}}YQ_0'Q(T \quad 0)' + C \Big] + Z_2 \Big[\Sigma^{\frac{1}{2}}YQ_0'BQ_0Y'\Sigma^{\frac{1}{2}}$$

$$+ \frac{1}{2}(0 \quad U)Q'Q_0Y'\Sigma^{\frac{1}{2}} + \frac{1}{2}\Sigma^{\frac{1}{2}}YQ_0'Q(0 \quad U)' + D \Big]$$

$$= Z_1 \left[\Sigma^{\frac{1}{2}}Y_1 EY_1'\Sigma^{\frac{1}{2}} + \frac{1}{2}TY_1'\Sigma^{\frac{1}{2}} + \frac{1}{2}\Sigma^{\frac{1}{2}}Y_1 T' + C \right]$$

$$+ Z_2 \left[\Sigma^{\frac{1}{2}}Y_2 FY_2'\Sigma^{\frac{1}{2}} + \frac{1}{2}UY_2'\Sigma^{\frac{1}{2}} + \frac{1}{2}\Sigma^{\frac{1}{2}}Y_2 U' + D \right] \tag{7.9.9}$$

and $Y\,(p \times n) = (Y_1 \quad Y_2 \quad Y_3)$, $Y_1\,(p \times t)$, $Y_2\,(p \times u)$, $Y_3\,(p \times (n - q))$, $q = t + u$. Furthermore,

$$(YQ_0' - \Sigma^{-\frac{1}{2}}M)(YQ_0' - \Sigma^{-\frac{1}{2}}M)' = (Y - \Sigma^{-\frac{1}{2}}MQ_0)(Y - \Sigma^{-\frac{1}{2}}MQ_0)'$$

$$= (Y - K)(Y - K)'$$

$$= (Y_1 - K_1)(Y_1 - K_1)' + (Y_2 - K_2)(Y_2 - K_2)'$$

$$+ (Y_3 - K_3)(Y_3 - K_3)', \tag{7.9.10}$$

where $K\,(p \times n) = \Sigma^{-\frac{1}{2}}MQ_0 = (K_1 \quad K_2 \quad K_3)$, $K_1\,(p \times t)$, $K_2\,(p \times u)$, $K_3\,(p \times (n-q))$. Now substituting (7.9.9) and (7.9.10) in (7.9.8), and integrating with respect to Y_3, we have

$$
M_{S_1,S_2}(Z_1, Z_2) = (2\pi)^{-\frac{1}{2}qp} \int_{Y_1 \in \mathbb{R}^{p \times t}} \int_{Y_2 \in \mathbb{R}^{p \times u}}
$$

$$
\mathrm{etr}\Big\{ Z_1\Big[\Sigma^{\frac{1}{2}}Y_1 EY_1'\Sigma^{\frac{1}{2}} + \frac{1}{2}TY_1'\Sigma^{\frac{1}{2}} + \frac{1}{2}\Sigma^{\frac{1}{2}}Y_1 T' + C\Big]
$$

$$
+ Z_2\Big[\Sigma^{\frac{1}{2}}Y_2 FY_2'\Sigma^{\frac{1}{2}} + \frac{1}{2}UY_2'\Sigma^{\frac{1}{2}} + \frac{1}{2}\Sigma^{\frac{1}{2}}Y_2 U' + D\Big]
$$

$$
-\frac{1}{2}(Y_1 - K_1)(Y_1 - K_1)' - \frac{1}{2}(Y_2 - K_2)(Y_2 - K_2)'\Big\}\, dY_1\, dY_2
$$

$$
= M_{S_1^*}(\Sigma^{\frac{1}{2}}Z_1\Sigma^{\frac{1}{2}})M_{S_2^*}(\Sigma^{\frac{1}{2}}Z_2\Sigma^{\frac{1}{2}}),
$$

where

$$
S_1^* = Y_1 EY_1' + \frac{1}{2}\Sigma^{-\frac{1}{2}}TY_1' + \frac{1}{2}Y_1 T'\Sigma^{-\frac{1}{2}} + \Sigma^{-\frac{1}{2}}C\Sigma^{-\frac{1}{2}},
$$

$$
S_2^* = Y_2 FY_2' + \frac{1}{2}\Sigma^{-\frac{1}{2}}UY_2' + \frac{1}{2}Y_2 U'\Sigma^{-\frac{1}{2}} + \Sigma^{-\frac{1}{2}}D\Sigma^{-\frac{1}{2}},
$$

$Y_1 \sim N_{p,t}(\Sigma^{-\frac{1}{2}}MQ_1, I_p \otimes I_t)$ and $Y_2 \sim N_{p,u}(\Sigma^{-\frac{1}{2}}MQ_2, I_p \otimes I_u)$. Evaluating $M_{S_1^*}(\Sigma^{\frac{1}{2}}Z_1\Sigma^{\frac{1}{2}})$ and $M_{S_2^*}(\Sigma^{\frac{1}{2}}Z_2\Sigma^{\frac{1}{2}})$ using Theorem 7.9.1 and writing E, F, T and U in terms of A, B, L and N, it can be seen that $M_{S_1^*}(\Sigma^{\frac{1}{2}}Z_1\Sigma^{\frac{1}{2}}) = M_{S_1}(Z_1)$ and $M_{S_2^*}(\Sigma^{\frac{1}{2}}Z_2\Sigma^{\frac{1}{2}}) = M_{S_2}(Z_2)$. Hence S_1 and S_2 are independent. This proves sufficiency.

Conversely if $S_1 = (s_{1ij})$ and $S_2 = (s_{2ij})$ are independent, then their diagonal elements are also independent, where

$$
s_{1ii} = x_i^{*'}Ax_i^* + \frac{1}{2}(\ell_i^{*'}x_i^* + x_i^{*'}\ell_i^*) + c_{ii},
$$

$$
s_{2ii} = x_i^{*'}Bx_i^* + \frac{1}{2}(n_i^{*'}x_i^* + x_i^{*'}n_i^*) + d_{ii},
$$

$x_i^{*'}$, $\ell_i^{*'}$ and $n_i^{*'}$ denote the ith row of the matrices X, L and N, and c_{ii} and d_{ii} are the $(i,i)^{\text{th}}$ diagonal elements of the matrices C and D respectively. Now from Lemma 7.9.3, s_{1ii} and s_{2ii} are independent if and only if $AB = 0$, $\ell_i^{*'}B = 0$, $n_i^{*'}A = 0$, and $\ell_i^{*'}n_i^* = 0$, $i = 1, \ldots, p$ or equivalently

$$
AB = 0, LB = 0, NA = 0 \text{ and diagonal elements of } LN' = 0. \tag{7.9.11}
$$

Further since $AB = 0$, we can find an orthogonal matrix $Q\,(n \times n) = (Q_1 \quad Q_2 \quad Q_3)$, $Q_1\,(n \times r)$, $Q_2\,(n \times s)$, $Q_3\,(n \times (n-q))$, $q = r + s$, such that

$$
Q_1'AQ_1 = \mathrm{diag}(\alpha_1, \ldots, \alpha_r) = D_\alpha, \tag{7.9.12}
$$

$$
Q_2'BQ_2 = \mathrm{diag}(\beta_1, \ldots, \beta_s) = D_\beta, \tag{7.9.13}
$$

where α_i, $i = 1, \ldots, r$ and β_i, $i = 1, \ldots, s$ are the nonzero characteristic roots of A and B respectively. Next using the transformation $Y = (Y_1 \quad Y_2 \quad Y_3) = \Sigma^{-\frac{1}{2}}XQ$,

Y_1 ($p \times r$), Y_2 ($p \times s$), Y_3 ($p \times (n - q)$) in (7.9.7) with the Jacobian $J(X \to Y) = \det(\Sigma)^{\frac{1}{2}n}$, and (7.9.11), (7.9.12), and (7.9.13), we get

$$
\begin{aligned}
M_{S_1, S_2}(Z_1, Z_2) &= (2\pi)^{-\frac{1}{2}np} \int_{Y \in \mathbb{R}^{p \times n}} \text{etr}\left\{ Z_1\left[\Sigma^{\frac{1}{2}}Y_1 D_\alpha Y_1' \Sigma^{\frac{1}{2}} + \frac{1}{2}LQ_1 Y_1' \Sigma^{\frac{1}{2}}\right.\right. \\
&\quad \left. + \frac{1}{2}\Sigma^{\frac{1}{2}}Y_1 Q_1' L' + C\right] + Z_2\left[\Sigma^{\frac{1}{2}}Y_2 D_\beta Y_2' \Sigma^{\frac{1}{2}} + \frac{1}{2}NQ_2 Y_2' \Sigma^{\frac{1}{2}}\right. \\
&\quad \left. + \frac{1}{2}\Sigma^{\frac{1}{2}}Y_2 Q_2' N' + D\right] + Z_1\left[\frac{1}{2}LQ_3 Y_3' \Sigma^{\frac{1}{2}} + \frac{1}{2}\Sigma^{\frac{1}{2}}Y_3 Q_3' L'\right] \\
&\quad \left. + Z_2\left[\frac{1}{2}NQ_3 Y_3' \Sigma^{\frac{1}{2}} + \frac{1}{2}\Sigma^{\frac{1}{2}}Y_3 Q_3' N'\right] - \frac{1}{2}(Y - K)(Y - K)'\right\} dY \\
&= M_{S_1^*}(\Sigma^{\frac{1}{2}}Z_1 \Sigma^{\frac{1}{2}})M_{S_2^*}(\Sigma^{\frac{1}{2}}Z_2 \Sigma^{\frac{1}{2}})M_{Y_3}(\Sigma^{\frac{1}{2}}(Z_1 L + Z_2 N)Q_3), \quad (7.9.14)
\end{aligned}
$$

where $K = \Sigma^{-\frac{1}{2}}MQ = (\ K_1 \ \ K_2 \ \ K_3\ )$, K_1 ($p \times r$), K_2 ($p \times s$), K_3 ($p \times (n - q)$),

$$
S_1^* = Y_1 D_\alpha Y_1' + \frac{1}{2}\Sigma^{-\frac{1}{2}}LQ_1 Y_1' + \frac{1}{2}Y_1 Q_1' L'\Sigma^{-\frac{1}{2}} + \Sigma^{-\frac{1}{2}}C\Sigma^{-\frac{1}{2}},
$$

$$
S_2^* = Y_2 D_\beta Y_2' + \frac{1}{2}\Sigma^{-\frac{1}{2}}NQ_2 Y_2' + \frac{1}{2}Y_2 Q_2' N'\Sigma^{-\frac{1}{2}} + \Sigma^{-\frac{1}{2}}D\Sigma^{-\frac{1}{2}},
$$

$Y_1 \sim N_{p,r}(K_1, I_p \otimes I_r)$, $Y_2 \sim N_{p,s}(K_2, I_p \otimes I_s)$ and $Y_3 \sim N_{p,n-q}(K_3, I_p \otimes I_{n-q})$. Now, from Theorem 2.3.2, the m.g.f. Y_3 is

$$
\begin{aligned}
M_{Y_3}(\Sigma^{\frac{1}{2}}(Z_1 L + Z_2 N)Q_3) &= \text{etr}\{(Z_1 L + Z_2 N)Q_3 Q_3' M' \\
&\quad + \Sigma(Z_1 L + Z_2 N)Q_3 Q_3'(Z_1 L + Z_2 N)'\}. \quad (7.9.15)
\end{aligned}
$$

Substituting from (7.9.15) in (7.9.14) and taking the logarithm we get

$$
\begin{aligned}
\ln M_{S_1, S_2}(Z_1, Z_2) &= \{\text{tr}(Z_1 LQ_3 Q_3' M' + \Sigma Z_1 LQ_3 Q_3' L' Z_1) \\
&\quad + \ln M_{S_1^*}(\Sigma^{\frac{1}{2}}Z_1 \Sigma^{\frac{1}{2}})\} + \{\text{tr}(Z_2 NQ_3 Q_3' M' \\
&\quad + \Sigma Z_2 NQ_3 Q_3' N' Z_2) + \ln M_{S_1^*}(\Sigma^{\frac{1}{2}}Z_2 \Sigma^{\frac{1}{2}})\} \\
&\quad + 2\,\text{tr}(\Sigma Z_1 LQ_3 Q_3' N' Z_2) \\
&= \ln M_{S_1}(Z_1) + \ln M_{S_2}(Z_2) \\
&\quad + 2\,\text{tr}(\Sigma Z_1 LQ_3 Q_3' N' Z_2). \quad (7.9.16)
\end{aligned}
$$

The last step follows from the Theorem 7.9.2. Therefore for independence of S_1 and S_2, we must have $\text{tr}(\Sigma Z_1 LQ_3 Q_3' N' Z_2) = 0$ *i.e.* $\text{tr}\{\Sigma Z_1 L(I_n - Q_1 Q_1' - Q_2 Q_2')N' Z_2\} = 0$ *i.e.* $\text{tr}(\Sigma Z_1 LN' Z_2) = 0$ for all symmetric matrices Z_1 and Z_2 and hence $LN' = 0$. This completes the proof of necessity. ∎

THEOREM 7.9.6. *Let $S_1 = XAX' + \frac{1}{2}(LX' + XL') + C$ and $S_2 = XBX' + \frac{1}{2}(NX' + XN') + D$, where $X \sim N_{p,n}(0, \Sigma \otimes \Psi)$, $A(n \times n) = A'$, $B(n \times n) = B'$, $C(p \times p) = C'$, $D(p \times p) = D'$, $L(p \times n)$, and $N(p \times n)$ are constant matrices. Then necessary and sufficient conditions for S_1 and S_2 to be stochastically independent are (i) $A\Psi B = 0$, (ii) $L\Psi B = 0$, and (iii) $N\Psi A = 0$ and (iv) $L\Psi N' = 0$.*

Proof: Note that the forms $S_1 = XAX' + \frac{1}{2}(LX' + XL') + C$ and $S_2 = XBX' + \frac{1}{2}(NX' + XN') + D$ are stochastically independent if and only if the quadratic forms $Y(\Psi^{\frac{1}{2}}A\Psi^{\frac{1}{2}})Y' + \frac{1}{2}(L\Psi^{\frac{1}{2}}Y' + Y\Psi^{\frac{1}{2}}L') + C$ and $Y(\Psi^{\frac{1}{2}}B\Psi^{\frac{1}{2}})Y' + \frac{1}{2}(N\Psi^{\frac{1}{2}}Y' + Y\Psi^{\frac{1}{2}}N') + D$, where $Y \sim N_{p,n}(M\Psi^{-\frac{1}{2}}, \Sigma \otimes I_n)$, are stochastically independent. Hence from Theorem 7.9.5, we get the required conditions. ∎

COROLLARY 7.9.6.1. *In the above notations S_1 and the linear form $NX' + XN'$ are stochastically independent if and only if (i) $A\Psi N' = 0$, and (ii) $L\Psi N' = 0$.*

COROLLARY 7.9.6.2. *In the above notations the linear forms $LX' + XL'$ and $NX' + XN'$ are stochastically independent if and only if $L\Psi N' = 0$.*

COROLLARY 7.9.6.3. *Let $X \sim N_{p,n}(M, \Sigma \otimes \Psi)$. Then the quadratic forms $(X + L_i)A_i(X + L_i)'$, $i = 1, \ldots, k$ are stochastically independent if and only if $A_i\Psi A_j = 0$, $i \neq j$, $i, j = 1, \ldots, k$.*

THEOREM 7.9.7. *Let $(X + L)A(X + L)' = \sum_{i=1}^{k}(X + L_i)A_i(X + L_i)'$, where $X \sim N_{p,n}(M, \Sigma \otimes I_n)$, $\text{rank}(A) = r \ (\geq p)$, and $\text{rank}(A_i) = r_i \ (\geq p)$, $i = 1, \ldots, k$. Consider the following four conditions:*

(i) $(X + L_i)A_i(X + L_i)' \sim W_p(r_i, \Sigma, \Sigma^{-1}(M + L_i)A_i(M + L_i)')$,

(ii) $(X + L_i)A_i(X + L_i)'$ and $(X + L_j)A_j(X + L_j)'$, $i \neq j$, are stochastically independent,

(iii) $(X + L)A(X + L)' \sim W_p(r, \Sigma, \Sigma^{-1}(M + L)A(M + L)')$, and

(iv) $r = \sum_{i=1}^{k} r_i$.

Then, (a) any two of the conditions (i), (ii), and (iii) imply the remaining, and (b) conditions (iii) and (iv) imply (i) and (ii).

Proof: The proof is similar to the proof of Theorem 7.8.7. ∎

7.10. WISHARTNESS AND INDEPENDENCE OF QUADRATIC FORMS OF THE TYPE $XAX' + L_1X' + XL'_2 + C$

Consider the polynomial of the type

$$S = XAX' + L_1X' + XL'_2 + C, \qquad (7.10.1)$$

where $A(n \times n) = A'$, $L_1(p \times n)$, $L_2(p \times n)$ and $C(p \times p)$ are constant matrices. For $L_1 = L_2 = \frac{1}{2}L$, $C = C'$, the polynomial (7.10.1) reduces to the polynomial (7.9.1) of Section 7.9. Conditions for Wishartness and independence, for this case, when $X \sim N_{p,n}(M, \Sigma \otimes \Psi)$, are given there. In this section we discuss such conditions for polynomials of type (7.10.1) when $X \sim N_{p,n}(M, \Sigma \otimes \Psi | r, s)$, $r \leq p$, $s \leq n$. First we derive its m.g.f.

Since $\Sigma \geq 0$ and $\Psi \geq 0$ are of ranks r and s respectively, we can write $\Sigma = B_1 B_1'$ and $Y = BB'$, where B_1 $(p \times r)$ and B $(p \times s)$ are of ranks r and s respectively. From Definition 2.4.1, we can write $X = M + B_1 Y B'$ where $Y \sim N_{r,s}(0, I_r \otimes I_s)$. Therefore S can be written as

$$S = B_1 Y A_{(1)} Y' B_1' + L_{(1)} Y' B_1' + B_1 Y L_{(2)}' + C_{(1)}, \qquad (7.10.2)$$

where

$$A_{(1)} = B'AB,$$

$$L_{(1)} = (MA + L_1)B,$$

$$L_{(2)} = (MA + L_2)B,$$

and

$$C_{(1)} = MAM' + L_1M' + ML_2' + C.$$

Now for any arbitrary matrix Z $(p \times p)$, and $Z_0 = \frac{1}{2}(Z + Z')$,

$$\text{tr}(ZS) = \text{tr}[Z_0 B_1 Y A_{(1)} Y' B_1' + (ZL_{(1)} + Z'L_{(2)})Y' B_1' + ZC_{(1)}]$$

$$= \text{tr}[B_1' Z_0 B_1 Y A_{(1)} Y' + B_1'(ZL_{(1)} + Z'L_{(2)})Y'] + \text{tr}(ZC_{(1)}). \quad (7.10.3)$$

Note that

$$\text{tr}(B_1' Z_0 B_1 Y A_{(1)} Y') = (\text{vec}(Y'))'(B_1' Z_0 B_1 \otimes A_{(1)}) \text{vec}(Y')$$

and

$$\text{tr}(LY') = (\text{vec}(Y'))' \text{vec}(L')$$

where $2L = B_1'(ZL_{(1)} + Z'L_{(2)})$. Hence, (7.10.3) can be written as

$$\text{tr}(ZS) = (\text{vec}(Y'))'(B_1' Z_0 B_1 \otimes A_{(1)}) \text{vec}(Y') + 2(\text{vec}(Y'))' \text{vec}(L')$$

$$+ \text{tr}(ZC_{(1)}), \qquad (7.10.4)$$

where $\text{vec}(Y') \sim N_{rs}(0, I_r \otimes I_s)$.

Using (7.10.4), the m.g.f. of S is

$$M_S(Z) = E[\exp\{(\text{vec}(Y'))'(B_1' Z_0 B_1 \otimes A_{(1)}) \text{vec}(Y') + 2(\text{vec}(Y'))' \text{vec}(L')$$

$$+ \text{tr}(ZC_{(1)})\}]$$

$$= \det(I_{rs} - 2(B_1' Z_0 B_1 \otimes A_{(1)}))^{-\frac{1}{2}} \text{etr}(ZC_{(1)}) E[\text{etr}\{2(\text{vec}(Y'))' \text{vec}(L')\}],$$

where now $\text{vec}(Y') \sim N_{rs}(\mathbf{0}, (I_{rs} - 2(B_1'Z_0B_1 \otimes A_{(1)}))^{-1})$. Therefore

$$M_S(Z) = \det(I_{rs} - 2(B_1'Z_0B_1 \otimes A_{(1)}))^{-\frac{1}{2}} \text{etr}(ZC_{(1)})$$

$$\exp\{2(\text{vec}(L'))'(I_{rs} - 2(B_1'Z_0B_1 \otimes A_{(1)}))^{-1} \text{vec}(L')\}. \quad (7.10.5)$$

Next let the spectral decomposition of $A_{(1)} = B'AB$ be $A_{(1)} = \sum_{j=1}^m \lambda_j E_j$, where E_j is symmetric, $E_j^2 = E_j$, $E_i E_j = 0$, $i \neq j$, and $\lambda_1 > \cdots > \lambda_m$ are the nonzero characteristic roots of $A_{(1)}$ with multiplicity $f_1, \ldots, f_m$ respectively. Let $E_0 = I_s - \sum_{j=1}^m E_j$, then $E_0^2 = E_0$, $E_0 E_j = 0$, and

$$(I_{rs} - 2(B_1'Z_0B_1 \otimes A_{(1)}))^{-1} = \left(I_{rs} - 2\left(B_1'Z_0B_1 \otimes \sum_{j=1}^m \lambda_j E_j\right)\right)^{-1}$$

$$= \left(I_r \otimes E_0 + \sum_{j=1}^m (I_r - 2\lambda_j B_1'Z_0B_1) \otimes E_j\right)^{-1}$$

$$= I_r \otimes E_0 + \sum_{j=1}^m (I_r - 2\lambda_j B_1'Z_0B_1)^{-1} \otimes E_j. \quad (7.10.6)$$

Now

$$(\text{vec}(L'))'(I_{rs} - 2(B_1'Z_0B_1 \otimes A_{(1)}))^{-1} \text{vec}(L')$$

$$= \text{tr}(LE_0L') + \sum_{j=1}^m \text{tr}\{(I_r - 2\lambda_j B_1'Z_0B_1)^{-1} LE_jL'\}$$

$$= \frac{1}{4}\sum_{j=0}^m \text{tr}\{(I_r - 2\lambda_j B_1'Z_0B_1)^{-1} B_1'(ZL_{(1)} + Z'L_{(2)})E_j(L_{(1)}'Z' + L_{(2)}'Z)B_1\},$$

$$(7.10.7)$$

where we define $\lambda_0 = 0$, and

$$\det(I_{rs} - 2(B_1'Z_0B_1 \otimes A_{(1)}))^{-\frac{1}{2}} = \prod_{j=1}^m \det(I_r - 2\lambda_j B_1'Z_0B_1)^{-\frac{1}{2}f_j}. \quad (7.10.8)$$

Substituting from (7.10.7) and (7.10.8) in (7.10.5), we get

$$M_S(Z) = \prod_{j=1}^m \det(I_r - 2\lambda_j B_1'Z_0B_1)^{-\frac{1}{2}f_j} \text{etr}(ZC_{(1)})$$

$$\text{etr}\left\{\frac{1}{2}\sum_{j=0}^m B_1(I_r - 2\lambda_j B_1'Z_0B_1)^{-1} B_1'(ZL_{(1)} + Z'L_{(2)})E_j(L_{(1)}'Z' + L_{(2)}'Z)\right\}$$

$$= \prod_{j=1}^m \det(I_p - 2\lambda_j \Sigma Z_0)^{-\frac{1}{2}f_j} \text{etr}(ZC_{(1)})$$

$$\text{etr}\left\{\frac{1}{2}\sum_{j=0}^m (I_p - 2\lambda_j \Sigma Z_0)^{-1}\Sigma(ZL_{(1)} + Z'L_{(2)})E_j(L_{(1)}'Z' + L_{(2)}'Z)\right\}. \quad (7.10.9)$$

If $L_{(1)} = L_{(2)} = L_{(0)}$ (say), and $C_{(1)}$ is symmetric, then $(L_{(1)} - L_{(2)})\Psi = 0$ and (7.10.9) reduces to

$$M_S(Z) = \text{etr}(Z_0 C_{(2)} + 2\Sigma Z_0 \Omega_0 Z_0) \prod_{j=1}^{m} \det(I_p - 2\lambda_j \Sigma Z_0)^{-\frac{1}{2} f_j}$$

$$\text{etr}\left\{ \sum_{j=1}^{m} \frac{1}{\lambda_j}(I_p - 2\lambda_j \Sigma Z_0)^{-1}\Omega_j Z_0 \right\}. \tag{7.10.10}$$

where $\Omega_j = L_{(0)} E_j L'_{(0)} = (MA + L_1)BE_j B'(MA + L_1)'$, $j = 0, 1, \ldots, m$ and $C_{(2)} = C_{(1)} - \sum_{j=1}^{m} \frac{\Omega_j}{\lambda_j}$. Now from (7.10.10), it is clear that S is distributed as $\sum_{j=1}^{m} \lambda_j W_j + \frac{1}{2}(Y + Y')$ where $W_1, \ldots, W_m$ and Y are independent, $Y \sim N_{p,p}(C_{(2)}, 4\Sigma \otimes \Omega_0)$ and if $f_j \geq p$, then $W_j \sim W_p(f_j, \frac{1}{\lambda_j^2}\Omega_j)$, $j = 1, \ldots, m$.

The results on Wishartness and independence of quadratic forms of the type (7.10.1) have also been given by Khatri (1980). Some of these are stated below without proof.

THEOREM 7.10.1. *Let* $S = XAX' + L_1X' + XL'_2 + C$, *where* $A(n \times n) = A'$, $L_1(p \times n)$, $L_2(p \times n)$ *and* $C(p \times p)$ *are constant matrices, and* $X \sim N_{p,n}(M, \Sigma \otimes \Psi | r, s)$. *Then* S *is distributed as* $\sum_{j=1}^{m} \lambda_j W_j$, *where* $W_1, \ldots, W_m$ *are independent,* $W_j \sim W_p(f_j, \Sigma, \frac{1}{\lambda_j^2}\Omega_j)$, *for distinct nonzero* $\lambda_1, \lambda_2, \ldots, \lambda_m$, *if and only if*

(i) $\lambda_1, \lambda_2, \ldots, \lambda_m$ *are distinct nonzero characteristic roots of* ΨA *with multiplicities* $f_1, f_2, \ldots, f_m$ *respectively such that* $f_j \geq p$, $j = 1, \ldots, p$,

(ii) $L_1\Psi = L_2\Psi$,

(iii) $(L_1 + MA)\Psi = L\Psi A\Psi$ *for some matrix* L,

(iv) $\Omega_j = (L_1 + MA)(BE_jB')(L_1 + MA)'$, *and*

(v) $MAM' + L_1M' + ML'_2 + L = \sum_{j=1}^{m} \frac{1}{\lambda_j}\Omega_j$.

The asymptotic distribution of $\sum_{j=1}^{m} \lambda_j W_j$, under the conditions of Theorem 7.10.1, is also given by Khatri (1980).

THEOREM 7.10.2. *Let* $S_i = XA_iX' + L_{1i}X' + XL'_{2i} + C_i$, *where* $A_i(n \times n) = A'_i$, $L_{1i}(p \times n)$, $L_{2i}(p \times n)$ *and* $C_i(p \times p)$ *are constant matrices,* $i = 1, 2$, *and* $X \sim N_{p,n}(M, \Sigma \otimes \Psi | r, s)$. *Then* S_1 *and* S_2 *are independently distributed if and only if*

(i) $\Psi A_1 \Psi A_2 \Psi = 0$,

(ii) $(L_{j1} + MA_1)\Psi A_2\Psi = (L_{j2} + MA_2)\Psi A_1\Psi = 0$, $j = 1, 2$, *and*

(iii) the coefficients of the elements of Z_1 *and* Z_2 *from* $\text{tr}(Z'_2 L_{(12)} + Z_2 L_{(22)})\Psi$ $(Z'_1 L_{(11)} + Z_1 L_{(21)})'$ *are zero where* $L_{(ji)} = L_{ji} + MA_i$, $i, j = 1, 2$.

PROBLEMS

7.1. Show that the Laplace transform of the density of $S = XAX'$, where $X \sim N_{p,n}(0, \Sigma \otimes I_n)$, is

$$L(Z) = \det(G)^{-\frac{1}{2}n} \sum_{k=0}^{\infty} \sum_{\kappa} \frac{(\frac{1}{2}n)_{\kappa}}{k!} \frac{C_{\kappa}(I_n - q^{-1}A)C_{\kappa}(I_n - G^{-1})}{C_{\kappa}(I_n)},$$

where q is a real positive quantity and $G = I_p + 2qZ\Sigma$. Hence derive the density (7.2.7).

(Shah, 1970)

7.2.   Derive the p.d.f. of $u = \operatorname{tr}(S)$ using the density (7.2.5) of S, as

$$f(u) = \left\{ 2^{\frac{1}{2}np} \Gamma\left(\frac{1}{2}np\right) \det(A\Psi)^{\frac{1}{2}p} \det(\Sigma)^{\frac{1}{2}n} \right\}^{-1} u^{\frac{1}{2}(np-2)}$$

$$_0F_0^{(np)}\left(\Sigma^{-1} \otimes A^{-\frac{1}{2}}\Psi^{-1}A^{-\frac{1}{2}}, -\frac{1}{2}u \right), \quad u > 0,$$

and then show that

$$P(u \le w) = \left\{ 2^{\frac{1}{2}np} \Gamma\left(\frac{1}{2}np + 1\right) \det(A\Psi)^{\frac{1}{2}p} \det(\Sigma)^{\frac{1}{2}n} \right\}^{-1} w^{\frac{1}{2}np}$$

$$_1F_1^{(np)}\left(\frac{1}{2}np; \frac{1}{2}np + 1; \Sigma^{-1} \otimes A^{-\frac{1}{2}}\Psi^{-1}A^{-\frac{1}{2}}, -\frac{1}{2}w \right).$$

(Hayakawa, 1966)

7.3.   Show that the limiting distributions of F_1 and F_2 as $q \to \infty$ are given by

$$\frac{\Gamma_p[\frac{1}{2}(m+n)]}{\Gamma_p(\frac{1}{2}m)\Gamma_n(\frac{1}{2}p)} \det(\Omega)^{-\frac{1}{2}n} \det(\Psi)^{-\frac{1}{2}p} \det(F_1)^{\frac{1}{2}(p-m-1)}$$

$$_1F_0^{(p)}\left(\frac{1}{2}(m+n); -\Omega^{-1}, \Psi^{-1}F_1 \right), \quad F_1 > 0,$$

and

$$\frac{\Gamma_p[\frac{1}{2}(m+n)]}{\Gamma_p(\frac{1}{2}m)\Gamma_p(\frac{1}{2}n)} \det(\Omega)^{-\frac{1}{2}n} \det(\Psi)^{-\frac{1}{2}p} \det(F_2)^{\frac{1}{2}(n-p-1)}$$

$$_1F_0^{(n)}\left(\frac{1}{2}(m+n); -\Psi^{-1}, \Omega^{-1}F_2 \right), \quad F_2 > 0,$$

respectively.

(Khatri, 1966)

7.4.   Let $S \sim Q_{p,n}(A, \Sigma, \Psi)$. Derive the p.d.f. of S^{-1} in the forms parallel to (7.2.1), (7.2.5) and (7.2.7).

7.5.   Let $S\ (p \times p)$ and $Y\ (p \times m)$ be independent, $S \sim Q_{p,n}(A, I_p, \Psi)$ and $Y \sim N_{p,m}(0, I_p \otimes I_m)$. Derive the density of $S^{-\frac{1}{2}}YY'S^{-\frac{1}{2}}$, for $p \le m$.

7.6.   Let the density of S is given by (7.2.1). Show that

$$E[\det(S)^h] = q^{(h+\frac{1}{2}n)p} \, 2^{ph} \frac{\Gamma_p(\frac{1}{2}n + h)}{\Gamma_p(\frac{1}{2}n)} \det(A\Psi)^{-\frac{1}{2}p} \det(\Sigma)^h$$

$$\quad {}_1F_0^{(n)}\left(\frac{1}{2}n + h; I_n - q\Psi^{-\frac{1}{2}}A^{-1}\Psi^{-\frac{1}{2}}, I_p\right)$$

$$= q^{(h+\frac{1}{2}n)p} \, 2^{ph} \frac{\Gamma_p(\frac{1}{2}n + h)}{\Gamma_p(\frac{1}{2}n)} \det(A\Psi)^{-\frac{1}{2}p} \det(\Sigma)^h$$

$$\quad {}_2F_1^{(n)}\left(\frac{1}{2}p, \frac{1}{2}n + h; \frac{1}{2}n; I_n - q\Psi^{-\frac{1}{2}}A^{-1}\Psi^{-\frac{1}{2}}\right).$$

7.7.   Let S be distributed as in Theorem 7.6.1. Then show that

$$E[\det(S)^h] = q^{-(h+\frac{1}{2}n)p} \, 2^{ph} \frac{\Gamma_p(\frac{1}{2}n + h)}{\Gamma_p(\frac{1}{2}n)} \det(A)^{-\frac{1}{2}p} \det(\Sigma)^h \, \text{etr}\left(-\frac{1}{2}\Sigma^{-1}MM'\right)$$

$$\sum_{k=0}^{\infty} \sum_{\kappa} \frac{(\frac{1}{2}n + h)_\kappa}{(\frac{1}{2}n)_\kappa \, k!} P_\kappa\left(\frac{1}{\sqrt{2}} \Sigma^{-\frac{1}{2}} M(I_n - qA)^{-\frac{1}{2}}, (qA)^{-1} - I_n\right),$$

where $\text{Re}(h) > -\frac{1}{2}(n - p + 1)$.

7.8.   Let $X \sim N_{p,n}(M, \Sigma \otimes \Psi)$, and $Q = (X - M)\Psi^{-1}(X - M)' - (X_1 - M_1)\Psi_{11}^{-1}$ $(X_1 - M_1)'$, where $X = (X_1 \quad X_2)$, $X_1 \, (p \times q)$, $M = (M_1 \quad M_2)$, $M_1 \, (p \times q)$ and $\Psi = \begin{pmatrix} \Psi_{11} & \Psi_{12} \\ \Psi_{21} & \Psi_{22} \end{pmatrix}$, $\Psi_{11} \, (q \times q)$, and $n \geq p + q$. Then show that $Q \sim W_p(n - q, \Sigma)$.

7.9.   Let $X \sim N_{p,n}(\Gamma W, \Sigma \otimes I_n)$ where $\Gamma \, (p \times r)$ and $W \, (r \times n)$ are constant matrices, $\text{rank}(W) = r \leq p$. Define $H = WW'$, $G = XW'H^{-1}$, and $Q = XX' - GHG'$. Then prove that G and Q are independent, and $Q \sim W_p(n - r, \Sigma)$, $n - r \geq p$.

7.10. Let $X \sim N_{p,n}(M, \Sigma \otimes \Psi)$, and A, $A_1, \ldots, A_{k-1}$, A_k be $n \times n$ real symmetric matrices so that $A = A_1 + \cdots + A_{k-1} + A_k$. Let XAX', $XA_1X', \ldots, XA_{k-1}X'$ have Wishart distributions and let A_k be positive semidefinite. Then XA_1X', $\ldots, XA_{k-1}X'$ and XA_kX' are stochastically independent, and XA_kX' has a Wishart distribution.

(Hogg, 1963)

7.11. Let $X \, (p \times n)$ be a real matrix. Let $A \, (n \times n)$ and $B \, (n \times n)$ be symmetric idempotent matrices of rank r and s respectively, $p \leq r < s$. Then prove that a necessary and sufficient condition for $B - A$ to be positive semidefinite is that $\det(XAX') \leq \det(XBX')$, for all $X \in \mathbb{R}^{p \times n}$.

(Hogg, 1963)

7.12. Let $XAX' = \sum_{i=1}^{k} XA_iX'$, where $X \sim N_{p,n}(M, \Sigma \otimes I_n)$. Then prove that any one of the following six conditions is a necessary and sufficient condition for

Theorem 7.8.7.

(i) $A_i^2 = A_i$, $i = 1, \ldots, k$, and XA_iX' and XA_jX', $i \neq j$, are stochastically independent,

(ii) $A_i^2 = A_i$, $i = 1, \ldots, k$, and $XAX' \sim W_p(\text{rank}(A), \Sigma, \Sigma^{-1}MAM')$,

(iii) $A_iA_j = 0$, $i \neq j$, and $XA_iX' \sim W_p(\text{rank}(A_i), \Sigma, \Sigma^{-1}MA_iM')$, $i = 1, \ldots, k$,

(iv) $A_iA_j = 0$, $i \neq j$, and $XAX' \sim W_p(\text{rank}(A), \Sigma, \Sigma^{-1}MAM')$,

(v) $A^2 = A$, and XA_iX' and XA_jX', $i \neq j$, are stochastically independent, and

(vi) $A^2 = A$, and $XA_iX' \sim W_p(\text{rank}(A_i), \Sigma, \Sigma^{-1}MA_iM')$, $i = 1, \ldots, k$.

7.13. Let $S_1 = XAX' + L_1X' + XL_2' + C$ and $S_2 = XBX' + N_1X' + XN_2' + D$, where $X \sim N_{p,n}(0, \Sigma \otimes \Psi)$. Then show that

(i) $E(S_1) = \text{tr}(A\Psi)\Sigma + C$,

(ii) $E(S_1GS_2) = \text{tr}(\Psi B'\Psi A')\,\text{tr}(G\Sigma)\Sigma + \text{tr}(A\Psi)\,\text{tr}(B\Psi)\Sigma G\Sigma$
$\quad + \text{tr}(A\Psi B'\Psi)\Sigma G'\Sigma + \text{tr}(A\Psi)\Sigma GD + L_1\Psi N_1'G'\Sigma + \text{tr}(G\Sigma)L_1\Psi N_2'$
$\quad + \text{tr}(L_2'GN_1\Psi)\Sigma + \Sigma G'L_2\Psi N_2' + \text{tr}(B\Psi)CG\Sigma + CGD$, and

(iii) $E(\text{tr}(S_2C)S_1) = \text{tr}(A'\Psi B\Psi)\Sigma G'\Sigma + \text{tr}(A\Psi)\,\text{tr}(B\Psi)\,\text{tr}(G\Sigma)\Sigma$
$\quad + \text{tr}(A'\Psi B'\Psi)\Sigma G\Sigma + \text{tr}(A\Psi)\,\text{tr}(DG)\Sigma + L_1\Psi N_1'G'\Sigma$
$\quad + \Sigma G'N_2\Psi L_2' + \text{tr}(\Sigma G)\,\text{tr}(\Psi B)C + \text{tr}(DG)C.$

7.14. Let $X \sim N_{p,n}(M, \Sigma \otimes \Psi)$ and define $S = D'XAX'D + \frac{1}{2}(LX'B + B'XL') + C$. Then show that

$$E(S) = D'MAM'D + \text{tr}(A\Psi)D'\Sigma D + \frac{1}{2}(LM'B + B'ML') + C$$

and

$$\begin{aligned}
\text{cov}(S) = M_p[\{2\,\text{tr}(A\Psi)^2\}(D'\Sigma D \otimes D'\Sigma D) + 4D'MA\Psi AM'D \otimes D'\Sigma D \\
+ L\Psi L' \otimes B'\Sigma B + 2L\Psi AM'D \otimes B'\Sigma D \\
+ 2(L\Psi AM'D)' \otimes (B'\Sigma D)']M_p.
\end{aligned}$$

(Brown and Neudecker, 1988)

7.15. Derive $E(S_1)$, $E(S_1GS_2)$ and $E(\text{tr}(S_2G)S_1)$ where S_1 and S_2 are defined in Problem 7.13 and $X \sim N_{p,n}(M, \Sigma \otimes \Psi)$.

7.16. Prove Theorem 7.8.9. (Hint: use Definition 2.4.1).

7.17. Let $XAX' + LX' + XL' + C = \sum_{i=1}^k (X+L_i)A_i(X+L_i)'$, where $X \sim N_{p,n}(M, \Sigma \otimes \Psi)$, $\text{rank}(A) = r\ (\geq p)$, and $\text{rank}(A_i) = r_i\ (\geq p)$, $i = 1, \ldots, k$. Consider the following conditions:

(a_1) $(X+L_i)A_i(X+L_i)' \sim W_p(r_i, \Sigma, \Sigma^{-1}(M+L_i)A_i(M+L_i)')$,

(a_2) $(X+L_i)A_i(X+L_i)'$ and $(X+L_j)A_j(X+L_j)'$, $i \neq j$, are stochastically independent,

(a_3) $(X+L)A(X+L)' \sim W_p(r, \Sigma, \Sigma^{-1}(M+L)A(M+L)')$,

(c_1) $A_i\Psi A_i = A_i$, $i = 1, \ldots, k$,

(c_2) $A_i\Psi A_j = 0$, $i \neq j$,

(c_3) $A\Psi A = A$,

and

(c_4) $r = \sum_{i=1}^{k} r_i$.

Then, prove that (a) any two of the three conditions (a_1), (a_2), and (a_3), or (b) any two of the three conditions (c_1), (c_2), and (c_3) or (c) any one set of (a_i) and (c_j), $i \neq j$, $i, j = 1, 2, 3$; or (d) conditions (c_3) and (a_3); or (e) conditions (c_3) and (c_4) are necessary and sufficient for all the remaining conditions.

(Khatri, 1962)

7.18. Let $S = XAX' + L_1X' + XL_2' + C$, where $X \sim N_{p,n}(0, \Sigma \otimes I_n)$. Show that the necessary and sufficient conditions for S to be distributed as $V + Y$, where $V \sim W_p(\text{rank}(A), \Sigma)$, Y ($p \times p$) is normal and is independent of V, are (i) $A^2 = A$, $\text{rank}(A) \geq p$ (ii) $(L_1 - L_2)A = 0$, and (iii) $L_1A = 0$.

7.19. Let $S_1 = XA_1X' + L_1X' + XL_2' + C_1$ and $S_2 = XA_2X' + N_1X' + XN_2' + C_2$, where $X \sim N_{p,n}(0, \Sigma \otimes I_n)$. Then show that S_1 and S_2 are stochastically independent if and only if (i) $A_1A_2 = 0$, (ii) $L_1A_2 = 0$, (iii) $N_1A_1 = 0$, (iv) $(N_1 - N_2)A_1 = 0$, (v) $(L_1 - L_2)A_2 = 0$, and (vi) $\begin{pmatrix} L_1N_1' & L_1(N_1 - N_2)' \\ (L_1 - L_2)N_1' & (L_1 - L_2)(N_1 - N_2)' \end{pmatrix} = 0$.

(Khatri, 1980)

7.20. Let $X \sim N_{p,n}(M, \Sigma \otimes I_n)$, $n \geq p$, $\Sigma > 0$ and $S = XAX'$ where $\text{rank}(A) = t$ ($p \leq t \leq n$). Further let $S = (S_{ij})$, $i, j = 1, \ldots, k$, S_{ij} ($q \times q$), and $kq = p$. Then show that the necessary and sufficient condition for principal minors $S_{11}, \ldots, S_{kk}$ to be distributed as the principal minors of a Wishart matrix is that A be idempotent.

(Gupta and Chattopadhyay, 1979)

CHAPTER 8

MISCELLANEOUS DISTRIBUTIONS

8.1. INTRODUCTION

In Chapters 1–7 we introduced the basic matrix variate distributions. These distributions, because of their wide applicability in multivariate statistical analysis and other fields, have been studied extensively. There are many other matrix variate distributions which have not been classified in the foregoing chapters.

In this chapter we give these distributions which, among others, have been studied by James (1954), Herz (1955), Khatri (1970a), Roux (1971), Downs (1972), van der Merwe and Roux (1974), Khatri and Mardia (1977), Mardia and Khatri (1977), de Waal (1979, 1983), and Chikuse (1990a, 1990b, 1991a, 1991b, 1993a, 1993b). However, the coverage here is not exhaustive. Patil, Boswell, Ratnaparkhi and Roux (1984) have also written a classified bibliography of statistical distributions which include matrix variate distributions.

8.2. UNIFORM DISTRIBUTION ON STIEFEL MANIFOLD

The uniform distribution on the Stiefel manifold, $O(p, n)$, has already been encountered in Chapter 1, while studying the Jacobian of a certain transformation involving semiorthogonal matrix X $(p \times n)$, $p \leq n$, $XX' = I_p$. Recall that

$$J((dX)X_0' \to (dX)) \, dX \qquad (8.2.1)$$

defines an invariant measure on the Stiefel manifold $O(p, n)$ and is denoted by $[(dX)X']$. Here X_0 $(n \times n) = (X' \quad X_1')$, $X_0'X_0 = I_n$ and

$$O(p, n) = \{X \ (p \times n) : XX' = I_p\}.$$

In Section 1.4, it was shown that

$$\text{Vol}(O(p, n)) \;=\; \int_{O(p,n)} [(dX)X']$$

279

$$= \frac{2^p \pi^{\frac{1}{2}np}}{\Gamma_p(\frac{1}{2}n)}.$$

Thus

$$\frac{1}{\text{Vol}(O(p,n))}[(dX)X'] = [dX] \qquad (8.2.2)$$

defines the probability element of the invariant distribution of random matrix X known as *uniform distribution on the Stiefel manifold*, $O(p,n)$, denoted by $\mathcal{U}_{p,n}$. Note that the random matrix X $(p \times n)$ has $np - \frac{1}{2}p(p+1)$ functionally independent and $\frac{1}{2}p(p+1)$ functionally dependent elements. Let X_I be the set of functionally independent elements and X_D be the set of functionally dependent elements of X. Let $J(XX' \to X_D)$ be the Jacobian of transformation from XX' to X_D at X_I (Roy, 1957, p. 170). Then the probability density function of X, with parameters p and n, is defined as

$$f_{p,n}(X) = \frac{\Gamma_p(\frac{1}{2}n)}{\pi^{\frac{1}{2}np}}\{J(XX' \to X_D)\}^{-1}, \; X \in O(p,n). \qquad (8.2.3)$$

The above density was given by Khatri (1970a). An alternative representation of density (8.2.3) in terms of generalized Eulerian angles θ_{ij}, $i = 1, \ldots, p$, $j = i+1, \ldots, n$ (Hoffman, Raffenetti and Ruedenberg, 1972; Girko and Gupta, 1996) is given by Khatri and Mardia (1977). Let $P_{ij}(\theta_{ij})$ be an $n \times n$ matrix with unities on the diagonal except in $(i,i)^{\text{th}}$ and $(j,j)^{\text{th}}$ positions which contain $\cos\theta_{ij}$, and all off-diagonal elements are zero except $(i,j)^{\text{th}}$ and $(j,i)^{\text{th}}$ elements which are $\sin\theta_{ij}$ and $-\sin\theta_{ij}$, respectively, $j > i$. Further, let

$$P = \prod_{i=1}^{p} \prod_{j=i+1}^{n} P_{ij}(\theta_{ij}),$$

where the product is written from right to left. The matrix P $(n \times n)$ is orthogonal and its first p rows can be chosen to represent X in polar coordinates, $p < n$. Then

$$[dX] = \frac{\Gamma_p(\frac{1}{2}n)}{2^p \pi^{\frac{1}{2}np}}\left\{\prod_{i=1}^{p} \prod_{j=i+1}^{n} \cos^{j-i-1}(\theta_{ij})\right\}\prod_{i<j} d\theta_{ij},$$

$$-\pi \le \theta_{j-1,j} \le \pi, \; j = 2,3,\ldots,n, \; -\frac{1}{2}\pi \le \theta_{ij} < \frac{1}{2}\pi, \; j \ne i+1. \quad (8.2.4)$$

Using (8.2.3), Khatri (1970a) has derived several results for uniform density. Here we state some of them without proof.

THEOREM 8.2.1. *Let $X \sim \mathcal{U}_{p,n}$. Then for $M \in O(p)$, and $N \in O(n)$, (i) $XN \sim \mathcal{U}_{p,n}$ (ii) $MX \sim \mathcal{U}_{p,n}$ and (iii) $MXN \sim \mathcal{U}_{p,n}$.*

THEOREM 8.2.2. *Let $X = (X_1 \; X_2)$ be distributed as (8.2.3) where X_i is $p \times n_i$, $i = 1,2$ with $n = n_1 + n_2$, $n_2 \ge p$. Then, all the elements of X_1 can be taken as random and the random matrices X_1 and $W = (I_p - X_1X_1')^{-\frac{1}{2}}X_2$ are independent, $W \sim \mathcal{U}_{p,n_2}$ and $X_1 \sim IT_{p,n_1}(n_2 + p + 1, 0, I_p, I_{n_1})$.*

For $n_1 = p$, above theorem was proved by Herz (1955, Lemma 3.7) using the uniqueness of Fourier transform and results on hypergeometric function of matrix argument.

THEOREM 8.2.3. *Let $X\,(p \times n) \sim \mathcal{U}_{p,n}$ and $R = (\,R_1 \quad R_2\,) \sim \mathcal{U}_{p,n+m}$, $R_1\,(p \times n)$, $R_2(p \times m)$ be independent and elements of R_2 be functionally independent. Then $W = (\,(R_1 R_1')^{\frac{1}{2}} X \quad R_2\,) \sim \mathcal{U}_{p,n+m}$.*

THEOREM 8.2.4. *Let $X \sim \mathcal{U}_{p,n}$ and $X' = (\,X_1' \quad X_2'\,)$, $X_1 = (\,X_{11} \quad X_{12}\,)$, $X_2 = (\,X_{21} \quad X_{22}\,)$ where X_{ij} is a matrix of order $p_i \times n_j$, $i, j = 1, 2$, $p_1 + p_2 = p$, $n_1 + n_2 = n$, $n_1 \geq p_1$. Define the matrices $T_1\,(n_1 \times (n_1 - p_1))$ of rank $n_1 - p_1$ such that $X_{11} T_1 = 0$, $T_1' T_1 = I_{n_1 - p_1}$ and $T_2 = (I_{n_1} + X_{12}'(X_{11} X_{11}')^{-1} X_{12})^{\frac{1}{2}}$. Then the random matrices $Y = (\,X_{21} T_1 \quad X_{22} T_2\,)$ and X_1 are independent, $Y \sim \mathcal{U}_{p_2,n-p_1}$ and $X_1 \sim \mathcal{U}_{p_1,n}$.*

THEOREM 8.2.5. *Let the random matrix $Y\,(p \times n)$, $n \geq p$ have a density with respect to the Lebesgue measure. Further let the distribution of Y be invariant under the transformation $Y \to YN$ for any orthogonal matrix $N\,(n \times n)$. Then YY' and $X = (YY')^{-\frac{1}{2}} Y$ are independently distributed, and $X \sim \mathcal{U}_{p,n}$.*

In the above theorem it is assumed that the distribution of YN does not depend on N. Without this condition X and YY' are not independent. Moreover, the distribution of Y may not be uniform, as shown by Chikuse (1990b)(also see Section 8.7).

Consider a random sample $X_i\,(p \times n)$, $i = 1, \ldots, N$, from the uniform distribution on the Stiefel manifold. Define the random matrices $Q = \frac{1}{N} \sum_{i=1}^{N} X_i' X_i$, and $D = \mathrm{diag}(\omega_1, \ldots, \omega_p)$ where ω_i, $i = 1, \ldots, p$ are the eigenvalues of Q. Then, Mardia and Khatri (1977) derived the distributions of Q as well as of D and also gave their asymptotic distributions.

8.3.   VON MISES-FISHER DISTRIBUTION

The von Mises-Fisher (or Langevin) matrix variate distribution defined in this section is useful in orientation statistics, Downs (1972), Khatri and Mardia (1977).

DEFINITION 8.3.1. *The random matrix $X\,(p \times n)$, $p \leq n$, is said to have von Mises-Fisher distribution with parameter $F\,(p \times n)$, if its probability element is given by*

$$a(F)\,\mathrm{etr}(FX')[dX], \; X \in O(p,n) \tag{8.3.1}$$

where $[dX]$ is the unit invariant measure on $O(p,n)$ and $a(F)$ is the normalizing constant.

If the probability element of a random matrix $X\,(p \times n)$ is given by (8.3.1), we will write $X \sim M_{p,n}(F)$. This distribution is a special case, for $C = I_p$, of Downs (1972) who studied the distribution of X when it lies on the Stiefel C-manifold : $S(C) = \{X\,(p \times n) : XX' = C > 0\}$. An alternate representation of (8.3.1) in terms of generalized Eulerian angles can be given by using (8.2.4) for $[dX]$. For $F = 0$, this distribution (8.3.1) reduces to uniform distribution on the Stiefel manifold. If the rank of F $(p \times n)$ is $r \leq p$, then the singular value decomposition of F can be written as

$$F = \Delta' D_\phi \Theta$$

where $\Delta \in O(r, p)$, $\Theta \in O(r, n)$, $D_\phi = \mathrm{diag}(\phi_1, \ldots, \phi_r)$, $\phi_i > 0$, $i = 1, \ldots, r$, and $\phi_1^2 > \phi_2^2 > \cdots > \phi_r^2 > 0$, are the nonzero eigenvalues of FF'. For the uniqueness of this decomposition we assume $\phi_1 > \phi_2 > \cdots > \phi_r > 0$, and the elements in the first column of Θ are positive. The matrices Δ and Θ indicate *orientations* and $\phi_1, \phi_2, \ldots, \phi_r$ are *concentration parameters* in the r directions determined by Δ and Θ. The distribution has *model orientation* $M = \Delta'\Theta$. It is rotationally symmetric around M (Chikuse, 1991b).

The normalizing constant $a(F)$ in (8.3.1) can be evaluated by using Theorem 1.6.4,

$$\{a(F)\}^{-1} = \int_{X \in O(p,n)} \mathrm{etr}(FX')\,[dX]$$

$$= {}_0F_1\left(\frac{1}{2}n; \frac{1}{4}FF'\right)$$

$$= {}_0F_1\left(\frac{1}{2}n; \frac{1}{4}F'F\right)$$

$$= {}_0F_1\left(\frac{1}{2}n; \frac{1}{4}D_\phi^2\right)$$

where $D_\phi^2 = \mathrm{diag}(\phi_1^2, \ldots, \phi_r^2)$. The m.g.f. of $X \sim M_{p,n}(F)$ is given by

$$M_X(Z) = \int_{X \in O(p,n)} \mathrm{etr}(FX')\,\mathrm{etr}(ZX')\,[dX]$$

$$= \frac{{}_0F_1(\frac{1}{2}n; \frac{1}{4}(F+Z)(F+Z)')}{{}_0F_1(\frac{1}{2}n; \frac{1}{4}FF')}.$$

The last step is obtained by using Theorem 1.6.4.

Now partition X $(p \times n) = \begin{pmatrix} X_1 \\ X_2 \end{pmatrix}$, X_i $(p_i \times n)$, $i = 1, 2$, and $F = \begin{pmatrix} F_1 \\ F_2 \end{pmatrix}$ similarly. The marginal distribution of X_1, when $X \sim M_{p,n}(F)$ can be obtained by using the decomposition of the unit invariant measure $[dX]$, as given in Chapter 1. For given X_1 $(p_1 \times n)$, we can find X_3 $((n - p_1) \times n) = G(X_1)$ and Y $(p_2 \times (n - p_1))$ such that X_0' $(n \times n) = (X_1' \ X_3')$ is orthogonal and $X_2 = YX_3$. The invariant measure $[dX]$, $X \in O(p, n)$, can be decomposed, Chikuse (1990a), as

$$[dX] = [dX_1][dY], \ X_1 \in O(p_1, n), \ Y \in O(p_2, n - p_1).$$

Further, when $X \sim M_{p,n}(F)$, using this factorization, the joint probability element of X_1 and Y is given by (Khatri and Mardia, 1977; Chikuse, 1990a)

$$\left\{ {}_0F_1\left(\frac{1}{2}n; \frac{1}{4}F'F\right) \right\}^{-1} \exp\{\mathrm{tr}(F_1'X_1) + \mathrm{tr}(X_3F_2'Y)\}\,[dX_1]\,[dY],$$

$$X_1 \in O(p_1, n), \ Y \in O(p_2, n - p_1). \tag{8.3.2}$$

The marginal probability element of X_1, after integrating with respect to Y, is

$$\left\{ {}_0F_1\left(\frac{1}{2}n; \frac{1}{4}F'F\right)\right\}^{-1} \exp\{\operatorname{tr}(F_1'X_1)\}$$

$$ {}_0F_1\left(\frac{1}{2}(n-p_1); \frac{1}{4}(I_n - X_1'X_1)F_2'F_2\right) [dX_1], \; X_1 \in O(p_1, n).$$

Note that when $F_2 = 0$, $X_1 \sim M_{p_1,n}(F)$. The conditional probability element of X_1 given X_2 is

$$\left\{ {}_0F_1\left(\frac{1}{2}(n-p_2); \frac{1}{4}(I_n - X_2'X_2)F_1'F_1\right)\right\}^{-1} \operatorname{etr}(F_1'X_1) [dX] [dX_2]^{-1},$$

where $[dX] [dX_2]^{-1}$ is the unit Haar measure of X_1 given X_2 subject to $X_1X_1' = I_{p_1}$ and $X_1X_2' = 0$. Hence the conditional distribution of X_1 given X_2 is essentially a von Mises-Fisher distribution.

If we partition $X\,(p \times n) = (\; X_1 \quad X_2\;)$, $X_i\,(p \times n_i)$, $i = 1, 2$, $n_2 \geq p$, and $F = (\; F_1 \quad F_2\;)$ similarly, then (Chikuse, 1990a) using Theorem 8.8.2, the joint probability element of X_1 and $W = (I_p - X_1X_1')^{-\frac{1}{2}}X_2$ is

$$\frac{\Gamma_p(\frac{1}{2}n)}{\pi^{\frac{1}{2}n_1 p}\Gamma_p(\frac{1}{2}n_1)} \left\{ {}_0F_1\left(\frac{1}{2}n; \frac{1}{4}FF'\right)\right\}^{-1} \exp\{\operatorname{tr}(F_1'X_1) + \operatorname{tr}(F_2'(I_p - X_1X_1')^{\frac{1}{2}}W)\}$$

$$\det(I_p - X_1X_1')^{\frac{1}{2}(n_2-p-1)} [dW] \, dX_1.$$

From above it is apparent that $W \sim M_{p,n_2}((I_p - X_1X_1')^{\frac{1}{2}}F_2)$ and the p.d.f. of X_1 is

$$\frac{\Gamma_p(\frac{1}{2}n)}{\pi^{\frac{1}{2}n_1 p}\Gamma_p(\frac{1}{2}n_1)} \left\{ {}_0F_1\left(\frac{1}{2}n; \frac{1}{4}F'F\right)\right\}^{-1} \exp\{\operatorname{tr}(F_1'X_1)\}$$

$$\det(I_p - X_1X_1')^{\frac{1}{2}(n_2-p-1)} \, {}_0F_1\left(\frac{1}{2}n_2; \frac{1}{4}F_2'(I_p - X_1X_1')F_2\right).$$

Thus, if $F_2 = 0$, W and X_1 are independently distributed. The matrix W is distributed uniformly on $O(p, n_2)$ and the p.d.f. of X_1 is

$$\frac{\Gamma_p(\frac{1}{2}n)}{\pi^{\frac{1}{2}n_1 p}\Gamma_p(\frac{1}{2}n_1)} \left\{ {}_0F_1\left(\frac{1}{2}n; \frac{1}{4}F'F\right)\right\}^{-1} \exp\{\operatorname{tr}(F_1'X_1)\} \det(I_p - X_1X_1')^{\frac{1}{2}(n_2-p-1)}.$$

Using the *sequential* decomposition of invariant measure on $O(p, n)$ into those for independent measures on component Stiefel manifolds and on subspaces of component rectangular matrices, Chikuse (1990a) has given further decomposition of (8.3.2).

Khatri and Mardia (1977) have given first two moments of X, and approximations to (8.3.1). For further insight into this distribution, and its special cases the reader is referred to Downs (1972), Khatri and Mardia (1977), Jupp and Mardia (1979), Chikuse (1990a, 1990b, 1991b, 1993a), and Bingham, Chang and Richards (1992).

8.4. BINGHAM MATRIX DISTRIBUTION

The Bingham matrix distribution defined in this section is the obvious analogue on the Stiefel manifold of Bingham's antipodally symmetric distribution on sphere (Bingham, 1974).

DEFINITION 8.4.1. *The random matrix X $(p \times n)$, $p \leq n$, is said to have Bingham matrix distribution with parameter $A\,(n \times n) = A'$, if its probability element is given by*

$$b(A)\,\text{etr}(XAX')\,[dX],\ X \in O(p,n), \tag{8.4.1}$$

where $[dX]$ is the unit invariant measure on $O(p,n)$ and $b(A)$ is the normalizing constant.

For identifiability of A we take $\text{tr}(A) = 0$. If the probability element of a random matrix X $(p \times n)$ is given by (8.4.1), we will write $X \sim B_{p,n}(A)$. A generalization of (8.4.1) may be given as

$$b_1(A,B)\,\text{etr}(BXAX')\,[dX],\ X \in O(p,n), \tag{8.4.2}$$

where B $(p \times p)$ is a symmetric matrix and $b_1(A,B)$ is the normalizing constant.

We shall denote this as $X \sim B_{p,n}(A,B)$. For $B = I_p$, (8.4.2) reduces to (8.4.1) and for $B = 0$, the matrix variate Bingham distribution reduces to the uniform distribution on the Stiefel manifold. An alternate representation of (8.4.1) in terms of generalized Eulerian angles can be given by using (8.2.4) for $[dX]$.

The Bingham matrix distribution (8.4.1) is a special case of the generalized von Mises-Fisher matrix variate distribution introduced by Khatri and Mardia (1977)(see Section 8.5).

The normalizing constant $b(A)$ in (8.4.1) can be evaluated by using Theorem 1.6.4,

$$\{b(A)\}^{-1} = \int_{X \in O(p,n)} \text{etr}(XAX')\,[dX]$$

$$= {}_1F_1^{(n)}\left(\frac{1}{2}n, \frac{1}{2}p; A\right).$$

Let us partition X $(p \times n) = \begin{pmatrix} X_1 \\ X_2 \end{pmatrix}$, $X_i\,(p_i \times n)$, $i = 1, 2$. For given X_1 $(p_1 \times n)$, we can find $X_3\,((n - p_1) \times n) = G(X_1)$ and Y $(p_2 \times (n - p_1))$ such that $X'_0\,(n \times n) = (\,X'_1\ \ X'_3\,)$ is orthogonal and $X_2 = YX_3$. Then using the factorization of invariant measure over the Stiefel manifolds, given in Section 1.3, the joint probability element of X_1 and Y is

$$\left\{ {}_1F_1^{(n)}\left(\frac{1}{2}n, \frac{1}{2}p; A\right) \right\}^{-1} \exp\{\text{tr}(X_1AX'_1) + \text{tr}(YX_3AX'_3Y')\}\,[dX_1]\,[dY],$$

$$X_1 \in O(p_1, n),\ Y \in O(p_2, n - p_1). \tag{8.4.3}$$

Now integrating (8.4.3) with respect to Y, using Theorem 1.6.4, we get the probability element of X_1 as

$$
\left\{ {}_1F_1^{(n)}\left(\tfrac{1}{2}n, \tfrac{1}{2}p; A\right)\right\}^{-1} \operatorname{etr}(X_1 A X_1') \, {}_1F_1\left(\tfrac{1}{2}(n - p_1); \tfrac{1}{2}p_2; X_3 A X_3'\right)[dX_1]
$$

$$
= \left\{ {}_1F_1^{(n)}\left(\tfrac{1}{2}n, \tfrac{1}{2}p; A\right)\right\}^{-1} \operatorname{etr}(X_1 A X_1') \, {}_1F_1\left(\tfrac{1}{2}(n - p_1); \tfrac{1}{2}p_2; (I_{p_1} - X_1'X_1)A\right)[dX_1],
$$

$$
X_1 \in O(p_1, n).
$$

since $X_0'X_0 = I_n$.

From (8.4.3), it is also seen that the conditional distribution of Y given X_1 is Bingham matrix distribution, $B_{p_2, n-p_1}(X_3 A X_3')$, $X_3 = G(X_1)$.

Using the *sequential* decomposition of invariant measure on $O(p, n)$ into those for independent measures on component Stiefel manifolds and on subspaces of component rectangular matrices, Chikuse (1990a) has given further decomposition of (8.4.3).

If we partition $X\,(p \times n) = (X_1 \quad X_2)$, $X_i(p \times n_i)$, $i = 1, 2$, $n_2 \geq p$, and $A = \begin{pmatrix} A_{11} & A_{12} \\ A_{21} & A_{22} \end{pmatrix}$, $A_{ij}\,(n_i \times n_j)$, $n_1 + n_2 = n$, then (Chikuse, 1990a) using Theorem 8.2.2, the joint probability element of X_1 and $W = (I_p - X_1 X_1')^{-\frac{1}{2}} X_2$ is

$$
\frac{\Gamma_p(\tfrac{1}{2}n)}{\pi^{\frac{1}{2}n_1 p}\Gamma_p(\tfrac{1}{2}n_1)} \left\{ {}_1F_1\left(\tfrac{1}{2}n; \tfrac{1}{2}p; A\right)\right\}^{-1} \operatorname{etr}\{X_1 A_{11} X_1' + (I_p - X_1 X_1')W A_{22} W'
$$

$$
+ 2(I_p - X_1 X_1')^{\frac{1}{2}} W A_{12}' X_1'\} \det(I_p - X_1 X_1')^{\frac{1}{2}(n_2 - p - 1)}[dW]\,dX_1.
$$

Thus the conditional probability element of W given X_1 is generalized Bingham-von Mises-Fisher matrix variate distribution (Khatri and Mardia, 1977) discussed in the next section.

The Bingham matrix distribution on the Stiefel manifold has been generalized by Prentice (1982). He has also obtained the large sample maximum likelihood estimators and uniformity test.

8.5.   GENERALIZED BINGHAM-VON MISES MATRIX DISTRIBUTION

Let $X \sim N_{p,n}(M, \Sigma \otimes \Psi)$, $p \leq n$. The conditional distribution of X on the Stiefel manifold $O(p, n)$ is known as generalized Bingham-von Mises-Fisher distribution. Khatri and Mardia (1977), using the density of $\Sigma^{-\frac{1}{2}} X X' \Sigma^{-\frac{1}{2}} = S$ given in (7.6.6), derived the probability element of X given $XX' = I_p$, as

$$
\{g(\Sigma, \Psi, \Delta)\}^{-1} \operatorname{etr}(\Sigma^{-1} X V X' + \Sigma^{-\frac{1}{2}} \Delta Q' X')[dX], \quad X \in O(p, n), \qquad (8.5.1)
$$

where $[dX]$ is the unit invariant measure on $O(p, n)$, $\Psi = QDQ'$, $Q'Q = I_n$, $D = \operatorname{diag}(\alpha_1, \ldots, \alpha_n)$, $V = \alpha I_n - \tfrac{1}{2}\Psi^{-1}$, $\Delta = \Sigma^{-\frac{1}{2}} MQD$, α is any arbitrary number, and $g(\cdot)$ is the normalizing constant.

de Waal (1979), using the density of XX' given in Theorem 7.6.2, derived the probability element of X given $XX' = I_p$, as

$$\{K(\Omega, \Phi, M)\}^{-1} \operatorname{etr}\{\Omega(X - M)\Phi(X - M)'\} [dX], \ X \in O(p, n), \qquad (8.5.2)$$

where

$$K(\Omega, \Phi, M) = \operatorname{etr}(\Omega M \Phi M') \operatorname{etr}(q\Omega) \sum_{k=0}^{\infty} \sum_{\kappa} \left\{ \left(\tfrac{1}{2}n\right)_\kappa k! \right\}^{-1}$$

$$P_\kappa\left(\iota \Omega^{\frac{1}{2}} M \Phi^{\frac{1}{2}} (I_n - q\Phi^{-1})^{-\frac{1}{2}}, \Phi - qI_n, -\Omega \right),$$

with $\Omega^{-1} = -2\Sigma$, $\Phi^{-1} = \Psi$ and $P_\kappa(\cdot)$ is the Hayakawa polynomial.
Next we give some special cases of (8.5.2).

(i) $M = 0$ (Generalized Bingham Distribution)
  The probability element of X is

$$\{K(\Omega, \Phi, 0)\}^{-1} \operatorname{etr}(\Omega X \Phi X') [dX], \ X \in O(p, n)$$

where

$$K(\Omega, \Phi, 0) = \operatorname{etr}(q\Omega) {}_0F_0(\Phi - qI_n, \Omega).$$

(ii) $M = 0$, $\Omega = I_p$ (Bingham Distribution)
  The probability element of X is

$$\{K(I_p, \Phi, 0)\}^{-1} \operatorname{etr}(X \Phi X') [dX], \ X \in O(p, n)$$

where

$$K(I_p, \Phi, 0) = \exp(pq) {}_1F_1\left(\tfrac{1}{2}p; \tfrac{1}{2}n; \Phi - qI_n\right).$$

(iii) $\Phi = I_n$ (von Mises-Fisher (or Langevin) Distribution)
  The probability element of X is

$$\{K(\Omega, I_n, M)\}^{-1} \operatorname{etr}(\Omega + \Omega M M') \operatorname{etr}(-2\Omega M X') [dX], \ X \in O(p, n)$$

where by evaluating the above density we obtain

$$K(\Omega, I_n, M) = \operatorname{etr}(\Omega + \Omega M M') {}_0F_1\left(\tfrac{1}{2}p; \tfrac{1}{2}n; \Omega^2 M M'\right).$$

(iv) $\Omega = I_p$ (Bingham-von Mises-Fisher Distribution)
  The probability element of X is

$$\{K(I_p, \Phi, M)\}^{-1} \operatorname{etr}\{(X - M)\Phi(X - M)'\} [dX], \ X \in O(p, n)$$

where

$$K(I_p, \Phi, M) = \operatorname{etr}(M \Phi M') \exp(pq) \sum_{k=0}^{\infty} \sum_{\kappa} \left\{ \left(\tfrac{1}{2}n\right)_\kappa k! \right\}^{-1}$$

$$P_\kappa\left(\iota M \Phi^{\frac{1}{2}} (I_n - q\Phi^{-1})^{-\frac{1}{2}}, \Phi - qI_n, -I_p \right).$$

(v) $\Omega = I_p$, $\Phi = I_n$ (von Mises-Fisher (or Langevin) Distribution)
The probability element of X is

$$\{K(I_p, I_n, M)\}^{-1} \operatorname{etr}((X - M)(X - M)')\,[dX],\ X \in O(p, n) \tag{8.5.3}$$

where, for $q = 0$,

$$K(I_p, I_n, M) = \operatorname{etr}(MM') \sum_{k=0}^{\infty} \sum_{\kappa} \left\{ \left(\tfrac{1}{2}n\right)_{\kappa} k! \right\}^{-1} L_{\kappa}^{\frac{1}{2}(n-p-1)}(-MM').$$

The form (8.5.3) for the von Mises-Fisher distribution is written in a different manner than (8.3.1).

8.6. MANIFOLD NORMAL DISTRIBUTION

Let $X \sim N_{p,n}(M, \Sigma \otimes \Psi)$, $p \le n$. The conditional distribution of X on the Stiefel C-manifold $S(C) = \{X\,(p \times n) : XX' = C > 0\}$ is known as the manifold normal distribution.

de Waal (1983), using the density of $XX' = S$ given in Theorem 7.6.2, derived the probability element of X given $XX' = S$ as

$$\{K(\Sigma, \Psi, M)\}^{-1} \operatorname{etr}\left(-\tfrac{1}{2}\Sigma^{-1} X \Psi^{-1} X' + \Sigma^{-1} X \Psi^{-1} M'\right)[dX]_c, \tag{8.6.1}$$

where

$$K(\Sigma, \Psi, M) = \sum_{k=0}^{\infty} \sum_{\kappa} \left\{ \left(\tfrac{1}{2}n\right)_{\kappa} k! \right\}^{-1} P_{\kappa}\left(\tfrac{1}{\sqrt{2}}\Sigma^{-1} M \Psi^{-\frac{1}{2}}, \Psi^{-\frac{1}{2}}, \tfrac{1}{2}\Sigma^{-\frac{1}{2}} S \Sigma^{-\frac{1}{2}}\right),$$

and $[dX]_c$ is the content element on $S(C)$. He has also given an approximation to (8.6.1). The m.g.f. of X is

$$M_X(Z) = \{K(\Sigma, \Psi, M)\}^{-1} \int_{S(C)} \operatorname{etr}(XZ') \operatorname{etr}\left(-\tfrac{1}{2}\Sigma^{-1} X \Psi^{-1} X'\right.$$
$$\left. + \Sigma^{-1} X \Psi^{-1} M'\right)[dX]_c$$
$$= \{K(\Sigma, \Psi, M)\}^{-1} \int_{S(C)} \operatorname{etr}\left(-\tfrac{1}{2}\Sigma^{-1} X \Psi^{-1} X'\right.$$
$$\left. + \Sigma^{-1} X \Psi^{-1}(M + \Sigma Z \Psi)'\right)[dX]_c$$
$$= \frac{K(\Sigma, \Psi, M + \Sigma Z \Psi))}{K(\Sigma, \Psi, M)}.$$

It may be noted that for $C = I_p$, $[dX]_c = [dX]$ and (8.6.1) reduces to (8.5.1). Hence manifold normal distribution can be regarded as a generalization of Bingham-von Mises-Fisher distribution discussed in Section 8.5.

8.7.   MATRIX ANGULAR CENTRAL GAUSSIAN DISTRIBUTION

In Sections 3 through 6, we have defined matrix von Mises-Fisher distribution and Bingham matrix distribution and their extensions. The Bingham matrix distribution is an antipodally symmetric distribution.

Chikuse (1990b) using polar decomposition of a random matrix, has proposed matrix angular central Gaussian distribution as an alternative to Bingham matrix distribution for modeling antipodally symmetric orientational data on Stiefel manifold (see also Tyler, 1987).

For any random matrix X $(p \times n)$ of rank $p \leq n$, the unique polar decomposition of X is defined (Chikuse, 1990b) as

$$X = S_X^{\frac{1}{2}} H_X \tag{8.7.1}$$

with $S_X = XX'$, and $H_X = (XX')^{-\frac{1}{2}}X$. So that $H_X \in O(p, n)$ and $S_X^{\frac{1}{2}}$ is the unique positive definite square root of S_X.

Let $f_X(X)$ be the density of X $(p \times n)$. Then the joint probability element of S_X and H_X is (Chikuse, 1990b),

$$\frac{\pi^{\frac{1}{2}np}}{\Gamma_p(\frac{1}{2}n)} f_X\left(S_X^{\frac{1}{2}} H_X\right) \det(S_X)^{\frac{1}{2}(n-p-1)} [dH_X] dS_X. \tag{8.7.2}$$

Integrating (8.7.2) with respect to S_X, the probability element of H_X is obtained as

$$\frac{\pi^{\frac{1}{2}np}}{\Gamma_p(\frac{1}{2}n)} [dH_X] \int_{S_X > 0} f_X\left(S_X^{\frac{1}{2}} H_X\right) \det(S_X)^{\frac{1}{2}(n-p-1)} dS_X. \tag{8.7.3}$$

Integrating (8.7.2) with respect to H_X, the denisity of S_X is given by

$$\frac{\pi^{\frac{1}{2}np}}{\Gamma_p(\frac{1}{2}n)} \det(S_X)^{\frac{1}{2}(n-p-1)} \int_{H_X \in O(p,n)} f_X\left(S_X^{\frac{1}{2}} H_X\right) [dH_X]. \tag{8.7.4}$$

From (8.7.2), it may be noted that if the density of XN, $N \in O(n)$ does not depend on N, then S_X and H_X are independent, the distribution of H_X is uniform, and the density of S_X is

$$\frac{\pi^{\frac{1}{2}np}}{\Gamma_p(\frac{1}{2}n)} f_X\left(S_X^{\frac{1}{2}}\right) \det(S_X)^{\frac{1}{2}(n-p-1)}, \; S_X > 0.$$

If $X \sim N_{p,n}(M, I_p \otimes \Psi)$, then the distribution of H_X is called matrix angular central Gaussian distribution (ACG) with parameters p, n and Ψ. This is denoted by $H_X \sim ACG_{p,n}(\Psi)$. From (8.7.3) the probability element of H_X is

$$\det(\Psi)^{-\frac{1}{2}p} \det(H_X \Psi^{-1} H_X')^{-\frac{1}{2}n} [dH_X], \; H_X \in O(p, n). \tag{8.7.5}$$

The density of $S_X = XX'$, from (7.2.1), is

$$\frac{\det(\Psi)^{-\frac{1}{2}p}}{2^{\frac{1}{2}np}\Gamma_p(\frac{1}{2}n)} \det(S_X)^{\frac{1}{2}(n-p-1)} \operatorname{etr}\left(-\frac{1}{2}S_X\right) {}_0F_0^{(n)}\left(I_n - \Psi^{-1}, \frac{1}{2}S_X\right), \ S_X > 0.$$

Note that the ACG distribution (8.7.5) is invariant under the transformation $H_X \to QH_X$, $Q \in O(p)$. In case $\Psi = I_n$, this distribution reduces to the uniform distribution over $O(p, n)$.

If the density of random matrix X $(p \times n)$ is of the form $g(XX')$, then the density of XN, $N \in O(n)$ obviously does not depend on N and therefore

(i) H_X and S_X are independent,
(ii) H_X is distributed uniformly over $O(p, n)$, and
(iii) the density of S_X has the form

$$\frac{\pi^{\frac{1}{2}np}}{\Gamma_p(\frac{1}{2}n)} \det(S_X)^{\frac{1}{2}(n-p-1)} g(S_X), \ S_X > 0.$$

Chikuse (1990b) has proved the necessity of conditions (i), (ii) and (iii) for the density of X to be of the form $g(XX')$. For $g(S_X) = \operatorname{etr}(-\frac{1}{2}S_X)$, the conditions (i), (ii) and (iii) provide characterization of matrix variate standard normal distribution. For $g(S_X) = \det(I_p + S_X)^{-\frac{1}{2}(n+m+p-1)}$, the conditions (i), (ii) and (iii) give a characterization of matrix variate t-distribution, $T_{p,n}(m, 0, I_p, I_n)$. For other relevant results the reader is referred to Chikuse (1990a, 1990b).

8.8. BIMATRIX WISHART DISTRIBUTION

The following distribution has been given by Roux and Raath (1973).

DEFINITION 8.8.1. *A random bimatrix* $X = (X_1, X_2)$, *where* X_i $(p \times p)$ *is symmetric,* $i = 1, 2$, *is said to have bimatrix Wishart distribution with parameters* $n_1 (\geq p)$, $n_2 (\geq p)$, $\Psi_1 (> 0)$, $\Psi_2 (> 0)$ *and* $\Theta(= \Theta')$ *if its p.d.f. is given by*

$$\prod_{i=1}^{2}\left[\left\{2^{\frac{1}{2}n_ip}\Gamma_p\left(\frac{1}{2}n_i\right)\det(\Psi_i)^{\frac{1}{2}n_i}\right\}^{-1}\operatorname{etr}\left(-\frac{1}{2}\Psi_i^{-1}X_i\right)\det(X_i)^{\frac{1}{2}(n_i-p-1)}\right]$$

$$\sum_{k=0}^{\infty}\sum_{\kappa}\frac{C_\kappa(\Theta)}{k!\,[C_\kappa(I_p)]^2}\prod_{i=1}^{2}\left\{\left(\frac{1}{2}n_i\right)_\kappa\right\}^{-1}L_\kappa^{\frac{1}{2}(n_i-p-1)}\left(\frac{1}{2}\Psi_i^{-1}X_i\right), \ X_1, X_2 > 0.$$

Here L_κ^γ is Laguerre polynomial of matrix argument defined in Chapter 1.

For $\Theta = 0$, it is easily seen that the matrices X_1 and X_2 are independent, and $X_i \sim W_p(n_i, \Psi_i)$, $i = 1, 2$.

THEOREM 8.8.1. *The joint m.g.f. of* $X = (X_1, X_2)$ *is given by*

$$M_{X_1,X_2}(Z_1, Z_2) = \prod_{i=1}^{2}\det(I_p - 2\Psi_iZ_i)^{-\frac{1}{2}n_i}\sum_{k=0}^{\infty}\sum_{\kappa}\frac{C_\kappa(\Theta)}{k!\,[C_\kappa(I_p)]^2}$$

$$\prod_{i=1}^{2}C_\kappa(2\Psi_iZ_i(I_p - 2\Psi_iZ_i)^{-1}).$$

Proof: The joint m.g.f. of X_1 and X_2 is

$$M_{X_1,X_2}(Z_1, Z_2) = \left[\prod_{i=1}^{2} \left\{ 2^{\frac{1}{2}n_i p} \Gamma_p\left(\frac{1}{2}n_i\right) \det(\Psi_i)^{\frac{1}{2}n_i} \right\}^{-1} \right] \sum_{k=0}^{\infty} \sum_{\kappa} \frac{C_\kappa(\Theta)}{k![C_\kappa(I_p)]^2}$$

$$\prod_{i=1}^{2} \left\{ \left(\frac{1}{2}n_i\right)_\kappa \right\}^{-1} \int_{X_i>0} \text{etr}\left(X_i Z_i - \frac{1}{2}\Psi_i^{-1} X_i \right) \det(X_i)^{\frac{1}{2}(n_i-p-1)}$$

$$L_\kappa^{\frac{1}{2}(n_i-p-1)}\left(\frac{1}{2}\Psi_i^{-1} X_i\right) dX_i.$$

Now transforming $Y_i = \Psi_i^{-\frac{1}{2}} X_i \Psi_i^{-\frac{1}{2}}$ with Jacobian $J(X_i \to Y_i) = \det(\Psi_i)^{\frac{1}{2}(p+1)}$, and using Theorem 1.7.1, the integral on the right hand side becomes

$$\int_{X_i>0} \text{etr}\left(X_i Z_i - \frac{1}{2}\Psi_i^{-1} X_i \right) \det(X_i)^{\frac{1}{2}(n_i-p-1)} L_\kappa^{\frac{1}{2}(n_i-p-1)}\left(\frac{1}{2}\Psi_i^{-1} X_i\right) dX_i$$

$$= 2^{\frac{1}{2}n_i p} \det(\Psi_i)^{\frac{1}{2}n_i} \left(\frac{1}{2}n_i\right)_\kappa \Gamma_p\left(\frac{1}{2}n_i\right) \det(I_p - 2\Psi_i Z_i)^{-\frac{1}{2}n_i}$$

$$C_\kappa(-2\Psi_i Z_i(I_p - 2\Psi_i Z_i)^{-1}), \ I_p - 2\Psi_i Z_i > 0.$$

Finally substituting from (8.8.2) in (8.8.1) we get the desired result. ∎

From the Definition 8.8.1, it can easily be shown that the marginal p.d.f. of X_i is $W_p(n_i, \Psi_i)$, $i = 1, 2$. The conditional density of X_1 given X_2 can easily be seen to be

$$\left\{ 2^{\frac{1}{2}n_1 p} \Gamma_p\left(\frac{1}{2}n_1\right) \det(\Psi_1)^{\frac{1}{2}n_1} \right\}^{-1} \text{etr}\left(-\frac{1}{2}\Psi_1^{-1} X_1 \right) \det(X_1)^{\frac{1}{2}(n_1-p-1)}$$

$$\sum_{k=0}^{\infty} \sum_{\kappa} \frac{C_\kappa(\Theta)}{k![C_\kappa(I_p)]^2} \prod_{i=1}^{2} \left[\left\{ \left(\frac{1}{2}n_i\right)_\kappa \right\}^{-1} L_\kappa^{\frac{1}{2}(n_i-p-1)}\left(\frac{1}{2}\Psi_i^{-1} X_i\right) \right], \ X_1, X_2 > 0.$$

8.9.   BETA-WISHART DISTRIBUTION

In this section we give two distributions of bimatrix $X = (X_1, X_2)$, with specified marginals and conditionals. First we define beta-Wishart type I distribution.

DEFINITION 8.9.1. *A random bimatrix* $X = (X_1, X_2)$, *where* $X_i\ (p \times p)$ *is symmetric,* $i = 1, 2$, *is said to have beta-Wishart type I distribution with parameters* $n_1 (\geq p)$, $n_2 (\geq p)$, $n_3 (\geq p)$, *and* $\Sigma > 0$ *if its p.d.f. is given by*

$$\left\{ 2^{\frac{1}{2}n_1 p} \Gamma_p\left(\frac{1}{2}n_1\right) \det \Sigma)^{\frac{1}{2}n_1} \beta_p\left(\frac{1}{2}n_2, \frac{1}{2}n_3\right) \right\}^{-1} \text{etr}\left(-\frac{1}{2}\Sigma^{-1} X_1 \right) \det(X_1)^{\frac{1}{2}(n_1-n_2-n_3)}$$

$$\det(X_2)^{\frac{1}{2}(n_2-p-1)} \det(X_1 - X_2)^{\frac{1}{2}(n_3-p-1)}, \ 0 < X_2 < X_1.$$

From Definition 8.9.1, it can be shown that

$$X_1 \sim W_p(n_1, \Sigma),$$

$$X_2 \sim CH_p^{II}\Big(\frac{1}{2}n_1, \frac{1}{2}n_3, \frac{1}{2}(n_1 - n_2 + p + 1), \frac{1}{2}\Sigma^{-1}; \text{kind } 1\Big), \text{ and}$$

$$X_2|X_1 \sim GB_p^I\Big(\frac{1}{2}n_2, \frac{1}{2}n_3; X_1, 0\Big),$$

where CH_p^{II} denotes the confluent hypergeometric function kind 2 and type II distribution defined in Section 8.11.

For $p = 1$, the above distribution reduces to the beta-Stacy distribution (Mihram and Hulquist, 1967).

Next we define beta-Wishart type II distribution.

DEFINITION 8.9.2. *A random bimatrix* $X = (X_1, X_2)$, *where* $X_i \, (p \times p)$ *is symmetric,* $i = 1, 2$, *is said to have beta-Wishart type II distribution with parameters* $n_1 \, (\geq p)$, $n_2 \, (\geq p)$ *if its p.d.f. is given by*

$$\Big\{2^{\frac{1}{2}(n_1+n_2)p}\Gamma_p\Big(\frac{1}{2}n_1\Big)\Gamma_p\Big(\frac{1}{2}n_2\Big)\Big\}^{-1} \operatorname{etr}\Big\{-\frac{1}{2}X_1(I_p + X_2)\Big\}$$

$$\det(X_1)^{\frac{1}{2}(n_1+n_2-p-1)}\det(X_2)^{\frac{1}{2}(n_2-p-1)}, X_1 > 0, X_2 > 0.$$

From Definition 8.9.2, it can be shown that

$$X_1 \sim W_p(n_1, I_p),$$

$$X_2 \sim B_p^{II}\Big(\frac{1}{2}n_1, \frac{1}{2}n_2\Big),$$

$$X_1|X_2 \sim W_p\Big(n_1 + n_2, (I_p + X_2)^{-1}\Big), \text{ and}$$

$$X_2|X_1 \sim W_p(n_1, X_1^{-1}).$$

It may be noted that if $X_1 \sim W_p(n_1, I_p)$ and $U \sim W_p(n_2, I_p)$ are independent, then the joint distribution of $X_2 = X_1^{-\frac{1}{2}}UX_1^{-\frac{1}{2}}$ and X_1 is beta-Wishatr type II distribution with parameters n_1 and n_2. This result is given in Chapter 5, in (5.2.9), where it is proved that $X_2 = X_1^{-\frac{1}{2}}UX_1^{-\frac{1}{2}} \sim B_p^{II}(\frac{1}{2}n_2, \frac{1}{2}n_1)$.

8.10.   CONFLUENT HYPERGEOMETRIC FUNCTION KIND 1 DISTRIBUTION

Here we define a matrix variate distribution in terms of the confluent hypergeometric function. This distribution arises in the study of ratios of certain random matrices.

DEFINITION 8.10.1. *A random symmetric positive definite matrix* $X \, (p \times p)$ *is said to have a confluent hypergeometric function kind 1 distribution if its p.d.f. is given by*

$$\frac{\Gamma_p(\alpha)\Gamma_p(\beta - n)}{\Gamma_p(n)\Gamma_p(\beta)\Gamma_p(\alpha - n)}\det(X)^{n-\frac{1}{2}(p+1)} {}_1F_1(\alpha; \beta; -X), X > 0, \tag{8.10.1}$$

where $\operatorname{Re}(\beta - n) > 0$, *and* $\operatorname{Re}(\alpha - n) > 0$. *The parameters* n, α, *and* β *are restricted to take values such that the density function is non-negative.*

We denote this distribution by $CH_p(n, \alpha, \beta, \text{kind } 1)$. By transforming $X = A^{\frac{1}{2}} Y A^{\frac{1}{2}}$, $(A > 0)$ with the Jacobian $J(X \to Y) = \det(A)^{\frac{1}{2}(p+1)}$, the density of Y is obtained as

$$\frac{\Gamma_p(\alpha)\Gamma_p(\beta - n)}{\Gamma_p(n)\Gamma_p(\beta)\Gamma_p(\alpha - n)} \det(A)^n \det(Y)^{n-\frac{1}{2}(p+1)} {}_1F_1(\alpha; \beta; -AY), \quad Y > 0. \quad (8.10.2)$$

We denote this distribution by $CH_p(n, \alpha, \beta; A, \text{kind } 1)$. When $\alpha = \beta$, the $CH_p(n, \alpha, \beta; A, \text{kind } 1)$ density simplifies to $G_p(n, A)$ defined in Chapter 3.

The c.d.f. of X, obtained from (8.10.1) is

$$P(X < \Omega) = \frac{\Gamma_p(\alpha)\Gamma_p(\beta - n)\Gamma_p[\frac{1}{2}(p+1)]}{\Gamma_p(\beta)\Gamma_p(\alpha - n)\Gamma_p[n + \frac{1}{2}(p+1)]} \det(\Omega)^n {}_2F_2\left(n, \alpha; n + \frac{1}{2}(p+1), \beta; -\Omega\right).$$

The confluent hypergeometric function kind 1 distribution arises as the distribution of ratio of beta and gamma matrices, as shown in the following theorem.

THEOREM 8.10.1. Let $W \sim G_p(n, I_p)$ and $U \sim B_p^I(a, b)$ be independent. Then $X = U^{-\frac{1}{2}} W U^{-\frac{1}{2}} \sim CH_p(n, a + n, a + b + n, \text{kind } 1)$.

Proof: The joint density of W and U is given by

$$\{\beta_p(a, b)\Gamma_p(n)\}^{-1} \det(U)^{a - \frac{1}{2}(p+1)} \det(I_p - U)^{b - \frac{1}{2}(p+1)}$$

$$\text{etr}(-W) \det(W)^{n - \frac{1}{2}(p+1)}, \quad 0 < U < I_p, W > 0.$$

Now transform $X = U^{-\frac{1}{2}} W U^{-\frac{1}{2}}$, with the Jacobian $J(W \to X) = \det(U)^{\frac{1}{2}(p+1)}$, the joint density of U and X is given by

$$\{\beta_p(a, b)\Gamma_p(n)\}^{-1} \det(X)^{n - \frac{1}{2}(p+1)} \text{etr}(-UX)$$

$$\det(U)^{a - \frac{1}{2}(p+1)} \det(I_p - U)^{b - \frac{1}{2}(p+1)}, \quad 0 < U < I_p, X > 0. \quad (8.10.3)$$

Integrating (8.10.3) with respect to U, using Corollary 1.6.3.1, we get the marginal p.d.f. of X as

$$\frac{\Gamma_p(a + b)\Gamma_p(a + n)}{\Gamma_p(a)\Gamma_p(n)\Gamma_p(a + b + n)} \det(X)^{n - \frac{1}{2}(p+1)} {}_1F_1(a + n; a + b + n; -X), \quad X > 0,$$

which completes the proof of the theorem. ∎

In the next theorem we derive the m.g.f. of $X \sim CH_p(n, \alpha, \beta, \text{kind } 1)$.

THEOREM 8.10.2. Let $X \sim CH_p(n, \alpha, \beta, \text{kind } 1)$. Then the m.g.f. of X is

$$M_X(Z) = \frac{\Gamma_p(\alpha)\Gamma_p(\beta - n)}{\Gamma_p(\beta)\Gamma_p(\alpha - n)} \det(I_p - Z)^{-n} {}_2F_1(n, \beta - \alpha; \beta; (I_p - Z)^{-1}), \quad (8.10.4)$$

where $I_p - Z > 0$.

Proof: The m.g.f. of X is

$$
\begin{aligned}
M_X(Z) &= \frac{\Gamma_p(\alpha)\Gamma_p(\beta - n)}{\Gamma_p(n)\Gamma_p(\beta)\Gamma_p(\alpha - n)} \int_{X>0} \operatorname{etr}(ZX) \det(X)^{n-\frac{1}{2}(p+1)} \, {}_1F_1(\alpha; \beta; -X) \, dX \\
&= \frac{\Gamma_p(\alpha)\Gamma_p(\beta - n)}{\Gamma_p(n)\Gamma_p(\beta)\Gamma_p(\alpha - n)} \int_{X>0} \operatorname{etr}\{-(I_p - Z)X\} \det(X)^{n-\frac{1}{2}(p+1)} \\
&\qquad\qquad\qquad {}_1F_1(\beta - \alpha; \beta; X) \, dX \\
&= \frac{\Gamma_p(\alpha)\Gamma_p(\beta - n)}{\Gamma_p(\beta)\Gamma_p(\alpha - n)} \det(I_p - Z)^{-n} \, {}_2F_1(n, \beta - \alpha; \beta; (I_p - Z)^{-1}),
\end{aligned}
$$

where the last two steps have been obtained using (1.6.9) and (1.6.4), respectively. ∎

From (8.10.4) the m.g.f. of $\operatorname{tr}(X)$ is easily obtained as

$$
M_{\operatorname{tr}(X)}(Z) = \frac{\Gamma_p(\alpha)\Gamma_p(\beta - n)}{\Gamma_p(\beta)\Gamma_p(\alpha - n)} \sum_{k=0}^{\infty} \sum_{\kappa} (1 - z)^{-(np+k)} \frac{(n)_\kappa (\beta - \alpha)_\kappa}{(\beta)_\kappa \, k!} C_\kappa(I_p), \qquad (8.10.5)
$$

for $|1 - z| > 0$. Next expanding $(1 - z)^{-(np+k)} = \sum_{s=0}^{\infty}(np + k)_s \frac{z^s}{s!}$, $|z| < 1$, and substituting in (8.10.5) and equating the coefficients of $\frac{z^s}{s!}$, we obtain

$$
E[(\operatorname{tr} X)^s] = \frac{\Gamma_p(\alpha)\Gamma_p(\beta - n)}{\Gamma_p(\beta)\Gamma_p(\alpha - n)} \sum_{k=0}^{\infty}(np + k)_s \sum_{\kappa} \frac{(n)_\kappa (\beta - \alpha)_\kappa}{(\beta)_\kappa \, k!} C_\kappa(I_p).
$$

Now we give certain properties of the confluent hypergeometric function kind 1 distribution.

THEOREM 8.10.3. *Let* $X \sim CH_p(n, \alpha, \beta, \text{kind } 1)$ *and* A *be any* $p \times p$ *constant nonsingular matrix. Then* $AXA' \sim CH_p(n, \alpha, \beta; (AA')^{-1}, \text{kind } 1)$.

THEOREM 8.10.4. *Let* $X \sim CH_p(n, \alpha, \beta, \text{kind } 1)$, *and* H $(p \times p)$ *be an orthogonal matrix whose elements are either constants or random variables distributed independently of* X. *Then, the distribution of* X *is invariant under the transformation* $X \to HXH'$, *and is independent of* H *in the latter case.*

Proof: The proof is similar to the proof of Theorem 3.3.2. ∎

THEOREM 8.10.5. *Let* $X \sim CH_p(n, \alpha, \beta, \text{kind } 1)$, *and partition* X *as*

$$
X = \begin{pmatrix} X_{11} & X_{12} \\ X_{21} & X_{22} \end{pmatrix} \begin{matrix} q \\ p - q \end{matrix}.
$$
$$
\qquad q \quad\ \ p - q
$$

Then X_{11} *and* $X_{22\cdot1} = X_{22} - X_{21}X_{11}^{-1}X_{12}$ *are independent and* $X_{11} \sim CH_q(n, \alpha - \frac{1}{2}(p - q), \beta - \frac{1}{2}(p - q), \text{kind } 1)$, *and* $X_{22\cdot1} \sim CH_{p-q}(n - \frac{1}{2}q, \alpha - \frac{1}{2}q, \beta - \frac{1}{2}q, \text{kind } 1)$.

Proof: The theorem can be proved by using the integral representation of $_1F_1$ in (8.10.1) and integrating with respect to X_{12}. ∎

THEOREM 8.10.6. *Let* $X \sim CH_p(n, \alpha, \beta, kind\ 1)$. *Then for a constant matrix* $A\,(q \times p)$, *with* $\mathrm{rank}(A) = q \leq p$, $AXA' \sim CH_q(n, \alpha - \frac{1}{2}(p-q), \beta - \frac{1}{2}(p-q); (AA')^{-1}, kind\ 1)$.

COROLLARY 8.10.6.1. *Let* $X \sim CH_p(n, \alpha, \beta, kind\ 1)$. *Then, for* $\boldsymbol{a} \neq \boldsymbol{0}$,

$$\frac{\boldsymbol{a}'X\boldsymbol{a}}{\boldsymbol{a}'\boldsymbol{a}} \sim CH_1\left(n, \alpha - \frac{1}{2}(p-1), \beta - \frac{1}{2}(p-1), kind\ 1\right).$$

In the above corollary it is clear that the distribution of $\frac{\boldsymbol{a}'X\boldsymbol{a}}{\boldsymbol{a}'\boldsymbol{a}}$ does not depend on $\boldsymbol{a}$. Thus, for any random vector $\boldsymbol{y}\,(p \times 1)$ distributed independently of X with $P(\boldsymbol{y} \neq \boldsymbol{0}) = 1, \frac{\boldsymbol{y}'X\boldsymbol{y}}{\boldsymbol{y}'\boldsymbol{y}} \sim CH_1(n, \alpha - \frac{1}{2}(p-1), \beta - \frac{1}{2}(p-1), kind\ 1)$.

THEOREM 8.10.7. *Let* $X \sim CH_p(n, \alpha, \beta, kind\ 1)$, *and* $A\,(q \times p)$ *be a constant matrix of rank* $q \leq p$. *Then* $(AX^{-1}A')^{-1} \sim CH_q(n - \frac{1}{2}(p-q), \alpha - \frac{1}{2}(p-q), \beta - \frac{1}{2}(p-q); AA', kind\ 1)$.

COROLLARY 8.10.7.1. *Let* $X \sim CH_p(n, \alpha, \beta, kind\ 1)$. *Then, for* $\boldsymbol{a} \neq \boldsymbol{0}$,

$$\frac{\boldsymbol{a}'\boldsymbol{a}}{\boldsymbol{a}'X^{-1}\boldsymbol{a}} \sim CH_1\left(n - \frac{1}{2}(p-1), \alpha - \frac{1}{2}(p-1), \beta - \frac{1}{2}(p-1), kind\ 1\right).$$

THEOREM 8.10.8. *Let* $X \sim CH_p(n, \alpha, \beta, kind\ 1)$. *Then*

(i) $E[C_\kappa(X)] = \dfrac{(n - \beta + \frac{1}{2}(p+1))_\kappa (n)_\kappa}{(n - \alpha + \frac{1}{2}(p+1))_\kappa} C_\kappa(I_p)$, $\mathrm{Re}(\alpha - n) > \dfrac{1}{2}(p-1) + k_1$,

and

(ii) $E[C_\kappa(X^{-1})] = \dfrac{(-1)^k (\alpha - n)_\kappa}{(\beta - n)_\kappa (-n + \frac{1}{2}(p+1))_\kappa} C_\kappa(I_p)$, $\mathrm{Re}(n) > \dfrac{1}{2}(p-1) + k_1$.

THEOREM 8.10.9. *Let* $X \sim CH_p(n, \alpha, \beta; \alpha^{-1}A, kind\ 1)$. *Then* $X \xrightarrow{\mathcal{D}} Y$ *as* $\alpha \to \infty$, *where the p.d.f. of* Y *is given by*

$$\frac{\Gamma_p(\beta - n)}{\Gamma_p(n)\Gamma_p(\beta)} \det(A)^n \det(Y)^{n - \frac{1}{2}(p+1)} {}_0F_1(\beta; -AY), \quad Y > 0,$$

where "$X \xrightarrow{\mathcal{D}} Y$" denotes convergence in distribution.

If, on the other hand, $X \sim CH_p(n, \alpha, \beta; \beta A, kind\ 1)$, then $X \xrightarrow{\mathcal{D}} Y$ as $\beta \to \infty$, where $Y \sim GB_p^{II}(n, \alpha - n; A^{-1}, 0)$. These two results were obtained by van der Merwe and Roux (1974) by using the confluence relations given in Chapter 1.

8.11. CONFLUENT HYPERGEOMETRIC FUNCTION KIND 2 DISTRIBUTION

In section 8.10 we defined confluent hypergeometric function kind 1 distribution. In this section we study certain distributions which correspond to the confluent hypergeometric function of kind 2 defined in Chapter 1.

DEFINITION 8.11.1. *A symmetric random matrix* X ($p \times p$) *is said to have a confluent hypergeometric function kind 2 and type I distribution, if its p.d.f. is given by*

$$\frac{\Gamma_p(\alpha)\Gamma_p[\alpha - \beta + \frac{1}{2}(p+1)]}{\Gamma_p(n)\Gamma_p(\alpha - n)\Gamma_p[n - \beta + \frac{1}{2}(p+1)]} \det(X)^{n - \frac{1}{2}(p+1)} \Psi(\alpha, \beta; X), \ X > 0,$$

where $\mathrm{Re}(n, \alpha - n) > \frac{1}{2}(p - 1)$ *and* $\mathrm{Re}(n - \beta) > -1$.

The parameters n, α, and β are restricted to take values such that the density function is non-negative. This distribution will be denoted by $CH_p^I(n, \alpha, \beta, \text{kind } 2)$. By transforming $X = A^{\frac{1}{2}}YA^{\frac{1}{2}}$, ($A > 0$) with the Jacobian $J(X \to Y) = \det(A)^{\frac{1}{2}(p+1)}$, the density of Y is

$$\frac{\Gamma_p(\alpha)\Gamma_p[\alpha - \beta + \frac{1}{2}(p+1)]}{\Gamma_p(n)\Gamma_p(\alpha - n)\Gamma_p[n - \beta + \frac{1}{2}(p+1)]} \det(A)^n \det(Y)^{n - \frac{1}{2}(p+1)} \Psi(\alpha, \beta; AY), \ Y > 0.$$

This distribution will be denoted by $CH_p^I(n, \alpha, \beta; A, \text{kind } 2)$.

THEOREM 8.11.1. *Let* $W \sim G_p(n, I_p)$ *and* $V \sim B_p^{II}(a, b)$ *be independent. Then* $X = V^{\frac{1}{2}}WV^{\frac{1}{2}} \sim CH_p^I(n, b + n, n - a + \frac{1}{2}(p + 1), \text{kind } 2)$.

Proof: Making the transformation $X = V^{\frac{1}{2}}WV^{\frac{1}{2}}$ with the Jacobian $J(W \to X) = \det(V)^{-\frac{1}{2}(p+1)}$, in the joint density of V and W, and then integrating with respect to V, we get

$$\frac{\det(X)^{n - \frac{1}{2}(p+1)}}{\beta_p(a, b)\Gamma_p(n)} \int_{V > 0} \mathrm{etr}(-VX) \det(V)^{b + n - \frac{1}{2}(p+1)} \det(I_p + V)^{-(a+b)} \, dV$$

$$= \frac{\Gamma_p(b + n)\Gamma_p(a + b)}{\Gamma_p(n)\Gamma_p(a)\Gamma_p(b)} \det(X)^{n - \frac{1}{2}(p+1)} \Psi\left(b + n, n - a + \frac{1}{2}(p + 1); X\right), \ X > 0.$$

The last step is obtained from the Definition 1.6.13. ∎

If $X \sim CH_p^I(n, \alpha, \beta, \text{kind } 2)$, then the Laplace transform of its density is

$$\frac{\Gamma_p(\alpha)\Gamma_p[\alpha - \beta + \frac{1}{2}(p+1)]}{\Gamma_p(n)\Gamma_p(\alpha - n)\Gamma_p[\alpha + n - \beta + \frac{1}{2}(p+1)]}$$

$$_2F_1\left(n - \beta + \frac{1}{2}(p + 1), n; \alpha + n - \beta + \frac{1}{2}(p + 1); I_p - Z\right), \quad (8.11.1)$$

where $\mathrm{Re}(I_p - Z) < I_p$, $\mathrm{Re}(n - \beta) > -1$ and $\mathrm{Re}(\alpha) > \frac{1}{2}(p - 1)$.

Now we give certain properties of the confluent hypergeometric function kind 2 and type I distribution.

THEOREM 8.11.2. *Let* $X \sim CH_p^I(n, \alpha, \beta, kind\ 2)$, *and* A *be any* $p \times p$ *constant nonsingular matrix. Then* $AXA' \sim CH_p^I(n, \alpha, \beta, (AA')^{-1}, kind\ 2)$.

THEOREM 8.11.3. *Let* $X \sim CH_p^I(n, \alpha, \beta, kind\ 2)$, *and* $H\ (p \times p)$ *be an orthogonal matrix whose elements are either constants or random variables distributed independently of* X. *Then, the distribution of* X *is invariant under the transformation* $X \to HXH'$, *and is independent of* H *in the latter case.*

THEOREM 8.11.4. *Let* $X \sim CH_p^I(n, \alpha, \beta, kind\ 2)$, *and partition* X *as*

$$X = \begin{pmatrix} X_{11} & X_{12} \\ X_{21} & X_{22} \end{pmatrix} \begin{matrix} q \\ p-q \end{matrix}.$$
$$\begin{matrix} q & p-q \end{matrix}$$

Then X_{11} *and* $X_{22 \cdot 1} = X_{22} - X_{21}X_{11}^{-1}X_{12}$ *are independent,* $X_{11} \sim CH_q^I(n, \alpha - \frac{1}{2}(p - q), \beta - \frac{1}{2}(p-q), kind\ 2)$, *and* $X_{22 \cdot 1} \sim CH_{p-q}^I(n - \frac{1}{2}q, \alpha - \frac{1}{2}q, \beta - \frac{1}{2}(p-q), kind\ 2)$.

THEOREM 8.11.5. *Let* $X \sim CH_p^I(n, \alpha, \beta, kind\ 2)$. *Then for a constant matrix* $A\ (q \times p)$, *with* $\mathrm{rank}(A) = q \leq p$, $AXA' \sim CH_q^I(n, \alpha - \frac{1}{2}(p - q), \beta - \frac{1}{2}(p - q); (AA')^{-1}, kind\ 2)$.

COROLLARY 8.11.5.1. *Let* $X \sim CH_p^I(n, \alpha, \beta, kind\ 2)$. *Then, for* $a \neq 0$,

$$\frac{a'Xa}{a'a} \sim CH_1^I\left(n, \alpha - \frac{1}{2}(p - 1), \beta - \frac{1}{2}(p - 1), kind\ 2\right).$$

In the above corollary it is noted that the distribution of $\frac{a'Xa}{a'a}$ does not depend on a. Thus, for any random vector $y\ (p \times 1)$ distributed independently of X with $P(y \neq 0) = 1, \frac{y'Xy}{y'y} \sim CH_1^I(n, \alpha - \frac{1}{2}(p - 1), \beta - \frac{1}{2}(p - 1), kind\ 2)$.

THEOREM 8.11.6. *Let* $X \sim CH_p^I(n, \alpha, \beta, kind\ 2)$, *and* $A\ (q \times p)$ *be a constant matrix of rank* $q \leq p$. *Then* $(AX^{-1}A')^{-1} \sim CH_q^I(n - \frac{1}{2}(p - q), \alpha - \frac{1}{2}(p - q), \beta - \frac{1}{2}(p - q); AA', kind\ 2)$.

COROLLARY 8.11.6.1. *Let* $X \sim CH_p^I(n, \alpha, \beta, kind\ 2)$. *Then, for* $a \neq 0$,

$$\frac{a'a}{a'X^{-1}a} \sim CH_1^I\left(n - \frac{1}{2}(p - 1), \alpha - \frac{1}{2}(p - 1), \beta - \frac{1}{2}(p - 1), kind\ 2\right).$$

Next, we define confluent hypergeometric function kind 2 and type II distribution and study its properties.

DEFINITION 8.11.2. *A random symmetric matrix* $X\ (p \times p)$ *is said to have a confluent hypergeometric function kind 2 and type II distribution, if its p.d.f. is given by*

$$\frac{\Gamma_p[\alpha - \beta + n + \frac{1}{2}(p + 1)]}{\Gamma_p(n)\Gamma_p[n - \beta + \frac{1}{2}(p + 1)]} \det(X)^{n - \frac{1}{2}(p+1)} \mathrm{etr}(-X)\ \Psi(\alpha, \beta; X),\ X > 0,$$

where $\mathrm{Re}(n, \alpha) > \frac{1}{2}(p - 1)$ *and* $\mathrm{Re}(n - \beta) > -1$.

The parameters n, α and β are restricted to take values such that the density function is non-negative. We wiil denote this distribution by $CH_p^{II}(n, \alpha, \beta, \text{kind 2})$. By transforming $X = A^{\frac{1}{2}} Y A^{\frac{1}{2}}$, $(A > 0)$ with the Jacobian $J(X \to Y) = \det(A)^{\frac{1}{2}(p+1)}$, we get the density of Y as

$$\frac{\Gamma_p[\alpha - \beta + n + \frac{1}{2}(p+1)]}{\Gamma_p(n)\Gamma_p[n - \beta + \frac{1}{2}(p+1)]} \det(A)^n \det(Y)^{n - \frac{1}{2}(p+1)} \operatorname{etr}(-AY) \, \Psi(\alpha, \beta; AY), \ Y > 0,$$

$$\operatorname{Re}(n) > \frac{1}{2}(p-1), \ \operatorname{Re}(\alpha) > \frac{1}{2}(p-1), \ \operatorname{Re}(n - \beta) > -1.$$

This distribution will be denoted by $CH_p^{II}(n, \alpha, \beta; A, \text{kind 2})$.

THEOREM 8.11.7. Let $W \sim G_p(n, I_p)$ and $U \sim B_p^I(a, b)$ be independent. Then $X = U^{\frac{1}{2}} W U^{\frac{1}{2}} \sim CH_p^{II}(n, a, n - b + \frac{1}{2}(p+1), \text{kind 2})$.

Proof: See Khattree and R. D. Gupta (1989). ∎

THEOREM 8.11.8. Let $X \sim CH_p^{II}(n, \alpha, \beta, \text{kind 2})$. Then its m.g.f. is

$$M_X(Z) = {}_2F_1\left(n - \beta + \frac{1}{2}(p+1), n; \alpha + n - \beta + \frac{1}{2}(p+1); Z\right).$$

Proof: The m.g.f. of X is given by

$$M_X(Z) = \frac{\Gamma_p[\alpha - \beta + n + \frac{1}{2}(p+1)]}{\Gamma_p(n)\Gamma_p[n - \beta + \frac{1}{2}(p+1)]}$$

$$\int_{X>0} \operatorname{etr}\{-(I_p - Z)X\} \det(X)^{n - \frac{1}{2}(p+1)} \, \Psi(\alpha, \beta; X) \, dX$$

$$= {}_2F_1\left(n - \beta + \frac{1}{2}(p+1), n; \alpha + n - \beta + \frac{1}{2}(p+1); Z\right),$$

$$\operatorname{Re}(I_p - Z) > 0, \ \operatorname{Re}(\alpha) > \frac{1}{2}(p-1), \ \operatorname{Re}(n - \beta) > -1.$$

The last step has been obtained by using the Laplace transform of confluent hypergeometric function given in Problem 1.23 in Chapter 1. ∎

The m.g.f. of $\operatorname{tr}(X)$, from the above theorem, is derived as

$$M_{\operatorname{tr}(X)}(Z) = {}_2F_1\left(n - \beta + \frac{1}{2}(p+1), n; \alpha + n - \beta + \frac{1}{2}(p+1); z I_p\right)$$

$$= \sum_{k=0}^{\infty} \sum_{\kappa} \frac{(n)_\kappa (n - \beta + \frac{1}{2}(p+1))_\kappa}{(n + \alpha - \beta + \frac{1}{2}(p+1))_\kappa} \frac{C_\kappa(I_p)}{k!} z^k, \ |z| < 1. \qquad (8.11.2)$$

From (8.11.2) the kth moment of $\operatorname{tr}(X)$ is obtained as

$$E[(\operatorname{tr} X)^k] = \sum_{\kappa} \frac{(n)_\kappa (n - \beta + \frac{1}{2}(p+1))_\kappa}{(n + \alpha - \beta + \frac{1}{2}(p+1))_\kappa} C_\kappa(I_p), \ k = 1, 2, \ldots, \ldots$$

THEOREM 8.11.9. *Let* $X \sim CH_p^{II}(n, \alpha, \beta, kind\ 2)$, *and* A *be any* $p \times p$ *constant nonsingular matrix. Then* $AXA' \sim CH_p^{II}(n, \alpha, \beta; (AA')^{-1}, kind\ 2)$.

THEOREM 8.11.10. *Let* $X \sim CH_p^{II}(n, \alpha, \beta, kind\ 2)$, *and* $H\ (p \times p)$ *be an orthogonal matrix whose elements are either constants or random variables distributed independent of* X. *Then, the distribution of* X *is invariant under the transformation* $X \to HXH'$, *and is independent of* H *in the latter case.*

Khattree and R. D. Gupta (1989) have derived many results on expectations using Theorem 8.11.8. They have shown that if $X \sim CH_p^{II}(n, a, n - b + \frac{1}{2}(p + 1), kind\ 2)$ then for $\boldsymbol{\delta}\ (p \times 1) \neq \mathbf{0}$,

$$\frac{\boldsymbol{\delta}'X\boldsymbol{\delta}}{\boldsymbol{\delta}'\boldsymbol{\delta}} \sim CH_1^{II}(n, a, n - b + 1, kind\ 2)$$

and

$$\frac{\boldsymbol{\delta}'\boldsymbol{\delta}}{\boldsymbol{\delta}'X^{-1}\boldsymbol{\delta}} \sim CH_1^{II}\left(n - \frac{1}{2}(p - 1), a, n - b + 1, kind\ 2\right).$$

8.12. HYPERGEOMETRIC FUNCTION DISTRIBUTIONS

In this section we give hypergeometric function distributions of two types. First we define hypergeometric function distribution of type I.

DEFINITION 8.12.1. *A random symmetric matrix* $X\ (p \times p)$ *is said to have hypergeometric function distribution of type I, if its p.d.f. is given by*

$$\frac{\Gamma_p(\gamma + n - \alpha)\Gamma_p(\gamma + n - \beta)}{\Gamma_p(\gamma)\Gamma_p(n)\Gamma_p(\gamma + n - \alpha - \beta)} \det(X)^{n - \frac{1}{2}(p+1)} \det(I_p - X)^{\gamma - \frac{1}{2}(p+1)}$$

$$_2F_1(\alpha, \beta; \gamma; I_p - X), 0 < X < I_p, \qquad (8.12.1)$$

where $\operatorname{Re}(\gamma + n - \alpha - \beta) > \frac{1}{2}(p - 1)$, $\operatorname{Re}(\gamma) > \frac{1}{2}(p - 1)$ *and* $\operatorname{Re}(n) > \frac{1}{2}(p - 1)$.

The parameters α, β, γ and n are restricted to take values such that the density function is non-negative. We will denote this distribution by $H_p^I(n, \alpha, \beta, \gamma)$. For $\alpha = \gamma$, the density (8.12.1) reduces to

$$\{\beta_p(\gamma, n - \beta)\}^{-1} \det(X)^{n - \beta - \frac{1}{2}(p+1)} \det(I_p - X)^{\gamma - \frac{1}{2}(p+1)}, 0 < X < I_p,$$

and for $\beta = \gamma$, hypergeometric function density of type I (8.12.1) reduces to

$$\{\beta_p(\gamma - \alpha, n)\}^{-1} \det(X)^{n - \alpha - \frac{1}{2}(p+1)} \det(I_p - X)^{\gamma - \frac{1}{2}(p+1)}, 0 < X < I_p.$$

THEOREM 8.12.1. *Let* $U \sim B_p^I(a, b)$ *and* $V \sim B_p^I(c, d)$ *be independent. Then* $Z = U^{\frac{1}{2}}VU^{\frac{1}{2}} \sim H_p^I(c, b, c + d - a, b + d)$.

Proof: The joint density of U and V is given by

$$\{\beta_p(a,b)\beta_p(c,d)\}^{-1} \det(U)^{a-\frac{1}{2}(p+1)} \det(I_p - U)^{b-\frac{1}{2}(p+1)}$$

$$\det(V)^{c-\frac{1}{2}(p+1)} \det(I_p - V)^{d-\frac{1}{2}(p+1)}, \; 0 < U < I_p, \; 0 < V < I_p.$$

Making the transformation $Z = U^{\frac{1}{2}}VU^{\frac{1}{2}}$, with Jacobian $J(V \to Z) = \det(U)^{-\frac{1}{2}(p+1)}$, and integrating out U from the joint density of U and Z, we get the marginal density of Z as

$$\{\beta_p(a,b)\beta_p(c,d)\}^{-1} \det(Z)^{c-\frac{1}{2}(p+1)} \int_{Z<U<I_p} \det(U)^{a-c-\frac{1}{2}(p+1)} \det(I_p - U)^{b-\frac{1}{2}(p+1)}$$

$$\det(I_p - U^{-1}Z)^{d-\frac{1}{2}(p+1)} \, dU. \quad (8.12.2)$$

Now substituting $A = (I_p - Z)^{-\frac{1}{2}}(I_p - U)(I_p - Z)^{-\frac{1}{2}}$, (8.12.2) becomes

$$\{\beta_p(a,b)\beta_p(c,d)\}^{-1} \det(Z)^{c-\frac{1}{2}(p+1)} \det(I_p - Z)^{b+d-\frac{1}{2}(p+1)}$$

$$\int_{0<A<I_p} \det(A)^{b-\frac{1}{2}(p+1)} \det(I_p - A)^{d-\frac{1}{2}(p+1)} \det(I_p - (I_p - Z)A)^{-(c+d-a)} \, dA$$

$$= \frac{\Gamma_p(a+b)\Gamma_p(c+d)}{\Gamma_p(a)\Gamma_p(c)\Gamma_p(b+d)} \det(Z)^{c-\frac{1}{2}(p+1)} \det(I_p - Z)^{b+d-\frac{1}{2}(p+1)}$$

$$_2F_1(b, c+d-a; b+d; I_p - Z), \; 0 < Z < I_p.$$

The last step has been obtained by using the Corollary 1.6.3.2. ∎

For $c = a+b$, the above theorem gives $Z \sim B_p^I(a, b+d)$, as proved in Theorem 5.3.25. The m.g.f. of $X \sim H_p^I(n, \alpha, \beta, \gamma)$ is given by

$$M_X(Z) = \frac{\Gamma_p(\gamma+n-\alpha)\Gamma_p(\gamma+n-\beta)}{\Gamma_p(\gamma)\Gamma_p(n)\Gamma_p(\gamma+n-\alpha-\beta)} \int_{0<X<I_p} \text{etr}(ZX) \det(X)^{n-\frac{1}{2}(p+1)}$$

$$\det(I_p - X)^{\gamma-\frac{1}{2}(p+1)} {}_2F_1(\alpha, \beta; \gamma; I_p - X) \, dX$$

$$= \frac{\Gamma_p(\gamma+n-\alpha)\Gamma_p(\gamma+n-\beta)}{\Gamma_p(\gamma)\Gamma_p(n)\Gamma_p(\gamma+n-\alpha-\beta)} \int_{0<Y<I_p} \text{etr}\{Z(I_p - Y)\} \det(Y)^{\gamma-\frac{1}{2}(p+1)}$$

$$\det(I_p - Y)^{n-\frac{1}{2}(p+1)} {}_2F_1(\alpha, \beta; \gamma; Y) \, dY$$

$$= \frac{\Gamma_p(\gamma+n-\alpha)\Gamma_p(\gamma+n-\beta)}{\Gamma_p(\gamma)\Gamma_p(n)\Gamma_p(\gamma+n-\alpha-\beta)} \sum_{k=0}^{\infty} \sum_{\kappa} \frac{1}{k!} \int_{0<Y<I_p} C_\kappa(Z(I_p - Y))$$

$$\det(Y)^{\gamma-\frac{1}{2}(p+1)} \det(I_p - Y)^{n-\frac{1}{2}(p+1)} {}_2F_1(\alpha, \beta; \gamma; Y) \, dY$$

$$= {}_2F_2(n, \gamma+n-\alpha-\beta; \gamma+n-\alpha, \gamma+n-\beta; \gamma; Z),$$

where the last step is derived by using the result of Problem 1.16.

van der Merwe and Roux (1974), using Gauss' hypergeometric function, have defined the following density.

DEFINITION 8.12.2. *A random symmetric matrix X $(p \times p)$ is said to have hypergeometric function distribution of type II, if its p.d.f. is given by*

$$\frac{\Gamma_p(\alpha)\Gamma_p(\beta)\Gamma_p(\gamma - n)\det(A)^n}{\Gamma_p(n)\Gamma_p(\gamma)\Gamma_p(\alpha - n)\Gamma_p(\beta - n)} \det(X)^{n-\frac{1}{2}(p+1)} {}_2F_1(\alpha, \beta; \gamma; -AX), \; X > 0, \quad (8.12.3)$$

where $A > 0$, $\mathrm{Re}(\gamma - n) > \frac{1}{2}(p-1)$, $\mathrm{Re}(\alpha - n) > \frac{1}{2}(p-1)$ and $\mathrm{Re}(\beta - n) > \frac{1}{2}(p-1)$.

The parameters n, α, β and γ are restricted to take values such that the density function is non-negative. Denote the above distribution by $H_p^{II}(n, \alpha, \beta, \gamma; A)$. The c.d.f. of the random matrix X can be shown to be

$$P(X < \Omega) = \frac{\Gamma_p(\alpha)\Gamma_p(\beta)\Gamma_p(\gamma - n)\Gamma_p[\frac{1}{2}(p+1)]}{\Gamma_p(n)\Gamma_p(\gamma)\Gamma_p(\alpha - n)\Gamma_p(\beta - n)\Gamma_p[n + \frac{1}{2}(p+1)]}$$

$$\det(A)^n \det(\Omega)^n {}_3F_2(\alpha, \beta, n; n + \frac{1}{2}(p+1), \gamma; -A\Omega), \quad (8.12.4)$$

where ${}_3F_2(\;)$ is the generalized hypergeometric function defined in Chapter 1.

For $\beta = \gamma$, the density (8.12.3) reduces to the density

$$\{\beta_p(n, \alpha - n)\}^{-1} \det(A)^n \det(X)^{n-\frac{1}{2}(p+1)} \det(I_p + AX)^{-a}, \; X > 0,$$

which is the density of generalized beta type II distribution $GB_p^{II}(n, \alpha - n; A^{-1}, 0)$. From the confluence relation (1.6.12) it follows that if $X \sim H_p^{II}(n, \alpha, \beta, \gamma; \beta^{-1}A)$, then $X \xrightarrow{\mathcal{D}} Y$ as $\beta \to \infty$, where the p.d.f. of Y is $CH_p(n, \alpha, \gamma; A, \text{kind } 1)$.

The following theorem gives the hypergeometric function distribution of type II as the distribution of ratio of two independent beta matrices.

THEOREM 8.12.2. *Let $U \sim B_p^I(a, b)$ and $V \sim B_p^{II}(c, d)$ be independent. Then $X = U^{-\frac{1}{2}}VU^{-\frac{1}{2}} \sim H_p^{II}(c, a + c, c + d, a + b + c; I_p)$.*

Proof: The joint density of U and V as given by

$$\{\beta_p(a, b)\beta_p(c, d)\}^{-1} \det(U)^{a-\frac{1}{2}(p+1)} \det(I_p - U)^{b-\frac{1}{2}(p+1)}$$

$$\det(V)^{c-\frac{1}{2}(p+1)} \det(I_p + V)^{-(c+d)}, \; 0 < U < I_p, \; V > 0.$$

Transforming $X = U^{-\frac{1}{2}}VU^{-\frac{1}{2}}$, with the Jacobian $J(V \to X) = \det(U)^{\frac{1}{2}(p+1)}$, and integrating out U from the joint density of U and X we get the marginal density of X as

$$\{\beta_p(a, b)\beta_p(c, d)\}^{-1} \det(X)^{c-\frac{1}{2}(p+1)} \int_{0<U<I_p} \det(U)^{a+c-\frac{1}{2}(p+1)} \det(I_p - U)^{b-\frac{1}{2}(p+1)}$$

$$\det(I_p + XU)^{-(c+d)} \, dU$$

$$= \frac{\Gamma_p(a+b)\Gamma_p(c+d)\Gamma_p(a+c)}{\Gamma_p(a)\Gamma_p(c)\Gamma_p(d)\Gamma_p(a+b+c)} \det(X)^{c-\frac{1}{2}(p+1)} {}_2F_1(a+c, c+d; a+b+c; -X).$$

The last step has been obtained by using Corollary 1.6.3.2. ∎

8.13. GENERALIZED HYPERGEOMETRIC FUNCTION DISTRIBUTIONS

Roux (1971), by multiplying Wishart, beta, and Dirichlet densities by generalized hypergeometric function, has defined a number of densities which are given in this section.

(i) The random symmetric matrix X $(p \times p)$ is said to have *Generalized Hypergeometric Function* (GHF) type I distribution if its density is

$$\frac{\text{etr}(-X)\det(X)^{n-\frac{1}{2}(p+1)}}{\Gamma_p(n)\,_{r+1}F_s(a_1,\ldots,a_r,n;b_1,\ldots,b_s;\Theta)}\,_rF_s(a_1,\ldots,a_r;b_1,\ldots,b_s;\Theta X), \quad X > 0,$$

$$(8.13.1)$$

where $n > \frac{1}{2}(p-1)$ and the parameters a_i, b_j, $i = 1,\ldots,r$, $j = 1,\ldots,s$ are restricted to take those values for which the density function is non-negative.

For $\Theta = 0$, we get the Wishart density, $W_p(2n, 2I_p)$, as a special case of the above density. When $r = 0$, $s = 1$, and $b_1 = n$ we get

$$\frac{\text{etr}(-\Theta)}{\Gamma_p(n)}\,\text{etr}(-X)\det(X)^{n-\frac{1}{2}(p+1)}\,_0F_1(n;\Theta X), \quad (8.13.2)$$

which is the noncentral Wishart density, $W_p(2n, 2I_p, \Theta)$. For $\Theta = I_p$, $r = s = 1$, the density (8.13.1) reduces to

$$\frac{\text{etr}(-X)\det(X)^{n-\frac{1}{2}(p+1)}}{\Gamma_p(n)\,_2F_1(a_1,n;b_1;I_p)}\,_1F_1(a_1;b_1;X), \quad X > 0.$$

Simplifying this density using the results

$$_2F_1(a_1,n;b_1;I_p) = \frac{\Gamma_p(b_1)\Gamma_p(b_1 - a_1 - n)}{\Gamma_p(b_1 - a_1)\Gamma_p(b_1 - n)}$$

and

$$_1F_1(a_1;b_1;X) = \text{etr}(-X)\,_1F_1(b_1 - a_1;b_1;-X),$$

it can be easily seen that $X \sim CH_p(n, b_1 - a_1, b_1, \text{kind } 1)$.

(ii) The random symmetric matrix X $(p \times p)$ is said to have *Generalized Hypergeometric Function* (GHF) type II distribution if its density is

$$\frac{\Gamma_p(m+n)\det(X)^{m-\frac{1}{2}(p+1)}\det(I_p - X)^{n-\frac{1}{2}(p+1)}}{\Gamma_p(m)\Gamma_p(n)\,_{r+1}F_{s+1}(a_1,\ldots,a_r,m;b_1,\ldots,b_s,m+n;\Theta)}$$

$$_rF_s(a_1,\ldots,a_r;b_1,\ldots,b_s;\Theta X), \quad 0 < X < I_p, \quad (8.13.3)$$

where $\Theta = \Theta'$, $m > \frac{1}{2}(p-1)$, $n > \frac{1}{2}(p-1)$ and the parameters a_i, b_j, $i = 1,\ldots,r$, $j = 1,\ldots,s$ are restricted to take those values for which the density function is non-negative.

For $\Theta = 0$, the GHF type II density (8.13.3) reduces to the matrix variate beta type I density with parameters m and n. For $r = s = 1$, $a_1 = m + n$, and $b_1 = m$ we get the density of X as

$$\frac{\text{etr}(-\Theta)}{\beta_p(m,n)} \det(X)^{m-\frac{1}{2}(p+1)} \det(I_p - X)^{n-\frac{1}{2}(p+1)} {}_1F_1(m+n; m; \Theta X), \ 0 < X < I_p,$$

which is the noncentral matrix variate beta type I(B) density with parameters m, n, and Θ.

(iii) The random symmetric matrix X ($p \times p$) is said to have *Generalized Hypergeometric Function* (GHF) type III distribution if its density is

$$\frac{\Gamma_p(m+n) \det(X)^{m-\frac{1}{2}(p+1)} \det(I_p + X)^{-(m+n)}}{\Gamma_p(m)\Gamma_p(n) \, {}_{r+1}F_{s+1}(a_1,\ldots,a_r,n;b_1,\ldots,b_s,m+n;\Theta)}$$

$$ {}_rF_s(a_1,\ldots,a_r;b_1,\ldots,b_s;\Theta(I_p + X)^{-1}), \ X > 0, \qquad (8.13.4)$$

where $\Theta = \Theta'$, $m > \frac{1}{2}(p-1)$, $n > \frac{1}{2}(p-1)$ and the parameters a_i, b_j, $i = 1,\ldots,r$, $j = 1,\ldots,s$ are restricted to take those values for which the density function is non-negative.

For $\Theta = 0$, the GHF type III density (8.13.4) reduces to the matrix variate beta type II density with parameters m and n. For $r = s = 1$, $a_1 = m + n$, and $b_1 = n$ we get the density of X as

$$\frac{\text{etr}(-\Theta)}{\beta_p(m,n)} \det(X)^{m-\frac{1}{2}(p+1)} \det(I_p + X)^{-(m+n)} {}_1F_1(m+n; n; \Theta(I_p + X)^{-1}), \ X > 0,$$

which is the noncentral matrix variate beta type II(A) density with parameters m, n, and Θ (see Theorem 5.5.3).

(iv) The $p \times p$ random symmetric matrices $X_1,\ldots,X_q$ are said to have *Generalized Hypergeometric Function* (GHF) type IV distribution if their joint density is

$$\frac{\Gamma_p(m+n) \prod_{k=1}^q \det(X_k)^{m_k-\frac{1}{2}(p+1)} \det(I_p - \sum_{k=1}^q X_k)^{n-\frac{1}{2}(p+1)}}{\prod_{k=1}^q \Gamma_p(m_k)\Gamma_p(n) \, {}_{r+1}F_{s+1}(a_1,\ldots,a_r,m;b_1,\ldots,b_s,m+n;\Theta)}$$

$$ {}_rF_s\Big(a_1,\ldots,a_r;b_1,\ldots,b_s;\Theta\sum_{k=1}^q X_k\Big), \ \sum_{k=1}^q X_k < I_p, \ X_k > 0, \ k = 1,\ldots,q, \qquad (8.13.5)$$

where $m = \sum_{k=1}^q m_k$, $\Theta = \Theta'$, $m_k > \frac{1}{2}(p-1)$, $k = 1,\ldots,q$, $n > \frac{1}{2}(p-1)$, and the parameters a_i, b_j, $i = 1,\ldots,r$, $j = 1,\ldots,s$ are restricted to take those values for which the density function is non-negative.

Note that for $q = 1$, the GHF type IV density (8.13.5) becomes GHF type II density (8.11.3). For $\Theta = 0$, the GHF type IV density (8.13.5) reduces to matrix variate Dirichlet type I density with parameters $m_1,\ldots,m_q$ and n. For $r = s = 1$, $a_1 = m + n$, and $b_1 = m$ we get the density of X as

$$\frac{\Gamma_p(m+n)}{\prod_{k=1}^q \Gamma_p(m_k)\Gamma_p(n)} \prod_{k=1}^q \det(X_k)^{m_k-\frac{1}{2}(p+1)} \det\Big(I_p - \sum_{k=1}^q X_k\Big)^{n-\frac{1}{2}(p+1)}$$

$$ {}_1F_1\Big(m+n; m; \Theta\sum_{k=1}^q X_k\Big), \ \sum_{k=1}^q X_k < I_p, \ X_k > 0, \ k = 1,\ldots,q,$$

which is the noncentral matrix variate Dirichlet type I density with parameters $m_1, \ldots,$ $m_q; n,$ and $\Theta.$

(v) The $p \times p$ random symmetric matrices $X_1, \ldots, X_q$ are said to have *Generalized Hypergeometric Function* (GHF) type V distribution if their joint density is

$$\frac{\Gamma_p(m+n) \prod_{k=1}^{q} \det(X_k)^{m_k - \frac{1}{2}(p+1)} \det(I_p + \sum_{k=1}^{q} X_k)^{-(m+n)}}{\prod_{k=1}^{q} \Gamma_p(m_k)\Gamma_p(n) \, _{r+1}F_{s+1}(a_1, \ldots, a_r, n; b_1, \ldots, b_s, m+n; \Theta)}$$

$$_rF_s\left(a_1, \ldots, a_r; b_1, \ldots, b_s; \Theta\left(I_p + \sum_{k=1}^{q} X_k\right)^{-1}\right), \quad X_k > 0, \ k = 1, \ldots, q, \quad (8.13.6)$$

where $m = \sum_{k=1}^{q} m_k$, $\Theta = \Theta'$, $m_k > \frac{1}{2}(p-1)$, $k = 1, \ldots, q$, $n > \frac{1}{2}(p-1)$, and the parameters a_i, b_j, $i = 1, \ldots, r$, $j = 1, \ldots, s$ are restricted to take those values for which the density function is non-negative.

Note that for $q = 1$, the GHF type V density (8.13.6) becomes GHF type III density (8.11.4). For $\Theta = 0$, the GHF type V density reduces to matrix variate Dirichlet type II density with parameters $m_1, \ldots, m_q$ and n. For $r = s = 1$, $a_1 = m+n$, and $b_1 = m$ we get the density of X as

$$\frac{\Gamma_p(m+n)}{\prod_{k=1}^{q} \Gamma_p(m_k)\Gamma_p(n)} \prod_{k=1}^{q} \det(X_k)^{m_k - \frac{1}{2}(p+1)} \det\left(I_p + \sum_{k=1}^{q} X_k\right)^{-(m+n)}$$

$$_1F_1\left(m+n; n; \Theta\left(I_p + \sum_{k=1}^{q} X_k\right)^{-1}\right), \quad X_k > 0, \ k = 1, \ldots, q,$$

which is the noncentral matrix variate Dirichlet type II density with parameters $m_1, \ldots, m_q; n,$ and $\Theta.$

8.14.   COMPLEX MATRIX VARIATE DISTRIBUTIONS

The complex multivariate distributions play an important role in various fields of research. The complex multivariate Gaussian distribution was introduced by Wooding (1956), Turin (1960), and Goodman (1963a). The complex Wishart distribution was derived by Goodman (1963a) to approximate the distribution of an estimate of the spectral density matrix for a vector valued stationary Gaussian process. In multiple time series analysis, complex multivariate distributions are used to describe estimators of frequency domain parameters. For applications of these distributions in time series analysis, reference may be made to Whaba (1968, 1971), Goodman and Dubman (1969), Hannan (1970), Priestly, Subba Rao and Tong (1973), Brillinger (1969, 1975), and Shaman (1980). This distribution has also been found useful in nuclear physics in studying the distribution of spacings between energy levels of nuclei in high excitation. For further details reference may be made to Dyson (1962a, 1962b, 1962c), Dyson and Mehta (1963a, 1963b), Bronk (1965), Porter (1965), and Carmeli (1974, 1983).

The complex multivariate elliptically symmetric distribution has been studied by Krishnaiah and Lin (1986), and Khatri and Bhavsar (1990). This family includes complex multivariate Gaussian and complex multivariate t-distributions.

The joint distributions of the roots of some complex random matrices have been derived by James (1964), Wigner (1965), and Khatri (1965). Parallel to the real case, substantial work in the complex case hase been carried out. The distributions of several test statistics in the complex case have been studied by several authors, *e.g.*, see Goodman (1963b) Khatri (1965, 1969, 1970b), Pillai and Jouris (1971), Nagarsenker and Das (1975), Chikuse (1976), Krishnaiah (1976), Gupta (1971a, 1973, 1976), Fang, Krishnaiah and Nagarsenkar (1982), Gupta and Rathie (1983a, 1983b), Gupta and Conradie (1987), Gupta and Nagar (1985, 1987, 1988, 1989, 1992), Nagar, Jain and Gupta (1985), and Nagar and Gupta (1993). A number of results on the distribution of complex random matrices has also been derived. Srivastava (1965) gave a derivation of complex Wishart distribution. A characterization of complex Wishart distribution has been given by Gupta and Kabe (1998). James (1964) and Khatri (1965) derived the complex central as well as the noncentral matrix variate beta distributions. Systematic treatment of the distributions of complex random matrices was given by Tan (1968) which included the Gaussian, Wishart, beta, and Dirichlet distributions. Kabe (1984) defined hyper complex matrix variate Gaussian distribution which includes Hamilton's quaternions, biquaternions, octonions, and bioctonions. He also studied the corresponding sampling distribution theory. Rautenbach and Roux (1985) have also derived the quaternion distribution and studied its properties.

PROBLEMS

8.1. Let the joint p.d.f. of the random matrices $X_1\,(p \times p)$ and $X_2\,(p \times p)$ be

$$\prod_{i=1}^{2} \left[\left\{ 2^{\frac{1}{2}n_i p}\, \Gamma_p\left(\tfrac{1}{2}n_i\right) \det(\Psi_i)^{\frac{1}{2}n_i} \right\}^{-1} \operatorname{etr}\left(-\tfrac{1}{2}\Psi_i^{-1} X_i \right) \det(X_i)^{\frac{1}{2}(n_i-p-1)} \right]$$

$$\sum_{k=0}^{\infty} \sum_{\kappa} \frac{(a_1)_\kappa \cdots (a_r)_\kappa}{(b_1)_\kappa \cdots (b_s)_\kappa} \frac{C_\kappa(\Theta)}{k!\, [C_\kappa(I_p)]^2} \prod_{i=1}^{2} \left\{ \left(\tfrac{1}{2}n_i\right)_\kappa \right\}^{-1} L_\kappa^{\frac{1}{2}(n_i-p-1)}\left(\tfrac{1}{2}\Psi_i^{-1} X_i \right),$$

$$X_1, X_2 > 0.$$

Then show that the joint m.g.f. of X_1 and X_2 is

$$M_{X_1,X_2}(Z_1, Z_2) = \prod_{i=1}^{2} \det(I_p - 2\Psi_i Z_i)^{-\frac{1}{2}n_i} \sum_{k=0}^{\infty} \sum_{\kappa} \frac{(a_1)_\kappa \cdots (a_r)_\kappa}{(b_1)_\kappa \cdots (b_s)_\kappa} \frac{C_\kappa(\Theta)}{k!\, [C_\kappa(I_p)]^2}$$

$$\prod_{i=1}^{2} C_\kappa\big(2\Psi_i Z_i (I_p - 2\Psi_i Z_i)^{-1} \big).$$

(Roux and Raath, 1973)

8.2. Prove that (8.10.1) is a density.

8.3.   Let $X \sim CH_p(m, n_1+m, n_1+n_2+m, \text{kind } 1)$, where n_1, n_2 and m are positive integers. Then show that

$$E(X) = \frac{m(n-p-1)}{n_1-p-1}I_p, \ n_1 - p - 1 > 0,$$

$$E(X^2) = \frac{m(m+1)(n-p-1)}{(n_1-p)(n_1-p-1)(n_1-p-3)}$$
$$\{(n_1-p)(n_1-p-3) + n_2(n_1-1)\}I_p, \ n_1 - p - 3 > 0,$$

$$E(X^{-1}) = \frac{2n_1}{(2m-p-1)n}I_p, \ 2m - p - 1 > 0,$$

$$E(X^{-2}) = \frac{4n_1[(2m-1)\{(n+1)+n_2(p+1)-2\}+2n_2-p(p+1)n_2]}{(2m-p)(2m-p-1)(2m-p-3)n(n-1)(n+2)}I_p,$$
$$2m - p - 3 > 0,$$

where $n_1 + n_2 = n$.

8.4.   Prove Theorem 8.10.4.

8.5.   Prove Theorem 8.10.5.

8.6.   Prove Theorem 8.10.6.

8.7.   Prove Theorem 8.10.7.

8.8.   Prove Theorem 8.10.8.

8.9.   Let $X \sim CH_p(n, \alpha, \beta, \text{kind } 1)$. Then show that

$$E[\det(X)^h] = \frac{\Gamma_p(\beta-n)\Gamma_p(n+h)\Gamma_p(\alpha-n-h)}{\Gamma_p(n)\Gamma_p(\alpha-n)\Gamma_p(\beta-n-h)},$$
$$-n + \frac{1}{2}(p-1) < \text{Re}(h) < \min[\text{Re}(\alpha-n), \text{Re}(\beta-n)] - \frac{1}{2}(p-1).$$

8.10.   Derive the Laplace transform of (8.11.1).

8.11.   Prove Theorem 8.11.3.

8.12.   Prove Theorem 8.11.4.

8.13.   Prove Theorem 8.11.5.

8.14.   Prove Theorem 8.11.6.

8.15.   Prove Theorem 8.11.7.

8.16.   Prove Theorem 8.11.9.

8.17.   Prove Theorem 8.11.10.

8.18. Let $X \sim CH_p^{II}(n, a, n - b + \frac{1}{2}(p + 1), \text{kind } 2)$. Then show that

(i) $E(C_\kappa(X^{-1})) = \prod_{j=1}^{p} \dfrac{(a + b - k_j - \frac{1}{2}(p - j))_{k_j}}{(b - k_j - \frac{1}{2}(p - j))_{k_j}(n - k_j - \frac{1}{2}(p - j))_{k_j}} C_\kappa(I_p),$

(ii) $E(\text{tr}(X^{-1})) = \dfrac{p(a + b - \frac{1}{2}(p + 1))}{(n - \frac{1}{2}(p + 1))(b - \frac{1}{2}(p + 1))},$

$$\text{Re}(n) > \frac{1}{2}(p + 1), \ \text{Re}(b) > \frac{1}{2}(p + 1),$$

(iii) $E(\text{tr}(X^{-2})) = \dfrac{\frac{1}{3}(a + b - \frac{1}{2}(p + 3))(a + b - \frac{1}{2}(p + 1))p(p + 2)}{(b - \frac{1}{2}(p + 3))(b - \frac{1}{2}(p + 1))(n - \frac{1}{2}(p + 3))(n - \frac{1}{2}(p + 1))}$

$$- \dfrac{\frac{1}{3}(a + b - \frac{1}{2}(p + 1))(a + b - \frac{1}{2}p)p(p - 1)}{(b - \frac{1}{2}(p + 1))(b - \frac{1}{2}p)(n - \frac{1}{2}(p + 1))(n - \frac{1}{2}p)},$$

$$\text{Re}(n) > \frac{1}{2}(p + 3), \ \text{Re}(b) > \frac{1}{2}(p + 3),$$

(iv) $E[\det(X)^h] = \dfrac{\Gamma_p(n + h)\Gamma_p(b + h)\Gamma_p(a + b)}{\Gamma_p(n)\Gamma_p(b)\Gamma_p(a + b + h)},$

$$\text{Re}(h) > \max\left[-b + \frac{1}{2}(p + 1), -n + \frac{1}{2}(p + 1)\right].$$

(Khattree and R. D. Gupta, 1989)

8.19. Let $X \sim CH_p^{II}(n, \alpha, \beta, \text{kind } 2)$, and partition X as

$$X = \begin{pmatrix} X_{11} & X_{12} \\ X_{21} & X_{22} \end{pmatrix} \begin{matrix} q \\ p - q \end{matrix}.$$
$$\quad\quad q \quad\ \ p - q$$

Then show that X_{11} and $X_{22 \cdot 1} = X_{22} - X_{21}X_{11}^{-1}X_{12}$ are independently distributed. Furthermore $X_{22 \cdot 1} \sim CH_{p-q}^{II}(n - \frac{1}{2}q, \alpha, \beta - \frac{1}{2}q, \text{kind } 2)$.

8.20. If $X \sim CH_p^{I}(n, \alpha, \beta, \text{kind } 2)$, then show that

$$E[\det(X)^h] = \dfrac{\Gamma_p(n + h)\Gamma_p(\alpha - n - h)\Gamma_p[n - \beta + \frac{1}{2}(p + 1) + h]}{\Gamma_p(n)\Gamma_p(\alpha - n)\Gamma_p[n - \beta + \frac{1}{2}(p + 1)]},$$

$$-n + \max\left[\text{Re}(\beta - 1), \frac{1}{2}(p - 1)\right] < \text{Re}(h) < \text{Re}(\alpha - n) - \frac{1}{2}(p - 1).$$

8.21. If $X \sim CH_p^{II}(n, \alpha, \beta, \text{kind } 2)$, then prove that that

$$E[\det(X)^h] = \dfrac{\Gamma_p(n + h)\Gamma_p[n - \beta + \frac{1}{2}(p + 1) + h]}{\Gamma_p(n)\Gamma_p[n - \beta + \frac{1}{2}(p + 1)]},$$

$$\text{Re}(h) > -n + \max\left[\text{Re}(\beta - 1), \frac{1}{2}(p + 1)\right].$$

8.22. Let $W \sim G_p(m, I_p)$ and $U \sim H_p^I(n, \alpha, \beta, \gamma)$ be independent. Then show that the p.d.f. of $X = U^{-\frac{1}{2}}WU^{-\frac{1}{2}}$ is

$$\frac{\Gamma_p(\gamma + n - \alpha)\Gamma_p(\gamma + n - \beta)\Gamma_p(n + m)\Gamma_p(\gamma + n + m - \alpha - \beta)}{\Gamma_p(n)\Gamma_p(\gamma + n - \alpha - \beta)\Gamma_p(\gamma + n + m - \alpha)\Gamma_p(\gamma + n + m - \beta)\Gamma_p(m)}$$

$$\det(X)^{m - \frac{1}{2}(p+1)} {}_2F_2(n + m, \gamma + n + m - \alpha - \beta;$$

$$\gamma + n + m - \alpha, \gamma + n + m - \beta; -X), \quad X > 0.$$

8.23. Let $X \sim H_p^I(n, \alpha, \beta, \gamma)$. Then show that

$$E[\det(X)^h] = \frac{\Gamma_p(\gamma - n + \alpha)\Gamma_p(\gamma + n - \beta)\Gamma_p(\gamma + h)}{\Gamma_p(\gamma)\Gamma_p(\gamma + n - \alpha - \beta)\Gamma_p(\gamma + n + h)}$$

$$ {}_3F_2(\alpha, \beta, \gamma + h; \gamma, \gamma + n + h; I_p), \quad \mathrm{Re}(h) > 0.$$

8.24. Let $X \sim CH_p^{II}(n, \alpha, \beta, \gamma; A)$. Then show that

$$E[\det(X)^h] = \frac{\Gamma_p(\gamma - n)\Gamma_p(n + h)\Gamma_p(\alpha - n - h)\Gamma_p(\beta - n - h)}{\Gamma_p(n)\Gamma_p(\alpha - n)\Gamma_p(\beta - n)\Gamma_p(\gamma - n - h)} \det(A)^{-h},$$

$$-n + \frac{1}{2}(p - 1) < \mathrm{Re}(h) < \min[\mathrm{Re}(\beta - n), \mathrm{Re}(\gamma - n)] - \frac{1}{2}(p - 1).$$

8.25. Show that the m.g.f. of the GHF type I distribution (8.13.1) is given by

$$M_X(Z) = \frac{{}_{r+1}F_s(a_1, \ldots, a_r, n; b_1, \ldots, b_s; \Theta(I_p - Z)^{-1})}{{}_{r+1}F_s(a_1, \ldots, a_r, n; b_1, \ldots, b_s; \Theta)} \det(I_p - Z)^{-n}, \quad Z < I_p.$$

8.26. Show that the hth moment of $\det(X)$, where X has GHF type I distribution (8.13.1), is given by

$$E[\det(X)^h] = \frac{\Gamma_p(n + h)}{\Gamma_p(n)} \frac{{}_{r+1}F_s(a_1, \ldots, a_r, n + h; b_1, \ldots, b_s; \Theta)}{{}_{r+1}F_s(a_1, \ldots, a_r, n; b_1, \ldots, b_s; \Theta)},$$

$$\mathrm{Re}(h) > -n + \frac{1}{2}(p - 1).$$

8.27. Show that the following results hold good for the generalized hypergeometric function density (8.13.3),

$$E[\det(X)^h] = \frac{\Gamma_p(m + n)\Gamma_p(m + h)}{\Gamma_p(m)\Gamma_p(m + n + h)}$$

$$\frac{{}_{r+1}F_{s+1}(a_1, \ldots, a_r, m + h; b_1, \ldots, b_s, m + n + h; \Theta)}{{}_{r+1}F_{s+1}(a_1, \ldots, a_r, m; b_1, \ldots, b_s, m + n; \Theta)},$$

$$\mathrm{Re}(h) > -m + \frac{1}{2}(p - 1).$$

and

$$E[\det(I_p - X)^h] = \frac{\Gamma_p(m+n)\Gamma_p(n+h)}{\Gamma_p(n)\Gamma_p(m+n+h)}$$

$$\frac{{}_{r+1}F_{s+1}(a_1,\ldots,a_r,m;b_1,\ldots,b_s,m+n+h;\Theta)}{{}_{r+1}F_{s+1}(a_1,\ldots,a_r,m;b_1,\ldots,b_s,m+n;\Theta)},$$

$$\mathrm{Re}(h) > -n + \frac{1}{2}(p-1).$$

(Roux, 1971)

8.28. For the density (8.13.4), show that $E[\det(X)^h]$ and $E[\det(I_p + X)^{-h}]$ are given by

$$E[\det(X)^h] = \frac{\Gamma_p(m+h)\Gamma_p(n-h)}{\Gamma_p(m)\Gamma_p(n)}$$

$$\frac{{}_{r+1}F_{s+1}(a_1,\ldots,a_r,n-h;b_1,\ldots,b_s,m+n;\Theta)}{{}_{r+1}F_{s+1}(a_1,\ldots,a_r,n;b_1,\ldots,b_s,m+n;\Theta)},$$

$$-m + \frac{1}{2}(p-1) < \mathrm{Re}(h) < n - \frac{1}{2}(p-1),$$

and

$$E[\det(I_p + X)^{-h}] = \frac{\Gamma_p(n+h)\Gamma_p(m+n)}{\Gamma_p(n)\Gamma_p(m+n+h)}$$

$$\frac{{}_{r+1}F_{s+1}(a_1,\ldots,a_r,n+h;b_1,\ldots,b_s,m+n+h;\Theta)}{{}_{r+1}F_{s+1}(a_1,\ldots,a_r,n;b_1,\ldots,b_s,n+h;\Theta)},$$

$$\mathrm{Re}(h) > -n + \frac{1}{2}(p-1),$$

respectively.

(Roux, 1971)

8.29. For the density (8.13.5), prove the following.

$$E\Big[\prod_{k=1}^{q}\det(X_k)^{h_k}\Big] = \frac{\prod_{k=1}^{q}\Gamma_p(m_k+h_k)\Gamma_p(m+n)}{\prod_{k=1}^{q}\Gamma_p(m_k)\Gamma_p(m+n+h)}$$

$$\frac{{}_{r+1}F_{s+1}(a_1,\ldots,a_r,m+h;b_1,\ldots,b_s,m+n+h;\Theta)}{{}_{r+1}F_{s+1}(a_1,\ldots,a_r,m;b_1,\ldots,b_s,m+n;\Theta)},$$

$$\mathrm{Re}(h_k) > -m_k + \frac{1}{2}(p-1),\ k = 1,\ldots,q$$

where $h = \sum_{k=1}^{q} h_k$, and

$$E\Big[\det\Big(I_p - \sum_{k=1}^{q} X_k\Big)^h\Big] = \frac{\Gamma_p(m+n)\Gamma_p(n+h)}{\Gamma_p(m+n+h)\Gamma_p(n)}$$

$$\frac{_{r+1}F_{s+1}(a_1, \ldots, a_r, m; b_1, \ldots, b_s, m+n+h; \Theta)}{_{r+1}F_{s+1}(a_1, \ldots, a_r, m; b_1, \ldots, b_s, m+n; \Theta)},$$

$$\mathrm{Re}(h) > -n + \frac{1}{2}(p-1),$$

respectively.

(Roux, 1971)

8.30. For the density (8.13.6), prove the following.

$$E\Big[\prod_{k=1}^{q} \det(X_k)^{h_k}\Big] = \frac{\prod_{k=1}^{q} \Gamma_p(m_k + h_k)\Gamma_p(n-h)}{\prod_{k=1}^{q} \Gamma_p(m_k)\Gamma_p(n)}$$

$$\frac{_{r+1}F_{s+1}(a_1, \ldots, a_r, m+h; b_1, \ldots, b_s, m+n; \Theta)}{_{r+1}F_{s+1}(a_1, \ldots, a_r, m; b_1, \ldots, b_s, m+n; \Theta)},$$

$$\mathrm{Re}(h_k) > -m_k + \frac{1}{2}(p-1), \ k = 1, \ldots, q,$$

$$\mathrm{Re}(h) < n - \frac{1}{2}(p-1),$$

where $h = \sum_{k=1}^{q} h_k$, and

$$E\Big[\det\Big(I_p + \sum_{k=1}^{q} X_k\Big)^{-h}\Big] = \frac{\Gamma_p(m+n)\Gamma_p(n+h)}{\Gamma_p(m+n+h)\Gamma_p(n)}$$

$$\frac{_{r+1}F_{s+1}(a_1, \ldots, a_r, m; b_1, \ldots, b_s, m+n+h; \Theta)}{_{r+1}F_{s+1}(a_1, \ldots, a_r, m; b_1, \ldots, b_s, m+n; \Theta)},$$

$$\mathrm{Re}(h) > -n + \frac{1}{2}(p-1),$$

respectively.

(Roux, 1971)

8.31. Let $W \sim G_p(n_1, I_p)$ and $U \sim CH_p(n_2, \alpha, \beta, \text{kind } 1)$ be independent. Let $X = W + U$ and $Y = (W + U)^{-\frac{1}{2}} U (W + U)^{-\frac{1}{2}}$. Then show that
(i) the random matrix X is distributed as GHF type I with density

$$\frac{\Gamma_p(\alpha)\Gamma_p(\beta - n_2)\Gamma_p(n_1)}{\Gamma_p(\beta)\Gamma_p(\alpha - n_2)\Gamma_p(n_1 + n_2)\Gamma_p(n_2)} \, \mathrm{etr}(-X)\det(X)^{n_1+n_2-\frac{1}{2}(p+1)}$$

$$_2F_2(n_2, \beta - \alpha; n_1 + n_2, \beta; X), \ X > 0,$$

and

(ii) the random matrix Y is distributed as GHF type II with density

$$\frac{\Gamma_p(\alpha)\Gamma_p(\beta - n_2)\Gamma_p(n_1 + n_2)}{\Gamma_p(\beta)\Gamma_p(\alpha - n_2)\Gamma_p(n_1)\Gamma_p(n_2)} \det(Y)^{n_2 - \frac{1}{2}(p+1)} \det(I_p - Y)^{n_1 - \frac{1}{2}(p+1)}$$

$$_2F_1(n_1 + n_2, \beta - \alpha; \beta; Y),\ 0 < Y < I_p.$$

CHAPTER 9

GENERAL FAMILIES OF MATRIX VARIATE DISTRIBUTIONS

9.1. INTRODUCTION

In Chapters 1 through 7 we have considered a number of models leading to matrix variate generalizations of well known continuous distributions. These distributions usually arise as sampling distributions, when the underlying population is multivariate normal.

In this chapter we study families of distributions which are defined through functional form assumption, either on density function, or on characteristic function, or invariance property.

9.2. MATRIX VARIATE LIOUVILLE DISTRIBUTIONS

In this section we study a family of distributions defined through functional form assumption on the density.

The random variables $x_1, \ldots, x_r$ are said to have Liouville distribution of the first kind if their joint p.d.f. is proportional to

$$\prod_{i=1}^{r} x_i^{a_i-1} g\left(\sum_{i=1}^{r} x_i\right), \ 0 < x_i < \infty, \ a_i > 0, \ i = 1, \ldots, r. \tag{9.2.1}$$

Here g is a measurable positive real valued function defined on the interval $(0, \infty)$ such that $\int_0^\infty g(\tau)\tau^{s-1} \, d\tau$ exists for all $s > 0$.

The random variables $y_1, \ldots, y_r$ are said to have Liouville distribution of the second kind if their joint p.d.f. is proportional to

$$\prod_{i=1}^{r} y_i^{b_i-1} g\left(\sum_{i=1}^{r} y_i\right), \ 0 < y_i < 1, \ \sum_{i=1}^{r} y_i < 1, \ b_i > 0, \ i = 1, \ldots, r, \tag{9.2.2}$$

where g is a measurable positive real valued function defined on the interval $(0, 1)$ such that $\int_0^1 g(\tau)\tau^{s-1}\, d\tau$ exists for all $s > 0$.

These families of distributions were defined by Marshall and Olkin (1979), and Sivazlian (1981). They include Dirichlet distributions, and have found applications in compositional data (Aitchison, 1986), and life time data (Barlow and Mendal, 1992). The distributional properties of these families and their extensions have been studied by Sivazlian (1981), R. D. Gupta and Richards (1987, 1990, 1991, 1992), Fang, Kotz and Ng (1990), Song and Gupta (1997), and Gupta and Song (1996).

In this section we give matrix variate generalizations of (9.2.1) and (9.2.2) studied by R. D. Gupta and Richards (1987).

DEFINITION 9.2.1. *The $p \times p$ symmetric positive definite random matrices $X_1, \ldots, X_r$ are said to have Liouville distribution of the first kind if their joint p.d.f. is proportional to*

$$\prod_{i=1}^{r} \det(X_i)^{a_i - \frac{1}{2}(p+1)} g\Big(\sum_{i=1}^{r} X_i\Big),\ X_i > 0,\ a_i > \frac{1}{2}(p-1),\ i = 1, \ldots, r, \qquad (9.2.3)$$

where $g(\cdot)$ is positive, continuous, supported on $S = \{X\,(p \times p) : X > 0\}$ such that

$$\int_{T>0} \det(T)^{a - \frac{1}{2}(p+1)} g(T)\, dT < \infty,$$

and $a = \sum_{i=1}^{r} a_i$.

This distribution will be denoted by $L_r^{(1)}(g, a_1, \ldots, a_r)$.

The normalizing constant of the density (9.2.3) depends on the function g and for given g, can be evaluated explicitly. In general, this constant can be written in terms of Weyl fractional integral defined below.

If a real valued continuous function f defined on the space of $p \times p$ symmetric positive definite matrices satisfies the condition

$$\int_{T>0} \det(T)^{\alpha - \frac{1}{2}(p+1)} g(T)\, dT < \infty, \qquad (9.2.4)$$

where $\alpha > \frac{1}{2}(p-1)$, then the Weyl fractional integral of order α of f is defined as

$$W^\alpha f(T) = \frac{1}{\Gamma(\alpha)} \int_{S>T} \det(S - T)^{\alpha - \frac{1}{2}(p+1)} f(S)\, dS. \qquad (9.2.5)$$

Properties of $W^\alpha f(T)$ are given by Gindikin (1964), and Rooney (1972) for $p = 1$, and by Richards (1984) for arbitrary p. There is one to one correspondence between $f(\cdot)$ and its Weyl fractional derivative $W^\alpha f(\cdot)$. The Weyl fractional integral W^α also satisfies the semigroup property $W^{\alpha+\beta} = W^\alpha W^\beta$, $\alpha > \frac{1}{2}(p-1)$, $\beta > \frac{1}{2}(p-1)$.

Now we turn to the evaluation of the normalizing constant A of the density (9.2.3),

$$\frac{1}{A} = \int_{X_1>0} \cdots \int_{X_r>0} \prod_{i=1}^{r} \det(X_i)^{a_i - \frac{1}{2}(p+1)} g\Big(\sum_{i=1}^{r} X_i\Big) \prod_{i=1}^{r} dX_i$$

$$= \frac{\prod_{i=1}^{r} \Gamma_p(a_i)}{\Gamma_p(a)} \int_{T>0} \det(T)^{a - \frac{1}{2}(p+1)} g(T)\, dT. \qquad (9.2.6)$$

The last step is obtained by using Theorem 1.4.4. From the definition of Weyl fractional integral (9.2.5), it follows that

$$\frac{1}{A} = \prod_{i=1}^{r} \Gamma_p(a_i) W^a g(0). \tag{9.2.7}$$

DEFINITION 9.2.2. *The $p \times p$ symmetric positive definite random matrices $Y_1, \ldots, Y_r$ are said to have Liouville distribution of the second kind if their joint p.d.f. is proportional to*

$$\prod_{i=1}^{r} \det(Y_i)^{b_i - \frac{1}{2}(p+1)} g\left(\sum_{i=1}^{r} Y_i\right), \quad 0 < Y_i < I_p, \quad \sum_{i=1}^{r} Y_i < I_p$$

$$b_i > \frac{1}{2}(p-1), \quad i = 1, \ldots, r, \tag{9.2.8}$$

where $g(\cdot)$ is positive, continuous, supported on $S = \{X \, (p \times p) : 0 < X < I_p\}$ such that

$$\int_S \det(T)^{b - \frac{1}{2}(p+1)} g(T) \, dT < \infty,$$

and $b = \sum_{i=1}^{r} b_i$.

This distribution will be denoted by $L_r^{(2)}(g, b_1, \ldots, b_r)$.

The normalizing constant of the density (9.2.8) is given in (9.2.7). Next we give some special cases of the above densities.

(i) In (9.2.3) taking $g(T) = \det(I_p + T)^{-\sum_{j=1}^{r+1} a_j}$, where $a_{r+1} > \frac{1}{2}(p-1)$, we get the matrix variate Dirichlet type II distribution with parameters $(a_1, \ldots, a_r; a_{r+1})$.

(ii) In (9.2.3) taking $g(T) = \text{etr}(-T)$, we get the product of Wishart densities.

(iii) In (9.2.8) taking $g(T) = \det(I_p - T)^{b_{r+1} - \frac{1}{2}(p+1)}$, where $b_{r+1} > \frac{1}{2}(p-1)$, we get the matrix variate Dirichlet type I distribution with parameters $(b_1, \ldots, b_r; b_{r+1})$.

Next we study some properties of above distributions. The first theorem gives relationship between the first kind and the second kind. The proof is similar to that of Theorem 6.3.1.

THEOREM 9.2.1. *(i) If $(Y_1, \ldots, Y_r) \sim L_r^{(2)}(g, b_1, \ldots, b_r)$ and*

$$X_i = \left(I_p - \sum_{j=1}^{r} Y_j\right)^{-\frac{1}{2}} Y_i \left(I_p - \sum_{j=1}^{r} Y_j\right)^{-\frac{1}{2}}, \quad i = 1, \ldots, r,$$

then $(X_1, \ldots, X_r) \sim L_r^{(1)}(f, b_1, \ldots, b_r)$ where

$$f(T) = \det(I_p + T)^{-b - \frac{1}{2}(p+1)} g(T(I_p + T)^{-1}), \quad T > 0.$$

Further there is one to one correspondence between $g(\cdot)$ and $f(\cdot)$.

(ii) If $(X_1, \ldots, X_r) \sim L_r^{(1)}(g, a_1, \ldots, a_r)$ and

$$Y_i = \left(I_p + \sum_{j=1}^{r} X_j\right)^{-\frac{1}{2}} X_i \left(I_p + \sum_{j=1}^{r} X_j\right)^{-\frac{1}{2}}, \quad i = 1, \ldots, r,$$

then $(Y_1, \ldots, Y_r) \sim L_r^{(2)}(f, a_1, \ldots, a_r)$, where

$$f(T) = \det(I_p - T)^{-a - \frac{1}{2}(p+1)} g(T(I_p - T)^{-1}), \ 0 < T < I_p.$$

Further there is one to one correspondence between $g(\cdot)$ and $f(\cdot)$.

In the following theorems we give stochastic representation of $X_1, \ldots, X_r$ where $(X_1, \ldots, X_r) \sim L_r^{(i)}(g, a_1, \ldots, a_r)$, $i = 1, 2$.

THEOREM 9.2.2. Let $(X_1, \ldots, X_r) \sim L_r^{(i)}(g, a_1, \ldots, a_r)$. Define $X_j = Y_r^{\frac{1}{2}} Y_j Y_r^{\frac{1}{2}}$, $j = 1, \ldots, r - 1$, and $X_r = Y_r^{\frac{1}{2}}(I_p - \sum_{j=1}^{r-1} Y_j) Y_r^{\frac{1}{2}}$. Then $(Y_1, \ldots, Y_{r-1})$ and Y_r are independent, $(Y_1, \ldots, Y_{r-1}) \sim D_p^I(a_1, \ldots, a_{r-1}; a_r)$ and $Y_r \sim L_1^{(i)}(g, \sum_{i=1}^r a_i)$.

From this result, in view of Theorem 6.3.4, it is easily seen that

$$\left(\sum_{i=1}^r X_i\right)^{-\frac{1}{2}} \left(\sum_{i=1}^s X_i\right) \left(\sum_{i=1}^r X_i\right)^{-\frac{1}{2}} \sim B_p^I\left(\sum_{i=1}^s a_i, \sum_{i=s+1}^r a_i\right), \ s < r.$$

THEOREM 9.2.3. Let $(X_1, \ldots, X_r) \sim L_r^{(i)}(g, a_1, \ldots, a_r)$. Define

$$X_1 = Y_r^{\frac{1}{2}} \prod_{j=1}^{r-1} Y_{r-j}^{\frac{1}{2}} \prod_{j=1}^{r-1} Y_j^{\frac{1}{2}} Y_r^{\frac{1}{2}},$$

$$X_2 = Y_r^{\frac{1}{2}} \prod_{j=2}^{r-1} Y_{r+1-j}^{\frac{1}{2}} (I_p - Y_1) \prod_{j=2}^{r-1} Y_j^{\frac{1}{2}} Y_r^{\frac{1}{2}},$$

$$\vdots$$

$$X_r = Y_r^{\frac{1}{2}} (I_p - Y_{r-1}) Y_r^{\frac{1}{2}}.$$

Then $Y_1, \ldots, Y_r$ are independent, $Y_k \sim B_p^I(\sum_{i=1}^k a_i, a_{k+1})$, $k = 1, \ldots, r - 1$, and $Y_r \sim L_1^{(i)}(g, \sum_{i=1}^r a_i)$.

THEOREM 9.2.4. Let $(X_1, \ldots, X_r) \sim L_r^{(i)}(g, a_1, \ldots, a_r)$. Define

$$X_1 = Y_r^{\frac{1}{2}} \prod_{j=1}^{r-1} (I_p + Y_{r-j})^{-\frac{1}{2}} \prod_{j=1}^{r-1} (I_p + Y_j)^{-\frac{1}{2}} Y_r^{\frac{1}{2}},$$

$$X_2 = Y_r^{\frac{1}{2}} \prod_{j=2}^{r-1} (I_p + Y_{r+1-j})^{-\frac{1}{2}} Y_1 \prod_{j=2}^{r-1} (I_p + Y_j)^{-\frac{1}{2}} Y_r^{\frac{1}{2}},$$

$$\vdots$$

$$X_r = Y_r^{\frac{1}{2}} (I_p + Y_{r-1})^{-\frac{1}{2}} Y_{r-1} (I_p + Y_{r-1})^{-\frac{1}{2}} Y_r^{\frac{1}{2}}.$$

Then $Y_1, \ldots, Y_r$ are independent, $Y_k \sim B_p^{II}(a_{k+1}, \sum_{i=1}^k a_i)$, $k = 1, \ldots, r - 1$, and $Y_r \sim L_1^{(i)}(g, \sum_{i=1}^r a_i)$.

THEOREM 9.2.5. *Let* $(X_1, \ldots, X_r) \sim L_r^{(i)}(g, a_1, \ldots, a_r)$. *Define*

$$X_1 = Y_r^{\frac{1}{2}} Y_1 Y_r^{\frac{1}{2}},$$

$$X_2 = Y_r^{\frac{1}{2}} (I_p - Y_1)^{\frac{1}{2}} Y_2 (I_p - Y_1)^{\frac{1}{2}} Y_r^{\frac{1}{2}},$$

$$\vdots$$

$$X_r = Y_r^{\frac{1}{2}} (I_p - Y_1)^{\frac{1}{2}} \cdots (I_p - Y_{r-1})^{\frac{1}{2}} (I_p - Y_{r-1})^{\frac{1}{2}} \cdots (I_p - Y_1)^{\frac{1}{2}} Y_r^{\frac{1}{2}}.$$

Then $Y_1, \ldots, Y_r$ *are independent,* $Y_k \sim B_p^I(a_k, \sum_{i=k+1}^r a_i)$, $k = 1, \ldots, r-1$, *and* $Y_r \sim L_1^{(i)}(g, \sum_{i=1}^r a_i)$.

THEOREM 9.2.6. *Let* $(X_1, \ldots, X_r) \sim L_r^{(i)}(g, a_1, \ldots, a_r)$. *Then*
 (i) $(X_1, \ldots, X_s) \sim L_s^{(i)}(g_s, a_1, \ldots, a_s)$, $s < r$,
where $g_s(T) = W^a g(T)$ *is the Weyl fractional integral of order* $a = \sum_{i=s+1}^r a_i$,
 (ii) $(X_{s+1}, \ldots, X_r)|(X_1, \ldots, X_s) \sim L_{r-s}^{(i)}(f_s, a_{s+1}, \ldots, a_r)$, $s < r$,
where $f_s(T) = \frac{g(\sum_{i=1}^s X_i + T)}{g_s(\sum_{i=1}^s X_i)}$.

Proof: (i) The joint p.d.f. of $X_1, \ldots, X_s$ $(s < r)$ is

$$A \int_{X_{s+1}>0} \cdots \int_{X_r>0} \prod_{i=1}^r \det(X_i)^{a_i - \frac{1}{2}(p+1)} g\left(\sum_{i=1}^r X_i\right) \prod_{i=s+1}^r dX_i, \qquad (9.2.9)$$

where A is given by (9.2.7). Integrating $X_{s+1}, \ldots, X_r$, using Theorem 1.4.4, from (9.2.9), we get

$$A \frac{\prod_{i=s+1}^r \Gamma_p(a_i)}{\Gamma_p(\sum_{i=s+1}^r a_i)} \prod_{i=1}^s \det(X_i)^{a_i - \frac{1}{2}(p+1)} \int_{Y>0} \det(Y)^{\sum_{i=s+1}^r a_i - \frac{1}{2}(p+1)} g\left(\sum_{i=1}^s X_i + Y\right) dY$$

$$= A_1 \prod_{i=1}^s \det(X_i)^{a_i - \frac{1}{2}(p+1)} W^{\sum_{i=s+1}^r a_i} g\left(\sum_{i=1}^s X_i\right)$$

$$= A_1 \prod_{i=1}^s \det(X_i)^{a_i - \frac{1}{2}(p+1)} g_s\left(\sum_{i=1}^s X_i\right).$$

The last step is obtained by using the definition of Weyl fractional integral.
 (ii) The proof is straightforward. ∎

9.3. MATRIX VARIATE SPHERICAL DISTRIBUTIONS

Sometimes it is desirable to study robustness of normal theory model under nonnormal situation. The class of elliptically contoured distributions in such studies is useful because the density functions of such distributions have the same elliptical shape as the normal density. For properties of these distributions one can refer to Kelker (1970), Chmielewski (1981), Cambanis, Huang and Simons (1981) and Muirhead (1982).

In this and subsequent sections we study the matrix variate generalizations of these families. We begin by defining matrix variate spherical distribution studied by Dawid (1977, 1978) and Fang and Chen (1984).

DEFINITION 9.3.1. *The random matrix $X\,(p \times n)$ is said to have*

(i) *right spherical distribution if $X \overset{d}{=} X\Lambda$, $\forall\, \Lambda \in O(n)$,*

(ii) *left spherical distribution if $X \overset{d}{=} \Gamma X$, $\forall\, \Gamma \in O(p)$, and*

(iii) *spherical distribution if $X \overset{d}{=} \Gamma X \Lambda$, $\forall\, \Gamma \in O(p)$, $\forall\, \Lambda \in O(n)$.*

It may be noted that if $X\,(p \times n)$ is right spherical, then, for $T\,(p \times n)$, its characteristic function is of the form $\phi(TT')$. We have

$$\Psi_X(T) = E[\mathrm{etr}(\iota T X')]$$

$$= E[\mathrm{etr}(\iota T \Lambda \Lambda' X')],\ \Lambda \in O(n)$$

$$= E[\mathrm{etr}(\iota (T\Lambda)(X\Lambda)')],\ \Lambda \in O(n)$$

$$= E[\mathrm{etr}(\iota (T\Lambda) X')],\ \text{ since } X \overset{d}{=} X\Lambda,\ \forall\, \Lambda \in O(n)$$

$$= \Psi_X(T\Lambda).$$

Hence $\Psi_X(T)$ is invariant under $O(n)$ and is a function of the maximal invariant under $O(n)$, *i.e.*, for some function ϕ,

$$\Psi_X(T) = \phi(TT').$$

If $X\,(p \times n)$ is right spherical with the characteristic function $\phi(TT')$, we will denote it by $X \sim RS_{p,n}(\phi)$. If $X\,(p \times n)$ is left spherical, we will write $X \sim LS_{p,n}(\phi)$.

THEOREM 9.3.1. *If $X\,(p \times n)$ is right spherical, then (i) X' is left spherical (ii) $-X$ is right spherical, $-X \overset{d}{=} X$.*

THEOREM 9.3.2. *Let $X \sim RS_{p,n}(\phi)$.*

(i) *For a constant matrix $A\,(q \times p)$, $AX \sim RS_{q,n}(\psi)$ where $\psi(TT') = \phi(A'TT'A)$, $T\,(q \times n)$.*

(ii) *For $X = (\,X_1\ \ X_2\,)$, where X_1 is $p \times m$, $X_1 \sim RS_{p,m}(\phi)$.*

In Chapter 8 we have defined uniform distribution over Stiefel manifold. This distribution belongs to the class of right spherical distributions, as shown in Theorem 8.2.1. Its converse is given in the next theorem.

THEOREM 9.3.3. *Let $X\,(p \times n)$ be right spherical and $XX' = I_p$, $p \le n$. Then $X \sim \mathcal{U}_{p,n}$.*

THEOREM 9.3.4. *The distribution of right spherical matrix $X\,(p \times n)$ is fully determined by that of XX'.*

Proof: Let Y $(p \times n)$ be another right spherical matrix such that $XX' \overset{d}{=} YY'$. Further let $U \sim \mathcal{U}_{n,n}$ with characteristic function $\omega(ZZ')$, Z $(n \times n)$. The characteristic function $\phi(TT')$, T $(p \times n)$, of X is

$$\phi_X(TT') = E_X[\text{etr}(\iota TX')]$$

$$= E_X\left\{ \int_{O(n)} \text{etr}(\iota TX') \, [dU] \right\}$$

$$= E_X\left\{ \int_{O(n)} \text{etr}(\iota TUU'X') \, [dU] \right\}$$

$$= E_X\left\{ \int_{O(n)} \text{etr}(\iota T'XU') \, [dU] \right\}$$

$$= E_X[\omega(T'XX'T)]$$

$$= E_Y[\omega(T'YY'T)]$$

$$= \phi_Y(TT').$$

Therefore $X \overset{d}{=} Y$. ∎

From the above theorem, it follows that the uniform distribution is the unique right spherical distribution over $O(p, n)$. For right spherical matrix, in general the density may not exist. However if X has a density with respect to a Lebesgue measure on $\mathbb{R}^{p \times n}$, then it is of the form $f(XX')$. Some examples of this distribution are given below.

(i) When $X \sim N_{p,n}(0, \Sigma \otimes I_n)$, the density of X is

$$(2\pi)^{-\frac{1}{2}np} \det(\Sigma)^{-\frac{1}{2}n} \text{etr}\left(-\frac{1}{2}\Sigma^{-1}XX' \right), \ X \in \mathbb{R}^{p \times n},$$

with the characteristic function $\text{etr}(-\frac{1}{2}\Sigma TT')$.

(ii) When $X \sim T_{p,n}(\delta, 0, \Sigma, I_n)$, the density of X is

$$\frac{\Gamma_p[\frac{1}{2}(\delta + n + p - 1)]}{(2\pi)^{\frac{1}{2}np} \det(\Sigma)^{\frac{1}{2}n} \Gamma_p[\frac{1}{2}(\delta + p - 1)]} \det(I_p + \Sigma^{-1}XX')^{-\frac{1}{2}(\delta+n+p-1)}, \ X \in \mathbb{R}^{p \times n},$$

with the characteristic function

$$\left\{ \Gamma_p\left[\frac{1}{2}(\delta + p - 1)\right] \right\}^{-1} B_{-\frac{1}{2}(\delta+p-1)}\left(\frac{1}{4}\Sigma TT' \right)$$

where $B_\delta(\cdot)$ is the Herz's Bessel function of second kind of order δ.

THEOREM 9.3.5. *If X $(p \times n)$ is right spherical and K $(n \times m)$ is a fixed matrix, then the distribution of XK depends on K only through $K'K$.*

Proof: Let the matrix H $(n \times m)$ be such that $H'H = K'K$. Then $H = \Gamma K$ for some $\Gamma \in O(n)$. Hence $XH = X\Gamma K \overset{d}{=} XK$, from which it follows that the distribution of XK depends on K only through $K'K$. ∎

In the above theorem, if $K'K = I_m$ then the distribution of XK is right spherical. This can be shown by evaluating the c.f. of XK.

Let $X = (X_1 \quad X_2)$, $X_1 (p \times (n - m))$, $X_2 (p \times m)$, $K' = (K_1' \quad K_2')$, $K_1 ((n - m) \times m) = 0$, $K_2 (m \times m) = I_m$. Then $K'K = I_m$, and therefore $XK = X_2$ is right spherical.

It is easy to show that if the distribution of X is mixture of right spherical distributions, then X is right spherical. It follows that if $X (p \times n)$, conditional on a random variable v, is right spherical and $Q (q \times p)$ is a function of v, then QX is right spherical.

The results given above have obvious analogues for left spherical distributions.

Now from the theory of spherical distributions many results for uniform distribution can be derived. In the next theorem we give some results for the uniform distribution.

THEOREM 9.3.6. *Let $U \sim \mathcal{U}_{p,n}$.*

(i) *Partition $U' = (U_1' \quad U_2')$, $U_1 (q \times n)$, $1 \leq q < p$. Then $U_1 \sim \mathcal{U}_{q,n}$.*

(ii) *For fixed $\Gamma \in O(q,p)$, $p \geq q$, $\Gamma U \sim \mathcal{U}_{q,n}$.*

(iii) *U is spherical.*

(iv) *If $n = p$, then $U' = U^{-1} \sim \mathcal{U}_{p,p}$.*

Proof: (i) Since U is also right spherical, U_1 is right spherical. From the fact that $U_1 U_1' = I_q$, and Theorem 9.3.2, the result follows.

(ii) Note that ΓU is right spherical and $(\Gamma U)(\Gamma U)' = I_q$. Therefore, from Theorem 9.3.3, $\Gamma U \sim \mathcal{U}_{q,n}$.

(iii) For $q = p$, $\Gamma \in O(p)$, and $\Gamma U \sim \mathcal{U}_{p,n}$. Hence U is left spherical. Since U is also right spherical, the result follows.

(iv) Since $UU' = I_p$, $U' = U^{-1}$, from (iii) U is left spherical and hence $U' = U^{-1}$ is right spherical. Now the result follows from Theorem 9.3.2, since $U^{-1}(U^{-1})' = I_p$. ■

We now study the stochastic representation of spherical distribution.

THEOREM 9.3.7. *Let $X \sim RS_{p,n}(\phi)$. Then there exists a random matrix $A (p \times p)$ such that*

$$X \stackrel{d}{=} AU \tag{9.3.1}$$

where $U \sim \mathcal{U}_{p,n}$ is independent of A.

Proof: For $X (p \times n)$, we can find $A (p \times p)$ such that $XX' \stackrel{d}{=} AA'$. Let $U \sim \mathcal{U}_{p,n}$ be independent of A. Define

$$Y = AU.$$

Then $YY' = AA' \stackrel{d}{=} XX'$. From Theorem 9.3.4 it follows that $X \stackrel{d}{=} Y = AU$. ■

The matrix A in stochastic representation (9.3.1) is not unique. One can take it to be lower (upper) triangular matrix with non-negative diagonal elements or right spherical matrix with $A \geq 0$. Further, in addition, if we assume that $P(\det(XX') = 0) = 0$, then the distribution of A is unique.

The next theorem proves the uniqueness of A when it is lower triangular.

THEOREM 9.3.8. *Let* $X \sim RS_{p,n}(\phi)$ *and* $P(\det(XX') \neq 0) = 1$. *Then for A, B lower triangular matrices with positive diagonal elements and* $U \sim \mathcal{U}_{p,n}$, $Q \sim \mathcal{U}_{p,n}$,

    (i) $X \stackrel{d}{=} AU$ *and* $X \stackrel{d}{=} BU \Rightarrow A \stackrel{d}{=} B$,

    (ii) $X \stackrel{d}{=} AU$ *and* $X \stackrel{d}{=} AQ \Rightarrow U \stackrel{d}{=} Q$.

Proof: (i) Note that $AA' = BB'$. Now consider one to one function $f(A) = AA'$. Then for any Borel measurable function $h(\cdot)$,

$$E\{h(A)\} = E\{h(f^{-1}(AA'))\}$$
$$= E\{h(f^{-1}(BB'))\}$$
$$= E\{h(B)\}$$

and hence $A \stackrel{d}{=} B$.

    (ii) Define the function $g(AQ) = (A, Q)$. Then $(A, Q) = g(AQ) \stackrel{d}{=} g(X) \stackrel{d}{=} g(AU) = (A, U)$, and hence $Q \stackrel{d}{=} U$. ∎

For studying the spherical distribution, singular value decomposition of the matrix X $(p \times n)$ provides a powerful tool. When $p \leq n$, let $X = G\Lambda H$, where $G \in O(p)$, $H \in O(p, n)$, $\Lambda = \text{diag}(\lambda_1, \ldots, \lambda_p)$, $\lambda_1 \geq \lambda_2 \geq \cdots \geq \lambda_p \geq 0$, and λ_i 's are the eigenvalues of $(XX')^{\frac{1}{2}}$.

THEOREM 9.3.9. *If* X $(p \times n)$, $p \leq n$, *is spherical, then*

$$X \stackrel{d}{=} U\Lambda V \qquad (9.3.2)$$

where $U \sim \mathcal{U}_{p,p}$, $V \sim \mathcal{U}_{p,n}$ *and* Λ *are mutually independent.*

Proof: Let $X = G\Lambda H$, $G \in O(p)$, $H \in O(p, n)$, and $\Lambda = \text{diag}(\lambda_1, \ldots, \lambda_p)$, be the singular value decomposition of the matrix X $(p \times n)$. Further let $U^* \sim \mathcal{U}_{p,p}$, $V* \sim \mathcal{U}_{n,n}$ be independent of (G, Λ, H), and define $U = U^*G$, $V = HV^*$, and $X^* = U^*XV^*$. Then, given (G, Λ, H), $U \sim \mathcal{U}_{p,p}$, $V \sim \mathcal{U}_{p,n}$ are independent. Hence U, V and Λ are independent. Now, since X is spherical, for given U^* and V^*, $X^* = U^*XV^* \stackrel{d}{=} X$. The proof is completed by noting that $X^* = U\Lambda V$. ∎

THEOREM 9.3.10. *If* X $(p \times n)$ *is spherical, then its characteristic function is of the form* $\phi(\lambda(TT'))$, *where* T $(p \times n)$, $\lambda(TT') = \text{diag}(\tau_1, \ldots, \tau_1)$, *and* $\tau_1 \geq \cdots \geq \tau_p \geq 0$ *are the eigenvalues of* TT'.

Proof: From the definition of spherical distribution, it follows that the characteristic function of X is

$$\Psi_X(T) = E[\text{etr}(\iota TX')]$$
$$= E[\text{etr}(\iota(\Gamma T\Delta)(\Gamma X\Delta)')], \quad \Gamma \in O(p), \Delta \in O(n)$$
$$= \Psi_X(\Gamma T\Delta).$$

Thus the characteristic function $\Psi_X(T)$ satisfies the equation $\Psi_X(T) = \Psi_X(\Gamma T \Delta)$ for every $\Gamma \in O(p)$ and $\Delta \in O(n)$. The maximal invariant of T in this case is $\lambda(TT')$. Hence $\Psi_X(T) = \phi(\lambda(TT'))$ for some function ϕ. ∎

From the above theorem it follows that, if the density of a spherical matrix X exists, then it is of the form $f(\lambda(XX'))$.

THEOREM 9.3.11. *Let $X \sim RS_{p,n}(\phi)$. If the second order moments of X exist then*

(i) $E(X) = 0$,
(ii) $\mathrm{cov}(X) = V \otimes I_n$, where $V = E(\boldsymbol{x}_1 \boldsymbol{x}_1')$, $X = (\boldsymbol{x}_1, \ldots, \boldsymbol{x}_n)$.

Proof: (i) Since $-X \overset{d}{=} X$, it follows that $E(X) = 0$.
(ii) See Fang and Zhang (1990, p. 104). ∎

THEOREM 9.3.12. *Let $X \sim RS_{p,n}(\phi)$ with the density $f(XX')$. Then the density of $S = XX'$, $n \geq p$, is*

$$\frac{\pi^{\frac{1}{2}np}}{\Gamma_p(\frac{1}{2}n)} \det(S)^{\frac{1}{2}(n-p-1)} f(S), \; S > 0.$$

Proof: Let $h(\cdot)$ be a non-negative Borel measurable function. Then

$$E[h(S)] = \int_{X \in \mathbb{R}^{p \times n}} h(XX') f(XX') \, dX$$

$$= \int_{S>0} \int_{XX'=S} h(XX') f(XX') \, dX \, dS$$

$$= \int_{S>0} h(S) f(S) \, dS \int_{XX'=S} dX$$

$$= \frac{\pi^{\frac{1}{2}np}}{\Gamma_p(\frac{1}{2}n)} \int_{S>0} h(S) \det(S)^{\frac{1}{2}(n-p-1)} f(S) \, dS.$$

The last step is obtained by using Theorem 1.4.10. Hence the density of S is

$$\frac{\pi^{\frac{1}{2}np}}{\Gamma_p(\frac{1}{2}n)} \det(S)^{\frac{1}{2}(n-p-1)} f(S), \; S > 0. \; ∎$$

From the above theorem, it follows that if $X \sim RS_{p,n}(\phi)$, with density $f(XX')$, then $XX' \sim L_1^{(1)}(f, \frac{1}{2}n)$, $n \geq p$.

COROLLARY 9.3.12.1. *Let $X \sim N_{p,n}(0, \Sigma \otimes I_n)$. Then*

$$f(XX') = (2\pi)^{-\frac{1}{2}np} \det(\Sigma)^{-\frac{1}{2}n} \mathrm{etr}\left(-\frac{1}{2}\Sigma^{-1}XX'\right), \; X \in \mathbb{R}^{p \times n},$$

and $S = XX' \sim W_p(n, \Sigma)$, $n \geq p$, with the density

$$\left\{ 2^{\frac{1}{2}np} \Gamma_p\left(\frac{1}{2}n\right) \det(\Sigma)^{\frac{1}{2}n} \right\}^{-1} \det(S)^{\frac{1}{2}(n-p-1)} \mathrm{etr}\left(-\frac{1}{2}\Sigma^{-1}S\right), \; S > 0.$$

COROLLARY 9.3.12.2. *Let* $X \sim T_{p,n}(\delta, 0, \Sigma, I_n)$. *Then*

$$f(XX') = \frac{\Gamma_p[\frac{1}{2}(\delta + n + p - 1)]}{(2\pi)^{\frac{1}{2}np}\Gamma_p[\frac{1}{2}(\delta + p - 1)]}$$

$$\det(\Sigma)^{-\frac{1}{2}n}\det(I_p + \Sigma^{-1}XX')^{-\frac{1}{2}(\delta+n+p-1)}, \quad X \in \mathbb{R}^{p \times n},$$

and $S = XX' \sim GB_p^{II}(\frac{1}{2}n, \frac{1}{2}(\delta + p - 1); \Sigma, 0)$ *with the density*

$$\left\{\beta_p\left(\frac{1}{2}n, \frac{1}{2}(\delta + p - 1)\right)\right\}^{-1}\det(\Sigma)^{-\frac{1}{2}n}$$

$$\det(S)^{\frac{1}{2}(n-p-1)}\det(I_p + \Sigma^{-1}S)^{-\frac{1}{2}(\delta+n+p-1)}, \quad S > 0.$$

The above results have also been studied in Chapters 3 and 5 respectively. Theorem 9.3.12 can be generalized as follows and gives the joint distribution of several quadratic forms in terms of Liouville distribution.

THEOREM 9.3.13. *Let* $X \sim RS_{p,n}(\phi)$ *with the density* $f(XX')$. *Partition* X *as* $X = (X_1, \ldots, X_r)$, X_i $(p \times n_i)$, $n_i \geq p$, $i = 1, \ldots, r$, $\sum_{i=1}^r n_i = n$. *Define* $S_i = X_iX_i'$, $i = 1, \ldots, r$. *Then* $(S_1, \ldots, S_r) \sim L_r^{(1)}(f, \frac{1}{2}n_1, \ldots, \frac{1}{2}n_r)$ *with p.d.f.*

$$\frac{\pi^{\frac{1}{2}np}}{\prod_{i=1}^r \Gamma_p(\frac{1}{2}n_i)} \prod_{i=1}^r \det(S_i)^{\frac{1}{2}(n_i-p-1)} f\left(\sum_{i=1}^r S_i\right), \quad S_i > 0, \ i = 1, \ldots, r.$$

Proof: The proof is similar to the proof of Theorem 9.3.12 and hence is not given here. ∎

The above theorem has been generalized further by Anderson and Fang (1987). Let $X \sim RS_{p,n}(\phi)$ with the density $f(XX')$, and A $(n \times n)$ be a symmetric matrix. Then

$$XAX' \sim L_1^{(1)}\left(f_1, \frac{1}{2}k\right) \tag{9.3.3}$$

where $f_1(T) = W^{\frac{1}{2}(n-k)}f(T)$, if and only if $A^2 = A$ and rank$(A) = k \geq p$. Further, let A_1 $(n \times n), \ldots, A_s$ $(n \times n)$ be symmetric matrices. Then

$$(XA_1X', \ldots, XA_sX') \sim L_s^{(1)}\left(f_1, \frac{1}{2}n_1, \ldots, \frac{1}{2}n_s\right), \tag{9.3.4}$$

where $f_1(T) = W^{\frac{1}{2}(n-n_1-\cdots-n_s)}f(T)$, if and only if $A_iA_j = \delta_{ij}A_i$, and rank$(A_i) = n_i$, $n_i \geq p$, $i, j = 1, \ldots, s$.

It may be mentioned that the class of matrix variate spherical distributions studied here are generalizations of multivariate spherical distributions. There are several other classes of matrix variate generalizations of multivariate spherical distributions studied by Jensen and Good (1981), Fang and Chen (1984, 1986) and Fang and Anderson (1990).

9.4. MATRIX VARIATE ELLIPTICALLY CONTOURED DISTRIBUTIONS

The class of matrix variate elliptically contoured distributions can be defined in many ways, *e.g.*, see Fang and Zhang (1990) and Gupta and Varga (1993).

DEFINITION 9.4.1. *Let $X\,(p \times n) \sim RS_{p,n}(\phi)$ and $M\,(p \times m)$, $B\,(n \times m)$ be constant matrices. Then the random matrix $Y\,(p \times m)$, where*

$$Y \stackrel{d}{=} M + XB, \; \Sigma = B'B$$

is said to have matrix variate elliptically contoured distribution, denoted by $Y \sim ERS_{p,m}(M, \Sigma, \phi)$.

The characteristic function of Y can be shown to be

$$\mathrm{etr}(\iota T'M)\phi(T\Sigma T'), \; T\,(p \times m). \tag{9.4.1}$$

If the density of $Y\,(p \times m)$ exists, then it has the form

$$\det(\Sigma)^{-\frac{1}{2}p} f((Y - M)\Sigma^{-1}(Y - M)'). \tag{9.4.2}$$

Next we give the distribution of a linear transformation of an elliptically contoured matrix.

THEOREM 9.4.1. *Let $Y\,(p \times m) \sim ERS_{p,m}(M, \Sigma, \phi)$ and $C\,(m \times q)$, $N\,(p \times q)$ be constant matrices. Then $Z = N + YC \sim ERS_{p,q}(N + MC, C'\Sigma C, \phi)$.*

Proof: The characteristic function of Z, evaluated at $T\,(p \times q)$ is

$$\begin{aligned}
\Psi_Z(T) &= E[\mathrm{etr}(\iota TZ')] \\
&= \mathrm{etr}(\iota TN')E[\mathrm{etr}(\iota(TC'Y'))] \\
&= \mathrm{etr}(\iota T(N' + C'M'))\phi(T(C'\Sigma C)T'). \tag{9.4.3}
\end{aligned}$$

The above expression of the characteristic function of Y is derived from the Definition 9.4.1 and the characteristic function (9.4.1). Now from (9.4.3) the desired result follows. ∎

COROLLARY 9.4.1.1. *Partition Y, M and Σ as $Y = (\,Y_1 \;\; Y_2\,)$, $Y_1\,(p \times q)$, $M = (\,M_1 \;\; M_2\,)$, $M_1\,(p \times q)$ and $\Sigma = \begin{pmatrix} \Sigma_{11} & \Sigma_{12} \\ \Sigma_{21} & \Sigma_{22} \end{pmatrix}$, $\Sigma_{11}\,(q \times q)$. Then $Y_1 \sim ERS_{p,q}(M_1, \Sigma_{11}, \phi)$.*

Proof: In the above theorem, let $N = 0$, and $C' = (\,I_q \;\; 0\,)$. Then $Y_1 = Z \sim ERS_{p,q}(M_1, \Sigma_{11}, \phi)$. ∎

THEOREM 9.4.2. *Let* $Y \sim ERS_{p,m}(M, \Sigma, \phi)$ *and partition* Y, M *and* Σ *as* $Y = \begin{pmatrix} Y_{11} & Y_{12} \\ Y_{21} & Y_{22} \end{pmatrix}$, $Y_{11}\,(q \times r)$, $M = \begin{pmatrix} M_{11} & M_{12} \\ M_{21} & M_{22} \end{pmatrix}$, $M_{11}\,(q \times r)$ *and* $\Sigma = \begin{pmatrix} \Sigma_{11} & \Sigma_{12} \\ \Sigma_{21} & \Sigma_{22} \end{pmatrix}$, $\Sigma_{11}\,(r \times r)$. *Then* $Y_{11} \sim ERS_{q,r}(M_{11}, \Sigma_{11}, \phi^*)$, *where* $\phi^*(A_{11}) = \phi(A)$, *for* $A\,(p \times p) = \begin{pmatrix} A_{11} & 0 \\ 0 & 0 \end{pmatrix}$, $A_{11}\,(q \times q)$.

Proof: The proof is straightforward and is left to the reader as an exercise. ∎

THEOREM 9.4.3. *Let* $Y \sim ERS_{p,m}(M, \Sigma, \phi)$. *If the second order moment of* Y *exist then*

    (i) $E(Y) = M$,

    (ii) $\mathrm{cov}(Y) = V \otimes \Sigma$, *where* $V = E(x_1 x_1')$, $X = (x_1, \ldots, x_n)$ *and* $Y \overset{d}{=} M + XB$, $\Sigma = B'B$.

For further results on matrix variate elliptically contoured distribution the reader is referred to Hayakawa (1986, 1987, 1989), Sutradhar and Ali (1989), Fang and Zhang (1990), Gupta and Verga (1991, 1994a, 1994c, 1994d, 1995b, 1997), Wong and Liu (1994), Li and Fang (1995), Girko and Gupta (1996), Gupta and Girko (1996) and Gupta (1998).

9.5. ORTHOGONALLY INVARIANT AND RESIDUAL INDEPENDENT MATRIX DISTRIBUTIONS

In the preceding chapters we have seen that the Wishart, gamma, beta type I and beta type II distributions are orthogonally invariant. That is, the distribution of $X\,(p \times p) > 0$ is same as that of $\Gamma X \Gamma'$, $\Gamma \in O(p)$. Many other properties follow from this fact, *i.e.*, the diagonal elements of X are identically distributed. Additionally for $X = TT'$, where T is a triangular matrix, the diagonal elements of T are independent. Motivated from these common properties, Khatri, Khattree and R. D. Gupta (1991) have defined the orthogonally invariant and residual independent, ORIARIM in short, family of distributions, $\mathcal{C}_p$.

DEFINITION 9.5.1. *The random symmetric positive definite matrix* $X\,(p \times p)$ *is said to have an ORIARIM distribution if*

    (i) for any $\Gamma \in O(p)$, *the distribution of* X *and* $\Gamma X \Gamma'$ *are identical, and*

    (ii) for any lower triangular factorization $X = TT'$, $T = (T_{ij})$, $T_{ii}\,(p_i \times p_i)$, $i = 1, \ldots, k$ *are independent, for any partition* $\{p_1, p_2, \ldots, p_k\}$ *of* p.

When X has ORIARIM distribution, we will write $X \in \mathcal{C}_p$. The matrix variate beta type I and type II, gamma $(C = I_p)$, Wishart $(\Sigma = I_p)$, and inverted Wishart $(\Psi = I_p)$ distributions belong to this class.

Next we give some properties of the class of ORIARIM distributions.

THEOREM 9.5.1. *Let $X \in C_p$. Partition X as*

$$X = \begin{pmatrix} X_{11} & X_{12} \\ X_{21} & X_{22} \end{pmatrix} \begin{matrix} q \\ p-q \end{matrix} .$$
$$\begin{matrix} q & p-q \end{matrix}$$

Then X_{11} and $X_{22\cdot1} = X_{22} - X_{21}X_{11}^{-1}X_{12}$ are independent and $X_{11} \in C_q$, and $X_{22\cdot1} \in C_{p-q}$.

THEOREM 9.5.2. *Let $X \in C_p$. Then, for $a\,(p \times 1) \neq 0$,*

(i) $\dfrac{a'Xa}{a'a}$ *has same distribution as x_{11} where $X = (x_{ij})$, and*

(ii) $\dfrac{a'\,a}{a'X^{-1}a}$ *has same distribution as x^{11} where $X^{-1} = (x^{ij})$.*

THEOREM 9.5.3. *Let $X \in C_p$ and $Y \in C_p$ be independent. Further let T and U be two different square roots of Y. Then TXT' and UXU' have identical distributions.*

THEOREM 9.5.4. *Let $X \in C_p$ and $Y \in C_p$ be independent. Then for any square root T of Y, the distribution of $Z = TXT'$ belongs to C_p.*

From the above theorem it follows that if $\Gamma = (\Gamma_1 \quad \Gamma_2)$, $\Gamma_i\,(p \times p_i)$, $i = 1, 2$, $p_1 + p_2 = p$ is a random orthogonal matrix independent of $Z \in C_p$, then $\Gamma_1'Z\Gamma_1 \in C_{p_1}$ and $(\Gamma_1'Z^{-1}\Gamma_1)^{-1} \in C_{p_2}$ are independent. Further if $E(Z)$, $E(Z^{-1})$, and $E(Z^\alpha)$, α an integer, exist, then

(i) $E(Z) = aI_p$

(ii) $E(Z^{-1}) = bI_p$

(iii) $E(Z^\alpha) = c_\alpha I_p$,

where $a = E(x_{11}y_{11})$, $b = E(x^{11})E(y^{11})$, and the constant c_α depends on moments of order less than or equal to α of X and Y.

Let $Z^{[i]}$ be any principal minor of Z of order i and $Y = TT'$, $X = UU'$ be lower triangular factorizations. Then

$$v_{ii} = \frac{\det(Z^{[i]})}{\det(Z^{[i-1]})} = t_{ii}^2 u_{ii}^2, \; i = 1, \ldots, p,$$

where $\det(Z^{[0]}) = 1$, are independent and

$$E(\det(Z)^\alpha) = \prod_{i=1}^{p} E(v_{ii}^\alpha)$$

provided the expectations involved exist.

Let $A_i \sim B_p^I(a_i, b_i)$, $B_i \sim B_p^{II}(c_i, d_i)$, $i = 1, 2$, $A \sim B_p^I(a, b)$, and $B \sim B_p^{II}(c, d)$ be independent. Define

$$Z_1 = A_1^{\frac{1}{2}}A_2(A_1^{\frac{1}{2}})' \tag{9.5.1}$$

$$Z_2 = B_1^{\frac{1}{2}}B_2(B_1^{\frac{1}{2}})' \tag{9.5.2}$$

$$Z_3 = A^{\frac{1}{2}} B (A^{\frac{1}{2}})' \tag{9.5.3}$$

and

$$Z_4 = B^{\frac{1}{2}} A (B^{\frac{1}{2}})'. \tag{9.5.4}$$

Then from Theorem 9.5.4, it follows that $Z_i \in C_p$, $i = 1, 2, 3, 4$. From Theorem 8.12.1, $Z_1 \sim H_p^I(a_2, b_1, a_2 + b_2 - a_1, b_1 + b_2)$ and its p.d.f. is

$$\frac{\Gamma_p(a_1 + b_2)\Gamma_p(a_2 + b_2)}{\Gamma_p(a_1)\Gamma_p(a_2)\Gamma_p(b_1 + b_2)} \det(Z_1)^{a_2 - \frac{1}{2}(p+1)} \det(I_p - Z_1)^{b_1 + b_2 - \frac{1}{2}(p+1)}$$

$$_2F_1(b_1, a_2 + b_2 - a_1; b_1 + b_2; I_p - Z_1),\ 0 < Z_1 < I_p.$$

The density of Z_2 can be shown to be

$$\frac{\beta_p(d_1 + c_2, c_1 + d_2)}{\beta_p(c_1, d_1)\beta_p(c_2, d_2)} \det(Z_2)^{c_2 - \frac{1}{2}(p+1)}$$

$$_2F_1(d_1 + c_2, c_2 + d_2; c_1 + c_2 + d_1 + d_2; I_p - Z_2),\ Z_2 > 0. \tag{9.5.5}$$

Next from the joint p.d.f. of A and B, by transforming $Z_3 = A^{\frac{1}{2}} B A^{\frac{1}{2}}$ with the Jacobian $J(A, B \to A, Z_3) = \det(A)^{-\frac{1}{2}(p+1)}$, and using the definition of $_2F_1$, the marginal p.d.f. of Z_3 is obtained as

$$\frac{\beta_p(b, a + d)}{\beta_p(a, b)\beta_p(c, d)} \det(Z_3)^{-d_2 - \frac{1}{2}(p+1)} {}_2F_1(c + d, a + d; a + b + d; -Z_3^{-1}),\ \dot{Z}_3 > 0.$$

Note that the distribution of Z_4 is same as that of Z_3.

Next let $X_i \sim G_p(m_i, I_p)$, $Y_i \sim IG_p(n_i + \frac{1}{2}(p + 1), I_p)$, $i = 1, 2$, $X \sim G_p(m, I_p)$, and $Y \sim IG_p(n + \frac{1}{2}(p + 1), I_p)$ be independent. Let

$$Z_5 = X_1^{\frac{1}{2}} X_2 (X_1^{\frac{1}{2}})' \tag{9.5.6}$$

$$Z_6 = Y_1^{\frac{1}{2}} Y_2 (Y_1^{\frac{1}{2}})' \tag{9.5.7}$$

$$Z_7 = X^{\frac{1}{2}} Y (X^{\frac{1}{2}})' \tag{9.5.8}$$

and

$$Z_8 = Y^{\frac{1}{2}} X (Y^{\frac{1}{2}})'. \tag{9.5.9}$$

Then the p.d.f. of Z_5 is

$$\{\Gamma_p(m_1)\Gamma_p(m_2)\}^{-1} \det(Z_5)^{m_1 - \frac{1}{2}(p+1)} B_{m_1 - m_2}(Z_5),\ Z_5 > 0.$$

Since $Z_6^{-1} = (Y_1^{\frac{1}{2}} Y_2 (Y_1^{\frac{1}{2}})')^{-1} = (Y_1^{-\frac{1}{2}})' Y_2^{-1} Y_1^{-\frac{1}{2}}$, $Y_1^{-1} = (Y_1^{-\frac{1}{2}})' Y_1^{-\frac{1}{2}} \sim G_p(n_1, I_p)$, $Y_2^{-1} \sim G_p(n_2, I_p)$, the p.d.f. of Z_6 obtained from the p.d.f. of Z_5 is

$$\{\Gamma_p(n_1)\Gamma_p(n_2)\}^{-1} \det(Z_6)^{-n_1 - \frac{1}{2}(p+1)} B_{n_1 - n_2}(Z_6),\ Z_6 > 0,$$

where $B_\delta(\cdot)$ is the Herz's Bessel function of type II. Note that $Z_7 \sim B_p^{II}(m, n)$, and also $Z_8 \sim B_p^{II}(m, n)$ which follows from Theorem 5.2.5.

Further define the following random matrices which again belong to the class $\mathcal{C}_p$:

$$Z_9 = A^{\frac{1}{2}} X (A^{\frac{1}{2}})' \tag{9.5.10}$$

$$Z_{10} = X^{\frac{1}{2}} A (X^{\frac{1}{2}})' \tag{9.5.11}$$

$$Z_{11} = B^{\frac{1}{2}} X (B^{\frac{1}{2}})' \tag{9.5.12}$$

$$Z_{12} = X^{\frac{1}{2}} B (X^{\frac{1}{2}})' \tag{9.5.13}$$

$$Z_{13} = A^{\frac{1}{2}} Y (A^{\frac{1}{2}})' \tag{9.5.14}$$

$$Z_{14} = Y^{\frac{1}{2}} A (Y^{\frac{1}{2}})' \tag{9.5.15}$$

$$Z_{15} = B^{\frac{1}{2}} Y (B^{\frac{1}{2}})' \tag{9.5.16}$$

and

$$Z_{16} = Y^{\frac{1}{2}} B (Y^{\frac{1}{2}})'. \tag{9.5.17}$$

The random matrices Z_9 and Z_{10} have the same density, given in Theorem 8.11.7. The random matrices Z_{11} and Z_{12} have the same density, given in Theorem 8.11.1. Similarly, the random matrices Z_{13} and Z_{14} have the same density as do the random matrices Z_{15} and Z_{16}, given by

$$\frac{\beta_p(a+n,b)}{\beta_p(a,b)\Gamma_p(n)} \det(Z_{13})^{-n-\frac{1}{2}(p+1)} \, {}_1F_1(a+n; a+b+n; -Z_{13}^{-1}), \ Z_{13} > 0,$$

and

$$\frac{\Gamma_p(c+n)}{\beta_p(c,d)\Gamma_p(n)} \det(Z_{15})^{-n-\frac{1}{2}(p+1)} \, \Psi\left(c+n; n-d+\frac{1}{2}(p+1); Z_{15}^{-1}\right), \ Z_{15} > 0,$$

respectively, where ${}_1F_1(\cdot)$ and $\Psi(\cdot)$ are confluent hypergeometric functions of kind 1 and 2.

The p.d.f.'s of random matrices Z_i, $i = 1, \ldots, 16$ have been studied by Khattree and R. D. Gupta (1992). However, in their paper the p.d.f.'s of Z_2, Z_5, Z_6, Z_{11} and Z_{13} seem to be in error. These p.d.f.'s have been given here in their corrected forms.

Next we give some properties of the random matrix $Z_1 \in \mathcal{C}_p$.

(i) For $\boldsymbol{a}\,(p \times 1) \neq \boldsymbol{0}$,

$$\frac{\boldsymbol{a}' Z_1 \boldsymbol{a}}{\boldsymbol{a}' \boldsymbol{a}} \sim H_1^I(a_2, b_1, a_2 + b_2 - a_1; b_1 + b_2)$$

and

$$\frac{\boldsymbol{a}' \boldsymbol{a}}{\boldsymbol{a}' Z_1^{-1} \boldsymbol{a}} \sim H_1^I\left(a_2 - \frac{1}{2}(p-1), b_1, a_2 + b_2 - a_1; b_1 + b_2\right).$$

(ii) Let $Z_1 = (z_{1ij})$ and $Z_1^{-1} = (z_1^{ij})$. Then $z_{1ii} \sim H_1^I(a_2, b_1, a_2 + b_2 - a_1; b_1 + b_2)$, $i = 1, \ldots, p$ and $z_1^{ii} \sim H_1^I(a_2 - \frac{1}{2}(p-1), b_1, a_2 + b_2 - a_1; b_1 + b_2)$, $i = 1, \ldots, p$.

(iii) Let

$$Z_1 = \begin{pmatrix} Z_{111} & Z_{112} \\ Z_{121} & Z_{122} \end{pmatrix} \begin{matrix} p_1 \\ p_2 \end{matrix} , \; p_1 + p_2 = p.$$
$$\qquad\quad p_1 \qquad p_2$$

Then Z_{111} and $Z_{122\cdot 1} = Z_{122} - Z_{121}Z_{111}^{-1}Z_{112}$ are independent, $Z_{111} \sim H_{p_1}^I(a_2, b_1, a_2 + b_2 - a_1; b_1 + b_2)$ and $Z_{122\cdot 1} \sim H_{p_2}^I(a_2 - \frac{1}{2}p_1, b_1, a_2 + b_2 - a_1; b_1 + b_2)$.

(iv) Let $Z_1^{[i]} = (z_{1jk})$, $1 \le j, k \le i$. Define

$$v_i = \frac{\det(Z_1^{[i]})}{\det(Z_1^{[i-1]})}, i = 1, \ldots, p \text{ and } \det(Z_1^{[0]}) = 1.$$

Then $v_1, \ldots, v_p$ are mutually independent and $v_i \sim H_1^I(a_2 - \frac{1}{2}(i-1), b_1, a_2 + b_2 - a_1; b_1 + b_2)$, $i = 1, \ldots, p$. Further the p.d.f. of $\det(Z_1)$ is the same as that of $\prod_{i=1}^p v_i$.

(v) For $p = 2$, $\det(Z_1)^{\frac{1}{2}} \sim H_1^I(a_2 - 1, b_1, a_2 + b_2 - a_1; b_1 + b_2)$

(vi) Using the representation $Z_1 = A_1^{\frac{1}{2}} A_2 (A_1^{\frac{1}{2}})'$, the following expected values can easily be obtained:

$$E(Z_1) = \frac{a_1 a_2}{(a_1 + b_1)(a_2 + b_2)} I_p,$$

$$E(Z_1^{-1}) = \frac{(2a_1 + 2b_1 - p - 1)(2a_2 + 2b_2 - p - 1)}{(2a_1 - p - 1)(2a_2 - p - 1)} I_p,$$

$$E(C_\kappa(Z_1)) = \frac{(a_1)_\kappa (a_2)_\kappa}{(a_1 + b_1)_\kappa (a_2 + b_2)_\kappa} C_\kappa(I_p),$$

$$E(C_\kappa(Z_1^{-1})) = \frac{(-a_1 - b_1 + \frac{1}{2}(p+1))_\kappa (-a_2 - b_2 + \frac{1}{2}(p+1))_\kappa}{(-a_1 + \frac{1}{2}(p+1))_\kappa (-a_2 + \frac{1}{2}(p+1))_\kappa} C_\kappa(I_p),$$

$$\mathrm{Re}(a_i) > k_1 + \frac{1}{2}(p-1), \; i = 1, 2,$$

$$E(Z_1^2) = \frac{a_1 a_2}{3(a_1 + b_1)(a_2 + b_2)} \left[\frac{(a_1 + 1)(a_2 + 1)(p + 2)}{(a_1 + b_1 + 1)(a_2 + b_2 + 1)} \right.$$
$$\left. - \frac{(2a_1 - 1)(2a_2 - 1)(p - 1)}{(2a_1 + 2b_1 - 1)(2a_2 + 2b_2 - 1)} \right] I_p,$$

$$E(Z_1^{-2}) = \frac{(2a_1 + 2b_1 - p - 1)(2a_2 + 2b_2 - p - 1)}{3(2a_1 - p - 1)(2a_2 - p - 1)} \left[\frac{(2a_1 + 2b_1 - p - 3)}{(2a_1 - p - 3)} \right.$$
$$\left. \frac{(2a_2 + 2b_2 - p - 3)(p + 2)}{(2a_2 - p - 3)} - \frac{(2a_1 + 2b_1 - p)(2a_2 + 2b_2 - p)(p - 1)}{(2a_1 - p)(2a_2 - p)} \right] I_p,$$

$$\mathrm{Re}(a_i) > \frac{1}{2}(p + 3), \; i = 1, 2.$$

The results (i) through (vi) given above are based on Khattree and R. D. Gupta

(1992) who also derived them for Z_2 and Z_3. Similar results for the random matrices Z_9, Z_{11} and Z_{15} are available in Khatri, Khattree and R. D. Gupta (1991). Using their approach results for the other random matrices can also be derived.

PROBLEMS

9.1.   Prove Theorem 9.2.1.

9.2.   Prove Theorem 9.2.2.

9.3.   Prove Theorem 9.2.3.

9.4.   Prove Theorem 9.2.4.

9.5.   Prove Theorem 9.2.5.

9.6.   Prove Theorem 9.2.6.

9.7.   Prove that if X $(p \times n)$ is spherical, then for given K $(q \times p)$ and Q $(n \times m)$, the distribution of KXQ depends only on KK' and $Q'Q$.

9.8.   Let $X \sim RS_{p,n}(\phi)$ with density function $f(XX')$. Partition X as $X = (X_1 \ X_2)$, X_i $(p \times n_i)$, $n_i \geq p$, $i = 1, 2$, $n_1 + n_2 = n$. Then prove that $(XX')^{-\frac{1}{2}} X_1 X_1' (XX')^{-\frac{1}{2}} \sim B_p^I(\frac{1}{2}n_1, \frac{1}{2}n_2)$.

9.9.   Let $X \sim RS_{p,n}(\phi)$ with density function $f(XX')$. Partition the random matrix X $(p \times n)$ as $X = (X_1, \ldots, X_r)$, X_i $(p \times n_i)$, $n_i \geq p$, $i = 1, \ldots, r$, $n_1 + \cdots + n_r = n$. Define $W_i = (XX')^{-\frac{1}{2}} X_i X_i' (XX')^{-\frac{1}{2}}$, $i = 1, \ldots, r - 1$. Then prove that $(W_1, \ldots, W_{r-1}) \sim D_p^I(\frac{1}{2}n_1, \ldots, \frac{1}{2}n_{r-1}; \frac{1}{2}n_r)$.

9.10. Let $X \sim RS_{p,n}(\phi)$ with density function $f(XX')$, $n \geq p$. Then prove that the density function of $W = (XX')^{-1}$ is

$$\frac{\pi^{\frac{1}{2}np}}{\Gamma_p(\frac{1}{2}n)} \det(W)^{-\frac{1}{2}(n+p+1)} f(W^{-1}), \ W > 0.$$

9.11. Let $S_i \sim W_p(n_i, I_p)$, $i = 1, 2$ be independent. Prove that $(S_1 + S_2)^{-1} S_i (S_1 + S_2)^{-1} \in C_p$, $i = 1, 2$.

9.12. For the random matrix Z_5 defined in (9.5.6), prove that
(i) for $\mathbf{a}$ $(p \times 1) \neq \mathbf{0}$, the p.d.f. of $v = \frac{\mathbf{a}' Z_5 \mathbf{a}}{\mathbf{a}' \mathbf{a}}$ is

$$2\{\Gamma(m_1)\Gamma(m_2)\}^{-1} v^{\frac{1}{2}(m_1+m_2-2)} K_{m_1-m_2}(2\sqrt{v}), \ v > 0,$$

(ii) for $\mathbf{a}$ $(p \times 1) \neq \mathbf{0}$, the p.d.f. of $u = \frac{\mathbf{a}' \mathbf{a}}{\mathbf{a}' Z_5^{-1} \mathbf{a}}$ is

$$\{\Gamma(m_1 - p + 1)\Gamma(m_2 - p + 1)\}^{-1} u^{\frac{1}{2}(m_1+m_2-2p)} K_{m_1-m_2}(2\sqrt{u}), \ u > 0,$$

where K_δ is the Bessel function of scalar argument of the third kind.

9.13. (contd.) Partition Z_5 as

$$Z_5 = \begin{pmatrix} Z_{511} & Z_{512} \\ Z_{521} & Z_{522} \end{pmatrix} \begin{matrix} p_1 \\ p_2 \end{matrix}, \quad p_1 + p_2 = p.$$

Then prove that the random matrices Z_{511} and $Z_{522 \cdot 1} = Z_{522} - Z_{521} Z_{511}^{-1} Z_{512}$ are independent. Further prove that the p.d.f. of Z_{511} is

$$\{\Gamma_p(m_1)\Gamma_p(m_2)\}^{-1} \det(Z_{511})^{m_1 - \frac{1}{2}(p_1+1)} B^{(p_1)}_{m_1-m_2}(Z_{511}), \quad Z_{511} > 0,$$

and the p.d.f. of $Z_{522 \cdot 1}$ is

$$\left\{\Gamma_{p_2}\left(m_1 - \frac{1}{2}p_1\right)\Gamma_{p_2}\left(m_2 - \frac{1}{2}p_1\right)\right\}^{-1}$$

$$\det(Z_{522 \cdot 1})^{m_1 - \frac{1}{2}(p_1+p_2+1)} B^{(p_2)}_{m_1-m_2}(Z_{522 \cdot 1}), \quad Z_{522 \cdot 1} > 0.$$

9.14. (contd.) Let $Z_5^{[i]} = (z_{5jk})$, $1 \le j, k \le i$. Define $v_i = \frac{\det(Z_5^{[i]})}{\det(Z_5^{[i-1]})}$, $i = 1, \ldots, p$ and $\det(Z_5^{[0]}) = 1$. Then prove that $v_1, \ldots, v_p$ are independent and the p.d.f. of v_i is

$$2\left\{\Gamma\left[m_1 - \frac{1}{2}(i-1)\right]\Gamma\left[m_2 - \frac{1}{2}(i-1)\right]\right\}^{-1} v_i^{\frac{1}{2}(m_1+m_2-i-1)} K_{m_1-m_2}(2\sqrt{v_i}), \quad v_i > 0,$$

9.15. (contd.) Prove that

$$E(Z_5) = m_1 m_2 I_p, \operatorname{Re}(m_i) > \frac{1}{2}(p-1), i = 1, 2,$$

$$E(Z_5^{-1}) = \frac{4}{(2m_1 - p - 1)(2m_2 - p - 1)} I_p,$$

$$E(C_\kappa(Z_5)) = \Gamma_p(m_1)\Gamma_p(m_2)C_\kappa(I_p),$$

$$E(C_\kappa(Z_5^{-1})) = \frac{1}{(-m_1 + \frac{1}{2}(p+1))_\kappa(-m_2 + \frac{1}{2}(p+1))_\kappa} C_\kappa(I_p),$$

$$\operatorname{Re}(m_i) > k_1 + \frac{1}{2}(p-1), i = 1, 2.$$

GLOSSARY OF NOTATIONS
AND ABBREVIATIONS

$A\,(p \times q)$	: matrix with p rows and q columns
$A = (a_{ij})$	: matrix with elements a_{ij}'s
A'	: transposed matrix of A
$A^{-1} = (a^{ij})$	: inverse of a nonsingular matrix A, with elements a^{ij}
$A^{[\alpha]}$	: $A^{[\alpha]} = (a_{ij})$, $1 \leq i,j \leq \alpha$
$A_{[a]}$	: $A_{[a]} = (a_{ij})$, $p - \alpha + 1 \leq i,j \leq p$
$A_{(\alpha)}$	: $A_{(\alpha)} = (a_{ij})$, $\alpha \leq i,j \leq p$
$D_a = \mathrm{diag}(a_1,\ldots,a_p)$	: diagonal matrix with elements $a_1,\ldots,a_p$ along the main diagonal
I_p	: unit matrix of order p
$\det(A)$	: determinant of a nonsingular square matrix
$\mathrm{tr}(A)$	: trace of a square matrix
$A \otimes B$	: Kronecker product (direct product) of the matrices A and B
$A > 0$	: A is positive definite
$A \geq 0$	: A is positive semidefinite
$A > B$	: $A - B$ is positive definite
$A \geq B$	: $A - B$ is positive semidefinite
$\boldsymbol{a}\,(p \times 1)$	: column vector with elements $a_1,\ldots,a_p$
$\boldsymbol{e}\,(p \times 1)$	: column vector with elements unity
$\boldsymbol{e}_i\,(p \times 1)$	: column vector with unity at i^{th} place and zero elsewhere
$\mathrm{vec}(X)$	: for a matrix $X\,(m \times n)$, $\mathrm{vec}(X)$ is an $mn \times 1$ vector defined as

$$\mathrm{vec}(X) = \begin{pmatrix} \boldsymbol{x}_1 \\ \vdots \\ \boldsymbol{x}_m \end{pmatrix},$$

where $\boldsymbol{x}_i$, $i = 1, \ldots, n$ is the ith column of X

$\text{vecp}(X)$: for a symmetric matrix X $(p \times p)$, $\text{vecp}(X)$ is a $\frac{1}{2}p(p+1)$ column vector formed from the elements above and including the diagonal, taken columnwise. In other words if

$$X = \begin{pmatrix} x_{11} & x_{12} & \cdots & x_{1p} \\ x_{21} & x_{22} & \cdots & x_{2p} \\ \vdots & & & \\ x_{p1} & x_{p2} & \cdots & x_{pp} \end{pmatrix}$$

then

$$\text{vecp}(X) = \begin{pmatrix} x_{11} \\ x_{12} \\ x_{22} \\ \vdots \\ x_{1p} \\ \vdots \\ x_{pp} \end{pmatrix}$$

$O(p,n)$: Stiefel manifold, $O(p,n) = \{H_1 \, (p \times n) : H_1 H_1' = I_p\}$

$O(p)$: orthogonal group, $O(p) = \{H \, (p \times p) : HH' = I_p\}$

ι : $\sqrt{-1}$

$\blacksquare$: end of the proof of a theorem (corollary)

$\sim$: is distributed as

$\overset{d}{=}$: equal in distribution

$E(x)$, $E(\boldsymbol{x})$, $E(X)$: expected values of random quantity x, $\boldsymbol{x}$ and X respectively

$\text{var}(x)$: variance of a random variable x

$\text{cov}(x,y)$: covariance of random variables x and y

$\text{corr}(x,y)$: correlation coefficient between random variables x and y

$\text{var}(\boldsymbol{x})$: covariance matrix of a random vector $\boldsymbol{x}$

$\text{cov}(\boldsymbol{x},\boldsymbol{y})$: covariance matrix of random vectors $\boldsymbol{x}$ and $\boldsymbol{y}$

$\text{cov}(X,Y)$: covariance matrix of random matrices X $(p \times n)$ and Y $(r \times s)$, $\text{cov}(X,Y) = \text{cov}(\text{vec}(X'), \text{vec}(Y'))$

$\text{etr}(A)$: $\exp\{\text{tr}(A)\}$

$J(X \to Y)$: Jacobian of the transformation $Y = F(X)$

δ_{ij} : Kronecker delta, $\delta_{ij} = \begin{cases} 1, & \text{if } i = j \\ 0, & \text{if } i \neq j \end{cases}$

$\Gamma_p(a)$: multivariate gamma function,

$$\Gamma_p(a) = \pi^{\frac{1}{4}p(p-1)} \prod_{j=1}^{p} \Gamma\left[a - \frac{1}{2}(j-1)\right], \ \mathrm{Re}(h) > \frac{1}{2}(p-1)$$

$\beta_p(a,b)$: multivariate beta function,

$$\beta_p(a,b) = \frac{\Gamma_p(a)\Gamma_p(b)}{\Gamma_p(a+b)}$$

$\Gamma_p^*(a_1,\ldots,a_p)$: generalized multivariate gamma function,

$$\Gamma_p^*(a_1,\ldots,a_p) = \pi^{\frac{1}{4}p(p-1)} \prod_{j=1}^{p} \Gamma\left[a_j - \frac{1}{2}(j-1)\right]$$

$\beta_p^*(a_1,\ldots,a_p;b_1,\ldots,b_p)$: generalized multivariate beta function,

$$\beta_p^*(a_1,\ldots,a_p;b_1,\ldots,b_p) = \frac{\Gamma_p^*(a_1,\ldots,a_p)\Gamma_p^*(b_1,\ldots,b_p)}{\Gamma_p^*(a_1+b_1,\ldots,a_p+b_p)}$$

$\int_{A>0} f(A)\,dA$: integral of $f(A)$ over the domain $\{A : A > 0\}$ where $dA = \prod_{i \le j} da_{ij}$

$\int_{G(X)=0} f(X)\,dX$: integral of $f(X)$ over the domain $\{X : G(X) = 0\}$ where $dX = \prod_{i,j} dx_{ij}$

$[(dH_1)H_1']$: invariant measure on Stiefel manifold

$[(dH)H']$: invariant measure on orthogonal group

$[dH_1]$: unit invariant measure on Stiefel manifold

$[dH]$: unit invariant measure on orthogonal group or Haar measure

p.d.f. : probability density function

c.d.f. : cumulative distribution function

m.g.f. : moment generating function

c.f. : characteristic function

c.g.f. : cumulant generating function

$\mathrm{Re}(h)$: real part of h

UNIVARIARE DISTRIBUTIONS

$N(\mu,\sigma^2)$: normal distribution; its probability density function is

$$\frac{1}{\sqrt{2\pi}\,\sigma} \exp\left\{-\frac{(x-\mu)^2}{2\sigma^2}\right\}, \ x \in \mathbb{R}, \ \mu \in \mathbb{R} \text{ and } \sigma \in \mathbb{R}^+$$

χ_n^2 : chi-square distribution; its probability density function is

$$\frac{1}{2^{\frac{1}{2}n}\Gamma(\frac{1}{2}n)}x^{\frac{1}{2}n-1}\exp\left(-\frac{1}{2}x\right),\; x \in \mathbb{R}^+, n > 0$$

t_n : Student's t-distribution; its probability density function is

$$\frac{\Gamma[\frac{1}{2}(n+1)]}{\sqrt{n\pi}\,\Gamma(\frac{1}{2}n)}\left(1+\frac{x^2}{n}\right)^{-\frac{1}{2}(n+1)}, \; x \in \mathbb{R}, \, n > 0$$

$B^I(a,b)$: beta type I distribution; its probability density function is

$$\frac{1}{\beta(a,b)}x^{a-1}(1-x)^{b-1}, \, 0 < x < 1, \, a > 0, \, b > 0$$

$B^{II}(a,b)$: beta type II distribution; its probability density function is

$$\frac{1}{\beta(a,b)}x^{a-1}(1+x)^{-(a+b)}, \, x > 0, \, a > 0, \, b > 0$$

MULTIVARIATE DISTRIBUTIONS

$N_p(\boldsymbol{\mu},\Sigma)$: multivariate normal distribution; its p.d.f. is

$$(2\pi)^{-\frac{1}{2}p}\det(\Sigma)^{-\frac{1}{2}}\operatorname{etr}\left\{-\frac{1}{2}\Sigma^{-1}(\boldsymbol{x}-\boldsymbol{\mu})(\boldsymbol{x}-\boldsymbol{\mu})'\right\},$$

$\boldsymbol{x} \in \mathbb{R}^p$, $\boldsymbol{\mu} \in \mathbb{R}^p$, and $\Sigma > 0$

$t_p(n,\omega,\mu,\Sigma)$: multivariate t-distribution; its p.d.f. is

$$\frac{\Gamma[\frac{1}{2}(n+p)]}{\pi^{\frac{1}{2}p}\Gamma(\frac{1}{2}n)}\det(\Sigma)^{-\frac{1}{2}}\omega^{-\frac{1}{2}p}\left(1+\frac{1}{\omega}(\boldsymbol{x}-\boldsymbol{\mu})'\Sigma^{-1}(\boldsymbol{x}-\boldsymbol{\mu})\right)^{-\frac{1}{2}(n+p)},$$

$\boldsymbol{x} \in \mathbb{R}^p$, $\boldsymbol{\mu} \in \mathbb{R}^p$, $\omega > 0$, and $\Sigma > 0$

$D^I(a_1,\ldots,a_r;a_{r+1})$: Dirichlet type I distribution; its p.d.f. is

$$\frac{\Gamma(\sum_{i=1}^{r+1}a_i)}{\prod_{i=1}^{r+1}\Gamma(a_i)}\prod_{i=1}^{r}u_i^{a_i-1}\left(1-\sum_{i=1}^{r}u_i\right)^{a_{r+1}-1}, 0 < u_i < 1, \sum_{i=1}^{r}u_i < 1,$$

where $a_i > 0$, $i = 1,...,r+1$

$D^{II}(b_1,\ldots,b_r;b_{r+1})$: Dirichlet type II distribution; its p.d.f. is

$$\frac{\Gamma(\sum_{i=1}^{r+1}b_i)}{\prod_{i=1}^{r+1}\Gamma(b_i)}\prod_{i=1}^{r}v_i^{b_i-1}\left(1+\sum_{i=1}^{r}v_i\right)^{-\sum_{i=1}^{r+1}b_i}, \; v_i > 0,$$

where $b_i > 0$, $i = 1,\ldots,r+1$

MATRIX VARIATE DISTRIBUTIONS

$N_{p,n}(M, \Sigma \otimes \Psi)$: matrix variate normal distribution; its p.d.f. is

$$(2\pi)^{-\frac{1}{2}np} \det(\Sigma)^{-\frac{1}{2}n} \det(\Psi)^{-\frac{1}{2}p}$$

$$\text{etr}\left\{-\frac{1}{2}\Sigma^{-1}(X-M)\Psi^{-1}(X-M)'\right\}, X \in \mathbb{R}^{p \times n},$$

where $M \in \mathbb{R}^{p \times n}$, $\Sigma > 0$ and $\Psi > 0$

$SN_{p,p}(M, B_p'(\Sigma \otimes \Psi)B_p)$: symmetric matrix variate normal distribution; its p.d.f. is

$$(2\pi)^{-\frac{1}{4}p(p+1)} \det(B_p'(\Sigma \otimes \Psi)B_p)^{-\frac{1}{2}}$$

$$\text{etr}\left[-\frac{1}{2}\Sigma^{-1}(X-M)\Psi^{-1}(X-M)\right], X = X' \in \mathbb{R}^{p \times p}$$

where $M = M' \in \mathbb{R}^{p \times p}$, $\Sigma > 0$ and $\Psi > 0$

$N_{p,n}(M, \Sigma \otimes \Psi | s, C)$: restricted matrix variate normal distribution; its p.d.f. is

$$(2\pi)^{-\frac{1}{2}(n-s)p} \det(\Psi)^{-\frac{1}{2}p} \det(C'\Psi C)^{\frac{1}{2}p} \det(\Sigma)^{-\frac{1}{2}(n-s)}$$

$$\text{etr}\left\{-\frac{1}{2}\Sigma^{-1}(X-M)\Psi^{-1}(X-M)'\right\}, XC = 0$$

where $M \in \mathbb{R}^{p \times n}$, $\Sigma > 0$ and $\Psi > 0$

$N_{p,n}(M, A, B, \theta)$: matrix variate θ-generalized normal distribution; its p.d.f. is

$$\left\{2\Gamma\left(1 + \frac{1}{\theta}\right)\right\}^{-np} \det(A)^{-n} \det(B)^{-p}$$

$$\exp\left\{-\sum_{i=1}^{p}\sum_{j=1}^{n}\left|\sum_{k=1}^{p}\sum_{\ell=1}^{n} a^{ik}(y_{k\ell} - m_{k\ell})b^{\ell j}\right|^{\theta}\right\}, X \in \mathbb{R}^{p \times n}$$

where $M \in \mathbb{R}^{p \times n}$, $A > 0$, $B > 0$, $A^{-1} = (a^{ik})$, $B^{-1} = (b^{\ell j})$, $M = (m_{k\ell})$, and $Y = (y_{k\ell})$

$W_p(n, \Sigma)$: Wishart distribution; its probability density function is

$$\left\{2^{\frac{1}{2}np} \Gamma_p\left(\frac{1}{2}n\right) \det(\Sigma)^{\frac{1}{2}n}\right\}^{-1} \det(S)^{\frac{1}{2}(n-p-1)} \text{etr}\left(-\frac{1}{2}\Sigma^{-1}S\right),$$
$$S > 0, \, n \geq p$$

$W_p(n, \Sigma, \Theta)$: noncentral Wishart distribution; its p.d.f. is

$$\left\{2^{\frac{1}{2}np} \Gamma_p\left(\frac{1}{2}n\right) \det(\Sigma)^{\frac{1}{2}n}\right\}^{-1} \text{etr}\left(-\frac{1}{2}\Theta\right) \text{etr}\left(-\frac{1}{2}\Sigma^{-1}S\right)$$

$$\det(S)^{\frac{1}{2}(n-p-1)} {}_0F_1\left(\frac{1}{2}n; \frac{1}{4}\Theta\Sigma^{-1}S\right), S > 0, \, n \geq p$$

where ${}_0F_1$ is the hypergeometric function (Bessel function)

$IW_p(m, \Psi)$: inverted Wishart distribution; its probability density function is

$$\frac{2^{-\frac{1}{2}(m-p-1)p} \det(\Psi)^{\frac{1}{2}(m-p-1)}}{\Gamma_p[\frac{1}{2}(m-p-1)]\det(V)^{\frac{1}{2}m}} \, \mathrm{etr}\left(-\frac{1}{2}V^{-1}\Psi\right), \, V > 0,$$

where $\Psi > 0$ and $m > 2p$

$IW_p(m, \Psi, \Omega)$: noncentral inverted Wishart distribution; its p.d.f. is

$$\frac{2^{-\frac{1}{2}(m-p-1)p} \det(\Psi)^{\frac{1}{2}(m-p-1)}}{\Gamma_p[\frac{1}{2}(m-p-1)]} \, \mathrm{etr}\left(-\frac{1}{2}\Theta\right)\mathrm{etr}\left(-\frac{1}{2}V^{-1}\Psi\right)$$

$$\det(V)^{-\frac{1}{2}m} {}_0F_1\left(\frac{1}{2}(m-p-1); \frac{1}{4}\Theta\Psi V^{-1}\right), \, V > 0,$$

where $\Psi > 0$ and $m > 2p$

$G_p(a, C)$: matrix variate gamma distribution; its probability density function is

$$\left\{\Gamma_p(a)\det(C)^{-a}\right\}^{-1} \mathrm{etr}(-CW)\det(W)^{a-\frac{1}{2}(p+1)}, \, W > 0,$$

where $C > 0$ and $a > \frac{1}{2}(p-1)$

$IG_p(m, C)$: inverted matrix variate gamma distribution; its probability density function is

$$\frac{\det(B)^{m-\frac{1}{2}(p+1)}}{\Gamma_p[m-\frac{1}{2}(p+1)]} \det(W)^{-m} \mathrm{etr}(-BW^{-1}), \, W > 0,$$

where $B > 0$, and $m > p$

$G_p(a, C, \Theta)$: noncentral matrix variate gamma distribution; its probability density function is

$$\left\{\Gamma_p(a)\det(C)^{-a}\right\}^{-1} \mathrm{etr}(-\Theta - CW)\det(W)^{a-\frac{1}{2}(p+1)}$$

$${}_0F_1(a; \Theta CW), W > 0, a > \frac{1}{2}(p-1)$$

$BG_p^I(a_1, \ldots, a_p; C)$: Bellman gamma type I distribution; its p.d.f. is

$$\left\{\Gamma_p^*(a_1, \ldots, a_p) \prod_{\alpha=1}^{p} \det(C_{(\alpha)})^{-m_\alpha}\right\}^{-1} \mathrm{etr}(-CW)$$

$$\det(W)^{a_p - \frac{1}{2}(p+1)} \prod_{\alpha=1}^{p-1} \det(W^{[\alpha]})^{-m_{\alpha+1}}, \, W > 0,$$

where $C > 0$, $a_j = m_1 + \cdots + m_j$, $a_j > \frac{1}{2}(j-1)$, $j = 1\ldots, p$

$BG_p^{II}(b_1, \ldots, b_p; B)$: Bellman gamma type II distribution; its p.d.f. is

$$\left\{ \Gamma_p^*(b_1, \ldots, b_p) \prod_{a=1}^{p} \det(B^{[a]})^{-k_\alpha} \right\}^{-1} \operatorname{etr}(-BW)$$

$$\det(W)^{b_p - \frac{1}{2}(p+1)} \prod_{\alpha=2}^{p} \det(W_{(a)})^{-k_{\alpha-1}}, \ W > 0,$$

where $B > 0$, $b_j = k_{p-j+1} + \cdots + k_p$, $b_j > \frac{1}{2}(j-1)$, $j = 1, \ldots, p$

$T_{p,m}(n, M, \Sigma, \Omega)$: matrix variate t-distribution; its p.d.f. is

$$\frac{\Gamma_p[\frac{1}{2}(n+m+p-1)]}{\pi^{\frac{1}{2}mp} \Gamma_p[\frac{1}{2}(n+p-1)]} \det(\Sigma)^{-\frac{1}{2}m} \det(\Omega)^{-\frac{1}{2}p}$$

$$\det(I_p + \Sigma^{-1}(T-M)\Omega^{-1}(T-M)')^{-\frac{1}{2}(n+m+p-1)}, \ T \in \mathbb{R}^{p \times m}$$

where $M \in \mathbb{R}^{p \times m}$, $\Omega(m \times m) > 0$, $\Sigma(p \times p) > 0$ and $n > 0$

$IT_{p,m}(n, M, \Sigma, \Omega)$: inverted matrix variate t-distribution; its p.d.f. is

$$\frac{\Gamma_p[\frac{1}{2}(n+m+p-1)]}{\pi^{\frac{1}{2}mp} \Gamma_p[\frac{1}{2}(n+p-1)]} \det(\Sigma)^{-\frac{1}{2}m} \det(\Omega)^{-\frac{1}{2}p}$$

$$\det(I_p - \Sigma^{-1}(T-M)\Omega^{-1}(T-M)')^{\frac{1}{2}(n-2)}, \ T \in \mathbb{R}^{p \times m}$$

where $I_p - \Sigma^{-1}(T-M)\Omega^{-1}(T-M)' > 0$, $M \in \mathbb{R}^{p \times m}$, $\Omega(m \times m) > 0$, $\Sigma(p \times p) > 0$ and $n > 0$

$\overline{D}T_{p,m}(n+p, M, \Sigma)$: upper disguised matrix variate t-distribution; its p.d.f. is

$$\{K(m, p, n+p)\}^{-1} \det(\Sigma)^{-\frac{1}{2}m}$$

$$\det(I_m + (T-M)'\Sigma^{-1}(T-M))^{-\frac{1}{2}(n+p-m-1)}$$

$$\prod_{i=1}^{m} \det((I_m + (T-M)'\Sigma^{-1}(T-M))^{[i]})^{-1}, \ T \in \mathbb{R}^{p \times m}$$

$\underline{D}T_{p,m}(n+p, M, \Sigma)$: lower disguised matrix variate t-distribution; its p.d.f. is

$$\{K(m, p, n+p)\}^{-1} \det(\Sigma)^{-\frac{1}{2}m}$$

$$\det(I_m + (T-M)'\Sigma^{-1}(T-M))^{-\frac{1}{2}(n+p-m-1)}$$

$$\prod_{i=1}^{m} \det((I_m + (T-M)'\Sigma^{-1}(T-M))_{[i]})^{-1}, \ T \in \mathbb{R}^{p \times m}$$

$T_{p,m}(n, M, \Sigma, \Omega | s, C)$: restricted matrix variate t-distribution; its p.d.f. is

$$\frac{\Gamma_p[\frac{1}{2}(n+m+p-s-1)]}{\pi^{\frac{1}{2}(m-s)p} \Gamma_p[\frac{1}{2}(n+p-1)]} \det(\Sigma)^{-\frac{1}{2}(m-s)} \det(C'\Omega C)^{\frac{1}{2}p}$$

$$\det(\Omega)^{-\frac{1}{2}p} \det(I_p + \Sigma^{-1}(T-M)\Omega^{-1}(T-M)')^{-\frac{1}{2}(n+m+p-s-1)}$$

$B_p^I(a, b)$: matrix variate beta type I distribution; its density is

$$\{\beta_p(a, b)\}^{-1} \det(U)^{a - \frac{1}{2}(p+1)} \det(I_p - U)^{b - \frac{1}{2}(p+1)}, \; 0 < U < I_p,$$

where $a > \frac{1}{2}(p - 1)$, $b > \frac{1}{2}(p - 1)$, and $\beta_p(a, b)$ is the multivariate beta function

$B_p^{II}(a, b)$: matrix variate beta type II distribution; its density is

$$\{\beta_p(a, b)\}^{-1} \det(V)^{a - \frac{1}{2}(p+1)} \det(I_p + V)^{-(a+b)}, \; V > 0,$$

where $a > \frac{1}{2}(p - 1)$ and $b > \frac{1}{2}(p - 1)$

$D_p^I(a_1, \ldots, a_r; a_{r+1})$: matrix variate Dirichlet type I distribution; its density is

$$\{\beta_p(a_1, \ldots, a_r; a_{r+1})\}^{-1} \prod_{i=1}^{r} \det(U_i)^{a_i - \frac{1}{2}(p+1)}$$

$$\det\left(I_p - \sum_{i=1}^{r} U_i\right)^{a_{r+1} - \frac{1}{2}(p+1)}, \; 0 < U_i < I_p, \; 0 < \sum_{i=1}^{r} U_i < I_p,$$

where $a_i > \frac{1}{2}(p - 1)$, $i = 1, \ldots, r + 1$, and

$$\beta_p(a_1, \ldots, a_r; a_{r+1}) = \frac{\prod_{i=1}^{r+1} \Gamma_p(a_i)}{\Gamma_p(\sum_{i=1}^{r+1} a_i)}$$

$D_p^{II}(b_1, \ldots, b_r; b_{r+1})$: matrix variate Dirichlet type II distribution; its density is

$$\{\beta_p(b_1, \ldots, b_r; b_{r+1})\}^{-1} \prod_{i=1}^{r} \det(V_i)^{b_i - \frac{1}{2}(p+1)}$$

$$\det\left(I_p + \sum_{i=1}^{r} V_i\right)^{-\sum_{i=1}^{r+1} b_i}, \; V_i > 0,$$

where $b_i > \frac{1}{2}(p - 1)$, $i = 1, \ldots, r + 1$

$Q_{p,n}(A, \Sigma, \Psi)$: the density of $S = XAX'$, $A > 0$, $X \sim N_{p,n}(M, \Sigma \otimes \Psi)$ is

$$\left\{2^{\frac{1}{2}np} \Gamma_p\left(\frac{1}{2}n\right)\right\}^{-1} \det(A\Psi)^{-\frac{1}{2}p} \det(\Sigma)^{-\frac{1}{2}n} \det(S)^{\frac{1}{2}(n-p-1)}$$

$$\mathrm{etr}\left(-\frac{1}{2}q^{-1}\Sigma^{-1}S\right) {}_0F_0^{(n)}\left(B, \frac{1}{2}q^{-1}\Sigma^{-1}S\right), \; S > 0,$$

where $B = I_n - qA^{-\frac{1}{2}}\Psi^{-1}A^{-\frac{1}{2}}$ and $q > 0$ is an arbitrary constant

$M_{p,n}(F)$: von Mises-Fisher distribution with parameter matrix $F\,(p \times n)$; its probability element is given by

$$a(F) \, \mathrm{etr}(FX') \, [dX], \; X \in O(p, n), \; p \le n,$$

where $[dX]$ is the unit invariant measure on $O(p, n)$ and $a(F)$ is the normalizing constant given by

$$\{a(F)\}^{-1} = {}_0F_1\left(\frac{1}{2}n; \frac{1}{4}FF'\right) = {}_0F_1\left(\frac{1}{2}n; \frac{1}{4}F'F\right)$$

$B_{p,n}(A)$: Bingham matrix distribution with parameter matrix $A = A'$; its probability element is given by

$$b(A)\,\mathrm{etr}(XAX')\,[dX],\ X \in O(p,n),\ p \le n,$$

where $[dX]$ is the unit invariant measure on $O(p,n)$ and $b(A)$ is the normalizing constant given by

$$\{b(A)\}^{-1} = {}_1F_1^{(n)}\Big(\frac{1}{2}n, \frac{1}{2}p; A\Big)$$

$B_{p,n}(A, B)$: generalized Bingham matrix distribution with parameter matrices $A = A'$ and $B = B'$; its probability element is given by

$$b_1(A, B)\,\mathrm{etr}(BXAX')\,[dX],\ X \in O(p,n),$$

where $b_1(A, B)$ is the normalizing constant

$ACG_{p,n}(\Psi)$: matrix angular central Gaussian distribution (ACG) with parameters p, n and $\Psi > 0$; its probability element is

$$\det(\Psi)^{-\frac{1}{2}p}\det(H\Psi^{-1}H')^{-\frac{1}{2}n}\,[dH],\ H \in O(p,n)$$

$CH_p(n, \alpha, \beta, \text{kind } 1)$: confluent hypergeometric function kind 1 distribution; its p.d.f. is given by

$$\frac{\Gamma_p(\alpha)\Gamma_p(\beta - n)}{\Gamma_p(n)\Gamma_p(\beta)\Gamma_p(\alpha - n)}\det(X)^{n-\frac{1}{2}(p+1)}{}_1F_1(\alpha; \beta; -X),\ X > 0,$$

where $\mathrm{Re}(\beta-n) > 0$, and $\mathrm{Re}(\alpha-n) > 0$. The parameters n, α, and β are restricted to take values such that the density function is non-negative

$CH_p^I(n, \alpha, \beta, \text{kind } 2)$: confluent hypergeometric function kind 2 and type I distribution; its p.d.f. is given by

$$\frac{\Gamma_p(\alpha)\Gamma_p[\alpha - \beta + \frac{1}{2}(p+1)]}{\Gamma_p(n)\Gamma_p(\alpha - n)\Gamma_p[n - \beta + \frac{1}{2}(p+1)]}$$
$$\det(X)^{n-\frac{1}{2}(p+1)}\,\Psi(\alpha, \beta; X),\ X > 0,$$

where $\mathrm{Re}(n, \alpha - n) > \frac{1}{2}(p - 1)$ and $\mathrm{Re}(n - \beta) > -1$. The parameters n, α and β are restricted to take values such that the density function is non-negative

$CH_p^{II}(n, \alpha, \beta, \text{kind } 2)$: confluent hypergeometric function kind 2 and type II distribution; its p.d.f. is given by

$$\frac{\Gamma_p[\alpha - \beta + n + \frac{1}{2}(p+1)]}{\Gamma_p(n)\Gamma_p[n - \beta + \frac{1}{2}(p+1)]}\det(X)^{n-\frac{1}{2}(p+1)}$$
$$\mathrm{etr}(-X)\,\Psi(\alpha, \beta; X),\ X > 0,$$

where $\text{Re}(n, \alpha) > \frac{1}{2}(p - 1)$ and $\text{Re}(n - \beta) > -1$. The parameters n, α and β are restricted to take values such that the density function is non-negative

$H_p^I(n, \alpha, \beta, \gamma)$: hypergeometric function distribution of type I; its p.d.f. is given by

$$\frac{\Gamma_p(\gamma + n - \alpha)\Gamma_p(\gamma + n - \beta)}{\Gamma_p(\gamma)\Gamma_p(n)\Gamma_p(\gamma + n - \alpha - \beta)} \det(X)^{n - \frac{1}{2}(p+1)}$$

$$\det(I_p - X)^{\gamma - \frac{1}{2}(p+1)} {}_2F_1(\alpha, \beta; \gamma; I_p - X), \; 0 < X < I_p,$$

where $\text{Re}(\gamma + n - \alpha - \beta) > \frac{1}{2}(p - 1)$, $\text{Re}(\gamma) > \frac{1}{2}(p - 1)$ and $\text{Re}(n) > \frac{1}{2}(p - 1)$. The parameters α, β, γ and n are restricted to take values such that the density function is non-negative

$H_p^{II}(n, \alpha, \beta, \gamma; A)$: hypergeometric function distribution of type II; its p.d.f. is given by

$$\frac{\Gamma_p(\alpha)\Gamma_p(\beta)\Gamma_p(\gamma - n) \det(A)^n}{\Gamma_p(n)\Gamma_p(\gamma)\Gamma_p(\alpha - n)\Gamma_p(\beta - n)} \det(X)^{n - \frac{1}{2}(p+1)}$$

$$ {}_2F_1(\alpha, \beta; \gamma; -AX), \; X > 0,$$

where $A > 0$, $\text{Re}(\gamma - n) > \frac{1}{2}(p - 1)$, $\text{Re}(\alpha - n) > \frac{1}{2}(p - 1)$ and $\text{Re}(\beta - n) > \frac{1}{2}(p - 1)$. The parameters n, α, β and γ are restricted to take values such that the density function is non-negative

$L_r^{(1)}(g, a_1, \ldots, a_r)$: matrix variate Liouville distribution of the first kind; its p.d.f. is proportional to

$$\prod_{i=1}^{r} \det(X_i)^{a_i - \frac{1}{2}(p+1)} g\left(\sum_{i=1}^{r} X_i\right), \; X_i > 0,$$

$$a_i > \frac{1}{2}(p - 1), \; i = 1, \ldots, r,$$

where $g(\cdot)$ is positive, continuous, supported on $\mathcal{S} = \{X \, (p \times p) : X > 0\}$ such that

$$\int_{T > 0} \det(T)^{a - \frac{1}{2}(p+1)} g(T) \, dT < \infty,$$

and $a = \sum_{i=1}^{r} a_i$

$L_r^{(2)}(g, b_1, \ldots, b_r)$: matrix variate Liouville distribution of the second kind; its p.d.f. is proportional to

$$\prod_{i=1}^{r} \det(Y_i)^{b_i - \frac{1}{2}(p+1)} g\left(\sum_{i=1}^{r} Y_i\right), \; 0 < Y_i < I_p, \; \sum_{i=1}^{r} Y_i < I_p,$$

$$b_i > \frac{1}{2}(p - 1), \; i = 1, \ldots, r,$$

where $g(\cdot)$ is positive, continuous, supported on $\mathcal{S} = \{X \ (p \times p) : 0 < X < I_p\}$ such that

$$\int_{\mathcal{S}} \det(T)^{b-\frac{1}{2}(p+1)} g(T) \, dT < \infty,$$

and $b = \sum_{i=1}^{r} b_i$.

REFERENCES

Abdi, W. H. (1968) Whittaker's $M_{k,\mu}$-function of a matrix argument, *Rend. Circ. Mat. Palermo*, 17, 333–342.

Aitchison, J. (1986) *The Statistical Analysis of Compositional Data*, Chapman and Hall, New York.

Amey, A. K. A. and Gupta, A. K. (1992) Testing sphericity under a mixture model, *Aust. J. Statist.*, 34, 451–460.

Anderson, T. W. (1946) The non-central Wishart distribution and certain problems of multivariate statistics, *Ann. Math. Statist.*, 17, 409–431. Correction (1964), 35, 923–24.

Anderson, T. W. (1984) *An Introduction to Multivariate Statistical Analysis*, 2nd ed., John Wiley & Sons, New York.

Anderson, T. W. and Fang, K. T. (1987) Cochran's theorem for elliptically contoured distributions, *Sankhyā*, A49(3), 305–315.

Anderson, T. W. and Girshick, M. A. (1944) Some extensions of the Wishart distribution, *Ann. Math. Statist.*, 15, 345–357.

Asoo, Y. (1969) On the Γ-distribution of matric argument and its related distributions, *Memoirs of Faculty of Literature and Sciences*, Shimane University, Natural Science, 2, 1–13.

Barlow, R. E. and Mendel, M. B. (1992) De Finetti-type representation for life distributions, *J. Amer. Statist. Assoc.*, 87, 1116–1122.

Bartlett, M. S. (1933) On the theory of statistical regression, *Proc. Roy. Soc. Edinb.*, 53, 260–283.

Basu, D. and Khatri, C. G. (1969) Some characterizations of statistics, *Sankhyā*, A31, 199–208.

Bekker, A. and Roux, J. J. J. (1990) Some characterizations of the matrix variate normal distribution, *South African Statist. J.*, 24, 45–54.

Bellman, R. (1956) A generalization of some integral identities due to Ingham and Siegel, *Duke Math. J.*, 23, 571–577.

Bellman, R. (1970) *Introduction to Matrix Analysis*, 2nd ed., McGraw-Hill, New York.

Bilodeau, M. and Srivastava, M. S. (1992) Estimation of the eigenvalues of $\Sigma_1\Sigma_2^{-1}$, *J. Multivariate Anal.*, 41, 1–13.

Bingham, C. (1974) An antipodally symmetric distribution on sphere, *Ann. Statist.*, 2(6), 1201–1225.

Bingham, C., Chang, T. and Richards, Donald St. P. (1992) Approximating the matrix Fisher and Bingham distributions: Applications to spherical regression and procrustes analysis, *J. Multivariate Anal.*, 41, 314–337.

Bochner, S. (1952) Bessel functions and modular relations of higher type and hyperbolic differential equations, *Communications der Séminaire Mathématique de l'Université Lund*, Tome Supplementaire dedié á Mercel Riez, 12–20.

Box, G. E. P. and Tiao, G. C. (1973) *Bayesian Inference in Statistical Analysis*, Addison-Wesley, Massachusetts.

Brillinger, D. R. (1969) Asymptotic properties of spectral estimate of second order, *Biometrika*, 56, 375–387.

Brillinger, D. R. (1975) *Time Series: Data Analysis and Theory*, Holt, Rinehart and Winston, New York.

Bronk, R. V. (1965) Exponential ensembles for random matrix, *J. Math. Phys.*, 6, 228–237.

Brown, M. W. (1974) Generalized least square estimators in the analysis of variance, *South African Statist. J.*, 8, 1–24.

Brown, M. W. and Neudecker, H. (1988) The covariance matrix of a general symmetric second degree matrix polynomial under normality assumption, *Linear Algebra Appl.*, 103, 113–120.

Cambanis, S., Huang, S. and Simons, G. (1981) On the theory of elliptically contoured distributions, *J. Multivariate Anal.*, 11, 368–385.

Carmeli, M. (1974) Statistical theory of energy levels and random matrices in physics, *J. Statist. Phys.*, 10, 259–297.

Carmeli, M. (1983) *Statistical Theory and Random Matrices in Physics*, Marcel Dekker, New York.

Chang, Y. C. (1972) Bayesian Analysis of Multivariate Regressions Subjected to Constraints, Ph. D. thesis, University of Wisconsin, Madison.

Chikuse, Y. (1976) Partial differential equations for hypergeometric functions of complex matrices and their applications, *Ann. Inst. Statist. Math.*, 28, 187–199.

Chikuse, Y. (1990a) Distributions of orientations on Stiefel manifolds, *J. Multivariate Anal.*, 33(2), 247–264.

Chikuse, Y. (1990b) The matrix angular central Gaussian distribution, *J. Multivariate Anal.*, 33(2), 265–274.

Chikuse, Y. (1991a) High dimensional limit theorems and matrix decompositions on the Stiefel manifold, *J. Multivariate Anal.*, 36(2), 145–162.

Chikuse, Y. (1991b) Asymptotic expansions for distributions of the large sample matrix resultant and related statistics on the Stiefel manifold, *J. Multivariate Anal.*, 39(2), 270–283.

Chikuse, Y. (1993a) High dimensional asymptotic expansions for the matrix Langevin distributions on the Stiefel manifold, *J. Multivariate Anal.*, 44(1), 82– 101.

Chikuse, Y. (1993b) Asymptotic theory for the concentrated matrix Langevin distributions on the Grassmann manifold, *Statistical Sciences and Data Analysis* (K. Matusita, M. L. Puri and T. Hayakawa, eds.), VSP, Netherlands, 237–245.

Chikuse, Y. and Davis, A. W. (1986) Some properties of invariant polynomials with matrix arguments and their applictions in econometrics, *Ann. Inst. Statist. Math.*, 38, 109–122.

Chmielewski, M. A. (1981) Elliptically symmetric distributions: A review and bibliography, *Int. Statist. Rev.*, 49, 67–74.

Constantine, A. G. (1963) Some noncentral distribution problems in multivariate analysis, *Ann. Math. Statist.*, 34, 1270–1285.

Constantine, A. G. (1966) The distribution of Hotelling's generalized T_0^2, *Ann. Math. Statist.*, 37, 215–225.

Cornish, E. A. (1954) The multivariate t-distribution associated with a set of normal sample deviates, *Aust. J. Phys.*, 7, 531–542.

Cornish, E. A. (1955) The sampling distributions of statistics derived from the multivariate t-distribution, *Aust. J. Phys.*, 8, 193–199.

Cornish, E. A. (1962) The multivariate t-distribution associated with the general multivariate normal distribution, Division of Mathematical Statistics, Commonwealth Scientific and Industrial Research Organization, Australia, Technical Report No. 13.

Crowther, N. A. S. (1975) The exact non-central distribution of a quadratic form in normal vectors, *South African Statist. J.*, 9, 27–36.

Das Gupta, S. (1968) Some aspects of discrimination function coefficients, *Sankhyā*, A30, 387–400.

Das Gupta, S. (1971) Nonsingularity of the sample covariance matrix, *Sankhyā*, A33, 475–478.

Das Gupta, S. (1972) Non-central matrix-variate beta distribution and Wilks' U distribution, *Sankhyā*, A34, 357–362.

Davis, A. W. (1979) Invariant polynomials with two matrix arguments extending the zonal polynomials: Applications to multivariate distribution theory, *Ann. Inst. Statist. Math.*, A31, 465–485.

Davis, A. W. (1980) Invariant polynomials with two matrix arguments extending the zonal polynomials, *Multivariate Analysis V* (P. R. Krishnaiah, ed.), North-Holland, 287–299.

Dawid, A. P. (1977) Spherical matrix distributions and a multivariate model, *J. Roy. Statist. Soc.*, B39, 254–261.

Dawid, A. P. (1978) Extendability of spherical matrix distributions, *J. Multivariate Anal.*, 8, 559–556.

Deemer, W. L. and Olkin, I. (1951) The Jacobians of certain matrix transformations useful in multivariate analysis, Based on lectures of P. L. Hsu at the University of North Carolina, 1947, *Biometrika*, 38, 345–367.

de Waal, D. J. (1968) An asymptotic distribution for the determinant of a non-central B statistic in multivariate analysis, *South African Statist. J.*, 2, 77–84.

de Waal, D. J. (1969) The non-central multivariate beta type 2 distribution, *South African Statist. J.*, 3, 101–108.

de Waal, D. J. (1970) Distributions connected with a multi-variate beta statistic, *Ann. Math. Statist.*, 41(3), 1091–1095. Correction (1971), 42(6), 2165–2166.

de Waal, D. J. (1972a) On the expected values of the elementary symmetric functions of a noncentral Wishart matrix, *Ann. Math. Statist.*, 43, 344–347.

de Waal, D. J. (1972b) An asymptotic distribution of noncentral multivariate Dirichlet variates, *South African Statist. J.*, 6, 31–40.

de Waal, D. J. (1978) The expected values of the elementary symmetric functions of some matrices, *South African Statist. J.*, 12, 75–82.

de Waal, D. J. (1979) On the normalizing constant for the Bingham-von Mises-Fisher matrix distribution, *South African Statist. J.*, 13, 103–112.

de Waal, D. J. (1983) Quadratic forms and manifold normal distributions, *Contributions in Statistics*, Essays in honour of Norman L. Johnson (P. K. Sen, ed.), North-Holland, 115–121.

de Waal, D. J. and Nel, D. G. (1973) On some expectations with respect to Wishart matrices, *South African Statist. J.*, 7, 61–67.

Dickey, J. M. (1967) Matricvariate generalizations of the multivariate t distribution and the inverted multivariate t distribution, *Ann. Math. Statist.*, 38(2), 511–518.

Dickey, J. M. (1976) A new representation of Student's t as a function of independent t's, with a generalization to matrix t, *J. Multivariate Anal.*, 6, 343–346.

Dickey, J. M., Dawid, A. P. and Kadane, J. B. (1986) Subjective-probability assessment methods for multivariate-t and matrix-t models, *Bayesian Inference and Decision Techniques* (P. Goel and A. Zellner, eds.), Elsevier Science Publishers B. V., Chapter 12, 177–195.

Downs, T. D. (1972) Orientation statistics, *Biometrika*, **59**, 665–676.

Dunnett, C. W. and Sobel, M. (1954) A bivariate generalization of student's t- distribution with tables for certain special cases, *Biometrika*, 41, 153–169.

Dykstra, R. L. (1970) Establishing the positive definiteness of the sample covariance matrix, *Ann. Math. Statist.*, 41, 2153–2154.

Dyson, F. J. (1962a) Statistical theory of the energy levels of complex systems I, *J. Math. Phys.*, 3, 140–156.

Dyson, F. J. (1962b) Statistical theory of the energy levels of complex systems II, *J. Math. Phys.*, 3, 157–165.

Dyson, F. J. (1962c) Statistical theory of the energy levels of complex systems III, *J. Math. Phys.*, 3, 166–175.

Dyson, F. J. and Mehta, M. L.(1963a) Statistical theory of the energy levels of complex systems IV, *J. Math. Phys.*, 4, 701–712.

Dyson, F. J. and Mehta, M. L.(1963b) Statistical theory of the energy levels of complex systems V, *J. Math. Phys.*, 4, 713–719.

Eaton, M. L. (1972) *Multivariate Statistical Analysis*, Institute of Mathematical Statistics, University of Copenhagen, Denmark.

Eaton, M. L. and Olkin, I. (1987) Best equivariant estimators of a Cholesky decomposition, *Ann. Statist.*, 15, 1639–1650.

Eaton, M. L. and Perlman, M. D. (1973) The nonsingularity of generalized sample covariance matrices, *Ann. Statist.*, 1, 710–717.

Eben, K. (1994) A generalization of Wishart density for the case when the inverse of the covariance matrixis a band matrix, *Math. Bohem.*, 119(4), 337–346.

Elfving, G. (1947) A simple method of deducing certain distributions connected with multivariate sampling, *Skandinavisk Aktuarietidskrift*, 30, 56–74.

Fang, C., Krishnaiah, P. R. and Nagarsenkar, B. N. (1982) Asymptotic distribution of the likelihood ration test statistics for covariance structures of the complex multivariate normal distributions, *J. Multivariate Anal.*, 12, 597–611.

Fang, K. T. and Anderson, T. W. (Eds.) (1990) *Statistical Inference in Elliptically Contoured and Related Distributions*, Allerton Press, New York.

Fang, K. T. and Chen, H. F. (1984) Relationships among classes of spherical matrix distributions, *Acta Mathematica Sinica* (English Series), 1(2), 138–148.

Fang, K. T. and Chen, H. F. (1986) On the spectral decompositions of spherical matrix distributions and some of their subclasses, *J. Math. Res. Exposition*, 1, 47–156.

Fang, K. T., Kotz, S. and Ng, K. W. (1990) *Symmetric Multivariate and Related Distributions*, Chapman and Hall, New York.

Fang, K. T. and Zhang, Y. T. (1990) *Generalized Multivariate Analysis*, Science Press, Beijing and Springer-Verlag, Berlin.

Farrell, R. H. (1985) *Multivariate Calculation: Use of the Continuous Groups*, Springer-Verlag, New York.

Fisher, R. A. (1915) Frequency distribution of the values of the correlation coefficient in samples from an indefinitely large population, *Biometrika*, 10, 507–521.

Geisser, S. (1965) Bayesian estimation in multivariate analysis, *Ann. Math. Statist.*, 36, 150–159.

Ghurye, S. G. and Olkin, I. (1969) Unbiased estimation of some multivariate probability densities and related functions, *Ann. Math. Statist.*, 40(4), 1261–1271.

Gindikin, S. G. (1964) Analysis in homogeneous domains, *Russian Math. Surveys*, 19, 1–90.

Giri, N. C. (1977) *Multivariate Statistical Inference*, Academic Press, New York.

Girko, V. L. and Gupta, A. K. (1996) Multivariate elliptically contoured linear models and some aspects of the theory of random matrices, *Multidimensional Statistical Analysis and Theory of Random Matrices* (A. K. Gupta and V. L. Girko, eds.), VSP, Netherlands, 327–386.

Goldberger, A. S. (1970) Criteria and constraints in multivariate regression, EME 7026, Social System Research Institute, University of Wisconsin, Madison, paper presented at the Second World Congress of the Econometric Society, Cambridge, England, September 1970.

Goodman, N. R. (1963a) Statistical analysis based on a certain multivariate complex Gaussian distribution (an introduction), *Ann. Math. Statist.*, 34, 152–177.

Goodman, N. R. (1963b) The distribution of the determinant of complex Wishart distributed matrix, *Ann. Math. Statist.*, 34, 178–180.

Goodman, N. R. and Dubman, M. R. (1969) Theory of time-varying spectral analysis and complex Wishart process, *Multivariate Analysis-II* (P. R. Krishnaiah, ed.), Academic Press, New York, 351–365.

Goodman, T. R. and Kotz, S. (1973) Multivariate θ-generalized normal distributions, *J. Multivariate Anal.*, 3, 204–219.

Graham, Alexander (1981) *Kronecker Products and Matrix Calculus with Applications*, Ellis Horwood, Chichester.

Graybill, F. A. (1983) *Matrices with Applications in Statistics*, Wadsworth, Belmont, California.

Graybill, F. A. and Marsaglia, G. (1957) Idempotent matrices and quadratic forms in general linear hypothesis, *Ann. Math. Statist.*, 28, 678–686.

Groves, T. and Rothenberg, T. (1969) A note on the expected value of an inverse matrix, *Biometrika*, 56, 690–691.

Gupta, A. K. (1971a) Distribution of Wilks' likelihood ratio criterion in the complex case, *Ann. Inst. Statist. Math.*, 23, 77–87.

Gupta, A. K. (1971b) Noncentral distribution of Wilks' statistic in MANOVA, *Ann. Math. Statist.*, 42, 1254–1261.

Gupta, A. K. (1971c) On a stochastic inequality for the Wilks' statistic, *Ann. Inst. Statist. Math.*, 27, 341–348.

Gupta, A. K. (1973) On a test for reality of the covariance matrix of a complex Gaussian distribution, *J. Statist. Comp. Simul.*, 2, 333–342.

Gupta, A. K. (1976) Nonnull distribution of Wilks' statistic for MANOVA in the complex case, *Commun. Statist.-Simul. Comp.*, 5, 177–188.

Gupta, A. K. (1977) On the distribution of sphericity test criterion in the multivariate Gaussian distribution, *Aust. J. Statist.*, 19, 202–205.

Gupta, A. K. (1998) Multivariate elliptically contoured and θ-generalized normal models, *Random Oper. Stochastic Equations*, 6, 281–290.

Gupta, A. K. and Chattopadhyay, A. K. (1979) Gammaization and Wishartness of dependent quadratic forms, *Commun. Statist.-Theory Meth.*, A8(9), 945–951.

Gupta, A. K., Chattopadhyay, A. K. and Krishnaiah, P. R. (1975) Asymptotic distributions of the determinants of some random matrices, *Commun. Statist.*, 4, 33–47.

Gupta, A. K. and Conradie, W. (1987) Quadratic forms in complex normal variates: Basic results, *Statistica*, 47, 37–84.

Gupta, A. K. and Girko, V. L. (Eds.) (1996) *Multidimensional Statistical Analysis and Theory of Random Matrices*, Proceedings of the Sixth Lukacs Symposium, VSP, Netherlands.

Gupta, A. K. and Javier, W. R. (1986) Nonnull distribution of the determinant of B-statistic in multivariate analysis, *South African Statist. J.*, 20, 87–102.

Gupta, A. K. and Kabe, D. G. (1998) Characterization of gamma and the complex Wishart densities, *Applied Statistical Science III* (E. Ahmed, M. Ahsanullah and B. K. Sinha, eds.), Nova Science, 393–400.

Gupta, A. K. and Nagar, D. K. (1985) Nonnull distribution of LR-statistic for testing $\mu = \mu_0; \Sigma = \sigma^2 I$ in complex multivariate normal model, *Statistica*, 45(4), 457–464.

Gupta, A. K. and Nagar, D. K. (1987) Distribution of the product of determinants of random matrices connected with non-central matric variate Dirichlet distribution, *South African Statist. J.*, 21, 141–153.

Gupta, A. K. and Nagar, D. K. (1988) Nonnull distribution of likelihood ratio criterion for testing multisample sphericity in the complex case, *Aust. J. Statist.*, 30(3), 307–318.

Gupta, A. K. and Nagar, D. K. (1989) Asymptotic nonnull distribution of likelihood ratio statistic for testing homogeneity of complex multivariate Gaussian populations, *J. Statist. Comp. Simul.*, 31, 83–91.

Gupta, A. K. and Nagar, D. K. (1992) Distribution of LR-statistic for testing $H : \mu = \mu_0; \Sigma = \sigma^2 I$ in multivariate complex Gaussian distribution, *Statistica*, 52(2), 255–267.

Gupta, A. K. and Nagar, D. K. (1994) A note on the distribution of $(a'S^{-1}a)$ $(a'S^{-2}a)^{-1}$, *Random Oper. Stochastic Equations*, 2(4), 331–334.

Gupta, A. K. and Nagar, D. K. (1998) Quadratic forms in disguised matrix t-vatiate, *Statistics*, 30, 357–374.

Gupta, A. K. and Offori-Nyarko, S. (1995) On disguised inverted Wishart distribution, *Proc. Amer. Math. Soc.*, 123, 2557–2562.

Gupta, A. K. and Rathie, P. N. (1983a) Nonnull distribution of Wilks' Λ in the complex linear case, *Statistica*, 43(3), 445–450.

Gupta, A. K. and Rathie, P. N. (1983b) On the noncentral distribution of the determinant of a complex Wishart matrix, *Metron*, 41, 109–116.

Gupta, A. K. and Song, D. (1990) Asymptotic expansion of the matrix variate Dirichlet distribution, Department of Mathematics and Statistics, Bowling Green State University, Technical Report No. 90-04.

Gupta A. K. and Song, D. (1996) Generalized Liouville distribution, *Comput. Math. Appl.*, 32(2), 103–109.

Gupta, A. K. and Tang, J. (1984) Distribution of likelihood ratio statistic for testing equality of covariance matrices of multivariate Gaussian models, *Biometrika*, 71, 555–559.

Gupta, A. K. and Tang, J. (1986a) Some properties of LR-tests for generalized variances of two multivariate normal populations, *Publications de l'Institut de Statistique de l'Universite de Paris*, 31, 59–69.

Gupta, A. K. and Tang, J. (1986b) Exact distribution of certain general test statistic in multivariate analysis, *Aust. J. Statist.*, 28, 104–114.

Gupta, A. K. and Tang, J. (1987) On testing equality of generalized variances of k multivariate normal populations, *Publications de l'Institut de Statistique de l'Universite de Paris*, 32, 29–42.

Gupta, A. K. and Tang, J. (1988) A general distribution theory for a class of likelihood ratio criteria, *Aust. J. Statist.*, 30, 359–366.

Gupta, A. K. and Varga, T. (1991) Rank of a quadratic form in an elliptically contoured matrix random variable, *Statist. Probab. Lett.*, 11, 131–134.

Gupta, A. K. and Varga, T. (1992) Characterization of matrix variate normal distribution, *J. Multivariate Anal.*, 41, 80–88.

Gupta, A. K. and Varga, T. (1993) *Elliptically Contoured Models In Statistics*, Kluwer Academic Publishers, Dordrecht.

Gupta, A. K. and Varga, T. (1994a) Some applications of the stochastic representation of matrix variate elliptically contoured distributions, *Random Oper. Stochastic Equations*, 2(1), 1–11.

Gupta, A. K. and Varga, T. (1994b) Characterization of matrix variate normality through conditional distributions, *Math. Methods Statist.*, 3(2), 163–170.

Gupta, A. K. and Varga, T. (1994c) A new class of matrix variate elliptically contoured distributions, *J. Italian Statist. Soc.*, 3, 255–270.

Gupta, A. K. and Varga, T. (1994d) Moments and other expected values for matrix variate elliptically contoured distributions, *Statistica*, 54, 361–373.

Gupta, A. K. and Varga, T. (1995a) Matrix variate θ-generalized normal distribution, *Trans. Amer. Math. Soc.*, 347(4), 1429–1437.

Gupta, A. K. and Varga, T. (1995b) Some inference problems for matrix variate elliptically contoured distributions, *Statistics*, 26, 219–229.

Gupta, A. K. and Varga, T. (1997) Characterization of matrix variate elliptically contoured distributions, *Advances in the Theory and Practice of Statistics: A Volume in Honor of Samuel Kotz* (N. L. Johnson and N. Balakrishnan, eds.), John Wiley & Sons, New York, 455–467.

Gupta, R. D. and Richards, Donald St. P. (1987) Multivariate Liouville distributions, *J. Multivariate Anal.*, 23, 233–256.

Gupta, R. D. and Richards, Donald St. P. (1990) The Dirichlet distributions and polynomial regression, *J. Multivariate Anal.*, 32, 95–102.

Gupta, R. D. and Richards, Donald St. P. (1991) Multivariate Liouville distributions, II, *Probab. Math. Statist.*, 12(2), 291–301.

Gupta, R. D. and Richards, Donald St. P. (1992) Multivariate Liouville distributions, III, *J. Multivariate Anal.*, 43, 29–57.

Haff, L. R. (1979) An identity for the Wishart distribution with applications, *J. Multivariate Anal.*, 9, 531–544.

Hannan, E. J. (1970) *Multiple Time Series*, John Wiley & Sons, New York.

Haq, M. S. and Rinco, S. (1976) β-expectation tolerance regions for a generalized multivariate model with normal error variables, *J. Multivariate Anal.*, 6, 414–421.

Hart, M. L. and Money, A. H. (1976) On Wilks' multivariate generalization of the correlation ratio, *Biometrika*, 63(1), 59–67.

Hayakawa, T. (1966) On the distribution of a quadratic form in a multivariate normal sample, *Ann. Inst. Statist. Math.*, 18, 191–200.

Hayakawa, T. (1969) On the distributions of the latent roots of a positive definite random symmetric matrix I, *Ann. Inst. Statist. Math.*, 21, 1–21.

Hayakawa, T. (1972) On the distribution of the multivariate quadratic form in multivariate normal sample, *Ann. Inst. Statist. Math.*, 25, 205–230.

Hayakawa, T. (1985) On the distribution of a quadratic form of a matrix t-variate, *Statistical Theory and Data Analysis* (K. Matusita, ed.), Elsevier Science Publishers B. V. (North-Holland), 249–256.

Hayakawa, T. (1986) On testing hypotheses of covariance matrices under an elliptical population, *J. Statist. Plan. Inf.*, 13, 193–202.

Hayakawa, T. (1987) Normalizing and variance stabilizing transformation of multivariate statistics under an elliptical population, *Ann. Inst. Statist. Math.*, 39, 299–306.

Hayakawa, T. (1989) On the distributions of the functions of the F-matrix under an elliptical population, *J. Statist. Plan. Inf.*, 21, 41–52.

Hayakawa, T. and Kikuchi, Y. (1979) The moments of a function of traces of a matrix with a multivariate symmetric normal distribution, *South African Statist. J.*, 13, 71–82.

Herz, C. S. (1955) Bessel functions of matrix argument, *Ann. Math.*, 61, 474–523.

Hoffman, D. K., Raffenetti, R. C. and Ruedenberg, K. (1972) Generalization of Euler angles to N-dimensional orthogonal matrices, *J. Math. Phys.*, 13(4), 528–532.

Hogg, R. V. (1963) On the independence of certain Wishart variables, *Ann. Math. Statist.*, 34, 935–939.

Hogg, R. V. and Craig, A. T. (1994) *Introduction to Mathematical Statistics*, 5th ed., MacMillan, New York.

Hsu, P. L. (1939a) On the distribution of the roots of certain determinantal equations, *Annals of Eugenics*, 9, 250–258.

Hsu, P. L. (1939b) A new proof of the joint product moment distribution, *Proc. Camb. Phil. Soc.*, 35, 336–338.

Hsu, P. L. (1940) An algebraic derivation of the distribution of rectangular coordinates, *Proc. Edinb. Math. Soc.*, 6, 185–189.

Hua, L. K. (1959) *Harmonic Analysis of Functions of Several Complex Variables in Classical Domains*, Moscow (in Russian), American Mathematical Society (English Translation).

Hudak, D. and Richter, G. (1996) Moments of special normally distributed matrices, *Statistics*, 27, 363–378.

Ingham, A. E. (1933) An integral which occurs in statistics, *Proc. Camb. Phil. Soc.*, 29, 271–276.

Jambunathan, M. V. (1965) A quick method of deriving Wishart's distribution, *Current Science*, 34, 78.

James, A. T. (1954) Normal multivariate analysis and the orthogonal group, *Ann. Math. Statist.*, 25, 40–75.

James, A. T. (1955) The noncentral Wishart distribution, *Proc. Roy. Soc. Lond.*, A229, 364–366.

James, A. T. (1960) The distribution of the latent roots of the covariance matrix, *Ann. Math. Statist.*, 31, 151–158.

James, A. T. (1961a) The distribution of noncentral means with known covariance matrix, *Ann. Math. Statist.*, 32, 874–882.

James, A. T. (1961b) Zonal polynomials of the real positive definite symmetric matrices, *Ann. Math.*, 74, 456–469.

James, A. T. (1964) Distributions of matrix variate and latent roots derived from normal samples, *Ann. Math. Statist.*, 35, 475–501.

Javier, W. R. (1982) On the Distributions of Certain Random Matric Variates and Their Functions, Ph. D. thesis, Department of Mathematics and Statistics, Bowling Green State University, Bowling Green, Ohio.

Javier, W. R. and Gupta, A. K. (1985a) On generalized matric variate beta distributions, *Mathematische Operationsforschung und Statistik*, 16(4), 549–558.

Javier, W. R. and Gupta, A. K. (1985b) On matric variate-t distribution, *Commun. Statist.-Theory Meth.*, 14(6), 1413–1425.

Jensen, D. R. (1970) The joint distribution of traces of Wishart matrices and some applications, *Ann. Math. Statist.*, 41(1), 133–145.

Jensen, D. R. and Good, I. J. (1981) Invariant distributions associated with matrix laws under structural symmetry, *J. Roy. Statist. Soc.*, B43, 327–332.

Joarder, A. H. and Ali, M. M. (1992) On some generalized Wishart expectations, *Commun. Statist.-Theory Meth.*, 21(1), 283–294.

Johnson, N. L. and Kotz, S. (1970) *Continuous Univariate Distributions-2*, Houghton Mifflin, New York.

Johnson, N. L. and Kotz, S. (1972) *Continuous Multivariate Distributions*, John Wiley and Sons, New York.

Jupp, P. E. and Mardia, K. V. (1979) Maximum likelihood estimators for the matrix von Mises-Fisher and Bingham distributions, *Ann. Statist.*, 7(3), 599–606.

Juritz, J. M. (1973) Aspects of Noncentral Multivariate t-distributions, Ph. D. thesis, University of Cape Town, RSA.

Juritz, J. M. and Troskie, C. G. (1976) Noncentral matrix T distributions, *South African Statist. J.*, 10, 1–7.

Kabe, D. G. (1965) Generalization of Sverdrup's Lemma and its applications to multivariate distribution theory, *Ann. Math. Statist.*, 36, 671–676.

Kabe, D. G. (1979) On Subrahmaniam's conjecture for an integral involving zonal polynomials, *Utilitas Math.*, 15, 245–248.

Kabe, D. G. (1984) Classical statistical analysis based on a certain hypercomplex multivariate normal distribution, *Metrika*, 31, 63–76.

Kang, C. and Kim, B. C. (1996) The Nth moment of matrix quadratic form, *Statist. Probab. Lett.*, 28, 291–297.

Kaufman, G. M. (1967) Some Bayesian moment formulae, Centre for Operations Research and Econometrics, Catholic University of Louvain, Heverlee, Belgium, Report No. 6716.

Kelker, D. (1970) Distribution theory of spherical distributions and a location-scale parameter generalization, *Sankhyā*, A32, 419–430.

Khatri, C. G. (1959a) On the mutual independence of certain statistics, *Ann. Math. Statist.*, 30, 1258–1262.

Khatri, C. G. (1959b) Conditions for the forms of the type XAX' to be distributed independently or to obey Wishart distribution, *Calcutta Statist. Assoc. Bull.*, 8, 162–168.

Khatri, C. G. (1962) Conditions for Wishartness and independence of second degree polynomials in a normal vector, *Ann. Math. Statist.*, 33, 1002–1007.

Khatri, C. G. (1963) Further contributions to Wishartness and independence of second degree polynomials in a normal vectors, *J. Indian Statist. Assoc.*, 1, 61–70.

Khatri, C. G. (1965) Classical statistical analysis based on a certain multivariate complex Gaussian distribution, *Ann. Math. Statist.*, 36, 98–114.

Khatri, C. G. (1966) On certain distribution problems based on positive definite quadratic functions in normal vectors, *Ann. Math. Statist.*, 37, 468–479.

Khatri, C. G. (1969) Non-central distributions of i-th largest characteristic roots of three matrices concerning complex multivariate normal populations, *Ann. Inst. Statist. Math.*, 21, 23–32.

Khatri, C. G. (1970a) A note on Mitra's paper "A density-free approach to the matrix variate beta distribution," *Sankhyā*, A32, 311–318.

Khatri, C. G. (1970b) On the moments of traces of two matrices in three situations for complex multivariate normal populations, *Sankhyā*, A32, 65–80.

Khatri, C. G. (1971) Series representation of distributions of quadratic form in the normal vectors and generalized variance, *J. Multivariate Anal.*, 1(2), 199–214.

Khatri, C. G. (1975) Distribution of a quadratic form in normal vectors (multivariate non-central case), *A Modern Course on Statistical Distributions in Scientific Work*, Volume 1, Models and Structures (G. P. Patil, S. Kotz, and J. K. Ord, eds.), D. Reidel, Dordrecht-Holland, 345–354.

Khatri, C. G. (1977) Distribution of a quadratic form in noncentral normal vectors using generalized Laguerre polynomials, *South African Statist. J.*, 11, 167–179.

Khatri, C. G. (1980) Quadratic forms in normal variables, *Handbook of Statistics*, Volume 1 (P. R. Krishnaiah, ed.). North-Holland, 443–469.

Khatri, C. G. (1989) Multivariate generalization of t'-statistic based on the mean square successive differences, *Commun. Statist.-Theory Meth.*, 18(5), 1983–1992.

Khatri, C. G. and Bhavsar, C. D. (1990) Some asymptotic inferential problems connected with complex elliptical distribution, *J. Multivariate Anal.*, 35, 66–85.

Khatri, C. G. and Mardia, K. V. (1977) The von Mises-Fisher matrix distribution in orientation statistics, *J. Roy. Statist. Soc.*, 39(1), 95–106.

Khatri, C. G. and Rao, C. R. (1987) Effects of estimated noise covariance matrix in optimal signal detection, *IEEE Trans. Acoustics, Speech, and Signal Processing*, 35, 671–679.

Khatri, C. G., Khattree, R. and Gupta, R. D. (1991) On a class of orthogonal invariant and residual independent matrix distributions, *Sankhyā*, B53(1), 1–10.

Khattree, R. and Gupta, R. D. (1989) Estimation of matrix valued realized signal to noise ratio, *J. Multivariate Anal.*, 30(2), 312–327.

Khattree, R. and Gupta, R. D. (1992) Some probability distributions connected with beta and gamma matrices, *Commun. Statist.-Theory Meth.*, 21(2), 369–390.

Kollo, T. and von Rosen, D. (1995) Approximating by the Wishart distribution, *Ann. Inst. Statist. Math.*, 47, 767–783.

Konishi, S., Niki, N. and Gupta, A. K. (1988) Asymptotic expansions for the distribution of quadratic forms in normal variables, *Ann. Inst. Statist. Math.*, 40(2), 279–296.

Konno, Y. (1988) Exact moments of the multivariate F and beta distributions, *J. Japan Statist. Soc.*, 18(2), 123–130.

Kotz, S., Johnson, N. L. and Boyd, D. W. (1967a) Series representations of distributions of quadratic forms in normal variables I: central case, *Ann. Math. Statist.*, 38, 823–837.

Kotz, S., Johnson, N. L. and Boyd, D. W. (1967b) Series representations of distributions of quadratic forms in normal variables II: non-central case, *Ann. Math. Statist.*, 38, 838–848.

Krishnaiah, P. R. (1976) Some recent developments on complex multivariate distributions, *J. Multivariate Anal.*, 6, 1–30.

Krishnaiah, P. R. and Lin, J. (1986) Complex elliptically symmetric distributions, *Commun. Statist.-Theory Meth.*, 15(12), 3693–3718.

Krishnamoorthy, A. S. and Parthasarathy, M. (1951) A multivariate gamma-type distribution, *Ann. Math. Statist.*, 22, 549–557. Correction (1960), 31, 229.

Krishnamoorthy, K. and Gupta, A. K. (1989) Improved minimax estimation of a normal covariance matrix, *Canadian J. Statist.*, 17, 91–102.

Kshirsagar, A. M. (1959) Bartlett decomposition and Wishart distribution, *Ann. Math. Statist.*, 30, 239–241.

Kshirsagar, A. M. (1961a) Some extensions of the multivariate t distribution and the multivariate generalization of the distribution of the regression coefficients, *Proc. Camb. Phil. Soc.*, 57, 80–85.

Kshirsagar, A. M. (1961b) The non-central multivariate beta distribution, *Ann. Math. Statist.*, 32, 104–111.

Kshirsagar, A. M. (1972) *Multivariate Analysis*, Marcel Dekker, New York.

Laha, R. G. (1956) On the stochastic independence of two second-degree polynomial statistics in normally distributed variates, *Ann. Math. Statist.*, 27, 790–796.

le Roux, N. J. (1978) The Algebra of Random Matrices, Ph. D. thesis, Faculty of Science, University of South Africa, RSA.

Leung, P. L. (1994) An identity for the noncentral Wishart distribution with application, *J. Multivariate Anal.*, 48, 107–114.

Li, R. Z. and Fang, K. T. (1995) Estimation of scale matrix of elliptically contoured matrix distribution, *Statis. Probab. Lett.*, 24, 289–297.

Lin, P. E. (1972) Some characterizations of the multivariate t distribution, *J. Multivariate Anal.*, 2, 339–344.

Loynes, R. M. (1966) On idempotent matrices, *Ann. Math. Statist.*, 37, 295–296.

Luke, Y. L. (1969) *The Special Functions and Their Approximations*, Volume I, Academic Press, New York.

Madow, W. A. (1938) Contribution to the theory of multivariate statistical analysis, *Trans. Amer. Math. Soc.*, 44, 454–495.

Magnus, J. R. and Neudecker, H. (1979) The commutation matrix: some properties and applications, *Ann. Statist.*, 7, 381–394.

Magnus, J. R. and Neudecker, H. (1988) *Matrix Differential Calculus with Applications in Statistics and Econometrics*, John Wiley & Sons, Chichester.

Mahalanobis, P. C., Bose, R. C. and Roy, S. N. (1937) Normalization of statistical variates and the use of rectangular coordinates in the theory of sampling distributions, *Sankhyā*, 3, 1–40.

Mardia, K. V. and Khatri, C. G. (1977) Uniform distribution on a Stiefel manifold, *J. Multivariate Anal.*, 7, 468–473.

Marshall, A. and Olkin, I. (1979) *Inequalities: Theory of Majorization and Its Applications*, Academic Press, New York.

Marx, D. G. (1981) Aspects of the Matric t-Distribution, Ph. D. thesis, University of The Orange Free State, RSA.

Marx, D. G. (1983) Quadratic forms of a matric-t variate, *Ann. Inst. Statist. Math.*, 35, 347–353. Correction (1985), 37, 567.

Marx, D. G. and Nel, D. G. (1982) A note on the distribution of linear combinations of matric-, vector- and scalar-t variates, Department of Mathematical Statistics, University of The Orange Free State, Technical Report No. 85.

Mathai, A. M. (1981) Distribution of the canonical correlation matrix, *Ann. Inst. Statist. Math.*, A33, 35–43.

Mathai, A. M. and Provost, S. B. (1992) *Quadratic Forms In Random Variables*, Marcel Dekker, New York.

Mathai, A. M. and Tan, W. Y. (1977) The non-null distribution of the likelihood ratio criterion for testing the hypothesis that the covariance matrix is diagonal, *Canadian J. Statist.*, 5(1), 63–74.

Mauchly, J. W. (1940) Significance test for sphericity of a normal n-variate distribution, *Ann. Math. Statist.*, 11, 204–209.

Mauldon, J. G. (1955) Pivotal quantities for Wishart's and related distributions, and a paradox in fiducial theory, *J. Roy. Statist. Soc.*, B17, 79–85.

Mehta, M. L. (1991) *Random Matrices*, 2nd ed., Academic Press, New York.

Mirham, G. A. and Hultquist, R. A. (1967) A bivariate warning-time/failure-time distribution, *J. Amer. Statist. Assoc.*, 62, 589–599.

Mitra, S. K. (1969) Some characteristic and noncharacteristic properties of the Wishart distribution, *Sankhyā*, A31(1), 19–22.

Mitra, S. K. (1970) A density-free approach to the matrix variate beta distribution, *Sankhyā*, A32, 81–88.

Muirhead, R. J. (1970) Asymptotic distributions of some multivariate tests, *Ann. Math. Statist.*, 41, 1002–1010.

Muirhead, R. J. (1982) *Aspects of Multivariate Statistical Theory*, John Wiley & Sons, New York.

Muirhead, R. J. (1986) A note on some Wishart expectations, *Metrika*, 33, 247–251.

Nagar, D. K. and Gupta, A. K. (1993) Asymptotic non-null distribution of likelihood ratio statistic for testing $\mu = \mu_0; \Sigma = \sigma^2 I_p$ in complex multivariate Gaussian model, *Statistica*, 53(4), 603–617.

Nagar, D. K., Jain, S. K. and Gupta, A. K. (1985) Distribution of LRC for testing sphericity of a complex multivariate Gaussian model, *Int. J. Math. & Math. Sci.*, 8(3), 555–562.

Nagarsenker, B. N. (1979) Noncentral distribution of Wilk's statistic for test of three hypotheses, *Sankhyā*, A41(1 & 2), 67–81.

Nagarsenkar, B. N. and Das, M. M. (1975) Exact distribution of sphericity criterion in the complex case and its percentage points, *Commun. Statist.*, 4(4), 362–374.

Nel, D. G. (1978) On the symmetric multivariate normal distribution and the asymptotic expansion of a Wishart matrix, *South African Statist. J.*, 12, 145–159.

Nel, D. G. and Groenewald, P. C. N. (1979) On a Fisher-Cornish type expansion of Wishart matrices, Department of Mathematical Statistics, University of the Orange Free State, Technical Report No. 47.

Nel, H. M. (1977) On distributions and moments associated with matrix normal distributions, Department of Mathematical Statistics, University of the Orange Free State, Technical Report No. 24.

Neudecker, H. (1985) On the dispersion matrix of a matrix quadratic form connected with the noncentral Wishart distribution, *Linear Algebra Appl.*, 70, 257–267.

Neudecker, H. and Wansbeek, T. (1983) Some results on commutation matrices with statistical applications, *Canadian J. Statist.*, 11, 221–231.

Neudecker, H. and Wansbeek, T. (1987) Fourth order properties of normally distributed random matrices, *Linear Algebra Appl.*, 97, 13–21.

Ogawa, J. (1953) On the sampling distributions of classical statistics in multivariate analysis, *Osaka Math. J.*, 5, 13–52.

Olkin, I. (1953) Note on the Jacobians of certain matrix transformations useful in multivariate analysis, *Biometrika*, 40, 43–46.

Olkin, I. (1959) A class of integral identities with matrix argument, *Duke Math. J.*, 26, 207–214.

Olkin, I. (1979) Matrix extensions of Liouville-Dirichlet-type integrals, *Linear Algebra Appl.*, 28, 155–160.

Olkin, I. and Roy, S. N. (1954) On multivariate distribution theory, *Ann. Math. Statist.*, 25, 329–33

Olkin, I. and Rubin, H. (1962) A characterization of the Wishart distribution, *Ann. Math. Statist.*, 33(4), 1272–1280.

Olkin, I. and Rubin, H. (1964) Multivariate beta distributions and independence properties of Wishart distribution, *Ann. Math. Statist.*, 35, 261–269. Correction (1966), 37(1), 297.

Parkhurst, A. M. and James, A. T. (1974) Zonal polynomials of order 1 through 12, *Selected Tables in Mathematical Statistics* (H. L. Harter and D. B. Owen, eds.), American Mathematical Society, Providence, R. I., 199–388.

Patil, G. P., Boswell, M. T., Ratnaparkhi, M. V. and Roux, J. J. J. (1984) *Dictionary and Classified Bibliography of Statistical Distributions in Scientific Work*, Volume 3, Multivariate Models, International Co-operative Publishing House, Fairland, Maryland.

Perlman, M. D. (1977) A note on the matrix-variate F distribution, *Sankhyā*, A39(3), 290–298.

Phillips, P. C. B. (1985) The distribution of matrix quotients, *J. Multivariate Anal.*, 16, 157–161.

Pillai, K. C. S. and Gupta A. K. (1967) On the distribution of the second elementary symmetric function of the roots of a matrix, *Ann. Inst. Statist. Math.*, 19, 167–179.

Pillai, K. C. S. and Gupta A. K. (1968) On the noncentral distribution of second elementary symmetric function of the roots of a matrix, *Ann. Math. Statist.*, 39, 833–839.

Pillai, K. C. S. and Gupta, A. K. (1969) On the exact distribution of Wilks' criterion, *Biometrika*, 56, 109–118.

Pillai, K. C. S. and Jouris, G. M. (1971) Some distribution problems in the multivariate complex Gaussian case, *Ann. Math. Statist.*, 42, 517–525.

Porter, C. E. (1965) *Statistical Theory of Spectra: Fluctuations*, Academic Press, New York.

Prentice, M. J. (1982) Antipodally symmetric distributions for orientation statistics, *J. Statist. Plan. Inf.*, 6, 205–214.

Press, S. J. (1972) *Applied Multivariate Analysis*, Holt, Rinehart and Winston, Inc., New York.

Priestly, M. B., Subba Rao, T. and Tong, H. (1973) Identification of the structure of multivariable stochastic systems, *Multivariate Analysis-III* (P. R. Krishnaiah, ed.), Academic Press, New York, 351–368.

Rainville, E. D. (1970) *Special Functions*, Macmillan, New York.

Rao, C. R. (1952) *Advanced Statistical Methods in Biometric Research*, John Wiley & Sons, New York.

Rao, C. R. (1973) *Linear Statistical Inference and Its Applications*, 2nd ed., John Wiley & Sons, New York.

Rasch, G. (1948) A functional equation for Wishart's distribution, *Ann. Math. Statist.*, 19, 262–266.

Rautenbach, H. M. and Roux, J. J. J. (1985) Statistical analysis based on quaternion normal random variables, Department of Statistics, University of South Affrica, RSA, Research Report No. 85-09.

Richards, Donald St. P. (1984) Hyperspherical models, fractional derivatives and exponential distributions on matrix spaces, *Sankhyā*, A46(2), 155–165.

Rinco, S. (1973) β-Expectation Tolerance Regions Based on the Structural Models, Ph. D. thesis, The University of Western Ontario, London, Canada.

Rogers, G. S. (1980) *Matrix Derivatives*, Marcel Dekker, New York.

Rooney, P. G. (1972) On the ranges of certain fractional integrals, *Canadian J. Math.*, 24, 1198–1216.

Roux, J. J. J. (1971) On generalized multivariate distributions, *South African Statist. J.*, 5, 91–100.

Roux, J. J. J. (1975) New families of multivariate distributions, *A Modern Course on Statistical Distributions in Scientific Work*, Volume 1, Models and Structures (G. P. Patil, S. Kotz, and J. K. Ord, eds.), D. Reidel, Dordrecht-Holland, 281–297.

Roux, J. J. J. and Becker, P. J. (1984) On prior inverted Wishart distribution, Department of Statistics and Operations Research, University of South Africa, Pretoria, Research Report No. 2.

Roux, J. J. J. and Raath, E. L. (1973) Generalized Laguerre series forms of Wishart distributions, *South African Statist. J.*, 7, 23–34.

Roy, J. (1966) Power of the likelihood-ratio test used in analysis of dispersion, *Multivariate Analysis* (P. R. Krishnaiah, ed.), Academic Press, New York, 105–127.

Roy, S. N. (1957) *Some Aspects of Multivariate Analysis*, John Wiley & Sons, New York.

Saw, J. G. (1973) Expectation of elementary symmetric functions of a Wishart matrix, *Ann. Statist.*, 1(3), 580–582.

Searle, S. R. (1971) *Linear Models*, John Wiley & Sons, New York.

Sen Gupta, A. (1987) Tests for standardized generalized variances of multivariate normal populations of possibly different dimensions, *J. Multivariate Anal.*, 23(2), 51–59.

Shah, B. K. (1970) Distribution theory of a positive definite quadratic form with matrix argument, *Ann. Math. Statist.*, 41(2), 692–697.

Shah, B. K. and Khatri, C. G. (1974) Proof of conjectures about the expected values of the elementary symmetric functions of a noncentral Wishart matrix, *Ann. Statist.*, 2(4), 833–836.

Shaman, P. (1980) The inverted complex Wishart distribution and its application to spectral estimates, *J. Multivariate Anal.*, 10, 51–59.

Siotani, M., Hayakawa, T. and Fujikoshi, Y. (1985) *Modern Multivariate Statistical Analysis: A Graduate Course and Handbook*, American Sciences Press, Columbus, Ohio, USA.

Sivazlian, B. D. (1981) On a multivariate extension of the gamma and beta distributions, *SIAM J. Appl. Math.*, 41, 205–209.

Song, D. and Gupta A. K. (1997) Properties of generalized Liouville distribution, *Random Oper. Stochastic Equations*, 5(4), 337–348.

Srivastava, M. S. (1965) On the complex Wishart Distribution, *Ann. Math. Statist.*, 36, 312–315.

Srivastava, M. S. and Khatri, C. G. (1979) *An Introduction to Multivariate Statistics*, North Holland, New York.

Stein, C. (1969) Multivariate Analysis I, Department of Statistics, Stanford University, USA, Technical Report No. 42.

Steyn, H. S. and Roux, J. J. J. (1972) Approximations for the non-central Wishart distributions, *South African Statist. J.*, 6, 165–173.

Styan, G. P. H. (1979) Three useful expressions for expectations involving Wishart matrices, *Statistical Data Analysis and Inference* (Y. Dodge, ed.), Elsevier Science Publishers B. V., 283–196

Subrahmaniam, K. (1973) On some functions of matrix argument, *Utilitas Math.*, 3, 83–106.

Subrahmaniam, K. (1976) Recent trends in multivariate normal distribution: On the zonal polynomials and other functions of matrix argument, *Sankhyā*, A38, 221–258.

Sugiura, N. (1973) Derivatives of the characteristic root of a symmetric or a Hermitian matrix with two applications in multivariate analysis, *Commun. Statist.*, 1(5), 393–417.

Sutradhar, B. C. and Ali, M. M. (1989) A generalization of the Wishart distribution for the elliptical model and its moments for the multivariate t model, *J. Multivariate Anal.*, 22, 155–162.

Sverdrup, E. (1947) Derivation of the Wishart distribution of the second order sample moments by straight forward integration of a multiple integral, *Skandinavisk Aktuarietidskrift*, 30, 151–166.

Tan, W. Y. (1964) Bayesian Analysis of Random Effect Models, Ph. D. thesis, University of Wisconsin, Madison.

Tan, W. Y. (1968) Some distribution theory associated with complex Gaussian distribution, *Tamkang J.*, 7, 263–301.

Tan, W. Y. (1969a) Some results on multivariate regression analysis, *Nanta Mathematica*, 3, 54–71.

Tan, W. Y. (1969b) The restricted matric-t distribution and its applications in deriving posterior distributions of parameters in multivariate regression analysis, Department of Statistics, University of Wisconsin, Madison, Technical Report No. 205.

Tan, W. Y. (1969c) Note on the multivariate and the generalized multivariate beta distributions, *J. Amer. Statist. Assoc.*, 64, 230–241.

Tan, W. Y. (1973) Multivariate studentization and its applications, *Canadian J. Statist.*, 1(2), 181–199.

Tan, W. Y. (1979) On the approximation of noncentral Wishart distribution by Wishart distribution, *Metron*, 37(3), 49–58.

Tan, W. Y. (1980) On approximating multivariate distributions, *Multivariate Statistical Analysis* (R. P. Gupta, ed.), North-Holland, Amsterdam, 237–249.

Tan, W. Y. and Gupta, R. P. (1982) On approximating the non-central Wishart distribution by Wishart distribution: A monte carlo study, *Commun. Statist.-Simul. Comp.*, 11(1), 47–64.

Tan, W. Y. and Gupta, R. P. (1983) On approximating a linear combination of central Wishart matrices with positive coefficients, *Commun. Statist.-Theory Meth.*, 12(22), 2589–2600.

Tan, W. Y. and Guttman, I. (1971) A disguised Wishart variable and a related theorem, *J. Roy .Statist. Soc.*, B33, 147–152.

Tiao, G. C. and Guttman, I. (1965) The inverted Dirichlet distribution with application, *J. Amer. Statist. Assoc.*, 60, 793–805.

Tiao, G. C., Tan, W. Y. and Chang Y. C. (1970) A Bayesian approach to multivariate regression subject to linear constraints, paper presented at the Second World Congress of the Econometric Society, Cambridge, England, September 1970.

Tiao, G. C. and Zellner, A. (1964) On the Bayesian estimation of multivariate regression, *J. Roy. Statist. Soc.*, B26, 277–285.

Tracy, D. S. and Sultan, S. A. (1993) Third moment of matrix quadratic form, *Statist. Probab. Lett.*, 16, 71–76.

Troskie, C. G. (1967) Noncentral multivariate Dirichlet distributions, *South African Statist. J.*, 1, 21–32.

Troskie, C. G. (1969) The generalised multiple correlation matrix, *South African Statist. J.*, 3(2), 109–122.

Troskie, C. G. (1972) The distributions of some test criteria depending on multivariate Dirichlet distributions, *South African Statist. J.*, 6, 151–163.

Tsai, M. T. (1995) A generalization of Wishart distribution, *Statist. Probab. Lett.*, 24, 67–70.

Turin, G. L. (1960) The characteristic function of Hermitian quadratic forms in complex normal variables, *Biometrika*, 47, 199–201.

Tyler, D. E. (1987) Statistical analysis for the angular central Gaussian distributionon the sphere, *Biometrika*, 74(3), 579–589.

Uhlig, H. (1994) On singular Wishart and singular multivariate beta distributions, *Ann. Statist.*, 22(1), 3950–405.

van der Merwe, C. A. (1980) Expectations of the traces of functions of a multivariate normal variable, Department of Mathematical Statistics, University of the Orange Free State, RSA, Technical Report No. 56.

van der Merwe, G. J. and Roux, J. J. J. (1974) On a generalized matrix-variate hypergeometric distribution, *South African Statist. J.*, 8, 49–58.

von Rosen, D. (1988a) Moments for the inverted Wishart distribution, *Scand. J. Statist.*, 15, 97–109.

von Rosen, D. (1988b) Moments of matrix normal variables, *Statistics*, 19, 575–583.

Weibull, M. (1953) The distribution of t- and F-statistics and of correlation and regression coefficients in stratified samples from normal populations with different means, *Skandinavisk Aktuarietidskrift*, 1-2, Supplement, 1–106.

Whaba, G. (1968) On the distributions of some statistics useful in the analysis of jointly stationary time series, *Ann. Math. Statist.*, 39, 1849–1862.

Whaba, G. (1971) Some tests of independence for stationary multivariate time series, *J. Roy. Statist. Soc.*, B33, 153–166.

Wigner, E. P. (1965) Distribution laws for the roots of a random Hermitian matrix, *Statistical Theory of Spectra: Fluctuations* (C. E. Porter, ed.), Academic Press, New York, 446–461.

Wigner, E. P. (1967) Random matrices in physics, *SIAM Review*, 9, 1–23.

Wijsman, R. A. (1957) Random orthogonal transformation and their use in some classical distribution problems in multivariate analysis, *Ann. Math. Statist.*, 28, 415–423.

Wilks, S. S. (1932) Certain generalizations in the analysis of variance, *Biometrika*, 24, 471–494.

Wishart, J. (1928) The generalized product moment distribution in samples from a normal multivariate population, *Biometrika*, A20, 32–52.

Wishart, J. and Bartlett, M. S. (1933) The generalized product moment distribution in a normal system, *Proc. Camb. Phil. Soc.*, 29, 260–270.

Wong, C. S. and Liu, D. (1994) Moments for left elliptically contoured random matrices, *J. Multivariate Anal.*, 49, 1–23.

Wooding, R. A. (1956) The multivariate distribution of complex normal variables, *Biometrika*, 43, 212–215.

Xu, Jian-Lun (1987) Inverse Dirichlet distribution and its applications, *Acta Mathematicae Applicatae Sinica*, 10, 91–100. Reprinted in *Statistical Inference in Elliptically Contoured and Related Distributions* (K. T. Fang and T. W. Anderson, eds.), Allerton Press, New York.

SUBJECT INDEX